Trace Elements in Abiotic and Biotic Environments

Trace Elements in Abiotic and Biotic Environments

Alina Kabata-Pendias
Barbara Szteke

CRC Press
Taylor & Francis Group
Boca Raton London New York

CRC Press is an imprint of the
Taylor & Francis Group, an **informa** business

CRC Press
Taylor & Francis Group
6000 Broken Sound Parkway NW, Suite 300
Boca Raton, FL 33487-2742

First issued in hardback 2019
First issued in paperback 2021

ISBN 13: 978-1-03-209877-7 (pbk)
ISBN 13: 978-1-4822-1279-2 (hbk)

Publisher's Note
The publisher has gone to great lengths to ensure the quality of this reprint but points out that some imperfections in the original copies may be apparent.

Visit the Taylor & Francis Web site at
http://www.taylorandfrancis.com

and the CRC Press Web site at
http://www.crcpress.com

Contents

List of Acronyms/Abbreviations

[X, Z]	(at the beginning of each subchapter) X—symbol of chemical element, Z—atomic number of chemical element
AAAc	ammonium acetic acid
AAP	American Academy of Pediatrics
AVS	acid volatile sulfide
AW	ash weight
BC	Before Christ
BCF	bioconcentration factor (ratio of content in plant to content in soil)
bw	body weight
C_w/C_c	ratio of an element concentration in water to its content in the Earth's crust
ca.	about (from Latin)
CEC	cation exchangeable capacity
COPR	chromite-ore processing residue
CRT	cathode ray tube
DNA	desoxyribonucleic acid, carrier of genetic information
DOC	dissolved organic carbon
DOM	dissolved organic matter
DM-x	dimethylated element
DTPA	diethylenetriaminepentaacetic acid
EDTA	ethylenediaminetetraacetic acid
EEC	European Economic Community
Eh	oxidation–reduction electrical potential (volts)
ERL	effects range low
ERM	effects range median
EU	European Union
FA	fulvic acid
FMI	free metal ion
FW	fresh weight
GM	geometric mean
HA	humic acid
IDD	Iodine Deficiency Disorders
MAC	maximum allowable concentration
MMT	methylocyclopentadienyl manganese tricarbonyl (controversial gasoline additive)
MM-x	monomethylated element
MTL	maximum tolerable level
NIOSH	National Institute for Occupational Safety and Health
OM	organic matter
PEC	probable effect concentration
PEL	probable effect level

PET	polyester (polyethylene terephthalate)
PGM	platinum group metal
pH	negative logarithm, base 10, of hydrogen ion concentration
PTE	periodic table of elements
RNA	ribonucleic acid, structural element of the cytoplasm and cell nucleus
SEL	safe element level
SEM	simultaneously extracted metal
SOM	soluble organic matter
TAV	trigger action value
TEL	threshold effects level
TET	total element level
UK	United Kingdom
US EPA	Environmental Protection Agency
WHO	World Health Organization
yr	year

List of Data Units

PRESENTATION OF DATA

The basic units of the International System (SI units) are used in this book. Mean contents refer to arithmetic mean values, unless otherwise stated.

The concentrations of a trace element in soil, plant, human, and animal samples are based on the total content by weight of the element in air-dried or oven-dried (at 40°C) material. Otherwise it is indicated as follows: AW (ash weight) and FW (fresh weight).

Trace element contents of soils are given for the top layer, unless otherwise indicated. If not identified, the content of an element in samples is given as so-called *total*, which means that it is measured in totally digested samples or measured directly in a sample.

UNITS

Bq	becquerel
Ci curie	1 Ci = 3.7×10^{10} Bq
g	gram (10^{-3} kg)
ha	hectare (10,000 m^2)
kBq	kilobecquerel
kg	kilogram (10^3 g)
kt	thousand metric ton (10^3 t)
L	liter (1 dm^3)
mCi	millicurie
µg	microgram (10^{-3} mg)
mg	milligram (10^{-3} g)
Mt	million metric ton (10^3 kt)
nCi	nanocurie
pCi	picocurie
t	metric ton (10^3 kg)

Preface

All substances are poisons, and the right dose differentiates a poison from a remedy

P.A. Paracelsus (1493–1541)

It has already become a truism to say that the quality of human life depends on the chemical composition of food and surroundings. Among chemical elements, trace elements are very special as their quantity ratio is of a crucial importance. There is a negligible difference between the physiological and harmful contents of most trace elements, and as always their proper balance plays significant role in the biological behavior. The bioavailability of trace elements is variable and is controlled by specific properties of living organisms, as well as by physical and chemical factors of the abiotic spheres.

Recent improvements and sophisticated methods in analytical chemistry, as well as increasing fields of investigations, have added substantially to the present state of the knowledge on the biogeochemistry of trace elements. During the last three decades, there has been a real *explosion* of research data and various publications on the biogeochemical behavior of, almost all, trace elements of known and unknown physiological functions.

Human activities are modifying trace element concentrations in all environmental compartments. It is especially pronounced in various industrial regions, as well as in urban and highways vicinities. Also, both geology and geochemistry of the environment have significant impact on the chemistry and health of plants, animals, and humans.

The authors' intention is to summarize an updated interdisciplinary fundamental data with a condensed presentation of knowledge on trace element transfer in the food chain from soil to humans. The information on the international legislation on trace elements for both micronutrients and contaminants in soil and plant food will be presented, and will be related to ecological and health risk assessments. A special attention is focused on human health effects of deficiency and excess of trace elements.

A better understanding of the geochemical processes and comprehensive dataset on the abundance of trace elements in abiotic and biotic environmental compartments will be a key to managing trace elements in the environment, which is a perquisite to sustainable land use and the possible diminishing of health risks due to trace inorganic pollutants.

Alina Kabata-Pendias
Institute of Soil Science and Plant Cultivation

Barbara Szteke
Institute of Agricultural and Food Biotechnology

The Periodic Table of Elements

1	2	3	4	5	6	7	8	9	10	11	12	13	14	15	16	17	18
1 H																	2 He
3 Li	4 Be											5 B	6 C	7 N	8 O	9 F	10 Ne
11 Na	12 Mg											13 Al	14 Si	15 P	16 S	17 Cl	18 Ar
9 K	20 Ca	21 Sc	22 Ti	23 V	24 Cr	25 Mn	26 Fe	27 Co	28 Ni	29 Cu	30 Zn	31 Ga	32 Ge	33 As	34 Se	35 Br	36 Kr
37 Rb	38 Sr	39 Y	40 Zr	41 Nb	42 Mo	43 Tc	44 Ru	45 Rh	46 Pd	47 Ag	48 Cd	49 In	50 Sn	51 Sb	52 Te	53 I	54 Xe
55 Cs	56 Ba	57 * La	72 Hf	73 Ta	74 W	75 Re	76 Os	77 Ir	78 Pt	79 Au	80 Hg	81 Tl	82 Pb	83 Bi	84 Po	85 At	86 Rn
87 Fr	88 Ra	89 ** Ac	104 Rf	105 Db	106 Sg	107 Bh	108 Hs	109 Mt	110 Ds	111 Rg	112 Cn	113 Uut	114 Fl	115 Uup	116 Lv	117 Uus	118 Uuo

* Lanthanide series	58 Ce	59 Pr	60 Nd	61 Pm	62 Sm	63 Eu	64 Gd	65 Tb	66 Dy	67 Ho	68 Er	69 Tm	70 Yb	71 Lu
** Actinide series	90 Th	91 Pa	92 U	93 Np	94 Pu	95 Am	96 Cm	97 Bk	98 Cf	99 Es	100 Fm	101 Md	102 No	103 Lr

The symbolic given in thin letters indicate the elements not occurring naturally in the environment.

Authors

Alina Kabata-Pendias, PhD, DSC, professor ordinary of environmental geochemistry, was head of the Trace Elements Laboratory of the Institute of Soil Science and Plant Cultivation in Puławy, Poland, for more than 30 years. She also worked, for 30 years, at the Polish Geological Institute in Warsaw, Poland. Today, after 65 years at the institute in Puławy, she holds the title of professor emeritus and is a member of the staff of the soil department.

Professor Kabata-Pendias' research interest has been always focused on the fate of trace elements in the rock–soil–plant chain. She has published about 320 articles and book chapters. She is author or coauthor of 11 books, including several editions (up to 4) of some titles.

Professor Kabata-Pendias was active in various national and international scientific committees, and was awarded several medals, of which the most prestigious is the Philippe Duchaufour Medal, bestowed (in 2007) by the European Geosciences Union.

Barbara Szteke, PhD, DSC, professor ordinary of agricultural sciences, was head of the Department of Food Analysis of the Institute of Agricultural and Food Biotechnology, Warsaw, Poland, for 25 years. Most of her activities were focused on foodstuff contaminants coming from the environment (elements, pesticides, mycotoxins, etc.). Professor Szteke is author or coauthor of more than 200 papers, chapters in books, and reports and lectures concerning scientific investigations, and coeditor of some books.

Professor Szteke was a member of many Polish and international scientific organizations, among them the Committee of Analytical Chemistry of the Polish Academy of Sciences and IUPAC Commission on Food Chemistry. For many years, Professor Szteke was Polish representative in Codex Alimentarius Commission FAO/WHO Committee on Food Additives and Contaminants.

Professor Szteke was the initiator of the cyclic IUPAC Symposium "Trace Elements in Food," the first of which was held in Warsaw, Poland, in 2000 and the fifth in Copenhagen, Denmark, in 2014.

1 Aluminum [Al, 27]

1.1 INTRODUCTION

Aluminum (Al) is a metal of the group 13 in the periodic table of elements. It is the third most abundant element in the Earth's crust, occurring at about 8%, and up to 5% in soils. It reveals lithophile properties, and is common in both igneous and sedimentary rocks. In coal, its contents vary from 1.5% to >10%. Due to its low content, but important functions in plants and humans, it is included in the trace elements group.

Aluminum occurs mainly at 3+ oxidation stage and is a common component of several minerals, especially silicates. It reveals amphoteric properties and thus may react with both alkaline and acid compounds. Minerals composed only of Al are boehmite, γ-AlOOH; diaspore, α-AlOOH; hydrargilite/gibbsite, γ-$Al(OH)_3$; and corundum, Al_2O_3. The host minerals of Al are feldspars, micas, and all layer silicates. Under special crystal system, Al_2O_3 may be gemstones (e.g., sapphire or ruby, where some Al_3^+ is replaced by Cr_3^+).

World production of Al in 2010 was 41,400 kt (USGS 2011). The highest production was in China, 16,800 kt. A great proportion of Al, especially in Europe and the United States, is obtained from old scraps and wastes. The most common Al ores are bauxites, which mainly include Al minerals such as gibbsite, boehmite, and diaspore, very often occur in a mixture with two Fe oxides (goethite and hematite).

Due to the versatile properties of Al, it is used in various industrial sectors such as metallurgy, construction, electricity, chemistry, and transportation and packing (especially food products). It is used for several chemical treatments, such as water purification, sugar refining, wood preservations, leather tanning, and several other processes.

1.2 SOILS

Aluminum is an abundant element in soils and a main component of several common minerals. Its content is inherited from parent rocks, and its species and distribution in soil profiles are governed by soil properties, of which soil pH is the most significant. Only easily mobile and exchangeable fractions of Al play a significant role in soil properties, among which the most important are pH and soluble organic matter (SOM). Very acidic soils, at pH < 5, contain mainly Al^{3+}; at pH 5–7 $Al(OH)_2^+$ and $Al(OH)^{2+}$ predominate; and at pH > 6 $Al(OH)^{4-}$ dominates. Rainwater containing sulfur (mainly SO_2) stimulates the formation of $Al(OH)SO_4$, which is very mobile in soils. Drainage water from acidic soils may contain Al up to 250 μg/L, whereas those from other soils is about 50 μg/L. Mobile Al species in acidic soils can be taken up by plants, which creates a chemical stress in plants.

Behavior of Al in soils is highly modified by SOM, due to the organically bound Al forms in both solid and liquid phases. Plants, and especially tree vegetation, may have an effective impact on the concentration and mobilization of Al in soils (Álvarez et al. 2002).

1.3 WATERS

Aluminum concentrations in ocean and seawater vary between 0.03 and 2 μg/L, whereas in river water, it may be much higher, up to about 1000 μg/L. At the surface layer of seawater, it may be at a higher level due to the atmospheric deposition. Its average content in river water is calculated at 32 μg/L, and the global river flux at 1200 kt/yr (Gaillardet et al. 2003). Al concentrations in rainwater vary highly from 3 μg/L at the remote region (Kola Peninsula) to 105 μg/L in polluted regions (Table 1.1).

Several parameters have an impact on the variable speciation of Al in the aquatic systems. Among them, the most important are: pH, dissolved organic carbon, SOM, as well as contents of several elements such as F, S, P, and Si. In acidic water (pH < 5), the most common species is Al_3^+, whereas in water of a higher pH value, other species such as $AlOH_2^+$, $Al(OH)^{2+}$, and $Al(OH)O_3$ are likely to occur. The lowest Al solubility is at pH > 6 (Yokel 2004). Increased level of Al in acidic water may be toxic to fish and other organisms (Barker and Pilbaem 2007); this is observed especially in freshwater systems—lakes and rivers (Gensemer and Playle 1999).

Al concentration in drinking water is under the control, and its acceptable (by the U.S. EPA) levels vary with different countries, from 50 to 200 μg/L (Yokel 2004). Median Al content in bottled water in the Europe is 1.95 μg/L, which is lower than

TABLE 1.1
Aluminum Contents in Water and Air

Environmental Compartment	Range
Water (μg/L)	
Rain	3–105
River	2–1000
Sea, ocean	0.03–2
Air (ng/m³)[a]	
Urban	150–1300
Rural areas	46–70
Antarctica	0.3–0.8

Sources: Data are given for uncontaminated environments from Kabata-Pendias, A. and Mukherjee, A.B., *Trace Elements from Soil to Human*, Springer, Berlin, Germany, 2007.

[a] Reimann, C. and de Caritat, P., *Chemical Elements in the Environment*, Springer, Berlin, Germany, 1998.

in tap water—2.47 μg/L (Birke et al. 2010). Contribution of Al in drinking water to the total oral exposure is calculated to be about 4%.

1.4 AIR

Aluminum contents in air (Table 1.1) from remote regions are estimated by Reimann and de Caritat (1998) at the range of 46–70 ng/m³, whereas in urban regions it is 150–1300 ng/m³, and the maximum may be up to 3500 ng/m³. Natural sources of Al are wind-blown earth dust and volcanic eruptions. Industrial emissions contain various Al compounds: Al_2O_3, AlF_3, and Na_3AlF_6. It is estimated that the Al industry releases about 0.2 kg Al per 1 t of Al mined (Kabata-Pendias and Mukherjee 2007).

1.5 PLANTS

Aluminum is a common constituent of all plants. Its content may vary highly, depending on both soil factors and plant species. Physiological functions of Al in plants are not described; however, there are some evidences that its low levels may have a beneficial effect on plant growth. The most significant problem is associated with the Al toxicity, which is considered to be one of the major factors limiting plant growth on acid minerals in soils (Matsumoto et al. 2001).

Plant species and even cultivars differ considerably in their uptake ability, translocation, and toleration of the excess Al. All these processes are related to the cell walls and plasma membrane properties. Citrates excluded in the rhizosphere protect plants against increased Al uptake. In acidic soils, however, citrates are absorbed by oxide minerals and thus the phytoavailability of Al is increased (Hashimoto 2007). The most significant mechanisms of Al tolerance depend on (1) exclusion of Al in the root–soil interface; (2) plant-induced pH barrier in the rhizosphere; (3) Al immobilization at cell walls and binding by proteins; and (4) activities of enzymes. According to Barker and Pilbaem (2007) Al is one of the most important factors that limit plant growth.

Complex Al toxicity in plants is reflected in several interactions, mainly in the reduction of the uptake of nutrients such as P, Ca, Mg, K, and N. Addition of these elements to soils, especially of Ca and Mg, may reduce the Al toxicity. Al toxicity may be associated with increased levels of Fe and Mn, and also other cations, which are easily available to plants grown on acidic soils.

The highest Al contents are noticed in clovers (85–3470 mg/kg) and grasses (10–3410 mg/kg). Mean Al contents in cereal grains are within the range 30–70 mg/kg, with the highest value for rye grains. Among vegetables, the highest Al content (about 100 mg/kg) has been found in spinach leaves. The lowest Al contents are noticed in various fruits, up to 15 mg/kg in orange fruits. Contents of Al in the most tea leaves are very high (average 2969 mg/kg), since tea bushes grow on very acidic soils (Houba and Uittenbogaard 1994). Concentrations of Al in mature tea leaves correlated significantly with exchangeable Al in soils. Levels of Al in mature leaves and young shoots were reduced by the application of large amounts of N fertilizers (Ruan et al. 2006).

1.6 HUMANS

Total body burden of Al in healthy individuals is 30–50 mg. Approximately, 50% of Al is found in the skeleton and 25% is in the lungs, and increases with age. Aluminum levels range from 5 to 10 mg/kg in the bone tissue and from 1 to 3 µg/L in the serum of healthy individuals (ATSDR 2008). However, there are also other estimations including adult body burden (70 kg) is 60 mg, of which (in mg/kg) in bones is between 4 and 27; in the soft tissue 1–28; and in the blood 0.4 mg/L (Emsley 2011). In the blood, Al is approximately equally distributed between plasma and erythrocytes.

Aluminum is a nonessential metal, to which humans are frequently exposed by systemic absorption of Al ingested from water, foods, drugs, and air. In lungs, it may be from environmental-derived particles, occupational exposure, and distribution from the blood. This metal, for a long time, was considered to be safe for human health. Until the early 1970s, the possible toxicity of Al was not considered. However, it was discovered that patients with kidney malfunction maintained by dialysis developed a serious neurological syndrome called *dialysis encephalopathy*. The finding of high concentration of Al in gray matter from the brains of patients dying with this disease lead to suggestion that dialysis encephalopathy is caused by Al intoxication. Dialysis encephalopathy is also associated with anemia and demineralization of the bones, leading to increased risk of fractures. Later work implicated Al in the pathogenesis of Alzheimer's disease, but there is no evidence that this disease is caused by Al, and even the original observation of elevated Al has been challenged in several laboratories (Halliwell and Gutteridge 2007).

It has been suggested that Al is implicated in the aetiology of Alzheimer's disease and is associated with other neurodegenerative diseases in humans. However, these hypotheses remain controversial (EFSA 2008).

After absorption, Al distributes to all tissues of organisms, and accumulates in some, especially in bones. The main carrier of the Al ion in plasma is the Fe-binding protein, transferrin. Aluminum can enter the brain and reach the placenta and fetus. It may persist for a very long time in various organs and tissues, before it is excreted in the urine. At high levels of exposure, some Al compounds may affect DNA damage, *in vitro* and *in vivo*, through indirect mechanisms. Database on the carcinogenicity of Al compounds is limited.

It is considered that the use of Al nanoparticles as food additives could be a reason for the rising incidence of autoimmune diseases. Nano Al has the capability to induce the production of free radicals, thereby resulting in the oxidative stress in cells. Nanotoxicological effects of Al nanoparticles (Al_2O_3 NPs) on some human cells may be mediated through an increase in oxidative stress (Alshatwi et al. 2013).

Foods are generally the major dietary sources of Al in most countries (Table 1.2). Aluminum in the food supply comes from natural sources, water used in food preparation, food ingredients, and utensils used during food preparation: cookware, beverages in Al cans, tap water, table salt, baking powders, processed cheese, and bleached flour. Other sources of Al for humans are antacids, antiperspirants, vaccines, and other medications, as well as occupational exposure. Because of new evidence that Al could have effects on reproductive system and developing nervous

TABLE 1.2
Aluminum in Foodstuffs (mg/kg FW)

Product	Content
Food of Plant Origin	
Bread[a]	3.59
Wheat flour[b]	16.5
Bread rye[c]	2.64
Miscellaneous cereals[a]	17.5
Cabbage[b]	21.2
Green vegetables[a]	1.12
Other vegetables[a]	2.84
Fresh fruits[a]	0.48
Cocoa powder[d]	165
Chocolate[d]	48
Curry[e]	15.8–195
Food of Animal Origin	
Carcass meat[a]	0.24
Meat products[a]	2.50
Pork[b]	0.22
Eggs[b]	0.26
Fish and seafood[c]	0.62
Milk[a]	0.01
Dairy products[a]	0.50
American cheese[f]	695[g]

Sources: [a] Rose, M. et al., *Food Addit. Contam.*, 27, 1380–1404, 2010.
[b] Deng, G.-F. et al., *Food Addit. Contam.*, 4, 348–353, 2011.
[c] Bratakos, S.M. et al., *Food Addit. Contam.*, 5, 33–44, 2012.
[d] Stahl, T. et al., *Environ. Sci. Eur.*, 23, 37, 2011.
[e] Gonzalvez, A. et al., *Food Addit. Contam.*, 1, 114–121, 2008.
[f] Pennington and Jones (1998).
[g] With emulgator, Na–Al phosphorate.
FW, fresh weight.

system, in 2006, the Joint FAO/WHO Expert Committee on Food Additives (JECFA) reduced the previously established provisional tolerable weekly intake (PTWI) values for all forms of Al in food from 7 to 1 mg kg/bw; it applied to all Al compounds in food, including additives (WHO 2007).

Total Al intake generally relates to total food intake. In Hong Kong, the average dietary exposure to Al for a 60 kg adult was estimated to be 0.60 mg/kg bw/week, which amounted to 60% of PTWI (Wong et al. 2010). In the United Kingdom, daily Al intake is 5.4 mg (Rose et al. 2010). The mean dietary exposure of European adults was estimated to be 1.6–13 mg/day, or 0.2–1.5 mg/kg bw/week (EFSA 2008).

2 Antimony [Sb, 51]

2.1 INTRODUCTION

Antimony (Sb), a metalloid of the group 15 in the periodic table of elements, occurs in the Earth's crust up to 0.9 mg/kg, and in igneous rocks within the range 0.1–1 mg/kg. Its content in sedimentary rocks, especially argillaceous, may be higher, up to 4 mg/kg. Sb content in coal varies within the range 1–3 mg/kg; it may be concentrated in fly ash (up to 100 mg/kg). Sb amount in crude oil is within the order of magnitude 0.0X–0.X mg/kg. Antimony exhibits chalcophilic properties, and occurs with sulfur and some metal deposits such as Cu, Pb, and Ag, and, therefore, is used as a pathfinder for the geochemical exploration. Antimony deposits often contain minor amounts of Ag, Au, and Hg.

It has various oxidation stages: –3, +3, +4, and +5. More than 100 minerals of Sb are found naturally. It occurs in three types of minerals, namely, sulfides, oxides, and mixed oxides-sulfides (e.g., kermesite, Sb_2S_2O). Its main minerals are stibnite, Sb_2S_3, and valentinite, Sb_2O_3. Antimony is a component of several other minerals, such as pyrargyrite (Ag_3SbS_3) and bourmite ($PbCuSbS_3$). It may easily form soluble thiocomplexes (SbS_2^{2-}, SbS_4^{3-}) in various aquatic environments.

Global (excluding the United States) Sb production was estimated at 135 kt in 2010, with the highest in China at 120 kt (USGS 2011). Sb production increased significantly in recent years. Antimony is also recovered from various recycled scraps. It is an important product of the secondary lead smelting and lead–acid batteries.

Antimony trioxide (stibnite; Sb_2O_3) is the most important Sb compound used in the industry. It is used mainly as an alloying metal for Pb in batteries (Pb-acid), for added strength and improved electrical properties. Sb alloys with Pb and Sn are used in bullets, and for solders and bearings. A variety of compounds containing Sb as the major constituents are used for other ammunition types and explosives. It is also applied in some microelectronics. This metal is used in fire retardant formulations for plastics, rubbers, textiles, paper, and paints. Antimony compounds, primarily sodium antimonate, are also used in decolorizing and refining agents for optical glass and cathode ray tube glass. In the electronics industry, the use of Sb grows for diodes. It is also applied in several chemical industries and in cosmetic production.

Antimony and its compounds are considered to be serious pollutants, and their concentration in sewage sludge is recently of concern. Especially, Sb released from coal combustion is an important pollutant due to its small-sized aerial particles, high solubility, and reactivity (Shotyk et al. 2005).

2.2 SOILS

Sb background level in soils ranges from 0.05 to 4.0 mg/kg (Table 2.1). Its average content in worldwide soils is estimated at 0.67 mg/kg. Mean content calculated for top soils in the United States is 0.66 mg/kg (Burt et al. 2003), whereas this value for

TABLE 2.1
Antimony Contents in Soils, Water, and Air

Environmental Compartment	Range/Mean
Soil (mg/kg)	
Light sandy	0.05–1.3
Medium loamy	0.05–2.0
Heavy loamy	0.3–4.0
Organic	0.08–0.6
Water (μg/L)	
Rain	0.02–0.14
River	0.005–0.98
Sea, ocean	0.15–0.44
Air (ng/m³)	
Urban	0.77–55
Spitsbergen	0.002–0.25
Antarctica	0.003

Sources: Data are given for uncontaminated environments from Kabata-Pendias, A. and Mukherjee, A.B., *Trace Elements from Soil to Human*, Springer, Berlin, Germany, 2007; Reimann, C. and de Caritat, P., *Chemical Elements in the Environment*, Springer, Berlin, Germany, 1998.

European soils is estimated at 1.04 mg/kg (FOREGS 2005). However, mean contents of Sb in soils of some countries differ significantly (in mg/kg), for example, China, 3.6; the United States, 2.4; Slovak Republic, 0.7; and Sweden, 0.25 (Kabata-Pendias and Mukherjee 2007).

Antimony in soils and sediments is slightly mobile and absorbed by clay and residual fractions and by hydrous oxides, especially of Al, Fe, and Mn, and at low pH values (Manaka 2006). Also, the organic matter reveals a capacity to bound Sb; Sb complexes with humic acid are easily mobile. Because it occurs mainly as sulfide (SbS) species, Sb is slightly available to plants. In low Fe-content soils, Sb mobility is increased. In soil Sb^{5+} predominates, mainly in the form of $Sb(OH)_6^-$ (Koch 1998).

Increased level of arsenic in soils resulted in a lower Sb mobility (Casado et al. 2007). The most soluble Sb in contaminated soils is under the treatment of citric acid solution (Telford et al. 2008). According to Gal et al. (2007), Sb, even at a high concentration in soils (up to 1200 mg/kg), is slightly available to earthworms.

Soil contamination with Sb is often reported, mainly for soils at hazardous waste sites and soils around mining and smelter areas. Soils around various Sb smelters contain this metal ranging from 109 to 2550 mg/kg (ATSDR 2002a). Top soils around various industries also contain elevated Sb levels (in mg/kg): Cu smelter, United States; Maury Island, Washington, 49–204; Au refining, United States, 280 (Flynn et al. 2003). Soils around abandoned mining area in Scotland contain Sb up to 1200 mg/kg (Gal et al. 2007). For soils, the maximum allowable concentration and the trigger action values are estimated at 10 mg/kg and 20 mg/kg, respectively. The following are the mean values of Sb content in various fertilizers (mean values in mg/kg): NPK, 0.03; P, 0.20; solid pig manure, 0.16; sewage sludge, 0.20; and fly ash, 3.8.

2.3 WATERS

According to Gaillardet et al. (2003), the world average river flux of Sb in seawater is 2.6 kt/yr (Filella et al. 2000). Antimony concentrations reportedly ranges from 0.15 to 0.44 μg/L and from 0.05 to 0.98 μg/L in seawater and river water, respectively (Table 2.1). In some surface water (e.g., lake in a national park in Poland), Sb concentration may be much higher, up to 2.15 μg/L (Niedzielski et al. 2000a). Water of some polluted aquatic environments contains Sb over 100 μg/L, and geothermal water may have up to 500 μg/L (Filella et al. 2000; Sabadell and Axtmann 1975). In the bottom sediments of uncontaminated water, Sb contents vary from <0.1 to 4.3 mg/kg (Bojakowska et al. 2007).

In water, Sb occurs in two oxidation states, 3+ and 5+, depending on the redox potential. There are some suggestions that Sb^{5+} in the form of $Sb(OH)_6^-$ predominates over Sb^{3+}. Common compounds in water at neutral pH are $HSbO_2$ and $Sb(OH)_3$. Organoantimony compounds in aqueous environment are also methylated forms: $MeSbCl_2$, Me_3SbBr_2, and $MeSbI_2$, which may change to $Me_3Sb(OH)_2$ and Me_3SbO, respectively (Koch 1998). Various methylated species of Sb(CH_3SbH_2, $[CH_3]Sb$, $[C_2H_5]_3Sb$) are found in river sediments. According to this author, volatile Sb compounds, SbH_3 and Me_3Sb, are also present is some aquatic environments. Niedzielski et al. (2000b) reported that common species of Sb in lake water are $Sb(OH)_3$ and $Sb(OH)^{6-}$, and their contents are as follows: Sb^{3+}, 0.3–0.4 μg/L; Sb^{5+}, <0.15 μg/L; and Sb organic, 0.1–0.15 μg/L.

Rainwater contains relatively low concentration of Sb, up to 0.14 μg/L (Table 2.1). Sb wet deposition during 1999 in Sweden was calculated to be 0.8 g/ha/yr (Eriksson 2001a).

Median Sb concentration in tap water of European countries is estimated at 0.07 μg/L, which is lower than in bottled water at 0.22 μg/L (Birke et al. 2010). Sb content in polyester (polyethylene terephthalate) bottles is about 224 mg/kg; however, its transfer to beverages is relatively low (Welle and Franz 2011). Guideline for Sb concentration in drinking water is established at 20 μg/L by the WHO (2011a).

2.4 AIR

Concentrations of Sb in air in Spitsbergen vary from 0.002 to 0.25 ng/m³ and may be considered as natural. Sb content in air around urban and industrial areas reaches up to 55 ng/m³ (Table 2.1), and may be elevated up to 1000 ng/m³ according to the data presented by ATSDR (2002b). Sn concentrations in air of Helsinki downtown and closed rural area were 0.77 ng/m³ and 0.36 ng/m³, respectively (Pakkanen et al. 2002).

Antimony released to the atmosphere is in the forms of oxides, sulfide, and elemental Sb. Natural sources of Sb and its compounds are volcanic eruption, sea salt spray, forest fires, and wind-blown dust. Sb anthropogenic emission is mainly from fossil fuel combustion, incineration of wastes and cement kilns, and various metallurgical industries. It is often fixed by some metallic particles, such as Pb, Cd, and Zn in aerosols and may be transported over long distances. Sb emission in the region of Bavaria, Germany, is estimated at 6–8 kt/yr (Wegenke et al. 2005).

2.5 PLANTS

Antimony is relatively easily taken up by plants, when it is present in soluble forms, although it is not considered to be essential to plants. In contaminated soils, Sb often occurs in the SbS form, which is easily available to plants. The common range in cultivated plants is between <1 and 30 μg/kg; however, in roots, it is usually higher. Spinach grown on Sb-contaminated soil contains this metal up to 1130 μg/kg, whereas in parsley it is below 420 μg/kg (Hammel et al. 2000). Lowest Sb content is reported for cereals grains at 0.5–1 μg/kg (Eriksson 2001a). Methylantimony compounds may occur in plants within the range of 100–200 μg/kg.

Pot experiment with high Sb doses (1000 mg/kg) shows that Sb^{3+} is more toxic than Sb^{5+} to rice plants (He and Yang 1999).

2.6 HUMANS

Sb amount in the human body (average 70 kg) is approximately 2.0 mg, in the bone, it is between 0.01 and 0.6; in the tissue, it ranges from 0.01 to 0.21 mg/kg (Emsley 2011); and in the rib bone, it is ≤0.0096 mg/kg (Zaichik et al. 2011). Sb distribution within the body and its excretion are a function of both the route of administration and the valence state of Sb. After inhalation, the trivalent form, Sb^{3+}, accumulates more rapidly than the pentavalent form, Sb^{5+}, in the liver, whereas pentavalent is found preferentially in the skeleton. In the blood, the trivalent form is lodged primarily in the red blood cells, whereas the pentavalent form is carried by the plasma.

Therapeutic use of trivalent Sb against parasites leads to its higher accumulation in the liver, the thyroid, and the heart. Antimony concentrations in breast milk in Italian women ranged from <0.05 to 12.9 μg/kg, with a mean of 3.0 ± 0.4 μg/kg (Health Canada 1999). Urinary Sb levels are thought to be a good indicator for the overall Sb exposure, regardless of the source (Makris et al. 2013).

Sb toxicity is a function of the water solubility and the oxidation state in Sb species. In general, Sb^{3+} is more toxic than Sb^{5+}, and the inorganic compounds are more toxic than the organic compounds, with stibin (SbH_3), a lipophilic gas, being most toxic (by inhalation). Soluble Sb salts, after oral uptake, exert a strong irritating effect on the gastrointestinal mucosa and trigger sustained vomiting. Other effects include abdominal cramps, diarrhea, and cardiac toxicity (WHO 2003).

Antimony toxicity occurs either due to occupational exposure or during therapy. Occupational exposure may cause respiratory irritation, pneumoconiosis, antimony spots on the skin, and gastrointestinal symptoms. As a therapeutic, Sb has been mostly used for the treatment of leishmaniasis and schistosomiasis. The major toxic side effects of antimonials as a result of therapy are cardiotoxicity (~9% of patients) and pancreatitis, which is seen commonly in HIV and visceral leishmaniasis coinfections. Historically, Sb has been known for its emetic properties. Oral exposure to Sb predominantly affects the gastrointestinal system. Sb amounts as low as 0.529 mg/kg can result in vomiting (Sundar and Chakravarty 2010).

There is inadequate evidence for the carcinogenicity of Sb trioxide and trisulfide in humans, but these compounds have been seen to cause lung tumours in rats. The greatest concern, with regard to the carcinogenicity of Sb, relates to the inhalation route. Antimony trioxide is classified as possibly carcinogenic to humans (Group 2B) and Sb trisulfide (Group 3), as reported by the IARC (1989).

Primary source of exposure to Sb_2O_3 for the general population in Canada is flame retardants used in furniture upholstery and mattress covers. Flame retardant polyester fabrics used in children's stuffed toys may be another source of exposure (Health Canada 2010a). Sb trioxide is a priority pollutant in the Europe and the United States.

Antimony does not bioaccumulate, so exposure to naturally occurring Sb through food is very low (WHO 2003). It is present in food, including vegetables grown on Sb-contaminated soils, mostly in low µg/kg FW values range (Table 2.2).

TABLE 2.2
Antimony in Foodstuffs (µg/kg FW)

Product	Content
Food of Plant Origin	
White rice[a]	0.3
Cereal products[b]	2.0
Green vegetables[b]	0.5
Other vegetables[b]	5.5
Fresh fruits[b]	0.4
Nuts[b]	0.7
Food of Animal Origin	
Carcass meat[b]	0.8
Meat products[b]	9.9
Fish[b]	2.6
Milk and dairy products[c]	0.4–4.0
Milk[b]	<0.1
Ready Meals	
Lasagna (with meat)[d]	1.7
Canneloni[d]	3.7
Fish filets[d]	3.0
Pasta (with meat)[d]	2.6

Sources: [a] Batista, B.L. et al., *Food Addit. Contam.*, 3, 253–262, 2010.
[b] Rose, M. et al., *Food Addit. Contam.*, 27, 1380–1404, 2010.
[c] Chung, S.W.C. et al., *Food Addit. Contam.*, 25, 831–840, 2008.
[d] Haldimann, M. et al., *Food Addit. Contam.*, 30, 587–598, 2013.
FW, fresh weight.

In the United States, the daily Sb intake from food is of approximately 4.6 μg; in Canada, it is 7.44 μg (Health Canada 1999); in France, it is 3 μg (Noël et al. 2003); and in the United Kingdom, it is 2.3 μg (Rose et al. 2010). These intakes are far below the tolerable limit, established as tolerable daily intake of 6 μg/kg bw/day (approximately 47 μg for 70 kg body) (WHO 2003).

3 Arsenic [As, 33]

3.1 INTRODUCTION

Arsenic (As) is a metalloid of the group 5 in the periodic table of elements and occurs in the Earth's crust at levels between 0.5 and 2.5 mg/kg. It is likely to concentrate in argillaceous sediment, up to about 15 mg/kg. In igneous rocks, As concentrations range from 1.5 to 3.0 mg/kg, and in sedimentary rocks, it ranges from 1.7 to 400 mg/kg (average is about 3 mg/kg). It is likely to concentrate in coal, up to 82 mg/kg. In some brown coal, As content reached 1500 mg/kg (Smith et al. 1998). It may be also accumulated in fly ash—up to 60 mg/kg. Arsenic is likely to be concentrated in shales and peats.

Global As production (without the United States) (arsenic trioxide, As_2O_3) in 2010 was 54,500 t, with China and Chile producing 25,000 t and 11,500 t, respectively (USGS 2011). Arsenic may be obtained from copper, gold, and lead smelter dust as well as from roasting arsenopyrite, the most abundant ore mineral of arsenic.

Arsenic occurs at various oxidation stages: –2, –3, +3, and +5. About 200 minerals of As are known. It easily forms complex anions (AsO_4^{3-}) and is easily combined with some metals (e.g., lellingit, $FeAs_2$; doineykite, Cu_3As; nikieline, NiAs; and sperrylite, $PtAs_2$). Its common minerals are arsenopyrite, FeAsS; orpiment, As_2S_3, reagal AsS; and arenolite, As_2O_3. Also, organic compounds of As occur in nature, such as dimethylarsenic, trimethylarsenic, and arsenobetaine (AB).

About 80% of produced As has been used for pesticides. Due to its toxicity, the amount of As in pesticide has been reduced, but it is still a dominating element in pesticides (Deschamps and Matschullat 2011). Arsenic also has varied applications such as wood preservation (e.g., chromate copper arsenate, a wood preservative, stopped in recent years, is still used to preserve wood for nonresidential applications, and the resulting As leaching from treated timber can increase the As content in the soil nearby); production of photoelectric devices, Pb-acid batteries, and glassware; and are used in Cu alloys to improve corrosion resistance. Arsenic was also used in medicine until 1950 (Matschullat 2000).

3.2 SOILS

Arsenic contents in soils vary from <0.1 to 67 mg/kg, with the lowest in light sandy soils and the highest in heavy loamy soils and organic soils (Table 3.1). Mean content calculated for top soils in the United States is 7.2 mg/kg (Burt et al. 2003), whereas this value for European soils is estimated at 11.6 mg/kg (FOREGS 2005). Average background As concentration does not exceed 15 mg/kg. However, it may considerably vary, depending on the geology of the parent materials, which are the main source of As in uncontaminated soils.

TABLE 3.1
Arsenic Contents in Soils, Water, and Air

Environmental Compartment	Range/Mean
Soil (mg/kg)	
Light sandy	<0.1–30
Medium loamy	1.5–27
Heavy loamy	2–35
Organic	<0.1–67
Water (μg/L)	
Rain	0.3–0.4
River	0.2–9.5
Sea, ocean	1.2–3.7
Air (ng/m³)	
Urban	2–53
Greenland	0.6
Antarctica	0.007

Sources: Data are given for uncontaminated environments, from Kabata-Pendias, A. and Mukherjee, A.B., *Trace Elements from Soil to Human*, Springer, Berlin, Germany, 2007; Reimann, C. and de Caritat, P., *Chemical Elements in the Environment*, Springer, Berlin, Germany, 1998.

Due to the release of As from various industrial sources, especially from coal burning, its content in many soils is increased, sometimes even up to 20,000 mg/kg (Smith et al. 1998). Other sources of As are (1) landfills and other disposals, especially sludges; (2) chemical and allied products; and (3) lumber and various wood products. Agricultural and orchard practices may be a significant source of As, as its contents are usually elevated in pesticides, fertilizer, sewage sludge, and manure. As Smith et al. (1998) reported, the highest As concentrations in orchard soils of Washington State was within the range of 106–2553 mg/kg. Elevated As contents are also noticed in the house garden and children yard soils. Especially, groundwater contamination by As is of great agricultural concern.

According to the criteria for the contaminated land (Dutch List 2013), the following As concentrations in soils and groundwater are established (in mg/kg, and μg/L): uncontaminated—20 and 10; medium contaminated—50 and 30; heavily contaminated—50 and 100.

Arsenic readily changes valence state and reacts with soils to form species with varying mobility. It is easily changed from As^{5+} to As^{3+} by Fe^{2+} and sorbed by Fe compounds (Lee et al. 2011). Therefore, various treatments are used for the soil remediation, from As stabilization to electrokinetics and phytoremediation. De Andrade et al. (2012) investigated the speciation of As in sulfide material from gold-mining areas and reported that most of As was in forms susceptible to dissolution in acid medium; fractions coprecipitated with amorphous and crystalline Fe oxides, and were fixed by MnO_2 and Al_2O_3. Recently, it has been concluded that bacteria are responsible for most of the redox As transformation in the environment (Santini 2012).

3.3 WATERS

Range of As concentration in seawater is estimated at 1.2–3.7 μg/L (Table 3.1). Low mean As content in the north Pacific Ocean is given by Nozaki (2005). Global average of As content in river water is calculated as 0.62 μg/L, ranging from 0.1 to 9.5 μg/L. However, contaminated water may contain much higher amount of As, with Lake Erie (United States) and Tocone River (Chile) containing up to 300 μg/L and 800 μg/L, respectively (Kabata-Pendias and Pendias 1999). Gaillardet et al. (2003) calculated the global riverine flux of As to seawater at 25 kt/yr.

In water, As is present as trivalent (dominated in reducing conditions) and pentavalent (dominated in oxidizing conditions). Also, its methylated form, resulted mainly from phytoplankton activities, constitutes about 10% of the total As concentration in most oceans. Arsenic in water is often in acid forms: H_3AsO_3 and H_3AsO_4 (Höll 2011). Also, organic compounds, often as products of microbial transformation, are present in water as monomethyl arsenic acid (MMAA) and dimethylarsenic acid (DMAA). As Niedzielski et al. (2000b) reported that several other As-organic compounds may occur in water. According to these authors, the range of As species contents in lake water includes (in μg/L): As^{3+}: 0.85–1.0; As^{5+}: 0.15–0.35; and As-organic: <0.15.

Groundwater may contain very high levels of As, which resulted in its elevated concentration in well water. As-enriched groundwater is present over large areas in some countries. Reported by Battacharya et al. (2002) and Burges and Ahmed (2006), high As concentrations in shallow well water are as follows (in μg/L): Argentina, 100–4800; Bangladesh, <1–3000; China, <100–1860; Mexico, 330–1100; and Thailand, 120–6700.

Arsenic content in bottom sediments is a good information on water pollution. Its average content in bottom sediments of San River (Poland) varies between <0.5 and 10 mg/kg, and is a bit higher in mule sediments than the sandy ones (Bojakowska et al. 2008). As content in stream-bottom sediments of National Park, Montgomery (Pennsylvania State), in 1995 ranged from 2 to 5 mg/kg (USGS 1997). Assessment limits for As in sediments are established as follows (in mg/kg): effects range low, 8.2; effects range median, 70; probable effect level (PEL), 17; 33; and 17 (EPA 2000, 2013). The Environment Canada Sediment Quality Guidelines (USGS 2001) gave other values for As in lake-bottom sediments (in mg/kg): threshold effects level, 5.9; PEL, 17; and probable effect concentration, 33. Sediment Quality Guidelines are also based on the acid volatile sulfide/simultaneously extracted metal ratios (Griethuysen et al. 2006).

According to a recent estimation, up to 200 million people in 70 countries are at risk from As-contaminated drinking water (Santini 2012). Removal of arsenic from water is of a big concern, and several methods are proposed and used based on mechanisms such as precipitation, adsorption, coagulation, and so on (Bang et al. 2011; Höll 2011). Nanotechnology may help alleviate water pollution, and as a result, 100 times more As can be captured using nanorust than with filtration systems using larger particles. Nanoparticles of Fe-Mn oxide is used for the stabilization of As in groundwater (An and Zhao 2012). Groundwater contains Fe-oxides that remove arsenic as coprecipitates (Lee et al. 2011).

Median As concentration in bottled water of the EU countries is 0.21 μg/L, which is a little bit higher than tap water (0.19 μg/L) (Birke et al. 2010). In Serbia, drinking water contains up to about 6.5 μg/L, in both bottled and tap water (Ristić et al. 2011). Papić et al. (2012) reported growing health hazard, especially cancer risk, with increased As contents in drinking water. The provisional guideline value for As in drinking water is 10 μg/L (WHO 2011a). (The guideline value is designated as provisional on the basis of treatment performance and analytical achievability.)

3.4 AIR

As concentrations in air vary from 0.007 ng/m^3 in Antarctica to >50 ng/m^3 in urban regions (Table 3.1). Its concentrations in remote areas should not exceed 1 ng/m^3 (Reimann and de Caritat 1998).

Natural As sources in air are volcanic eruption (20%–40% of the total natural emission), sea salt aerosol (<10% of the total natural emission), soil-derived dusts, and forest fires. The greatest As emission in the atmosphere are industrial sources, mainly from coal combustion. Zevenhoven et al. (2006) calculated that the total anthropogenic As emission in EU-25 countries in 2000 as 260 t, whereas Pacyna and Pacyna (2001) estimated its worldwide emission in the same year at 5011 t.

3.5 PLANTS

As concentration in plants is a function of both its total and soluble species contents in soils. This suggests that it is taken by plants passively, with the water flow. However, various plants reveal different capability to As uptake, what is illustrated by variable As contents among plants from the As-contaminated region; the highest As content was in sarghina (*Corrigiola telephiifolia*), 1350 and 2110 mg/kg, and the lowest in Spanish foxglove (*Digitalis thapsi*), 94 and 356 mg/kg, in roots and tops, respectively (Garcia-Salgado et al. 2012).

Excessive As uptake, in all species (organic, As^{3+}, and As^{5+}) is toxic to plants. In general, As tolerance of various plants ranges from 2 to 8 mg/kg. However, the critical values for many plants are much higher, up to 100 mg/kg in tops and 1000 mg/kg in roots. Arsenic reacts with several enzymes, disrupts energy flow in cells, and inhibits root growth (Mukherjee and Bhattacharya 2001). As complexes with glutathione and phytochelatins, apparently to minimize the toxic effects, were identified in plants from As-contaminated sites (Jedynak et al. 2012).

Plants reveal various capabilities to uptake As from the growth media. For example, the average content of As in rice variety *Boro* is 0.29 mg/kg, whereas in the variety *Aman*, from the same site, it is 0.15 mg/kg (Ahmed et al. 2011). However, according to these authors, As contents in rice grains depends on environmental conditions, which affect As contents of all food plants (Table 3.2). Some plants may accumulate great amounts of As; for example, huisache (*Acacia farnesiana*) and smooth mesquites (*Prosopis laevigata*) from the same site contain 225 mg/kg and 83 mg/kg of As, respectively (Armienta et al. 2008). The authors concluded that hazaiches might be useful for the phytoremediation. Also, Ma et al. (2001) reported that Chinese brake fern (*Pteris vittate*) reveals a potential to clean up As-contaminated soils.

In some Asian countries, excessive As concentration in groundwater is a crucial problem associated with the cultivation of plants, especially rice. Load of As with water used for the irrigation varies from 1.36 to 5.5 kg/ha/yr (Huq and Naidu 2005). Arsenic from irrigation water is accumulated mainly in plant roots and decreases rice yield.

Arsenic has been used for a long time (in China since about 1000 years ago) for pesticides and herbicides. Recently, the most commonly used forms of arsenic are monosodium methyl arsenate, disodium methyl arsenate, and calcium acid methyl arsenate. These treatments resulted in increased levels of As in food plants. Use of As-containing pesticides has been forbidden since 1970s, especially in European countries.

TABLE 3.2
Arsenic Contents in Food Plants Grown in Bangladesh and China

Country and Plants	Mean Content (μg/kg)
Bangladesh[a]	
Carrot	10.1
Radish	15.7
Broccoli	14.7
Cabbage	60.8
Potatoes	9.2
China[b]	
Rice, grain	<930
Rice husk	4110
Maize	120
Soybean	790
Peanut	270
Beans	1330

Sources: [a] Al Ramalli, S.W. et al., *Sci. Total Environ.*, 337, 23–30, 2005, plants imported to England from Bangladesh.
[b] Liu, H. et al., *Sci. Total Environ.*, 339, 153–166, 2005, plants form Pb/Zn mining area.

3.6 HUMANS

Total As content in the human body is 7 mg but varies between 0.5 and 15 mg and tends to increase with age; in the blood, it varies between 2 and 9 µg/L; in tissues and bones, it ranges from 0.1 to 1.6 mg/kg (Emsley 2011). Mammals tend to accumulate As in keratin-rich tissues such as hair and nails. Inorganic As can easily pass through the placenta.

Arsenic has long been associated with criminal activity and still is an emotionally highly charged topic, as large homicidal doses can cause cholera-like symptoms (acute poisoning) and death. Arsenic-containing medicines have little place in modern medical practice, but at one time were important in the treatment of syphilis and other diseases (Salvarsan and Fowler's solution). However, As-containing herbicides (MMAA and DMAA) and veterinary products are still used. It is likely that they are ultimately degraded to inorganic As or volatile arsines by microbial activity. As absorption depends on chemical species and its solubility as well as the matrix in which it is present. Soluble arsenicals in water are highly bioavailable. Long-term health effects of exposure to As are skin lesions, skin cancer, internal cancers, bladder kidney, lung, neurological effects, hypertension and cardiovascular disease, pulmonary disease, peripheral vascular disease, and diabetes mellitus (Smith et al. 2000).

As toxicity depends largely on its chemical form and oxidation states. Generally, lower oxidation of As has higher toxic effects. As toxicity decreases with increasing complexity of the organic structure of As compounds:

$$AsH_3 \rightarrow {}_{\text{inorganic}}As^{3+} \rightarrow {}_{\text{organic}}As^{3+} \rightarrow {}_{\text{inorganic}}As^{5+} \rightarrow {}_{\text{organic}}As^{5+}$$

$$\rightarrow \text{arsenic compounds} \rightarrow {}_{\text{metalic}}As$$

First symptoms of long-term exposure to high levels of inorganic As are usually observed in the skin, and include pigmentation changes, skin lesions, and hard patches on the palms and soles of the feet (hyperkeratosis). These occur after a minimum exposure of approximately five years and may be a precursor to skin cancer. In addition to skin cancer, long-term exposure to As may also cause cancers of the bladder and lungs. The International Agency for Research on Cancer (IARC) has classified As and As compounds as carcinogenic to humans (Group1), and has also stated that As in drinking water is carcinogenic to humans (IARC 2013).

Other adverse health effects that may be associated with the long-term ingestion of inorganic As include developmental effects, neurotoxicity, diabetes, and cardiovascular disease. In China (province of Taiwan), As exposure has been linked to the *blackfoot disease*, which is a severe disease of blood vessels leading to gangrene. However, this disease has not been observed in other parts of the world, and it is possible that malnutrition contributes to its development (WHO 2011b).

Dietary intake from food, beverages, and drinking water are the primary sources of As exposure in humans (Table 3.3). In comparison with drinking water, food can contain a number of As species with varying levels of toxicity. Organic As in seafood and inorganic As in water, beverages, and drugs have been shown to be readily absorbed (70%–90%) by the gastrointestinal tract. Arsenic species commonly found in food include inorganic As such as arsenite (As^{3+}) and arsenate (As^{5+}), which

TABLE 3.3
Total and Inorganic Arsenic Content in Foodstuffs (mg/kg FW)

Product	Content	
	As—Inorganic	As—Total
Food of Plant Origin		
Bread[a]	<0.01	<0.005
Miscellaneous cereals[a]	0.012	0.018
Rice, different variety[b]	0.11	0.20
Rice[c]	0.082	0.1144
Rice, white[d]	<0.150	0.210
Vegetables[a]	<0.01	0.005
Fresh fruits[a]	<0.01	0.001
Grape juice[e]	0.0054	0.0059
Red wine[e]	0.0028	0.0039
White wine[e]	0.0048	0.0050
Beer[e]	0.0035	0.0036
Rice beer[e]	0.0057	0.0071
Food of Animal Origin		
Meat, poultry, game, and their products[f]	0.004	0.029
Fish[a]	0.015	3.99
Fish, different species[g]	<0.002–0.006	0.3–110
Fish and seafood and their products[f]	0.015	1.9
Poultry[a]	<0.01	0.022
Eggs and their products[f]	0.023	0.037
Dairy products[a]	<0.01	<0.003

Sources: [a] Rose, M. et al., *Food Addit. Contam.*, 27, 1380–1404, 2010.
[b] Jorhem et al. (2008a).
[c] Liang et al. (2010).
[d] Heitkemper et al. (2009).
[e] Huang et al. (2012).
[f] Chung, S.W. et al., *Food Addit. Contam.*, 31, 650–657, 2014.
[g] Julshamn et al. (2012).
FW, fresh weight.

are acutely toxic and carcinogenic. Methylated forms of As^{5+}, namely, DMAA and MMAA are significantly less toxic, but have been implicated as cancer promoters (IARC 1987). Additional species such as AB, arsenocholine (AC), and a variety of arsenosugars are found in fish, shellfish, and edible seaweeds. They have historically been considered to be nontoxic; however, much less is known regarding the biotransformation of these species and arsenosugars, in particular, than inorganic As species (Conklin et al. 2006).

Consumption of seafood contributes largely to human As intake. Because organic As compounds (e.g., AB) are the predominant As species in seafood, As toxicity from seafood consumption is not significant. General consensus in the literature is that about 85%–90% of the As in the edible parts of marine fish and shellfish is organic As (e.g., AB, AC, and DMA) and that approximately 10% is inorganic As. However, the inorganic As content in seafood may be highly variable. For example, a study in the Netherlands reported that inorganic As comprised of 0.1%–41% of the total As in seafood (Vaessen and van Ooik 1989).

The greatest threat to public health from As originates from contaminated groundwater. Inorganic As is naturally present at high levels in the groundwater of a number of countries, including Argentina, Bangladesh, Chile, China, India, Mexico, and the United States of America. Drinking water, crops irrigated with contaminated water, and food prepared with contaminated water are the sources of As exposure. It is particularly devastating in Bangladesh. Millions of tube wells were installed to provide *pure water* to prevent morbidity and mortality from gastrointestinal disease, but the water was not tested for As contamination. According to survey data from 2000 to 2010, an estimated 35 to 77 million people in the country have been chronically exposed to As in their drinking water, in what has been described as the largest mass poisoning in history. The tube wells draw the As-containing groundwater from a shallow depth of 10–70 m. Groundwater from depths >150 m usually contains less As and can be a sustainable drinking water source. Between 2000 and 2003, 4.94 million tube wells throughout Bangladesh were tested for As and were marked as safe or unsafe. Areas showing high proportions of unsafe wells (i.e., wells in which As concentration in water is >50 μg/L, the Bangladesh drinking-water quality standard) are largely the same areas experiencing the highest As concentrations (often >200 μg/L) (Flanagan et al. 2012).

Studies in other countries where the population has had long-term exposure to As in groundwater indicate that 1 in 10 people who drink water containing 500 μg of As/L may ultimately die from cancers caused by As, including lung, bladder, and skin cancers. It is estimated that the lifetime risk of dying from cancers of the liver, lung, kidney, or bladder while drinking 1 liter a day of As-containing water at concentration 50 μg/L could be as high as 13 per 1000 persons exposed; the risk estimate for 500 μg/L of As in drinking water would be 13 per 100 people (Smith et al. 2000).

The Joint FAO/WHO Expert Committee on Food Additives (JECFA) established the provisional tolerable weekly intake (PTWI) of 0.015 mg/kg bw for inorganic As (equivalent to 0.0021 mg/kg bw/day) (WHO 1989). But focused on more recent data showing effects at lower doses of inorganic This PTWI for inorganic As was withdrawn by JECFA because it was not possible to establish a safe exposure (WHO 2011b).

Cereals, fruits, vegetables, meat, poultry, fish, and dairy products can also be dietary sources of As (Table 3.4), although exposure from these foods is generally much lower compared to the exposure through contaminated groundwater. In seafood, As is mainly found in its less toxic organic form.

After Heitkemper et al. (2009), the average daily intake of inorganic As from rice in the United States would be approximately 0.003 mg. Ethnic groups that consume higher amounts of rice daily would have a daily intake of approximately 0.0105 mg of inorganic As. In the United Kingdom, the population dietary exposure to total As

TABLE 3.4
Arsenic in Animal Tissues (mg/kg FW)

Animals	Muscles	Liver	Kidney
	Tissues		
Cattle, agricultural area[a]	0.023	0.017	0.043
Cattle, industrial area[a]	0.010	0.037	0.093
Cattle[b]	0.007	0.016	0.020
Pigs[b]	0.003	0.007	0.007
Cross-bred bisons[b]	0.008	0.015	0.026
Wild pigs, industrial area[c]	0.018	0.034	0.032
Deer, industrial area[c]	0.088	0.059	0.038
Hens[b]	0.006	0.009	–
Carps[b]	0.010	0.014	–
Cods[b]	0.310	3.230	–

Sources: [a] Waegeneers, N. et al., *Food Addit. Contam.*, 26, 326–332, 2009a.
[b] Żmudzki and Szkoda (1994).
[c] Kucharczak et al. (2005).
FW, fresh weight.

was estimated to be 0.061 mg/day but to inorganic As, it was between 0.0014 and 0.007 mg/day (Rose et al. 2010).

Total estimated daily dietary intake of As may vary widely, mainly because of wide variations in the consumption of fish and shellfish. Most data reported are for total As intake and do not reflect the possible variation in the intake of the more toxic inorganic As species. Mean daily intake of As in food for adults is estimated to range from 0.0167 to 0.129 mg (WHO 2011b).

3.7 ANIMALS

Arsenic enters into the human and animal bodies through ingestion, inhalation, or skin absorption. After entering into the body, it is distributed into a large number of organs, including the lungs, liver, kidney, and skin. Arsenic residues in animal tissues are often in relation to their surrounding habitat, especially coming from industrial area (Table 3.4). Moreover, arsenicals are still used as antiparasitic agents in veterinary medicine.

4 Barium [Ba, 56]

4.1 INTRODUCTION

Barium (Ba) is an alkaline metal of the group 2 in the periodic table of elements and occurs in the Earth's crust at levels 550–668 mg/kg (mean 425 mg/kg). It has a lithophilic affinity and is likely to be concentrated in acidic igneous rocks (400–1200 mg/kg) and argillaceous sediments (400–800 mg/kg). Barium follows the K fate in geochemical processes and may be concentrated in phosphate rocks, and thus in P fertilizers, often as barium phosphate, $Ba_3(PO_4)_2$. In coal, Ba contents vary within the range 150–500 mg/kg, and in brown coal, it may concentrate up to about 1500 mg/kg. According to Finkelman (1999), Ba concentration in some coals may be up to 22,000 mg/kg. In fly ash, its average amount is estimated at 398 mg/kg.

Barium occurs at +2 oxidation state. Its common minerals are barite, $BaSO_4$; whiterite, $BaCO_3$; and hollandite, $Ba_2Mn_8O_{16}$.

Total barite production in 2010 was 6 900 kt, of which (in kt) China, 3600; India, 1000; and the United States, 670 (USGS 2011). Ba has varied industrial uses, and mainly in the production of bricks, ceramics, and glasses. It is widely applied in chemical industries, including rubber, paints, and plastics production. Nearly 95% of the barite sold in the United States is used as a weighting agent in gas- and oil-well drilling fluids. Barite is used in X-ray medical treatments, because it significantly blocks X-ray and gamma-ray emissions. It is also used as aggregate in high-density concrete for radiation shielding around X-ray units in hospitals, nuclear power plants, and various nuclear research facilities.

4.2 SOILS

Barium contents in soils vary from 10 to 1500 mg/kg, with the lowest in organic soils and the highest in loamy soils (Table 4.1). Mean content calculated for top soils in the United States is 580 mg/kg (Burt et al. 2003), and this value for European soils is 400 mg/kg (FOREGS 2005). Average Ba background concentration for worldwide soils is estimated at 460 mg/kg. However, it may considerably vary, depending on the geology of the parent materials.

The criteria for the contaminated land (Dutch List 2013), following Ba concentrations in soils and groundwater, are established (in mg/kg, and μg/L) as follows: uncontaminated, 200 and 50; medium contaminated, 400 and 100; and heavily contaminated, 2000 and 500. An interim soil-quality criterion for Ba content in agricultural soils of Canada has been established at 750 mg/kg (Jaritz 2004).

During weathering processes, Ba precipitates easily with sulfates and carbonates, and is also strongly adsorbed by clays, oxides, and hydroxides. It is relatively mobile, especially in acidic soils, although it may be strongly adsorbed, especially by Mn nodules. Also, some minerals, for example, mica, reveal a great sorption capacity

TABLE 4.1
Barium Contents in Soils, Water, and Air

Environmental Compartment	Range/Mean
Soil (mg/kg)	
Light sandy	85–780
Medium loamy	200–1500
Heavy loamy	200–1500
Organic	10–700
Water (μg/L)	
Rain	0.33–1.2
River	4.0–73
Sea, ocean	2.0–15
Air (ng/m³)	
Urban	>100
Rural areas	0.2–90
Antarctica	0.02–0.73

Sources: Data are given for uncontaminated environments, from Kabata-Pendias, A. and Mukherjee, A.B., *Trace Elements from Soil to Human*, Springer, Berlin, Germany, 2007; Reimann, C. and de Caritat, P., *Chemical Elements in the Environment*, Springer, Berlin, Germany, 1998.

for Ba. Often, Ba occurs as hollandite, an easily soluble mineral, which affected its mobility, and, in aridic soils, its upward migration.

Barium source may be from some aerial deposition, especially from the surroundings of P-fertilizer plants. In most cases, Ba input from aerial sources and with P fertilizers increases its output (leaching and plant uptake), and the Ba budget in rural soils is slowly increasing. Barium is found in most land soils at low levels; however, it may be at higher levels at hazardous waste sites.

4.3 WATERS

The range of Ba concentration in seawater is estimated at 2–15 μg/L, and in river and stream water at 4–73 μg/L (Table 4.1). Global average of Ba in river water is calculated as 23 μg/L (Gaillardet et al. 2003). A high concentration of Ba, up to 15,000 μg/L, is reported for polluted surface water (Jaritz 2004). However, in most aquatic environments, there are sufficient concentrations of sulfate anions to bond the Ba cations, and to keep its restively low Ba levels.

Water of some coal mines is enriched in Ba, which resulted in increased Ba contents (up to 1%) in bottom sediments of mine ponds (Pluta 2001). Bottom sediments of uncontaminated water contain Ba within the ranges 9–81 and 37–355 mg/kg, in sandy and mule sediments, respectively (Bojakowska et al. 2007).

Median Ba concentration in bottled water of the EU countries is 30.6 μg/L, which is about the same in tap water at 30.1 μg/L (Birke et al. 2010). In Poland, drinking water contains Ba up to 60 μg/L. The guideline value of Ba concentration in drinking water is established at 700 μg/L by the WHO (2011a), whereas this value estimated by the European Commission is 1000 μg/L (EU 1998a).

4.4 AIR

The range of Ba concentrations in air of remote area (Antarctica) is 0.02–0.73 ng/m^3 (Table 4.1). Ba contents in air of rural and urban regions increase up to 90 ng/m^3 and above 100 ng/m^3, respectively. Its main sources are mainly oil and coal combustion (ATSDR 2007a).

4.5 PLANTS

Barium is easily uptake by all plants, especially from acidic soils. Its mean contents in the most plants range from 2 to 14 mg/kg. Its content may be higher in some shrubs and trees, especially in areas of arid climate. Local conditions have an important impact on Ba in plants, which is illustrated by various contents, for example, blueberries sampled in Russia contain Ba at the range 8–60 mg/kg, whereas those sampled in Germany contain at the range 32–181 mg/kg (Markert and Vtorova 1995).

Mean Ba concentration in food plants is as follows (in mg/kg): wheat grain, 3.2; cereal grain, 5.5; carrot, 13; tomato, 2.1; and apple, 1.5.

High Ba content (about 2000 mg/kg) inhibited the growth of some plants. Addition of Ca and S compounds reduces the Ba toxicity, possible due to the formation of slightly soluble $BaCO_3$ and $BaSO_4$.

4.6 HUMANS

Amount of Ba found in the human body (average 70 kg) is 22 mg, of which in the blood 70 μg/L, in bones 3–70 mg/kg, in other tissues 0.1 mg/kg (Emsley 2011), and in the rib bone 2.5 mg/kg (Zaichik et al. 2011).

Barium is primarily distributed to the bone and teeth; it is not known if Ba distributed to these tissues would result in toxicity. However, there are no data correlating Ba levels in tissues and fluids with exposure levels (ATSDR 2007a).

Humans are exposed to Ba from a variety of sources such as the environment, where it naturally occurs; drinking water; and the diet. It may also be absorbed in working places, as it is used in many industrial applications. After absorption, Ba accumulates in the skeleton; an accumulation also takes place in the pigmented parts of the eye.

Toxicity of Ba compounds depends on their solubility. In humans, ingestion of high levels of soluble Ba compounds may cause gastroenteritis (vomiting, diarrhea, and abdominal pain), hypopotassemia, hypertension, cardiac arrhythmias, and skeletal muscle paralysis. The free ion is readily absorbed from the lung or gastrointestinal tract.

Oral absorption of soluble Ba^{2+} is highly variable, in both animals and humans. There is evidence that gastrointestinal absorption of Ba in humans is in the range 3%–60% of the administered dose, although higher spotted values are also reported.

Studies in rodents and dogs estimated an oral absorption between 7% and 85%. In humans, once absorbed, 90% of the Ba in the body is deposited into the bones. There is no evidence that Ba undergoes biotransformation other than as a divalent cation (SCHER 2012).

Soluble Ba compounds would generally expect to be of greater health concern than insoluble Ba compounds, because of their greater potential for absorption. Various Ba compounds have different solubilities in water and body fluids, and therefore serve as variable sources of the Ba^{2+} ion. This ion and the soluble compounds of Ba (notably chloride, nitrate, and hydroxide) are toxic to humans. Although Ba carbonate is relatively insoluble in water, it is toxic to humans, because it is soluble in the gastrointestinal tract. However, Ba sulfate or other insoluble Ba compounds may potentially be toxic when it is introduced into the gastrointestinal tract under conditions where there is colon cancer or perforations of the gastrointestinal tract,

TABLE 4.2
Barium in Foodstuffs (mg/kg FW)

Product	Content
Food of Plant Origin	
Bread	0.81
Miscellaneous cereals	0.74
Potatoes	0.17
Green vegetables	0.465
Root vegetables[a]	0.92
Fresh fruits	0.422
Fruits[a]	1.29
Sweeteners, honey, and confectionery[b]	0.045–5.10
Beer, different country of origin[c]	0.016–0.068
Curry, samples[d]	0.632–5.85
Food of Animal Origin	
Carcass meat	0.03
Meat products	0.33
Fish and fish products[b]	0.045–0.945
Eggs	0.33
Milk[e]	0.026
Dairy products	0.22

Sources: Rose, M. et al., *Food Addit. Contam.*, 27, 1380–1404, 2010, unless otherwise stated.

[a] Howe, A. et al., *Geochem. Health*, 27, 19–30, 2005.
[b] Millour, S. et al., *J. Food Compos. Anal.*, 25, 108–129, 2012.
[c] Pohl, P., *Food Addit. Contam.*, 25, 693–703, 2008.
[d] Gonzalvez, A. et al., *Food Addit. Contam.*, 1, 114–121, 2008.
[e] Gabryszuk, M. et al., *J. Elem.*, 15, 259–267, 2010.

FW, fresh weight.

and Ba is able to enter the bloodstream. Pulmonary baritosis (or simply known as baritosis) is a type of benign nonfibrotic pneumoconiosis, which can be precipitated by the inhalation of Ba particles (usually $BaSO_4$). Inhaled Ba particles can lie in the lungs for years, without producing symptoms or causing impairment in lung function (ATSDR 2007a).

Chronic exposure to Ba is often associated with effects on the cardiovascular system. There may be a connection between its high levels in drinking water and high blood pressure, as well as cardiac mortality (Reilly 2002).

The most important route of exposure to Ba appears to be the ingestion of Ba through drinking water and food. Particles containing Ba may be inhaled into the lungs, but little is known regarding the absorption of Ba by this route. However, there are data that suggest that Ba may interact with other cations and certain drugs.

The cations K, Ca, and Mg also interact with Ba. Barium exposure, for example, may cause a buildup of K inside the cell, resulting in extracellular hypokalemia, which is believed to mediate Ba-induced paralysis. In fact, K is a powerful antagonist of the cardiotoxic and paralyzing effects of Ba in animals, and is used as an antidote in cases of acute Ba poisoning. Ca and Mg suppress the uptake of Ba by pancreatic islets, *in vitro*. Conversely, Ba, in low concentrations, stimulates Ca uptake in these cells (ATSDR 2007a).

Barium is present in all food groups, at varying concentration, with the highest in nuts (Table 4.2). Tolerable daily intake of 0.2 mg Ba/kg bw/day is set by the WHO. Mean daily intake of Ba was estimated from 0.75 to 1.33 mg/person, including both food and fluids (WHO 2001b). In the United Kingdom, in 2006, the population dietary exposure to Ba was 0.847 mg/day, and this increased from 0.58 mg/day reported in 1994 (Rose et al. 2010).

5 Beryllium [Be, 4]

5.1 INTRODUCTION

Beryllium (Be) is an alkaline metal (the lightest of the alkaline Earth elements) of the group 2 in the periodic table of elements and reveals lithophilic affinity. Its average content of the Earth's upper crust is calculated within the range 4–6 mg/kg. It is likely to concentrate in acidic igneous rocks (2–6.5 mg/kg) and in argillaceous sediments (2–6 mg/kg). Contents of Be in coal vary from 10 to 330 mg/kg, and concentrations up to 2000 mg/kg are also reported (Veselý et al. 2002). Its average content in fly ash is about 10 mg/kg, but may be much higher.

Beryllium occurs at +2 oxidation state. It may easily substitute Ca, Al, and Si in some minerals. Among its many minerals, the most common are beryl, $Be_3Al_2Si_6O_{18}$; bertrandite, $Be_4Si_2O_7 \cdot H_2O$; and chrysoberyl, $BeAl_2O_3$. The first two minerals are of commercial importance.

Beryllium occurs in several isotopes, of which ^{9}Be is stable, and ^{7}Be and ^{10}Be are cosmogenic, and also produced by nuclear reactions.

World mine production of Be in 2010 was about 190 t, of which 170 t was mined in the United States (USGS 2011). Industrial uses of Be are of great importance as it is very light and very resistant metal. Cu–Be alloy is broadly used in the metallurgy, and especially for the construction of aircrafts, rockets, atomic reactors, and so on. Beryllium is also used in various electrical components.

5.2 SOILS

Beryllium contents in worldwide soils vary from 0.1 to 5 mg/kg (Table 5.1). Its highest contents are in heavy loamy soils, apparently due to its easy fixation by clay minerals, mainly montmorillonite. It is also easily bonded by soluble organic matter (SOM). The mean Be content in the U.S. soils is estimated as 0.6 mg/kg, within the range of 0.46–1.14 mg/kg (ATSDR 2002a). In Japanese soils, Be contents vary from 0.5 to 1.5 mg/kg in Andosols and Gleysols, respectively (Takeda et al. 2004). Swedish arable soils contain Be in the range from <0.5 to 1.8 mg/kg (Eriksson 2001a).

Recently, most data for Be in soils are related to contaminated soils. Increased beryllium levels in soil might be due to geological influence; for example, soils over Be-rich granite in Czech Republic contain up to 15 mg Be/kg (Veselý et al. 2002). Anthropogenic impact might result in much higher Be concentrations, up to 55 mg/kg in soils, in the neighborhood of the Be factories in Japan (Asami 1988).

Beryllium in soils occurs mostly as divalent cation; however, several anionic complexes are noticed, such as $Be(OH)CO_3^-$ and $Be(CO_3)_2^{2-}$. It is rather immobile in soils, but some of its salts (e.g., $BeCl_2$ and $BeSO_4$) are easily soluble and toxic to plants. Its concentrations in soil solution ranges from 0.2 to 1.1 μg/L, mainly as cations, for example, Be^{2+} and $BeOH^+$, and as anions, for example, BeO_3^{2-} and $Be(OH)_3^-$

TABLE 5.1
Beryllium Contents in Soils, Water, and Air

Environmental Compartment	Range/Mean
Soil (mg/kg)	
Light sandy	0.1–1.7
Medium loamy	0.7–2
Heavy loamy	2.5–5.0
Water (μg/L)	
Rain	<0.01–0.22
River	<0.008–0.6
Ocean	0.07–0.11
Air (ng/m³)	
Urban	0.1–0.5
Rural areas	0.1–0.4
Antarctica	0.1–0.4

Sources: Data are given for uncontaminated environments, from Kabata-Pendias, A. and Mukherjee, A.B., *Trace Elements from Soil to Human*, Springer, Berlin, Germany, 2007; Reimann, C. and de Caritat, P., *Chemical Elements in the Environment*, Springer, Berlin, Germany, 1998.

(Kabata-Pendias and Sadurski 2004). Most Be in soils does not dissolve easily in soil solution and remains in bound forms, especially at higher pH levels. SOM have a great sorption capacity for Be, that increases at high pH.

5.3 WATERS

The common range of Be concentration in river water ranges from <0.008 to 0.6 μg/L (Table 5.1). Its average dissolved load in surface water is calculated as 0.009 μg/L, and its world riverine flux is 0.33 kt/yr (Gaillardet et al. 2003). The highest Be concentration in water is at pH 4. In acidic conditions, Be is easily mobilized and leached out from soils, fly ash, slag, and various wastes. Increased Be levels in wastewater might be of a real concern.

Median Be concentration in bottled water of the EU countries is 0.003 μg/L, and is similar to its median contents in tap water (Birke et al. 2010). Vaessen and Szteke (2000) presented mean contents of Be in tap water of various countries as follows (in μg/L): United States, 0.013; Germany, 0.008; Netherlands, <0.1; and Saudi Arabia, 1.24. According to the data of the ATSDR (2002a), drinking water in the United States contain Be at an average value of 0.19 μg/L. Beryllium is rarely found in drinking water at concentrations of health concern, therefore a guideline value is not established (WHO 2011a). Its threshold value for wastewater is 820 μg/L (ATSDR 2002a).

5.4 AIR

The range of Be concentrations in air in the remote area (Antarctica) is 0.02–0.73 ng/m³ (Table 5.1). Its content in air of rural and urban regions increase up to 90 and above 100 ng/m³, respectively. Mean Be content of ambience air in the United States is 0.03 ng/m³, whereas in cities, it is 0.2 ng/m³ (ATSDR 2002a).

Main sources of beryllium are mainly oil and coal combustion. Be emissions resulting from coal combustion in EU countries in 1990 was 90 t (Pacyna and Pacyna 2001). At the same time, 450 t of Be compounds was released in the United States (Kabata-Pendias and Mukherjee 2007).

Increased levels of Be and its compounds in air are a great concern, because their inhalation is highly toxic to humans and animals (Rossmann 2004).

5.5 PLANTS

Beryllium uptake by plants is easy and fairly similar to the uptake of other divalent cations, such as Ca^{2+} and Mg^{2+}. These elements reveal antagonistic interactions and often Be may substitute Mg, and limits the availability of Ca and P, which disturbs metabolic processes. Most Be taken up by plants remains in roots, and only its small proportion is translocated to above-ground parts.

Mean contents of Be in food plants vary broadly within the range from <0.07 to 42 µg/kg FW in beans and green pepper, respectively (Vaessen and Szteke 2000). Increased level of As, up to 25 µg/kg FW, has been noticed in corn (ATSDR 2002a). In sites where rocks and soils contain higher amounts of Be, as, for example, in Bohemit region (Czech Republic), its amounts in needles and bark of spruce increase up to 310 and 78 µg/kg, respectively (Veselý et al. 2002). Be toxicity in plants is observed at its high levels (10–50 mg/kg) in soils. But sensitive plants, for example, lettuce, may be affected at 2 mg Be/kg in soil (Hlušek 2000).

Increased Be levels in plants can be a good indicator of pollution, mainly industrial. Moss samples from Norway (collected during the period 1990–1995) contain Be within the very broad range, from <0.4 to 370 µg/kg, indicating an impact of aerial pollution (Berg and Steinnes 1997).

5.6 HUMANS

The amount of Be found in the human body (average 70 kg) is approximately 35 µg (Emsley 2011). Average concentrations of Be were measured in human organs as follows (in µg/kg): 0.21 in lungs; 0.08 in the brain; 0.07 in both kidneys and the spleen; 0.04 in the liver, muscles, and vertebrae; and 0.03 in the heart (ATSDR 2002a).

Beryllium is one of the most toxic elements to living organisms. Toxicologically relevant exposure to Be and its compounds are almost exclusively confined to the workplace. Before the introduction of improved emission control, and hygiene measures in Be plants, several *neighbourhood* cases of chronic Be diseases were reported. In 1930s and 1940s, several hundred cases of acute Be diseases occurred, particularly among workers in Be-extraction plants of many countries (Vaessen and Szteke 2000).

General population can be exposed to trace amounts of Be through inhalation of air, consumption of food and water, and skin contact with air, water, or soil that contain Be. Beryllium and its compounds are not biotransformated, but soluble Be salts are partially converted to less soluble forms in the lungs (ATSDR 2002a).

Exposure to Be and its compounds is much more hazardous by inhalation than by ingestion, because they are poorly absorbed after oral and dermal exposure. The respiratory tract in humans and animals is the primary target of Be toxicity following inhalation exposure. Occupational exposure to high concentrations of soluble Be compounds can result in acute Be disease, whereas exposure to relatively low concentrations (0.5 μg/m^3) of soluble or insoluble Be compounds can result in chronic Be disease. Acute Be disease is characterized by an inflammation of the respiratory tract tissues and is usually resolved within several months of exposure termination. In contrast, chronic beryllium disease or berylliosis is immune response to Be, and is only observed in individuals who are sensitive to Be (usually <15% of exposed population). Other systemic effects, that have been observed in individuals with severe cases of chronic Be disease include damage to the right heart ventricle, hepatic necrosis, kidney stones, and weight loss. Dermal contact with Be can result in an allergic response, especially skin granulomas, in certain individuals. Dermatitis, which may be due to direct irritation rather than an immune response, has also been observed in workers exposed to high concentrations of airborne Be. Tobacco smoke is another potential source of exposure to Be in the general population. Beryllium levels of 0.47, 0.68, and 0.74 μg/cigarette were found in three brands of cigarettes. Between 1.6% and 10% of the Be content, or 0.011–0.074 μg/cigarette, was reported to pass into the smoke during smoking (WHO 2001a).

Inhaled and parenterally administered Be salts lead to its accumulation in the skeletal system. Following oral exposure, Be accumulates mainly in bones, but is also found in several other organs and soft tissues. Systemic distribution of the more soluble Be compounds is greater than that of the insoluble compounds. Transportation of Be across the placenta has been shown in rats and mice treated by intravenous injection (ATSDR 2002a).

Soluble Be compounds are absorbed to a greater degree (~20% of the initial lung burden) than sparingly soluble compounds (e.g., Be oxide). Ingested Be is poorly absorbed (<1%) from the gastrointestinal tract. Dermal exposure of Be and its compounds in humans can result in a delayed-type (cell-mediated) hypersensitivity skin response.

Beryllium is chemically similar to Mg and therefore can displace it from enzymes, which causes them to malfunction. It can deplete the store of Mg in the body, lodge in vital organs, and prevent them from performing their functions, and interfere with a number of enzyme systems of the body. IARC (1993) designates Be and its compounds as carcinogenic to humans (Group 1). Tolerable daily intake of 2 μg/kg bw was estimated (WHO 2011a). The content of Be in food is small (Table 5.2). There is a surprising lack of information on levels of Be in foods and beverages, in spite of the considerable interest in its environmental and toxicological importance (Reilly 2002).

TABLE 5.2
Beryllium in Foodstuffs (μg/kg FW)

Product	Content
Food of Plant Origin	
Banana pulps	4.2
Apple juice	10.8–43.6
Cabbages	0.2
Mushrooms, European wild	<5–36
Rice	3–5
Potatoes	33–59
Tomatoes	0.2
Wine[a]	0.01–1.0
Food of Animal Origin	
Meat	4.0
Fish	11.6–19
Clams	2
Crabs	15
Eggs	0.06
Milk	0.2
Milk[b]	1.0

Sources: Vaessen, H.A.M.G. and Szteke, B., *Food Addit. Contam.*, 17, 149–159, 2000, unless otherwise stated.
[a] Marengo, E. and Aceto, M., *Food Chem.*, 81, 621–630, 2003.
[b] Ikem, A. et al., *Food Chem.*, 77, 439–447, 2002.
FW, fresh weight.

The U.S. Environmental Protection Agency estimated the total daily Be intake as 0.423 μg, with the largest contributions from food (0.12 μg/day) and drinking water (0.3 μg/day), and smaller from air (0.0016 μg/day) (WHO 2001a). Daily Be exposure of the U.S. citizens was estimated to be 0.00006 μg by inhalation of air, 0.4 μg from drinking water, and 12 μg from food. Thus, its sources are mainly from food (ATSDR 2002a). The total Be intake from various sources is estimated to range from 0.12 to 100 μg/day (Emsley 2011; Vaessen and Szteke 2000).

6 Bismuth [Bi, 83]

6.1 INTRODUCTION

Bismuth (Bi) is a metal of the group 15 in the periodic table of elements and reveals both lithophilic and chalcophilic properties. Its contents on the Earth's upper crust range from 0.06 to 0.17 mg/kg. In sedimentary rocks, Bi contents are a little bit higher, up to 0.5 mg/kg. It is likely to concentrate in coal and graphite shales, up to 5 mg/kg. Its increased levels in volcanic emissions are reported (Ferrari et al. 2000).

Bismuth occurs mainly at +3 oxidation state. At the +5 oxidation state, it is likely to form stable compounds such as $NaBiO_6$ and BiF_5. Its common minerals are bismuthinite (Bi_2S_3) and bismite (Bi_2O_3). Bismuth may be concentrated at host minerals, such as galena, sphalerite, and chalcopirite. It naturally occurs as metal, and its crystals are present in the sulfide ores of some metals such as Ni, Co, Sn, and Ag.

World mine production (without the United States) of Bi in 2010 was 7600 t, of which China produced 5100 t (USGS 2011). Bismuth is also obtained from old scraps and as a by-product from Pb and Cu smelting, especially in the United States. Bi has varied industrial uses, especially in the production of low-melting shot, shotguns, glass, and ceramics. Bi compounds are used as catalysts in manufacturing synthetic fibers and rubbers. Some of its compounds are used in pharmaceutical and cosmetic production. Bismuth is an environmental-friendly substitute for Pb in plumbing and many other applications.

6.2 SOILS

Average Bi content in worldwide soils is estimated at 0.42 mg/kg; however, there are not much data on its occurrence in soils. Bismuth content in Polish soils is within the range of 0.03–0.45 mg/kg, and is associated mainly with the clay-soil fraction (Pasieczna 2012). Therefore, its content in heavy loamy soils is much higher than in medium loamy soils (Table 6.1).

In some soils derived from rocks of increased Bi contents, its levels may be higher. In ferralitic soils, it may be concentrated up to 10 mg/kg (Kabata-Pendias and Mukherjee 2007). Background Bi content in Slovakian soils is given as 1.3 mg/kg, but also concentration up to 37.2 mg/kg is reported (Čurlik and Šefčik 1999). In raw humus of some soils, Bi concentration at 92 mg/kg is reported (Tyler 2005). According to Hou et al. (2006), Bi is fixed mainly by Fe and Mn oxides and soluble organic matter. Thus, an increase of Bi in soil horizons rich in these compounds should be expected.

Sewage sludge and some P fertilizers are often a significant source of Bi in soils. According to Eriksson (2001a), Bi concentrations in materials used for soil amendments are as follows (in mg/kg): sewage sludge, 0.73; manure, 0.72; P fertilizers, 0.18. Additional source of Bi in soils, in Canada, are pellet, for which production is used, Bi instead of Pb (Fahey et al. 2008).

TABLE 6.1
Bismuth Contents in Soils and Water

Environmental Compartment	Range/Mean
Soil (mg/kg)[a]	
Medium loamy	0.1–0.4
Heavy loamy	1.3–1.5
Water (μg/L)[b]	
Rain	<0.01–0.05
River	0.006–0.04
Sea, ocean	0.02–0.04

Sources: [a] Kabata-Pendias, A. and Mukherjee, A.B., *Trace Elements from Soil to Human*, Springer, Berlin, Germany, 2007.
[b] Reimann, C. and de Caritat, P., *Chemical Elements in the Environment*, Springer, Berlin, Germany, 1998.

Radionuclides of Bi (212 and 214) may occur in some phosphate rocks, and thus may be added with P fertilizers to soil.

6.3 WATERS

Bismuth concentration in seawater and ocean water range from 0.02 to 0.04 μg/L, and its mean concentration is estimated at 0.02 μg/L (Reimann and de Caritat 1998). In river water, it may be even at lower level, at 0.006 (Table 6.1). Species of Bi in seawater are mainly BiO^+ and $BiCl^-$. Bismuth is easily fixed by Fe hydrous oxides, and therefore is likely to be concentrated in bottom sediments.

Mean concentration of Bi in rainwater collected in Sweden during 1999 was 0.003 μg/L, and its wet deposition was estimated at 33 mg/ha/yr (Eriksson 2001a).

6.4 AIR

Global atmospheric concentration of Bi is estimated to be up to 3 ng/m^3 (Kabata-Pendias and Pendias 1999). Natural sources of Bi are geothermal power plants and volcanic eruptions. Its anthropogenic sources are mainly from smelting metal ores, coal combustion, and waste incineration.

6.5 PLANTS

Apparently, Bi is taken up by plants passively with water and from aerial deposits. Standard Bi content in plants is estimated at 10 μg/kg (Markert 1992). However, in some plants, especially in trees, its contents may be highly elevated to about 800 μg/kg (Kabata-Pendias and Pendias 1999).

Berg and Steinnes (1997) reported mean Bi amount as 33 μg/kg, within the range 1–800 μg/kg, in mosses collected in Norway in 1995. Its mean content of mosses decreased to 33 μg/kg in 2000. Increased Bi levels resulted mainly from aerial pollution. Grass (*Agrostis scabra*), plots from where Bi pellets were used, contains elevated Bi amounts, up to 95 μg/kg (Fahey et al. 2008).

Wheat and barley grains contain Bi at the mean values of 3 and 5 μg/kg, respectively (Eriksson 2001a).

6.6 HUMANS

Amount of Bi found in the human body (average 70 kg) is less than 500 μg (Emsley 2011); in the rib bone is 15 μg/kg (Zaichik et al. 2011). Normal concentration of Bi in the blood is between 1 and 15 μg/L. Distribution of Bi in the organs is either largely independent of the compound administered or the route of administration: the concentration in the kidneys is always highest and retained for a long time. It is bound to a Bi metal-binding protein in the kidneys, the synthesis of which can be induced by the metal itself (Slikkerver and de Wolf 1989).

A number of toxic effects have been attributed to Bi compounds in humans: nephropathy, encephalopathy, osteoarthropathy, gingivitis, stomatitis, and colitis. Each of these adverse effects is associated with certain Bi compounds. Bismuth encephalopathy occurred in France as an epidemic of toxicity and was associated with the intake of inorganic salts including Bi subnitrate, subcarbonate, and subgallate (Slikkerver and de Wolf 1989).

Bismuth and its salts can cause kidney damage, although the degree of such damage is usually mild. Large doses can be fatal. Due to the low stability in aqueous solutions of Bi compounds, with +5 oxidation state, Bi with the +3 oxidation state is regarded as the only relevant Bi species in biological systems. It is seen as the least toxic metal for humans, and is widely used in medical applications for its good antibacterial properties, mainly because of their low uptake into human cells.

Compounds containing Bi are, therefore, widely used in medical applications. Bismuth-containing pharmaceuticals, partially in synergy with antibiotics, are already used or are being considered in the treatment of infections caused by certain bacteria, especially to eradicate *Helicobacter pylori*, *Pseudomonas aeruginosa*, and others. However, careless use of Bi-containing pharmaceuticals can result in encephalopathy, renal failure, and other adverse effects. Both the benefits and the adverse effects of Bi are based on the same property of the metal, that is, its strong affinity to thiols groups of proteins. It is important that the concentration of Bi applied in medication does not cause an increased accumulation of the metal in the cytoplasm of human cells (Thomas et al. 2012).

Microbial methylation of Bi by the human gut microbiota resulting in more mobile and presumably more harmful derivatives has recently been reported. Transformation of several elements such as Bi, As, Sb, Se, and Te into volatile derivatives by methylation or hydridization plays an important role in the spreading and cycling of these elements in natural and anthropogenetically modified environment. Because many of these volatile derivatives are more toxic than their (mostly inorganic) precursors, these processes may have an impact on human health. A high risk can be expected from scenarios in which these derivatives are accumulated in closed systems, such as compartments of all living organisms (Michalke et al. 2008).

TABLE 6.2
Bismuth in Foodstuffs (μg/kg FW)

Product	Content
Food of Plant Origin	
Bread	<1
Miscellaneous cereals	<1
Sugar and preserves	5
Vegetables	0.5
Potatoes	0.4
Fruits	0.3
Food of Animal Origin	
Meat products	<0.5
Poultry	<0.5
Fish	0.6
Milk	2.0
Dairy products	6.4

Source: Rose, M. et al., *Food Addit. Contam.*, 27, 1380–1404, 2010. With permission.

FW, fresh weight.

After the ingestion of inorganic Bi, the intestinal microbiota, in particular methanoarchaea, are capable of methylating inorganic Bi to soluble, partially methylated compounds, such as monomethyl (MMBi) and dimethylbismuth (DMBi), as well as volatile trimethylbismuth (TMBi). TMBi is characterized by a higher volatility and hydrophobicity in comparison with inorganic Bi, and can be therefore easily distributed inside the human body, and is able to pass the blood–brain barrier (Bialek et al. 2011).

Bismuth is not considered a human carcinogen. Its toxicity from foods has not been described and levels of Bi in foods are normally low, ranging from 0.3 to 6.4 μg/kg FW in all the food groups (Table 6.2).

Estimates by different authors' daily intake of Bi from food and water ranges from 2 to 30 μg/day. The dietary exposure of population in the United Kingdom was 2 μg Bi/day (Rose et al. 2010).

7 Boron [B, 5]

7.1 INTRODUCTION

Boron (B), a metalloid of the group 13 in the periodic table of elements, occurs in the Earth's crust at an average amount of 10 mg/kg. It is likely to be concentrated to about 30 mg/kg in acidic igneous rocks. In sedimentary rocks, its contents are higher than in igneous rocks, and are closely associated with the clay fraction; for example, in quaternary clay, its amounts are up to 96 mg/kg. The highest B contents, up to 145 mg/kg, are in limestones. It may also be concentrated in some carboniferous sediments and coals, as well as in fly ash, where its mean content is estimated at 509 mg/kg (Llorens et al. 2000).

Boron is widely distributed in the hydrosphere, and its water solubility highly controls its distribution in the environment. The largest quantities of B are accumulated in marine evaporites and marine argillaceos sediments. Borate minerals are usually deposited as evaporates (e.g., borax and kernite). Boron may also be increased in some volcanic emissions, mainly as orthoboric acid, H_3BO_3, and thus it occurs in higher amounts in volcanic rocks.

Boron oxidation state is +3. It is a constituent of several minerals, and some of its common minerals are borax, $Na_2B_4O_7{\cdot}10H_2O$; colemanite, $Ca_2B_6O_{11}{\cdot}5H_2O$; ulexite, $NaCaB_5O_9{\cdot}8H_2O$; kernite, $Na_2B_4O_6{\cdot}3H_2O$; and tourmaline, $NaFe^{2+}3Al_6(BO_3){\cdot}3Si_6O_{18}(OH)_4$. Boron is often associated with feldspars and micas.

World (without the United States) production of B, mainly as boric oxide (B_2O_3), in 2010 was 3500 kt (rounded), of which 1200 kt (18% of the global B demand) was produced in Turkey, as colemanite (USGS 2011). The United States supplies about 70% of the global B demand.

The major industrial-scale uses of B compounds are in the production of fiberglass and heat-resistant borosilicate glass and ceramics. Addition of B increases resistance to the thermal shock of various products. It is used in the synthesis of organic chemicals. Boron is widely used in fertilizers and pesticides, cosmetics, antiseptics, leather tanning, and laundry products. Also, a few boron-containing organic pharmaceuticals are produced. Borax ($Na_2B_4O_7{\cdot}10H_2O$) is a commonly used compound. Boron is a good neutron absorber and is used to control nuclear reactions.

7.2 SOILS

Boron contents in soils vary between 10 and 100 mg/kg, being enriched in calcisols (Table 7.1). World B average for soils is estimated at 42 mg/kg. Average B contents in various soils of the United States are within the range 20–55 mg/kg. A high B content, up to 1622 mg/kg (average 65 mg/kg), is reported for some soils in the Slovak Republic (Čurlik and Šefčik 1999). Boron is inherited mainly from the parent rocks, but industrial pollution, sludge, coal combustion, fly ash, and B fertilizers are

TABLE 7.1
Boron Contents in Soils and Water

Environmental Compartment	Range/Mean
Soil (mg/kg)[a]	
Light sandy	10–23
Medium loamy	15–35
Heavy loamy	15–40
Calcerous (Calcisols)	10–100
Water (μg/L)[b]	
Rain	0.5–2.9
River	1.5–150
Sea, ocean	4500

Sources: [a] Data are given for uncontaminated environments, Kabata-Pendias, A. and Mukherjee, A.B., *Trace Elements from Soil to Human*, Springer, Berlin, Germany, 2007.
[b] Reimann, C. and de Caritat, P., *Chemical Elements in the Environment*, Springer, Berlin, Germany, 1998.

its significant sources. It is more strongly sorbed in soils than other anions, by both organic and inorganic soil components. Its sorption increases with pH value, and the highest is at alkaline pH, about 9. Adsorbed B on soil minerals is rather easily leachable, but irreversibility of B sorption may also occur.

During weathering of rocks, B goes into the soil solution, where it occurs mainly as anions: BO_2^-, $B_2O_7^{2-}$, $H_2BO_3^-$, and $B(OH)_4^-$, and may be concentrated up to about 3000 mg/L, which is about 5% of its total soil contents (Kabata-Pendias and Pendias 1999). Boron movement in soils follows the water flux, and thus in cool humid-zone soils, it is leached downward in soil profiles, whereas in warm-zone soils, and especially in arid and semiarid regions, it is likely to be enriched in the surface soil layers. In these soils, excess of B may create agricultural problems, whereas in other soils, it is rather a deficient micronutrient. Water contaminated with B used for the irrigation resulted in its increased concentration in soils, up to 124 mg/kg (Lucho-Constantino et al. 2005).

Excess contents of B are associated mainly with solonetz soils and contaminated soils. There are several procedures for soil remediation: (1) light acidic soils—irrigation with water; (2) heavy sodic soils—amended with gypsum; and (3) various soils—phytoremediation, especially with grass and *Brassica* species.

7.3 WATERS

Mean boron concentration in worldwide seawater is estimated at 4500 μg/L (Table 7.1). Its content in river water is variable, ranging from 1.5 to 150 μg/L, whereas its average abundance is given as 10 μg/L. World average, annual river flux of seawater is calculated as 380 kt (Gaillardet et al. 2003). Boron concentration in surface water of the United States ranges from 1 to 5000 μg/L (Butterwick et al. 1989). Fairly similar broad ranges of B in river water are given for several other countries. Both natural and anthropogenic factors are responsible for this phenomenon. Anthropogenic sources of B in water are mainly sewage sludge, in which its maximum concentration is in the United States, 4000 μg/L, and in Europe, 5000 μg/L (Butterwick et al. 1989).

B concentration of groundwater is of an agricultural concern, especially in warm and arid climatic countries. Worldwide B contents in water of these regions range from <300 to >100,000 μg/L. The highest B concentrations in the water of the European countries are noticed in Italy and Spain (ECETOC 1997).

The most common B species in natural water are boric acid, $B(OH)_3$, and its dissociation products, such as $B(OH)_4^-$ and other borates. Soluble Al compounds react with B in water resulting in the formation of Al–B insoluble complex. Several other complex compounds, such as polyborates and fluoroborates, may also be formed in water, and decrease B solubility.

Boron concentration in rainwater is relatively low (Table 7.1). Rainwater sampled in Sweden, in 1999, contained B within the range of 1.3–2.9 μg/L, and its wet deposit was calculated at 18 g/ha/yr (Eriksson 2001a).

Median B concentration in bottled water of the EU countries is 48 μg/L, and is higher than those in tap water, 16 μg/L (Birke et al. 2010). Guideline value for B concentration in drinking water is established by the WHO at 2400 μg/L (WHO 2011a).

Boron is an essential micronutrient to aquatic plants, fish, and amphibians, but may be toxic, especially borate compounds, at concentrations above 10,000 μg/L. Some species of salmon and trout are sensitive to the increased B levels in water. Aquatic plants of rivers with polluted water from industrial or agricultural sources may contain B up to 170 mg/kg. However, it does not biomagnify in fish and other aquatic food chain (Kabata-Pendias and Mukherjee 2007). Marin and Oron (2007) suggested using duckweek (*Lemna gibba*) accumulating B from water up to 1900 mg/kg, which may reduce B concentrations significantly.

7.4 AIR

B concentrations in air commonly range between 0.5 and 80 ng/m³. Its natural sources are emissions from sea surface and volcanic activities. Industrial sources include various mining operation, glass, ceramic, and chemical productions. The highest amounts of B are released from coal combustions, especially fired power plants. The global anthropogenic B emission was estimated at 180–650 kt/yr, whereas from marine sources at 800–4000 kt/yr (Anderson et al. 1994). According to these authors, the total global emission of B may be up to 72,000 kt/yr. Boron from the atmosphere is easily deposited on the Earth, with rain and dry particles.

7.5 PLANTS

Soluble species of B are easily available to plants and may be taken up passively or actively. Properties of boric acid to complex with polysaccharides stimulate passive sorption of B by plants. In most cases, however, B absorption follows the water flow through the roots. Thus, its uptake is proportional to its concentration in water of growth media. All B species are available, but the most easily available to plants is boric acid, as well as all water-soluble B compounds (Kabata-Pendias 2011). Several soil properties have an impact on B availability to plants, but the most important seems to be the Ca:B ratio (Azarenko 2007).

Boron is an essential element for plant growth and its biochemical role is still under investigation. Its significant functions as a structural element of plant cell walls are closely related to cell metabolic processes (Goldbach and Wimmer 2007). It has a complexing ability to trigger the movement of sugars and other compounds, especially those involved in biocomplexing membrane bindings. Thus, an adequate B supply to sugar beets is necessary for sugar synthesis. Functions of B have not been identified with any specific enzyme; however, its activities are related to some basic processes as follows:

- Carbohydrate metabolism and transport of sugars through membranes
- Nucleic acids (DNA and RNA) and phytohormone syntheses
- Structural integrity of the plasma membrane (boric acid is involved in the linking of some cell components)
- Tissue and cell-wall developments
- Formation of B complexes with constituents of cell walls and plasma membranes

B deficiency has been reported in over 130 crops from 80 countries. Its deficiency may occur in all countries, but most often in Asian and African countries. This deficiency affects some commercial crops such as sugar beets, celery, sunflower, legumes, and apples. Critical B levels for most plants range at 5–30 m/kg. Mycorrhizal plants have a greater need for B supply than do nonmycorrhizal plants. It is related to a beneficial effect of B on the nodule formation and nitrogenase activity.

Enhanced B levels in soils can be harmful to crop plants, especially lemon and orange trees. Several other crops, such as cereals and cotton, when cultivated in irrigated areas might be affected by increased B contents. B toxicity is usually more common in arid and semiarid regions. B toxicity in some crops (mainly cereals and sunflower) occurs on soils contaminated due to (1) irrigation with municipal wastewater or B-enriched river water, in aridic climatic zones; (2) amendment with fly ash; and (3) foliar application of B, mainly in citrus or apple orchards. Toxic B content in most plants range from 330 to 350 mg/kg, for spinach and corn, respectively (Kabata-Pendias 2011).

Highest B contents are in fodder plants within the following ranges (in mg/kg): alfalfa, 26–40; clover, 14–33; and grass, 5–26. Mean B contents in cereal grains are within the range 0.69–3.3 mg/kg, with the highest value for oat grains (Eriksson 2001a).

There is often a narrow range of toxic and deficient B levels to plants. Boron levels in the most food plants cultivated in humid zones do not exit 18 mg/kg, whereas in fodder plants it is higher. In cotton grown on irrigated land, B contents may be up to 422 mg/kg. For animal nutrition, B level above 150 mg/kg is not recommended.

Soil contamination with B is considered now as a widespread agro-environmental problem. Therefore, the phytoremediaton of such soils has been investigated. The highest B removal from soils by crop plants is with lucerne (350 g/ha) and sugar beet (300/ha) (Shorrocks 1997). Poplar (*Populus* sp.) accumulated up to 845 mg B/kg, and is suggested by Robinson et al. (2007) to remove excess B from soils.

Boron interacts in the uptake of other nutrients by plants because it has an influence on the membrane permeability and cell colloids (Goldbach et al. 2007). Mechanisms of these reactions are still not well understood. The possible antagonisms with some elements, such as Cu, Cr, Mo, and Mn, may be the indirect effects of increased demands for these micronutrients, due to the increased plant growth. Other interactions are partly explained as follows:

- B–Zn: At low levels of available Zn plants may uptake B to toxic levels in plants, especially in roots.
- B–Si: Antagonism is an effect of competition of silicate ions for adsorption sites of B.
- B–Ca: Interrelationship is relatively often observed. It is mainly lime-induced B deficiency in acid soils. Therefore, toxic effects of B may be reduced by adding Ca to soils, especially in the form of $CaSO_4$.
- B–P: Interactions occur in soil media where P ions decrease B mobility. The uptake and distribution of P is dependent on the B concentration because B increases P immobility in roots. These two elements influence the integrity of cell membrane, thus their imbalances may lead to the aberration in their uptake.

Other interactions are also observed with elements such as K, N, and Al. It may be, however, a secondary effect associated with some physiological disorders in plants.

7.6 HUMANS

Amount of B found in the human body (average 70 kg) is 18 mg (Emsley 2011). In the human tissues, its concentrations are as follows (in mg/kg FW): in the kidneys, 0.6; in the lungs, 0.6; in the lymph nodes, 0.6; in the blood, 0.4; in the liver, 0.2; in the muscles, 0.1; in the testes, 0.09; in the brain, 0.06 (Health Canada 1991), and in the rib bone, ≤0.65 (Zaichik et al. 2011).

Boron, in intracellular and extracellular spaces, is categorized as a possible essential nutrient for humans, but this has not been yet directly proven (Coughlin and Nielsen 1999; EPA 2008a; Meacham et al. 2010). Nielsen (2008) suggests that in recent years the focus on B has shifted from toxicological effects and dietary recommendations to nutritional essentiality and biochemical mechanisms of action.

Boron plays an important role in mineral and hormonal metabolisms, cell membrane functions, and enzyme reactions. Boron also affects osteoporosis (is involved in Ca and bone metabolism), heart trouble, paralysis, diabetes, and senility. Its effects are more marked when cholecalciferol (vitamin D_3) and Mg are deficient. Boron is involved in biochemical indices associated with the metabolism of other nutrients, including Ca, Cu, N, and cholecalciferol, in the synthesis of extracellular matrix and is helpful in wound healing. It may be involved in cerebral function due to its effects on the transport across membranes. Compounds of B have been shown to be potent antiosteoporotic, anti-inflammatory, hypolipemic, anticoagulant, and antineoplastic agents, both *in vitro* and *in vivo* in animals. Improvements in brain function, bone density, immunity, cardiovascular effects, development and reproduction, and the incidence and prevalence of cancer biomarkers have been stated. However, the cellular mechanism of B has not yet been identified.

It is stated that in areas of the world where B intakes usually were 1.0 mg B or less a day, the estimated incidence of arthritis ranged from 20 to 70%, whereas in areas of the world where B intakes were usually 3 to 10 mg B, the estimated incidence of arthritis ranged from 0 to 10% (Meacham et al. 2010).

Dietary B has a low level of toxicity. A number of acute poisonings in human from boric acid or borax have been reported following ingestion, parenteral injection, lavage of serous cavities, and enemas and application of dressings, powders, or ointments to large areas of burned or abraded skin. Symptoms of acute poisoning by high doses of B include nausea, vomiting, diarrhea, headache, desquamation, and evidence of central nervous system stimulation followed by depression, dermatitis, lethargy, and other symptoms. In the extreme cases, kidney damage occurs, followed by circulatory collapse and death. It is not absorbed across intact skin. However, there is evidence that B can be absorbed through more severely damaged skin, especially from an aqueous vehicle and during inhalation exposure.

Ingested B is rapidly taken up from the gastrointestinal tract, with possibly more than 90% absorbed and is rapidly distributed throughout body liquids. A small amount appears to be retained in body tissues, especially bone and spleen (Reilly 2002). Boron is mostly present in body tissues and fluids as boric acid $B(OH)_3$, and in lesser amounts as anion $B(OH)_4^-$.

Due to the lack of human and the limited animal data, B is not classified as human carcinogen (WHO 1998). For humans, B exposure occurs primarily through the diet and drinking water. Boron is a notable contaminant or major ingredient of many nonfood personal-care products (Hunt et al. 1991). The amount of B intake by people is dependent on the types of food consumed. The richest sources of B are fruits, vegetables, pulses, legumes, and nuts. Dairy products, fish, meats, and most grains are poor sources of boron (Table 7.2). The tolerable daily intake for B is estimated at 0.17 mg/kg bw, rounded to 0.2 mg/kg bw (WHO 2011a). For the general population, the greatest B exposure comes from the oral intake of food (65%) and water (30%). According to the WHO, it is expected that the amounts of B taken are 0.44 μg/day through air, 0.2–0.6 mg/day through drinking water, and 1.2 mg/day through diet (WHO 1998). Dietary intake of B in the United Kingdom ranges from 0.8 to 1.9 mg/day. However, increased consumption of specific foods, with high B content, will increase B intake significantly; for example, one serving of wine or avocado provides 0.42 and 1.11 mg, respectively. Other B significant sources may be from cosmetics.

According to Meacham et al. (2010), B from dietary supplements and personal-care products is almost 100% absorbable through epithelial membranes. Some

TABLE 7.2
Boron in Foodstuffs (mg/kg FW)

Product	Content
Food of Plant Origin	
Bread	0.5
Miscellaneous cereals	0.9
Cornflakes[a]	0.31
Fresh fruits	3.4
Green vegetables	2.0
Potatoes	1.4
Nuts[a]	14–18
Peanut butter[b]	5.9
Wine[b]	3.6
Avocado[b]	14.3
Beer[a]	1.8
Honey[a]	7.2
Food of Animal Origin	
Beef, round, raw[a]	<0.015–<0.05
Chicken, breast, raw[a]	<0.015–0.09
Meat products	0.4
Eggs	<0.4
Fish	0.5
Milk	<0.4
Milk[c]	0.091

Source: Ysart, G. et al., *Food Addit. Contam.*, 16, 391–403, 1999, unless otherwise indicated.
[a] Hunt, C.D. et al., *J. Am. Diet. Assoc.*, 91, 558–568, 1991.
[b] Meacham, S. et al., *Open Mineral. Process. J.*, 3, 36–53, 2010.
[c] Gabryszuk, M. et al., *J. Elem.*, 15, 259–267, 2010.
FW, fresh weight.

B-containing products for adults are authorized to be used for specific structure-function claims as follows:

- Boron is important for bone metabolism and the calcification of bones. It affects Ca, Mg, and P levels.
- Boron and vitamin D facilitate the utilization of Ca.
- Boron is involved with the efficient absorption of Ca in the body.
- Calcium plus elements such as B and Mg are needed for optimum bone mineralization.

8 Bromine [Br, 35]

8.1 INTRODUCTION

Bromine (Br) is a chemical element of the group 17 (halogen) in the periodic table of elements. Its total contents of the Earth's crust is estimated at 1–2.4 mg/kg (mean 2 mg/kg) and is relatively uniformly distributed among igneous rocks, whereas in argillaceous sediments, it may be accumulated up to 10 mg/kg. Bromine is a very volatile element and its salts are readily soluble and mobile in most environments. It may, however, be concentrated in coal, at an average of 9.1 mg/kg, and accumulated up to 160 mg/kg (Finkelman 1999).

Elemental Br is a fuming red-brown liquid at room temperature, which is corrosive and toxic. Its chemical properties are similar to Cl and I. It is very reactive chemically and can have several valence states. Its most common oxidation state is −1, but may be from +1 to +7. However, as an one-electron atom, it can form only compounds with oxygen: Br_2O, BrO_2, BrO_3. Minerals containing Br are not common, owing to both its low abundance and high solubility. Its more often occurring minerals are bromyrite, AgBr; embolite, Ag(Cl,Br); and iodobromite, Ag(Cl,Br,I).

World total production (without the United States) of Br in 2010 was 380 kt, of which China and Israel produced 140 kt and 130 kt, respectively (USGS 2011). Commercially, Br is extracted from brine pools, mostly in the United States, Israel, and China. Bromine compounds are used for various purposes: in well-drilling fluids, film photography, flame-retardant compounds such as polybrominated diphenyl ethers (PBDEs), and as an intermediate in the manufacture of organic chemicals. Considerable amounts of Br are used in agriculture, in various pesticides.

Ethylene bromide ($C_2H_4Br_2$) was an additive in gasoline (especially used in some aircraft engines) in order to form volatile lead bromide ($PbBr_2$), which is exhausted from the engines. This Br application has declined since 1970s. There has been also a proposal of using Br to mitigate Hg emissions at the power plants, because Br compound bonds with Hg are more easily captured in flue-gas scrubbers than in $HgCl_2$ produced at many facilities. Now, half of Br produced is used for fire retardants brominated flame retardant (BFR) (West 2008).

Reactive Br species of anthropogenic origin affect ozone by directly destroying it and by reducing its production (von Glasow et al. 2004). As a result, many organobromide compounds that were formerly in common use (e.g., methyl bromide in pesticides) have been abandoned.

8.2 SOILS

Boron content in soils is closely associated with soil types. Its lowest amounts, below 10 mg/kg, are in acid light sandy soils (Table 8.1). In heavy loamy soils, Br may be up to 100 mg/kg. The highest Br concentrations, above 100 mg/kg are in volcanic ash

TABLE 8.1
Bromine Contents in Soils, Water, and Air

Environmental Compartment	Range/Mean
Soil (mg/kg)	
Light sandy	<0.1–8.0
Medium loamy	10–40
Heavy loamy	50–100
Calcareous	70–130
Water (mg/L)	
Rain	<0.2–15
River	0.002–4.5
Sea, ocean	67.7
Air (ng/m^3)	
Industrial regions	150–500
Greenland	7–20
Antarctica	0.4–1.4

Source: Data are given for uncontaminated environments, from Kabata-Pendias, A. and Mukherjee, A.B., *Trace Elements from Soil to Human*, Springer, Berlin, Germany, 2007; Reimann, C. and de Caritat, P., *Chemical Elements in the Environment*, Springer, Berlin, Germany, 1998.

soils. Its increased level is also observed in soils close to sea coasts, which is an effect of Br evaporation from sea. Soils in Japanese seas accumulated Br up to 495 mg/kg (Yuita 1983). The reference Br range in soils in the United States is between 1.4 and 7.8 mg/kg (Govindaraju 1994).

Presently, the anthropogenic sources of Br in soils are associated mainly with bromomethane (CH_3Br) application for the fumigation of soils with crop plants. Due to these practices, Br in soil may increase significantly, especially in soil solution, because Br^- anionic forms are very mobile. Carbon, hydroxides of Al and Fe, organic matter and clay fraction may accumulate a great proportion of Br, and thus affect its increased level in soils and sediments. Despite the great sorption capacity for Br in some soils, it is the element most easily leached out from soil profiles and transported to water basins.

Increased Br levels due to the bromomethane fumigation of soils can result in almost complete eradication of population of a wide variety of soil microflora and fauna, and thus alter the trophic structure of soils (Pavelka 2004).

Criteria used in the Netherlands for contaminated land (Dutch List 2013) are following for Br concentrations in soils and groundwater (in mg/kg, and μg/L): uncontaminated, 20 and 100; medium contaminated, 50 and 500; and heavily contaminated, 300 and 2000.

8.3 WATERS

Mean bromine concentration in ocean water is estimated at 67.7 mg/L (Table 8.1). In stream water, its contents are much lower, and do not exceed 4.5 mg/L. In brine and salt lake water, its concentration may be elevated, for example, up to about 7000 mg/L in the Dead Sea. The high solubility of bromide ions has caused its accumulation in water.

Some seaweeds absorbed Br in very high amounts, and might be a source of this element. Organobromide compounds are needed for the metabolism of sea plants, especially of some algae.

A great difference between Br in rainwater collected from Kola Peninsula (<0.2 mg/L) and from the United Kingdom (15 mg/L) clearly indicates anthropogenic impact (Reimann et al. 1999).

8.4 AIR

Bromine concentration in air vary highly. In remote areas, it is within the range 0.4–20 ng/m^3, whereas over industrialized regions it ranges from 150 to 500 ng/m^3, and might increase up to about 2500 ng/m^3. Industrial activities and decomposition of Br pesticides are the most significant Br sources in the atmosphere. Common Br compounds in the air are Br_2O_3, BrCl, $BrONO_2$, and CH_3Br.

At high temperatures, and especially under sunlight, organobromine compounds are readily converted to free Br atoms in the atmosphere, and an unwanted side effect of this process is ozone depletion.

8.5 PLANTS

Neither mechanisms of Br uptake by plants nor its function in plants are known. Most probably, Br is taken up by plants with the water flow from soils and by aerial parts from the air. Plants grown in Br-enriched soils always contain more of this element. It is especially visible in plants grown in volcanic ash soils, containing up to 2000 mg Br/kg. Increased levels of Br in plants cultivated in greenhouses, after a fumigation with CH_3Br (at the rate of 75 g/m^3) are a real evidence of aerial Br adsorption by plants. Methyl bromide and other Br organic compounds used as fumigants in greenhouses may be a serious source of Br in human diets.

Natural Br contents in crop plants do not exceed 30 mg/kg, but in forage plants, it might be much higher, up to 120 mg/kg in grasses. Mean B contents of cereal grains are 5.5 mg/kg in barley and 3.1 mg/kg in oats (Eriksson 2001a). Highly elevated Br contents, up to 300 mg/kg, are reported in some food plants (Pavelka 2004).

Usually, Br is present more in shoots than in roots and tubers. Its contents in plants do not correlate with any soil properties, but with its amounts in soils. Thus, plants have more Br when grown in soils enriched with Br. Apparently, it might be taken up also by the leaves from air. Especially, Br pesticides have a significant impact on its contents in plants.

Plants differ in their tolerance to Br in soil. Some vegetables and flowers such as potato, spinach, sugar beet, onion, carnation, and chrysanthemum are known to be sensitive to high Br levels. Resistant to Br toxicity are carrot, tomato, celery, melon, and tobacco. These plants may accumulate Br at about 2000 mg/kg, without showing any effects. Common symptoms of Br toxicity are chlorosis, necrosis, and dark green color of young leaves. Citrus seedlings are the most sensitive to excess Br in soils, and may be used as an indicator of excess Br in soils.

8.6 HUMANS

According to Emsley (2011), the Br content in human body is 260 mg, and averages in the blood at 5 mg/L, in the bones and tissues at 7 mg/kg, and in the rib bones at ≤3.9 mg/kg (Zaichik et al. 2011).

Exposure to Br may occur via inhalation, oral, dermal, or ocular routes. It is a gas and, therefore, inhalation exposure is the most relevant route to humans. Bromine is corrosive to human tissue in a liquid state, and its vapors irritate eyes and throat. Inhalation of Br vapors is very toxic.

Pure Br is a serious health hazard to humans. As a gas, it can burn the eyes, throat, and lungs. As a liquid spilled on skin, it produces painful blisters and burns. Swallowed Br can damage the nerves and brain.

Bromine metabolism in most living organisms is that of the bromide ion. Species differences in tissue Br concentrations are small and the element does not accumulate to any marked degree in any particular organ or tissue, which might indicate a specific physiological function of this ion. After oral ingestion, Br is rapidly and completely absorbed in the gastrointestinal tract, and analogously to chloride, distributed almost exclusively in the extracellular fluid (with the exception of erythrocytes) (Pavelka 2004).

Following ingestion, Br liquid is absorbed from the intestine by passive, paracellular transport. Bromine crosses blood cell membranes in an electrically neutral form and is distributed widely into various tissues, and mainly into the extracellular fluid of the body. Following inhalation, Br is absorbed by the lungs, and its deposition in the lungs is primarily determined by the water solubility of Br. High inhalation of Br concentrations, for example, in confined spaces, may also cause marked irritant effects on lungs.

There are no data regarding the metabolism of inhaled Br; however, Br has been known to quickly form bromide in living tissue. Bromide is partitioned in the body similar to Cl and acts by replacing itself. Bromide ion is a central nervous system depressant, and its adverse effects are results of overdoses.

The most important health effects that can be caused by Br-containing organic contaminants are malfunctioning of the nervous system and disturbances in genetic materials. Br organic can also cause damage to organs such as liver, kidneys, lungs, and milt; they can cause stomach and gastrointestinal malfunctioning, too. Some forms of Br organic compounds, such as ethylene bromide, can even cause cancer.

Humans can absorb organic Br through the skin, with food and during breathing. Organic Br is widely used as sprays to kill insects and other unwanted pests. However, they are poisonous not only to the organisms, which they are used against, but also to larger animals, as well as to humans.

Through food and drinking water, humans absorb high doses of inorganic Br. These bromines can damage the nervous system and the thyroid gland. Bromine disrupts the thyroid gland and interferes with the production of thyroid hormones. Thyroid gland relies on I obtained through the foods to produce thyroid hormones, which are essential for normal growth, development, and metabolism. Because Br is very similar to I, it can take the place of I, which results in less I for the thyroid gland, and may also increase the elimination of I from the body.

One type of Br compounds, diphenyl ether (PBDE), appears to be building up in people living in the United States. PBDEs are used in electronic equipment and furniture. The United States is the world's largest producer of PBDEs. Americans have 10 to 20 times more PBDEs in the body than the Europeans. Europeans have double the amount of PBDEs than the Japanese (West 2008).

BFRs, including PBDEs, have been used at increasing levels in home furnishings and electronics over the past 25 years. BFRs are ubiquitous industrial chemicals, and many of them are produced in large volumes. Due to this fact, several BFRs are found in quantifiable levels in wildlife, as well as in humans (Darnerud 2003). High PBDE levels have been detected in food, household dust, and indoor air, with subsequent appearance in animal and human tissues. When these data are considered in the combination with the stage-dependent effects of thyroid hormone on brain development, which are also implicated in autistic brains, a hypothesis that PBDEs might also serve as autism risk factors emerges to seem to be true (Messer 2010).

Some Br compounds were classified by International Agency for Research on Cancer (IARC) as potentially carcinogens: K bromate group 2B; bromochloroacetic acid group 2B; bromochloroacetonitrile group 3; bromodichloromethane group 2B; bromoethane group 3; and bromoform group 3 (IARC 2013).

Methyl bromide is widely used for fumigating postharvest commodities, such as wheat and cereals, spices, nuts, dried and fresh fruits, and tobacco. Methyl bromide concentrations usually decrease rapidly after aeration and residues are not detectable after some weeks. Some foods, such as nuts, seeds, and fatty foods, such as cheese, tend to retain methyl bromide and inorganic bromide.

In postharvest fumigation, it is possible that methyl bromide itself, as well as bromide and other possible reaction products, may be found in food (Table 8.2).

TABLE 8.2
Bromine in Products from Retail Outlets in the United Kingdom (mg/kg FW)

Product	Content
Apple	0.2
Aubergine	11
Cabbage	2
Cucumber	7
Orange	0.2
Tomato	3–11
Rice	0.72
White flour	1.22
Pasta	2.40

Source: WHO, *Methyl Bromide*, WHO, Geneva, Switzerland, 1995.
FW, fresh weight.

Bromine is used as a food additive in flour and in some fruit-flavored soft drinks. Bromine is used as an additive in the forms of K bromate and brominated vegetable oil. In citrus-flavored beverages, brominated vegetable oil keeps citrus flavors suspended throughout the fluid. When it is added to flour, $KBrO_3$ strengthens bread dough and improves the texture of finished products.

The amount of Br allowed in food and beverages is limited by the U.S. Food and Drug Administration. Acceptable daily intake (ADI) for Br ion is 1 mg/kg bw/day (WHO 1995). Estimated median daily intake of Br worldwide, from food and water, is 2–5 mg/day. In Japan, the total intake of Br was 8–12 mg, which is about 20% of ADI (Matsuda et al. 1994).

Bromoform (tribromomethane, $CHBr_3$) and dibromochloromethane ($ClCHBr_2$) are formed as by-products when Cl is added to water supply systems. Then, exposure to bromoform may occur from the consumption of chlorinated drinking water. Trihalomethanes (THMs) in swimming pools using Br agents indicate that in swimming pools, besides inhalation, dermal absorption is a relevant route for the incorporation of THMs, particularly those with lower degree of bromination (Lourencetti et al. 2012).

9 Cadmium [Cd, 48]

9.1 INTRODUCTION

Cadmium (Cd) is a metal of the group 12 in the periodic table of elements, and its mean content in the Earth's upper crust is within the range of 0.1–0.2 mg/kg. In igneous and sedimentary rocks, its contents are fairly similar, 0.2–0.3 mg/kg. The lowest Cd contents are in sandstones. Coals may contain variable amounts of Cd, broadly ranging from 0.5 to 170 mg/kg. Especially Zn-bearing coals, of Carboniferous age, are resources of Cd.

Cadmium is considered as one of the most ecotoxic metals that exhibit adverse effects on all biological processes. It reveals very harmful impact on the environment and the quality of food.

Cadmium occurs at +2 oxidation state, and easily forms complex cations and anions, especially with Cl, S, and OH. It exhibits chalcophillic properties and behaves very similar to Zn, but has a stronger affinity for S, and is more mobile in acidic environments. During weathering processes, Cd forms simple, mobile compounds, such as CdO, $Cd(OH)_2$, $CdCl_2$, and CdF_2. They are also isotypic with corresponding compounds of cations such as Co^{2+}, Ni^{2+}, Fe^{2+}, Mg^{2+}, and Ca^{2+}. Common minerals of Cd are greenockite, CdS; octavite, CdSe; and monteponite, CdO. They all are associates with Zn and Pb ore deposits. The host minerals of Cd are sphalerite (αZnS), wurtzite (βZnS), biotite, smithsonite, and amphibolites. Sphalerite is the most common source of Cd.

World production (rounded) of Cd in 2010 was 22,000 t of which China produced 5600t; Republic of Korea produced 3200t; and the United States produced 650t (USGS 2011). Secondary Cd production takes place mainly at NiCd battery recycling facilities. Industrial uses of Cd are varied, especially for NiCd battery production (about 70% of the total use). Cadmium is also added to nonferrous alloys. The other traditional uses of Cd, such as coatings, pigments (yellow), and stabilizers for plastics, have gradually decreased, owing to environmental and health concerns. Concern over Cd toxicity has spurred various recent legislative efforts, especially in the EU countries, to restrict the use of Cd in the most applications.

9.2 SOILS

Average Cd contents in soil worldwide is estimated at 0.41 mg/kg, within the range of mean values for various countries from 0.06 to 1.1 mg/kg. Parent materials are the main factor determining the Cd contents in soils. Its highest contents are in calcerous and organic soils, <0.8 and <2.5 mg/kg, respectively (Table 9.1). In some soils derived from bed rocks enriched in Cd, its concentration may be up to 8.9 mg/kg (Čurlik and Šefčik 1999). Mean Cd content in organic surface soils of Norway is 0.57 mg/kg, ranging from 0.05 to 6.7 mg/kg (Nygard et al. 2012).

TABLE 9.1
Cadmium Contents in Soils, Water, and Air

Environmental Compartment	Range/Mean
Soil (mg/kg)	
Light sandy	0.01–0.2
Medium loamy	0.08–0.3
Heavy loamy	0.2–0.5
Calcerous (calcisols)	0.4–0.8
Organic	0.2–2.5
Water (μg/L)	
Rain	<0.01–0.05
River	0.02–0.05
Sea, ocean	0.07–0.11
Air (ng/m³)	
Urban areas	0.2–150
Rural areas	0.1–4.0
Greenland	0.1–6.0
Antarctica	0.015

Source: Data are given for uncontaminated environments, from Kabata-Pendias, A. and Mukherjee, A.B., *Trace Elements from Soil to Human*, Springer, Berlin, Germany, 2007; Reimann, C. and de Caritat, P., *Chemical Elements in the Environment*, Springer, Berlin, Germany, 1998.

During weathering, Cd goes into soil solutions, where its concentrations vary highly from 0.2 to 300 μg/L. It may occur as both cationic and anionic species, as well as in organic chelates (Kabata-Pendias and Sadurski 2004). According to Sposito and Page (1984), Cd speciation in soil solutions varies depending on soil type, in oxic and acid soils: Cd^{2+}, $CdSO_4^\circ$ and $CdCl^+$; and in alkaline soil: $CdHCO_3^+$. However, Cd in soil solution is present as free cation Cd^{2+} between 55% and 90% and is readily available to plants (Taylor and Percival 2001). Especially, interactions with dissolved organic carbon affect the Cd speciation in soil solution (Ge et al. 2005).

Under anaerobic conditions, Cd in soil solution is governed by sulfide precipitation. The highest proportion of Cd in soil occurs in residual fraction (~40%) and in fraction adsorbed by hydrous oxides (~20%). After sludge application, more Cd became fixed by metal hydrous oxides (Kabata-Pendias 2011). Wang et al. (2012) reported that Cd in contaminated irrigated desert soil occurs mainly in the Fe–Mn oxide-bound fractions. Cd fixation by metal hydrous oxides controls its behavior in soils at low pH (<5.5), whereas at higher pH, mainly clays control the Cd fixation. In alkaline soil, at pH > 7.5, two compounds, $CdCO_3$ and $Cd_3(PO_4)_2$, are mainly responsible for Cd behavior. There are several evidences that Eh–pH controls the formation of these compounds, and thus

Cd mobility in soils. According to Matusik et al. (2008) Cd–phosphate compounds are formed under the influence of soil pH. Also, Cd sorption by goethite increases with soil pH, and reaches the maximum at pH about 7–9 (Violante 2013).

To summarize results of several studies on the Cd fixation in soils, it may be generalized that (1) in all soils, Cd mobility is strongly affected by pH; (2) in acidic soils, the soluble organic matter and sesquioxides largely control Cd activity; (3) in alkaline soils, precipitations with carbonates and phosphates have a great impact on Cd behavior.

Soil microbial activities play a significant role in Cd behavior in soils. Microorganisms may bond with Cd and later on release to the soil solution. Microorganisms have a great capability to accumulate Cd (up to 1120 mg/kg) and thus, the remediation technique based on inoculating soils with microorganisms has been proposed for Cd-contaminated soils (Burd et al. 2000). Some actinomycete strains may absorb much Cd and are proposed for the bioremediation of contaminated soils (Amoroso and Abate 2012). Increasing alkalinity of soils is also suggested for soil remediation, because it decreases Cd adsorption, probably due to the competition from Ca and Mg (Laxen 1985).

Soil contamination with Cd is believed to be the most serious health risk. Elevated Cd contents in soils are mainly human-induced, but sometimes it may also be of geogenic soil origin. For example, soils derived from altered pyritized andesine in Slovakia contain up to 222 mg Cd/kg (Čurlik and Šefčik 1999). Other Cd-enriched soils, up to about 8 mg/kg, are derived from Jurassic or Cretaceous limestone or from bauxite, up to 192 mg/kg (Davies et al. 2003).

Major sources of Cd contamination are atmospheric deposition, and sludge and P fertilizers. About 10%–15% of total airborne Cd emissions arise from natural sources, mainly volcanic activities. It has been calculated that the annual Cd natural emission may range between 150 and 2600 t (Ursinyova and Hladikova 2000). However, the highest Cd sources in soils are wet and dry depositions from metal processing, mainly smelters (Table 9.2). P fertilizers and sewage sludge are adding to

TABLE 9.2
Cadmium Contamination of Surface Soils (mg/kg)

Site and Pollution Source	Country	Maximum Content
Mining areas	England	468
	Belgium	144
	Poland	160
Metal-processing industries	Belgium	1781
	Japan	88
	Poland	270
	United States	1500
Sludged or irrigated farmland	England	167
	Netherlands	57
	Poland	107

Source: From Kabata-Pendias, A., *Trace Elements in Soils and Plants*, 4th ed., CRC Press, Boca Raton, FL, 2011.

soils also significant amounts of Cd. The output of Cd from agricultural soils is less than its input, which affects in its yearly increases in surface soils. The balance of Cd in various European soils was calculated as follows (in g/ha/yr): for forest ecosystems, 3.1–23, and for agricultural soils, 1.3–40 (Kabata-Pendias 2011).

According to the EU regulations, the use of compost materials on agricultural lands is advisable when their Cd contents vary between 0.2 and 1.3 mg/kg (Eckel et al. 2005). The criteria for contaminated land (Dutch List 2013), following Cd concentrations in soils and groundwater, are established (in mg/kg, and μg/L) as follows: uncontaminated, 1 and 1; medium contaminated, 5 and 2.5; and heavily contaminated, 20 and 10. Acceptable Cd levels in soils are estimated, based on various criteria, within the range of 3–3.7 mg/kg (Siebielec et al. 2012).

P fertilizers are also significant sources of Cd in soils; its doses to agricultural soils increased from about 5 g/ha in 1920 to about 100 g/ha in 1980. Manure may also be enriched in Cd, and according to Lourekari et al. (2000), about 50% of Cd added to rural soils may be from this source. Sewage sludge and various composts may add Cd up to 150 g/ha/yr (Kabata-Pendias 2011).

Cadmium contents in phosphate rocks may be very high (in mg/kg P_2O_5), up to 180, but the common range (rounded) is 20–60. Its concentrations in P fertilizers range from 22 to 90 mg/kg P_2O_5, in Finland and Belgium, respectively. Thus, its input to agro-soils varies from 1.8 to 150 g/ha/yr. The guidelines for maximum Cd levels in soils are from (in mg/kg) 0.5 (Finland) to 3.0 (United Kingdom) (Oosterhuis et al. 2000).

Several methods, based on adsorption and desorption processes, solubility and natural attenuation, as well as phytoremediation, have been applied for the remediation of soil contaminated with Cd, as well as with other trace metals (Hamon et al. 2006). Liming is an old practice to reduce phytoavailability of Cd, and it is still effective (Adriano et al. 2004). Addition of a brown coal preparation (composed of coal, fly ash, and peat) decreased Cd phytoavailability (Kwiatkowska-Malina and Maciejewska 2013). Some plant species reveal an unusual ability to uptake Cd and transport it to upper parts. Among several plants, sugarcane is a candidate for Cd phytoremediation; it concentrates much Cd in shoots, up to 451 mg/kg, without symptoms of toxicity (Sereno et al. 2007).

9.3 WATERS

Concentration of Cd in ocean water is estimated at mean values from 0.07 to 0.11 μg/L (Table 9.1) Baltic Sea contains Cd within the range of 0.06–1.99 μg/L (Szefer 2002). Average Cd concentration in river water is calculated as 0.08 μg/L, and its riverine flux is 3 kt/yr (Gaillardet et al. 2003). The annual input of Cd to oceans with stream water is also given as 7.5 kt/yr (Nriagu 1980).

Cadmium contents of surface water may vary highly, especially ranging from 0.05 to 1.0 μg/L. However, much higher values have also been reported (e.g., 6–1000 μg/L) for rivers and estuaries (Kabata-Pendias and Mukherjee 2007). The average range of Cd in European rivers is <0.01–0.1 μg/L (Pan et al. 2010). However, some European rivers contain much higher amounts of Cd, up to 1000 μg/L, as results of sewage sludge inputs (Kabata-Pendias and Pendias 1999). Atmospheric inputs may also be a significant source of Cd, especially in seawater. According to Nriagu and Pacyna (1988), Cd atmospheric input to the oceans is 2.4 kt/yr.

Cadmium is more mobile in seawater than in other water, where it is present in various forms such as Cd^{2+}, $Cd(OH)^+$, and $CdCO_3^\circ$, and as organic and inorganic complexes (Sundby et al. 2004).

Cadmium in water is easily absorbed by aquatic biota, which sometimes may be used for biomonitoring. In most marine biota, Cd contents range from 0.03 to 1.8 mg/kg, in flounder and oyster, respectively (Jeng et al. 2000). However, it might be much higher, as in flounder liver, 2.5 mg/kg (Voigt 2004). Growth of algae is stopped when Cd contents range from 0.05 to 2.0 mg/kg, depending on their sensitivity. Planktons from various lakes contain Cd from 5.6 to 15 mg/kg, indicating water pollution (Kabata-Pendias and Mukherjee 2007). Some aquatic bacteria may absorb high amounts of Cd, up to 55% of its concentration in water. Abyar et al. (2012) suggested that *Achromobacter denitrificans* may be used for Cd removal from polluted water, at a relatively low cost. The great biosorption of Cd reveals Nordmann fir cones (*Abies nordmannaana*), which is optimal at water pH 6.5 (Ozel 2012).

Cadmium content of bottom sediments is a good information on water contamination by this metal. Its average contents in bottom sediments of San River (Poland) vary between <0.5 and 1.7 mg/kg, and is higher in mule sediments (mean 0.24 mg/kg) than in the sandy ones (mean <0.1 mg/kg) (Bojakowska et al. 2007). Its content (in 1995) in stream-bottom sediments of National Park, Montgomery (Pennsylvania State) was <1 mg/kg (Reif and Sloto 1997). Cadmium content of surface-bottom sediments of a harbor in Klaipeda (Lithuania) depends on granulometric composition, and is (in mg/kg, average and maximum, respectively) as follows: in sand, 0.4 and 0.9, and in mud, 0.8 and 1.4 (Galkus et al. 2012).

Assessment limits for Cd in sediments are established as follows (in mg/kg): effects range low, 1.2; effects range median, 9.6; probable effect level (PEL), 3.5; 10; and 3 (EPA 2000, 2013). The Environment Canadian Sediment Quality Guidelines (USGS 2001) gave other values for Cd in lake-bottom sediments (in mg/kg): threshold effects level, 0.6, PEL, 3.5, and probable effect concentration, 4.98.

Median Cd concentration in bottled water of the EU countries is 0.0032 μg/L, and a little bit lower than in tap water, 0.0083 μg/L (Birke et al. 2010). The guideline Cd concentration in drinking water is established at 3.0 μg/L by the WHO (2011a).

9.4 AIR

Cadmium concentration in the atmosphere changes (in ng/m³) from 0.015 in Antarctica to over 4 in rural regions, and to 150 in urban areas (Table 9.1). This clearly shows that anthropogenic Cd emission should be of a real concern. A great proportion of Cd in the air is emitted mainly from smelting and refining of nonferrous metals, fossil fuel combustion, and municipal waste incineration. The natural source of Cd is volcanic emissions. Earlier estimations of anthropogenic Cd versus natural Cd emissions indicated that approximately 8–10 kt/yr is from anthropogenic sources compared to 0. 8 to 1 kt/yr from natural emissions (Nriagu and Pacyna 1988). In the European countries, and worldwide, approximately 85%–90% of total airborne Cd emissions arise from anthropogenic sources.

In the European countries, atmospheric emission of Cd in 2001 was 24 t, of which about 14 t was from metal production and refining, to comparing 158 t of Cd released in 1990 in these countries, which is the evident result of good effects of the technology in nonferrous smelters.

Cadmium and its compounds may be in the atmosphere in both forms, suspended particulate matter and vapor. Therefore, it is very easily transported, even for a long distance. Cd particles greater than 10 μm fall on soils around their released sources. However, there is still Cd deposition from the long, transboundary air transport.

9.5 PLANTS

Cadmium is taken up by plants passively, but may be also absorbed metabolically. Often, there is a linear relationship between Cd in plants and its concentration in the growth media. Soil pH is considered to be the major soil factor controlling Cd uptake. In most cases, the most phytoavailable Cd is in soils at pH ranging from 4.5 to 5.5. However, when Cd is present in complexes and/or chelates, its uptake is not controlled by soil pH. Also, low Eh values (about 20 V) of soils decrease solubility and thus the availability of Cd. Not only soluble Cd species are taken up by plants but also Cd fixed by the Fe–Mn oxide fractions is easily available to some plants (Wang et al. 2012).

Several soil properties control Cd solubility, and therefore predicting its phytoavailability is not simple. Nevertheless, most experiments have indicated that plant Cd is a function of soil Cd (Verma et al. 2007). Liming of soils gives variable results in minimizing Cd uptake. P fertilizers always increase Cd uptake. Also, N fertilizers increase Cd concentration in soil solution and thus its content in wheat grains (Wangstrand et al. 2007). Soil salinity stimulates the formation of several Cd–Cl complexes, which are easily soluble and easy phytoavailable. Cadmium from some composts and sewage sludge may also be easily available to plants (Chaudri et al. 2007). Its availability is under the variable impact of rhizosphere bacteria (Dell'Amino et al. 2008).

Distribution of Cd within plant organs is different, but most often illustrates its transport from roots to tops, in particular, to leaves. However, it may also be an effect of the atmospheric Cd deposition. Significant source of Cd from air is illustrated by its higher content in caps (5.22 mg/kg) than in stalks (1.86 mg/kg) of mushrooms (*Boletus edulis*) from Magurski National Park in Poland (Sembratowicz and Rusinek-Prystupa 2012). The mushrooms, common chanterelles (*Cantharellus cibarius*), grown in mountains contain higher amounts of Cd, 0.88 mg/kg, than those grown in the Baltic Sea coast, 0.43 mg/kg (Falandysz et al. 2012).

Mean Cd contents of cereal grains range from 5.6 to 32 μg/kg, of barley and wheat, respectively (Eriksson 2001a). Fodder plants contain much more Cd, within the range of mean values (in μg/kg): in clover, 80–460 and in grass, 70–400.

Biochemical functions of Cd are variable; it has a strong affinity to sulfhydryl and phosphate groups, to concentrate in proteins compounds, and to accumulate in phytochelatin complexes. It has a toxic impact on plants metabolism, disturbing enzyme activities. It reduces ascorbate and glutathione levels, inhibits chlorophyll formation and affects cell acidosis (Nocito et al. 2008). In general, symptoms induced by increased Cd contents of plants are growth retardation, root damage, inhibition of photosynthesis, disturbed permeability of cell membranes, chlorosis of leaves, and reddish-brown coloration of leaf margin and veins. At excess Cd, inhibitions of microorganism functions and disturbed symbiosis between microbes and plants are also observed.

Some native plants, in particular trees, shrubs, and mosses, may accumulate considerable amounts of Cd (up to about 3 mg/kg) without toxicity symptoms. The highest concentration of Cd (560 mg/kg) is reported for *Thlaspi caerulescens*, which is suggested for the phytoremediation (Felix et al. 1999).

Several elements are known to interact with Cd both in elements uptake by plants and biochemical processes. The most commonly observed interactions are as follows:

- Cd–Zn, variable effects (depressing and enhancing): most often there are synergistic interactions, and final impacts are increased uptake of both metals under their increased levels.
- Cd–Cu: inhibitory effect of Cu on Cd absorption.
- Cd–Mn/Ni: both metals may be replaced by Cd during the uptake processes.
- Cd–Fe: interactions are related do disturbance in the photosynthesis processes.
- Cd–Se: antagonistic effects due to Se–urea complex with Cd, which decreases Se availability.
- Cd–P: both increased and decreased uptakes of Cd are reported. Apparently, it depends on their ratio, as well as on soil properties.
- Cd–Ca: relationship is cross-linked with soil pH. Ions like Ca^{2+} are able to replace Cd^{2+} in some processes and thus inhibits Cd absorption.
- Cd interactions with several macro elements (e.g., Mg and K) are observed and may be related to an impaired effect of Cd on cell membranes.

Cadmium concentration in plants is of a great concern as a pathway of Cd to humans and animals. Therefore, tolerance and plant adaptation to higher Cd levels may create a health risk. Thus, Cd contents of food and fodder plants have been studied widely. This clearly indicates that, in general, plants from contaminated sites contain at about 1000 times higher amounts of Cd than plants grown on uncontaminated soils.

Cd background levels of common food plants that are reported for various countries are fairly similar and low. Mean Cd contents of cereal grains do not exceed 70 μg/kg. Much higher Cd amounts, up to 300 μg/kg, are present in some potato tubers. In potato tubers produced in Quebec, its contents vary from 40 to 200 μg/kg; the proposed tolerance content is 250 μg/kg (Fan et al. 2009). Leafy vegetables, lettuces, and spinach may contain elevated Cd levels between 150 and 400 μg/kg.

All plants grown in soils around metal mines and industrial regions contain highly elevated Cd levels (Table 9.3). Even grass and clover from the sites at the ancient abandoned mine contain 1–5 mg Cd/kg. Among various native plants grown in the polluted areas of the Northern Europe (Kola Peninsula), crowberry (*Empetrum nigrum*) has the highest capacity to accumulate Cd (0.058 mg/kg) (Reimann et al. 2001). Very high Cd levels, especially in lettuces and mushrooms, from areas close to metal industries should be of a real concern. Also, its elevated contents in lettuces and cabbages from some fertilized farmlands should be under the control.

TABLE 9.3
Cadmium Contents of Plants Grown in Contaminated Sites (mg/kg)

Pollution Source	Plant/Part	Range/Mean	Country
Ancient mining areas	Grass	1–2	United Kingdom
	Clover	4.9	United Kingdom
Metal-processing industry	Lettuce	45	Australia
	Lettuce	5–14	Poland
	Spinach	6.4	Zambia
	Grass	8.2	United Kingdom
	Mushrooms	3–56	Finland
Sludged, irrigated, or fertilized farmland	Cereal grains	0.1–1.1	Finland
	Lettuce	8–37	Germany
	Lettuce	0.5–23	United States
	Cabbage	130	Russia
	Sudan grass	0.3–3	Hungary

Source: After Kabata-Pendias, A., *Trace Elements in Soils and Plants*, 4th ed., CRC Press, Boca Raton, FL, 2011.

9.6 HUMANS

Amount of Cd in the human body (average 70 kg) increases with age, and at the age of 50, it is around 20 mg (Emsley 2011). In humans, about 50% of the Cd body burden is in the kidneys. Other major bioaccumulating organs or tissues contributing to the body burden are the liver (15%) and muscles (20%). Quantity of Cd in bones is small (WHO 2011b); in the rib bone, it is 0.044 mg/kg (Zaichik et al. 2011).

Mean Cd contents in women tissues are apparently higher than those in men (in μg/kg): in the lungs, 0.17 and 0.11; in the liver, 1.36 and 0.78; and in the kidneys, 18.1 and 14.6, respectively (Satarug et al. 2010).

Absorption of Cd, following oral exposure, depends on the physiologic status of an organism (age; body stores of Fe, Ca, and Zn; pregnancy history; etc.) and on the presence and levels of other ions and dietary components ingested together with Cd. Absorption of Cd from the diet is about 1%–10% for adults. Individuals with low body stores of Fe may have a higher absorption than those with adequate Fe stores (Concha et al. 2013). Cadmium absorption appears to be higher in newborns and infants, and in contrast to adults, independent of Fe status.

Slow excretion of Cd results in a long biological half-life, which has been estimated to be between 10 and 33 years. Cadmium in blood is used as an indicator of both recent and cumulative exposures, and urinary Cd predominantly reflects cumulative exposure and the concentration of Cd in the kidney. In the general population, normal blood Cd concentrations are in the range of 0.4–1.0 μg/L for nonsmokers, and 1.4–4 μg/L for smokers. Much higher Cd levels have been reported for environmental exposure (above 10 μg/L), and occupational exposure (up to 50 μg/L) (WHO 2011b).

Cadmium absorption after dietary exposure in humans is relatively low (3%–5%), but it is efficiently retained in the kidney and liver, where it is bound to metallothioneinen (MT, a low molecular weight cysteine-rich, intracellular protein) (Pappas et al. 2011).

Cadmium is primarily toxic to the kidney, especially to the proximal tubular cells, where it accumulates over time, and may cause a decrease in the glomerular filtration rate, and eventually renal failure. It can also cause bone demineralization, either through direct bone damage or indirectly, as a result of renal dysfunction (EFSA 2012a). Increased levels of Cd measured in blood or urine have been found to be associated with various cardiovascular end-points, including myocardial infarction, stroke, heart failure, hypertension, and changes in arterial function (aortic pulse wave velocity and carotid, brachial, and femoral pulse pressures). However, the epidemiological evidence for an association between cardiovascular diseases and Cd is weak (WHO 2011b).

Cadmium and its compounds have been classified by the International Agency for Research on Cancer (IARC) as carcinogenic to humans (group 1), with sufficient evidence for lung cancer. Also, positive associations have been observed between exposure to Cd and Cd compounds and cancers of the kidney and prostate. Most of the evidence is derived from high Cd exposure of exposed workers through inhalation (IARC 2012a).

Cadmium in the environment has been a concern since the 1960s, when a painful bone disease was reported to have been caused by Cd industrial pollution in an area in Japan (mine Zn–Pb). People who consumed the polluted rice and drunk the river water over a period of 30 years were found to have accumulated a large amount of Cd in their bodies, which leads to a serious osteoporosis—such as bone disease, known to the Japanese as *itai-itai disease*.

Toxicity of Cd is significantly enhanced for children, who are exposed to even low levels. Adverse effect of Cd exposure on child development resulted from Cd-induced neurotoxicity. Exposure to Cd may have a negative effect on fetal growth. Recently, in a large, population-based, longitudinal mother–child cohort in Bangladesh ($n = 1.616$) an association between maternal Cd exposure and birth size in girls (but not in boys) was observed. Recently, based on data from the same cohort ($n = 1.305$), an association between early-life Cd exposure and lower child intelligence scores is argued (Concha et al. 2013).

Children and infants may have higher exposure to metals, because they consume more food in relation to their body weight and absorb metals more readily than adults. Children with higher Cd levels are three times more likely to have learning disabilities, and participate in special education.

The Joint FAO/WHO Expert Committee on Food Additives (JECFA) noted that the existing health-based guidance value for Cd was expressed on a weekly basis (provisional tolerable weekly intake, or PTWI, but, owing to Cd exceptionally long half-life, considered that a monthly value was more appropriate. The Committee, therefore, withdrew the PTWI of 7 μg/kg bw and established a provisional tolerable monthly intake of 25 μg/kg bw (WHO 2011b), whereas the European Food Safety

Authority nominated a PTWI of 2.5 μg/kg bw, to ensure sufficient protection of all consumers (EFSA 2012a).

Dietary exposure, across European countries, was estimated to be 1.15–7.84 μg/kg bw/week (EFSA 2012a). Mean, monthly dietary exposures in some countries and regions of the world was estimated at (in μg/kg bw/month) 2.2 (Australia), 4.6 (United States), and 12 (Japan) (WHO 2011b).

For protecting the health of the consumers and ensuring fair practices in the food trade. The maximum permissible levels (ML) for Cd (and other contaminants) in certain foodstuffs were set by JECFA and EU (EC 2006; EU 2014) to protect the health of the consumers and ensure fair practices in the food trade. In dietary exposure to Cd, the highest quantity is given by potatoes, miscellaneous cereals, and bread (Table 9.4.). These high levels can result from a combination of both a high concentration in the food and a high level of consumption of this food group (Rose et al. 2010).

TABLE 9.4
Cadmium in Foodstuffs (mg/kg FW)

Product	Range or Mean	Maximum Level[a]
	Food of Plant Origin	
Wheat[b]	0.074–0.111	0.20
Rye[b]	0.030–0.049	0.10
Rice, different type[c]	0.020–0.025	0.20
Bread[d]	0.024	–
Miscellaneous cereals[d]	0.021	–
Cabbage[b]	0.005–0.009	0.20
Carrot[b]	0.024–0.047	0.10
Potatoes[b]	0.021–0.031	0.10
Apple[b]	0.003–0.007	0.05
Strawberry[b]	0.014–0.021	0.05
Cultivated mushroom[e]	0.098	0.20
Honey[e]	0.004	–
	Food of Animal Origin	
Meat[d]	<0.003	0.05
Offal[d]	0.084	0.5–1.0[f]
Eggs[e]	0.001	–
Milk[d]	<0.001	–
Fish[e]	0.031	0.05–0.1; 0.15–0.25[f]

Source: [a] EU, *J. Eur. Union*, L 138/75, 2014.
[b] Szteke, B., *J. Environ. Stud.*, 15, 2a, 189–194, 2006.
[c] Jorhem, L. et al., *Food Addit. Contam.*, 25, 284–292, 2008a.
[d] Rose, M. et al., *Food Addit. Contam.*, 27, 1380–1404, 2010.
[e] Vrooman, V. et al., *Food Addit. Contam.*, 27, 1665–1673, 2010.
[f] Depending on tissue or species.
FW, fresh weight.

TABLE 9.5
Cadmium in Animal Tissues (mg/kg FW)

Animal	Tissue		
	Muscles	**Liver**	**Kidney**
Cattle, agricultural area[a]	0.002	0.191	1.142
Cattle, industrial area[a]	0.004	0.446	2.862
Cattle[b]	0.005	0.200	0.557
Pigs[b]	0.002	0.049	0.301
Horses < 2 years[c]	0.023	2.70	31.82
Horses < 2 years[c]	0.771	40.51	166.79
Hunting animals[b]	0.006	0.225	2.127
Carps[b]	0.007	0.038	–

Source: [a] Waegeneers, N. et al., *Food Addit. Contam.*, 26, 326–332, 2009a.
[b] Żmudzki, J. and Szkoda, J., *Cadmium in the Environment*, The Polish Academy of Sciences, Warsaw, Poland, 2000.
[c] Szkoda, J. and Żmudzki, J., *Cadmium in the Environment*, The Polish Academy of Sciences, Warsaw, Poland, 2000.
FW, fresh weight.

9.7 ANIMALS

Contents of Cd in animal tissues is diverse, being the highest in kidney and the lowest in muscles. Cadmium concentrations in bovine kidneys and livers, from industrially contaminated areas, may be much higher compared to animals from rural areas, and are higher in the tissues of hunting animals and horses (Table 9.5). Cadmium levels in tissue samples increase with age of cattle, especially in horse tissues (Szkoda and Żmudzki 2000; Rudy 2009; Waegeneers et al. 2009b).

Supplementation of chicken diet with Se not only increases Se concentration but also reduces Cd concentration in the tissues. Cadmium is also negatively correlated with Zn and exposes weak correlation with Cu and Fe (Pappas et al. 2011).

10 Cesium [Cs, 55]

10.1 INTRODUCTION

Cesium (Cs) is an alkali metal of the group 1 in the periodic table of elements. Its concentration in the Earth's upper crust is within the (rounded) range 4–6 mg/kg. In igneous rocks, its average content is 4 mg/kg, and increases up to 5–10 mg/kg in argillaceous sedimentary rocks. Its content in coal is about 1 mg/kg.

The main oxidation state of Cesium is +1. It reveals lithophilic properties and a great affinity to aluminosilicates. Its common mineral is pollucite, $(Cs,Na)_2Al_2Si_4O_{12} \cdot H_2O$, which contains 5%–32% Cs_2O, and is of commercial importance.

Data on resources and mine production of Cs are presently not available. There are given information on Cs reserves, based on occurrences of pollucite and other pegmatite minerals (mainly lepidolite-Li). Cesium reserves, in Canada (and world), are estimated at 70 kt. There are information that the next largest Cs source, potentially economic, is in Zimbabwe. Cesium is used in drilling fluid (petroleum exploration), chemical production, electric power generation, electronics, and medical applications.

Cesium has several radioactive isotopes, but only ^{137}Cs has a long half-life of about 30 years, and is used in medical treatments (cancer). However, it is of a high health risk if released into the environment. Great amounts of this isotope were released during the Chernobyl accident in 1986.

10.2 SOILS

Cesium contents in various soils vary from <1 to 30 mg/kg (Table 10.1). Its mean concentration in soils of Europe is 5.6 mg/kg (FOREGS 2005). Its levels in Bulgarian soils are 2.2–16.7 mg/kg, being the highest in surface forest soils and chernozems. This may suggest that soluble organic matter (SOM) has the capability to absorb Cs. Cesium contents in soils from various countries are reported as follows (in mg/kg): Japan, 1–11; Switzerland, 0.3–4.5; Bulgaria, 2.2–16.7; and the United States, 0.4–5.1 (Kabata-Pendias 2011).

Cesium and its compounds are easily water soluble, but in soils, they are strongly adsorbed by clay minerals, especially by illites and apatite. Dust samples from roadside in the vicinity of metal factories contain Cs within the range from 1.53 to 3.63 mg/kg (ATSDR 2002b).

10.3 WATERS

Cesium concentrations in water are low, usually do not exceed (0.6 μg/L), and are higher in seas and oceans, than in rivers (Table 10.1). The worldwide mean Cs content in river water is estimated at 0.011 μg/L, and its river flux at 0.4 kt/yr (Gaillardet et al. 2003).

TABLE 10.1
Cesium Contents in Soils, Water, and Air

Environmental Compartment	Range/Mean
Soil (mg/kg)	
Light sandy	0.8–30
Medium loamy	2.5–20
Heavy loamy	5–20
Calcareous	5–15
Water (μg/L)	
Rain	0.01–0.5
River	0.003–0.02
Sea, ocean	0.3–0.6
Air (ng/m³)	
Industrial regions	14–18
Greenland	0.02–0.4
Antarctica	0.06–0.2

Source: Data are given for uncontaminated environments, from Kabata-Pendias, A. and Mukherjee, A.B., *Trace Elements from Soil to Human*, Springer, Berlin, Germany, 2007; Reimann, C. and de Caritat, P., *Chemical Elements in the Environment*, Springer, Berlin, Germany, 1998.

Cesium in water forms CsOH and is adsorbed by suspended solid particles, and is finely present in bottom sediments, which may contain up to >50 mg Cs/kg. Its content in surface sediments of the Baltic Sea ranges from 3 to 10 mg/kg, and in soft tissues of mussels, 0.001–0.05 mg/kg (Szefer 2002).

In rainwater collected in Sweden during 1999, Cs concentrations are in the range of 0.004–0.011 μg/L (Eriksson 2001a).

10.4 AIR

Cesium concentrations in air from remote regions do not exceed 0.4 ng/m^3, whereas in industrial areas it is up to 18 ng/m^3 (Table 10.1).

There is an observed decrease of Cs contents in lichens collected in Greece in 1996; this contained much less ^{137}Cs than the lichens collected during earlier years (ATSDR 2002b).

10.5 PLANTS

Cesium is relatively active in soils and is thus easily taken up by plants, especially from sandy, acidic soils. Addition of lime or peat to soils inhibited its phytoavailability.

Common Cs range in plants is estimated at <0.1–3 mg/kg. Tea leaves contain Cs within the range 0.5–1.0 mg/kg AW (Aidiniyan *vide* Kabata-Pendias 2011). Roots contain more Cs than tops of plants, as well as older plants have more Cs than younger ones. Cesium distribution in plants is similar to K, which suggests that it may compete with K in uptake processes and binding sites in cells (Isaure et al. 2005). Its toxicity may be related to these interactions. According to White et al. (2003), the transport of Cs and K to xylem is different, and thus, the Cs/K concentration ratio differs among plants.

Cesium is easily absorbed from the atmosphere by mosses; mosses collected in Norway during 1990–1995 contained cesium in the range 0.02–3.1 mg/kg (Berg and Steinnes 1997). Its deposition in the United Kingdom in 1972–1981 was 0.6–1.2 g/ha (Cawse 1987). The highest Cs content is reported in mosses from Germany (Reimann and de Caritat 1998).

10.6 CESIUM ISOTOPES

10.6.1 Introduction

Concerns with radiocesium (two radioisotopes ^{134}Cs and ^{137}Cs) in soils and food chain began since the accident in the Chernobyl nuclear power plant, in 1986. Some amounts of these two radioisotopes have been released into the environment during nearly all nuclear weapon tests, and some nuclear accidents. Also, ash from coal burning and municipal waste incinerations may be a source of ^{137}Cs. However, there is much less concern about ^{134}Cs, because it was released in small quantity and has short half-life ($t_{1/2}$ is 2.1 years) compared to great emission of ^{137}Cs ($t_{1/2}$ is 30 years). Deposition of both radionuclides was not only in surroundings of the reactor but also in other European countries, and in other continents.

10.6.2 Soils

The behavior of ^{137}Cs in soils is very similar to those of the stable Cs. Thus, when deposited in soils, it is absorbed by clay fractions and SOM, and, therefore, migrates slowly. Especially SOM, of a higher molecular weight, is likely to accumulate a high proportion of this radionuclide. Also Fe and Mn sesquioxides decreased its mobility in soils. These processes are also noticed in fresh alluvial sediments (Kaplan et al. 2005). Although clay minerals, in general, immobilize ^{137}Cs, hydromicas increase its mobility in soils (Korobova and Chizhikova 2006).

All soils contain a certain amounts of ^{137}Cs. Levels of this radionuclide in agricultural soils of Japan are 1–37 Bq/kg (Tsukada and Nakamura 1999); in wetland of South Carolina, United States, 0.2–10.1 Bq/g (Kaplan et al. 2005); in soils of Bulgaria, 0.9–21.2 kBq/m^2 (Zhiyanski et al. 2006); in soils of Kazakhstan, 82–120 mCi/km^2 (Panin 2004); in soils of France, >10 kBq/m^2 (Renaud et al. 2003); in soils of Belarus, 103–1500 kBq/m^2 (Kagan et al. 1996). The impact of industrial pollution on levels of ^{137}Cs, within the range 15–25 kBq/m^2, in forest soils of Finland has also been reported (Outola et al. 2003).

The March 2011 accident at the nuclear power plant in Fukushima in Japan, caused by the Tōhoku earthquake and tsunami, released both Cs radionuclides to the environment. Radionuclides deposited on the ground surface were washed off to rivers and finely transported to the sea, where they are absorbed by bottom sediments (Onishi and Yokuda 2013).

10.6.3 Waters

After the accident in Chernobyl, ^{137}Cs concentration in the Black Sea was 2.7–8.1 pCi/L, whereas in the Mediterranean Sea was 0.13 pCi/L (ATSDR 2002b). Typical present background of ^{137}Cs content in Russian river and lakes is within 1–10 Bq/m^3 (Krylova et al. 2011).

This radionuclide is likely to concentrate in bottom sediments. Its content in sediments of the pond close to Chernobyl was up to 11 Bq/g, whereas in surface layer of sediments of the Baltic Sea, it ranges between 100 and >200 mBq/g (Kabata-Pendias and Mukherjee 2007). The mean annual decrease in ^{137}Cs levels in bottom sediments of some Masurian lakes (Poland) during 1992–1995 was from 4.2% to 7.8% (Kapala et al. 2008).

Fishes from both freshwater and seawater may contain elevated amounts of ^{137}Cs. Its highest concentration, up to 8 Bq/kg, has been reported for fish from ponds in Chernobyl (Jagoe et al. 1997). Decreased levels of this radionuclide was reported in fish from the lakes of Finland, from 66,297 to 18,567 pCi/kg during 1988 and 1992 (ATSDR 2002b).

Concentrations of ^{134}Cs and ^{137}Cs in drinking water was established at 900 and 1000 pCi/L, respectively (ATSDR 2002b).

10.6.4 Air

Concentrations of ^{137}Cs in air in the South Pole in the 1970s was within the range of 0.072–0.14 pg/m^3, and in the North Pole in the 1990s was 10–60 pg/m^3. Its maximum concentration in air in New York City, after the Chernobyl accident, was reported as 0.26 pCi/m^3 (ATSDR 2002b).

10.6.5 Plants

Similarly, as stable Cs, its radionuclides are easily absorbed by plants. Experiments with ^{137}Cs addition to various soils have indicated that the highest amounts of ^{137}Cs were taken up from sandy acid soils (in nCi/g) by cabbage—leaves, 34.9; cucumber—fruits, 8.1; carrots—roots, 6.5; potato—tubers, 5.9; and oat—straw, 10–19 (Kabata-Pendias 2011). Mushrooms reveal a great ability to accumulate this radionuclide, which might be its serious source in food products (Yordanova et al. 2007). A broad range of ^{137}Cs, from 0.005 to 1.6 mg/kg, was found in mushrooms grown in Japan (Tsukada et al. 1998).

Tsukada et al. (2002) reported that up to 77% of ^{137}Cs was accumulated in nonedible (old) leaves of cabbage.

10.7 HUMANS

The amount of Cs found in the human body (average 70 kg) is approximately 6 mg, of which in the blood, 4 μg/L; in the bone, 10–50 μg/kg; and in other tissues, up to 1 mg/kg (Emsley 2011). Its content in the rib bone is about 7.7 μg/kg (Zaichik et al. 2011). Humans may be exposed to Cs by breathing, drinking, or eating.

The monovalent alkali metal Cs behaves similar to K, after absorption in blood and is accumulated in all cells. Higher concentrations of Cs have been reported in muscle than in other tissues, but the differences are small. It is generally assumed that Cs is distributed uniformly between and within body organs and tissues. After entering the blood, it is cleared rapidly and deposited more or less uniformly in the tissues. It is now well established that the rate of clearance is faster in women than in men, and that clearance may be further accelerated during pregnancy. The mean equivalent biological half-time of Cs ranges from about 47–152 days in men and from about 30–141 days in women. Rate of clearance of Cs is faster in children (IARC 2001).

The transfer of Cs isotopes to the fetus has been followed in both humans and animals. The concentrations of ^{137}Cs arising from exposure to fall-out from nuclear weapons were measured in nine newborn children, within three days of birth and in their mothers; concentrations were similar (IARC 2001).

Exposure to radioisotopes of Cs is of a great human health concern. Energy released by radioactive isotopes can result in significant damage to living cells. Both ^{134}Cs and ^{137}Cs emit β particles and γ rays, which may ionize molecules within cells; penetrations by these emissions resulted in tissue damage and disruption of cellular functions. The most important exposure routes for radioisotopes of Cs are external exposure to the radiation released by the radioisotopes and ingestion of radioactive Cs-contaminated food sources. Inhalation and dermal exposure routes may also present a health hazard. The hazards of external exposure to ^{134}Cs and ^{137}Cs are similar to those of other γ- and β-emitting radionuclides. Signs and symptoms of acute toxicity from external and internal exposure to high levels of radiation from ^{134}Cs or ^{137}Cs are typical of those observed in cases of high exposure to ionizing radiation in general. Depending on the radiation dose, symptoms may include typical of acute radiation syndrome (vomiting, nausea, and diarrhea), skin and ocular lesions, neurological signs, chromosomal abnormalities, compromised immune function, and death.

Acute or repeated exposure of humans or animals to ionizing radiation may result in reduced male fertility, abnormal neurological development following exposure during critical stages of fetal development, and genotoxic effects such as increased frequencies of chromosomal aberrations, T-lymphocyte point mutations, dominant lethal mutations, and reciprocal translocations. Animal studies have indicated increased risk of cancer following external or internal exposure to relatively high doses of radiation from ^{137}Cs sources. These studies showed that Cs is of relatively low toxicity (ATSDR 2004a).

The amount of Cs in foods and drinks depends upon the emission of radioactive Cs through nuclear power plants, mainly due to accidents. Impact of Chernobyl nuclear explosion on radionuclides in the U.K. foods is illustrated by the average level of the ^{137}Cs detected in Scottish milk. In 1985, it was 0.1 Bq/L, and after the

Chernobyl accident it has risen to 7.8 Bq/L. In the following year, it decreased to 5.4 Bq/L, and by 1990, it decreased to 0.2 Bq/L. Similar rises, and subsequent falls, were detected in several other foodstuffs, including mutton and lamb (Reilly 2002). The content of radioactive Cs isotopes in daily food rations, in different regions of Poland, in 1987 and 1988, was 1.80 and 0.93 Bq/kg, respectively (Skibniewska et al. 1993).

A survey on long-lived artificial radionuclides ^{137}Cs in vegetables in Finland, carried out in 2009–2010, showed that traces of ^{137}Cs deposition are still seen in vegetables. But it can be concluded that the committed effective dose from the ingestion of ^{137}Cs in vegetables is very low and the radiation-induced health detriment caused by this radionuclide is most probably negligible (Kostiainen and Turtainen 2013).

11 Chlorine [Cl, 17]

11.1 INTRODUCTION

Chlorine is a common element of the halogen group 17 in the periodic table of elements. Its total content in the upper Earth's crust is estimated at 640 mg/kg. In igneous rocks, its concentrations vary between 80 and 200 mg/kg, being the highest in granites. It is likely to accumulate in shales (200 mg/kg) and in some limestones (150 mg/kg). Its lowest mean contents are in sandstones, 10 mg/kg. High Cl amounts are in coal, within the range 200–1000 mg/kg, but contents up to about 9000 mg/kg are also reported (Finkelman 1999).

Under standard condition, Cl is a yellowish-green gas, of a specific irritating odor. Its chemical properties are similar to Br and I. It is chemically very reactive and can have several valence states. Most commonly its oxidation state is −1. It is a strong oxidizing agent. Chlorine is a lithophilic and a very mobile element; it may be a minor constituent of several minerals, mainly amphiboles, apatite, and micaceous and clay minerals. During weathering processes, Cl is released from its compounds and is transferred to water, mainly in oceans and seas.

Chlorine forms some minerals such as cerargyrite, AgCl; carnallite, $KMg(H_2O)_6Cl_2$; and sylvite, KCl. However, halite—commonly known as *rock salt*—NaCl, is the most abundant in both terrestrial and sea environments, and is the most important in human activities.

World total salt production in 2010 was about 280 Mt, of which China and the United States produced 22% and 15%, respectively (USGS 2010). Salt is obtained either from solid sedimentary deposits or from seawater and brines.

Major uses of salt (in %) are chemicals, 42%; road ice control, 38%; distributors (groceries, etc.), 8%; agricultural and food processing, 4% each; general industrial and primary water treatment, 2% each; and other uses less than 1%. These proportions of salt use may differ from countries; in some countries the use of NaCl as road salt exceeds 50% of its total use.

Chlorine is broadly used as ClO_2 in bleaching materials, for paper pulp and cellulose bleaching. Chloride dioxide is used in sterilization in hospitals. It is also used in supplied drinking water and in food preservations. Chlorine is a component of some pesticides and various chlorinated organic compounds. Chlorine dioxide load of various environmental compartments may infiltrate the food chain and is of a health concern (Coenen 2004).

11.2 SOILS

The mean content of Chlorine in worldwide soils is estimated at 300 mg/kg. Its contents in loamy soils, up to 1000 mg/kg, is much higher than in sandy soils, <120 mg/kg (Table 11.1). In humid climate zones, Cl^- is easily leached down in soil profiles, and is transported with drainage water. In soils of arid and semiarid climatic zones, it is concentrated in surface soil horizons. In solonetz and solonchaks, NaCl is inherited from parent material, and in addition accumulated due to climatic conditions. Concentrations of Cl in these soils vary within the order of 0.00X%–0.X%.

Amounts of Cl in soils, similar to other halogens, decreased with increasing distance from the sea coast. Forest soils of Norway contain Cl up to 1806 and 375 mg/kg, when located close to and far away from the sea, respectively. Mean accumulation of Cl in soils in Japan was 228 and 114 mg/kg, from the coastal plain and uplands fields, respectively (Kabata-Pendias 2011).

Higher concentrations of Cl in soils, and thus in soil solutions influence behavior of several cations. It increases mobility of several metals (e.g., Cu, Zn, Hg, and Pb). However, the greatest concern is associated with Cl affinity to form soluble and phytoavailable complexes with Cd (Weggler et al. 2004). Increasing Cl levels inhibit microbial growth and disturb their biological activities (Dinesh et al. 1995).

Increased amounts of Cl in soils result from anthropogenic activities. Main sources of salt are from (1) de-icing salt used for roadways; (2) field irrigation with mineralized water; (3) salt water spilled at oil extraction from deep deposits; (4) ammonium chloride used as N fertilizers; (5) coal burning; and (6) some pesticides application.

TABLE 11.1
Chlorine Contents in Soils, Water, and Air

Environmental Compartment	Range/Mean
Soil (mg/kg)	
Light sandy	80–120
Heavy loamy	<1000
Water (mg/L)	
Rain	<0.1–2.1
River	1–10
Sea, ocean	<20,000
Air (μg/m³)	
Urban areas	4–10
Greenland	0.04–0.1
Antarctica	0.003

Source: Data are given for uncontaminated environments, from Kabata-Pendias, A. and Mukherjee, A.B., *Trace Elements from Soil to Human*, Springer, Berlin, Germany, 2007; Reimann, C. and de Caritat, P., *Chemical Elements in the Environment*, Springer, Berlin, Germany, 1998.

11.3 WATERS

The median Cl concentration in seawater is estimated at 19,400 mg/L, and its mean contents in stream water vary within the range of 1.4–5.3 mg/L (Reimann and de Caritat 1998). Chlorine contents in precipitation are much lower, and do not exceed 2.1 mg/kg (Table 11.1). Rainwater over Kola Peninsula contain Cl between <0.1 and 2.1 mg/L, whereas rainwater over the United Kingdom has, on an average, 3 mg Cl/L (Fuge 1988).

Chlorine is easily transported along with stream and river water to all water basins. Chlorine concentration in water is usually measured as salinity (NaCl) content. Throughout the world, rivers carry about four billion tons of dissolved salts to the oceans and seas annually. Similar amounts of salt are deposited in bottom sediments. Thus, there is a balanced salt input and outgo. Salinity of the total ocean water is 3.5%, and of some salty lakes may be up to 27%. The highest salt content, about 34%, is in water of the Dead Sea (Salt Sea).

Chlorine is produced in large amounts and widely used, both industrially and domestically, as an important disinfectant and bleach. In particular, it is widely used as a disinfectant in swimming pools and is the most commonly used disinfectant and oxidant in drinking water treatment. In water, Cl reacts to form hypochlorous acid and hypochlorites. Concentrations of chlorate and some perchlorates increase in hypochlorite solutions, upon storage at high ambient temperatures or when new hypochlorite is added to old hypochlorite.

No health-based guideline value is proposed for Cl in drinking water. However, its concentrations in excess of about 250 mg/L can give rise to detectable taste in water. Presence of Cl in the most disinfected drinking water is within the concentrations of 0.2–1 mg/L (WHO 2011a).

11.4 AIR

World median Cl concentrations vary from about 0.023 μg/m^3 in air of remote world sites to 0.24 μg/m^3 in polluted areas (Reimann and de Caritat 1998). Its concentration in the atmospheric air is (in μg/m^3) in Antarctica 0.003, and up to around 10 in some urban areas (Table 11.1). Chlorine easily evaporates from the surface of seas and oceans, and therefore its Cl content increases in air. Additional sources of Cl are volcanic exhalations, most often in the form of hydrochloric acid (HCl).

Chlorine in air is of important health risk to human. Acute exposure to Cl is when its concentration is about 200 μg/m^3 (OEHHA 2000). Air in enclosed swimming pools may contain highly elevated Cl amounts, up to about 4000 μg/m^3. Children exposed to Cl gas have acute respiratory distress and eye irritation.

11.5 PLANTS

Soluble Cl species in soils are passively taken up by roots and easily transported in plants. Plant leaves are capable of absorbing Cl directly from air. Chlorine as a compound of monovalent cations (e.g., NaCl) is more readily absorbed than from other compounds. Some plants may accumulate relatively high amounts of Cl, especially when grown in coastal regions and semiarid climatic zones. Some experiments have shown that above half of Cl contents in radish and lettuce are absorbed by aboveground parts from evaporates of a soil arable layer (Kashparov et al. 2005).

Plants require relatively low levels of Cl for metabolic processes. It is concentrated in chloroplasts; however, its function in the photosynthesis is not fully known (Marschner 2005). Chlorine stimulates activity of some enzymes, such as cytochromeoxidase and phosphorylase, and plays a role in the osmotic system of plants. Because Cl is supplied to plants from various sources, more concern is about its toxicity than deficiency. However, Cl may be deficient to some plants grown in soils with Cl contents below 2 mg/kg. Especially some crops (e.g., tobacco) require Cl level in soils within the range 10–15 mg/kg. The threshold Cl content in soils is estimated at 2700 mg/kg. Higher Cl contents, especially in soil solution, may be toxic to some plants. The less tolerant plants (e.g., beans and apple trees) are inhibited by Cl concentrations from 460 to 673 mg/L, whereas resistant plants (e.g., tobacco, cereals, cotton, beets, and spinach) can grow at Cl contents within 887–3546 mg/L (Kabata-Pendias 2011). HCl and Cl_2, gases emitted from industrial sources, are the most toxic to plants. Very high Cl contents up to 3300 and 3550 mg/kg in tree leaves and vegetable and fruits, respectively, are affected by industrial pollution (Nersyan 2007). Sometimes, trees may secrete excess Cl, mainly with the falling of leaves in autumn.

Chlorine contents in crop plants are variable (in mg/kg) and are as follows: cereal plants, 10–20; sugar beet leaves, 100–200; potato tubers, 1300–5500.

11.6 HUMANS

Chlorine is present in the body of humans as chloride ions. On an average, an adult human body contains approximately 95 g of Cl, 900 mg/kg in the bone, average of 0.03% in the blood, and 0.2%–0.5% in the tissue (Emsley 2011).

The most harmful route of exposure is from breathing Cl gas. Exposure may also result from skin or eye contact with Cl gas or by swallowing Cl-contaminated food or water.

When Cl gas enters the body, as a result of breathing, swallowing, or skin contact, it reacts with water to produce acids. The acids are corrosive and damage the cells in the body. Most harmful Cl exposures are from its inhalation. Health effects typically begin within seconds to minutes. The health effects of Cl gas in humans come from accounts of soldiers exposed during gas attacks in World War I. Other Cl exposures are at work; accidental exposures occur due to leaks or explosions of storage tanks or due to the mishandling of bleach solutions or swimming pool chemicals. Effects of exposure to low Cl concentrations may be limited to irritation of the eyes, upper respiratory tract, and skin, whereas exposure to high concentrations may cause serious pulmonary effects and death. There is limited information on neurological effects in humans exposed to Cl gas. Oral exposure is not a relevant route of exposure to Cl gas in humans or animals (ATSDR 2010a).

Chlorine is used to clean swimming pool water to prevent contamination and bacterial overgrowth. It is not the safest method, but is probably the most common. It is confirmed that long-term exposure to chlorinated pools can cause symptoms of asthma in swimmers. Additionally, eye and skin irritation in swimmers has been hypothesized to originate from Cl exposure.

Chlorine is the main monovalent anion of extracellular fluids; with Na or K, it forms salts, and with hydrogen it forms HCl. Chlorine ions constitute about 65% of the total anions of the blood plasma and other extracellular fluids within the mammalian body. Chlorine ions are also known to be present in the red blood cells. Chlorine is essential for the regulation of osmotic pressure and acid–base balance and plays a specific role in the transportation of oxygen and carbon dioxide in the blood. It is necessary for the absorption of proteins and metallic minerals, as well as vitamin B_{12}.

Chlorine is stored, to a limited extent, in the skin, subcutaneous tissues, and skeleton; it constitutes two-thirds of the negatively charged ions in the blood. It is readily absorbed and excreted according to the body requirements. Gastric secretion is composed of chlorides in the form of HCl and salts. Chlorine is readily absorbed during digestion. Its rate of excretion through sweat, urine, and intestinal expulsion is high.

Chlorine provides the acid medium for the activation of the gastric enzymes and digestion in the stomach. HCl made in the stomach has two main purposes: to help destroy germs that arrived with the food and to help pepsin, an enzyme, break down the proteins of foodstuffs, ensuring that essential nutrients are made available to the body.

In the immune system, which is charged with fighting off the daily invasion of germs, Cl is very useful. When infections take place, hypochlorite, a Cl-containing compound, which is known as disinfectant, forms in white blood cells. Hypochlorite itself attacks the germs, or helps to activate other agents that do this work.

The requirement for Cl has not been ascertained, but if NaCl is taken liberally, it ensures the adequate intake of chloride as well. Thus, most of the chloride taken is from the common salt used during food processing and preparation. Sodium chloride (table salt) is the most common Cl compound associated with foods. Chlorine, in the form of NaCl, is used in many processed foods to prevent the growth of microorganisms, reduce the perception of dryness and enhance the flavor of the product. The minimum requirement for Cl by adults is about 700 mg per day. Tolerable daily intake for Cl is estimated for 150 μg/kg bw (WHO 2011a).

Common salt helps the body to normally meet the requirements of Na and Cl. The body's supplies of Cl are rapidly depleted during hot weather, when excessive perspiration reduces the fluid content of the body. Stored chlorides may become dangerously low in periods of severe vomiting and diarrhea, and in diseases that produce severe alkalosis, an accumulation of base or loss of acid in the body. Excessive chloride levels, on the other hand, can result in water retention and in elevated blood pressure, as well as increased risk of developing cancer. Also, excess chloride becomes free-radical initiator, resulting in the damage of the arterial walls, leading to arteriosclerosis.

12 Chromium [Cr, 24]

12.1 INTRODUCTION

Chromium (Cr), a metal of the group 6 in the periodic table of elements, has strong lithophillic tendencies. Its content on the Earth's crust is within the range of 102–185 mg/kg. The highest Cr contents in ultramafic rocks may be up to 3800 mg/kg, which decreased in acidic ones (e.g., granites) to 10–50 mg/kg. Its distribution in sedimentary rocks is as follows (in mg/kg): argillaceous (shales), 80–120; sandstones, 20–40; and limestones, 5–16. Coal contains Cr within the range 10–40 mg/kg, whereas fly ash may have up to about 150 mg/kg.

Chromium occurs at several oxidation states, of which the most common are +3 and +6. It is precipitated from magma mainly in the Cr spinel mineral groups of the following formula: (Mg,Fe)O $(Cr,Al,Fe)_2$ O_4. Its relatively common minerals are chromite ($FeCr_2O_4$) and crocoite ($PbCrO_4$). These minerals are likely to be associated with pyroxenes, amphibolites, and micas. The geochemical association of Cr with Fe and Mn resulted in its increased contents in Fe concretions of soils, and in ferromanganese nodules in sea bottom sediments.

World mine Cr production (rounded) in 2010 was 22 Mt, of which (in Mt) South Africa, 8.5; India, 3.8; and the United States, 3.8 (USGS 2011). Chromium is mined mainly as chromite ore, which is converted into ferrochromium, broadly used in metallurgical industries, and most of that is consumed to make stainless and heat-resisting steel. It is also used for chromium chemicals, applied in paints, varnishes, glazes, and inks. A great proportion of Cr compounds is utilized for leather tanning. It is also used in paper manufactories and various refractory products.

The main sources of Cr pollution are from dyestuffs and leather tanning. Also chromite ore processing residue is a significant source of Cr, mainly in surface water, which resulted in bottom sediments contamination.

12.2 SOILS

The world soils average Cr content has been estimated at 60 mg/kg. However, soils derived from parent rocks with high Cr levels may contain much higher amounts. For example, soils in some Carpathian rocks have about 6000 mg Cr/kg (Čurlik and Šefčik 1999). Soils developed from serpentines contain especially high Cr levels, sometimes above 100,000 mg/kg (Kabata-Pendias 2011). The most common average Cr contents are within the range 20–80 mg/kg, the lowest in organic soils and the highest in heavy loamy soils (Table 12.1). Mean Cr content in organic surface soils of Norway is 7.8 mg/kg, within the range <0.05–191 mg/kg (Nygard et al. 2012). Chromium contents of soils in Spitsbergen range from 0.1 to 0.8 mg/kg (Gulińska et al. 2003).

The oxidation states of Cr are the most important factors controlling Cr behavior. Cr^{3+} is slightly mobile in very acidic soils, and at about pH 5.5, it forms very stable

TABLE 12.1
Chromium Contents in Soils, Water, and Air

Environmental Compartment	Range	Mean
	Soil (mg/kg)	
Light sandy	2–350	40
Medium loamy	10–300	60
Heavy loamy	30–1100	80
Calcerous (calcisols)	2–100	50
Organic	2–40	20
	Water (μg/L)	
Rain	0.2–2.7	0.3
River	0.3–2.1	0.7
Sea, ocean	0.15–0.3	0.2
	Air (ng/m^3)	
Urban/industrial areas	1–1100	40
Greenland	0.6–0.8	0.5
Antarctica	0.003–0.01	–

Source: Data are given for uncontaminated environments, from Kabata-Pendias, A. and Mukherjee, A.B., *Trace Elements from Soil to Human*, Springer, Berlin, Germany, 2007; Reimann, C. and de Caritat, P., *Chemical Elements in the Environment*, Springer, Berlin, Germany, 1998.

compounds in both acidic and alkaline soils. At the high oxidation state, Cr^{6+} is very mobile in both, acidic and alkaline soils. Nevertheless, above 80% of total Cr contents of most soils is in immobile residual fraction. This explains, partly, elevated Cr levels in heavy loamy soils (Table 12.1). Organic soil horizons, and especially peat deposits exhibit a high sorption capacity for Cr. Peat capacity for Cr, mainly as insoluble metal complexes, is very high, within the values of 250–52,800 mg/kg (Kyzioł 2002). On the other hand, organic soils of Norway contain less Cr (mean 4.91 mg/kg) than all investigated soils (mean 7.86 mg/kg) (Nygard et al. 2012).

Formations of complex anionic and cationic species, as well as organic complexes of Cr is possible due to its highly variable oxidation states. The Cr^{6+} forms are more mobile than Cr^{3+}, especially under both, very acidic and/or alkaline ranges of pH. Mobile Cr species are as follows: $CrOH^{2+}, CrO_4^{2-}, HCrO_4^-, HCrO_3^{2-}, Cr(OH)_4^-$, and $Cr(CO_3)_3^{3-}$. $Cr(H_2O)_6^{3+}$ species is relatively inert (Kabata-Pendias and Sadurski 2004). All ionic Cr species are highly susceptible to oxidation–reduction processes. Its behavior is controlled also by the formation of various organic complexes. The impact of soluble organic matter on the conversion and, in particular, reduction of Cr is of a great environmental significance, because when it is bound to both humic acid and fulvic acid, it became easily mobilized, at various soil pH values. Both, biological reduction or reduction by organic molecules stimulates the formation of soluble Cr^{3+}-organic complexes.

However, under natural soil conditions, the Cr oxidation–reduction potential seems to be directly associated with Fe and Mn-oxides contents (Chung and Sa 2001).

Increased levels of mobile Cr species may be toxic to both microorganisms and crop plants. Activities of microbial enzymes decrease under elevated Cr contents, especially sensitive are nitrification processes. More toxic is Cr^{6+} than Cr^{3+}. However, toxic effects of Cr at various oxidation states may differ depending on species of the microorganisms (Wyszkowska and Kucharski 2004). Some microorganisms are resistant to Cr^{6+} due to chromate reduction processes (Ramirez-Diaz et al. 2008).

Cr contents of surface soils have increased due to pollution from various sources, of which the main are chromite-ore processing residue, municipal, and tannery and leather-manufacturing wastes. A number of studies have been recently carried out to understand Cr behavior in contaminated soils, and to apply effective remedial treatments (Bini et al. 2008; Geelhoed et al. 2003).

Municipal and industrial wastes may contain Cr up to 10,200 mg/kg. Its contents of sewage sludge applied to agricultural soils usually vary between 100 and 200 mg/kg (Maján et al. 2001). Its deposition to agricultural soils is estimated at 0.5–46 g/ha/yr, being the lowest in Finland and the highest in Italy (Nicholson et al. 2003). According to these authors, the total annual input of Cr to agricultural soils of the United Kingdom in 2000 was 327 t, of which 126 t was from inorganic fertilizers (mainly phosphate), 83 t from atmospheric deposition, and 78 t from sewage sludge (Maján et al. 2001).

Reported Cr contaminations of soils from industrial sources in various countries are as follows:

- *Italy, leather tannery:* range 50–10,000 and mean 210 mg/kg (Bini et al. 2008)
- *Portugal, electroplating plant:* Up to 27,132 and mean 1,000 mg/kg (Morgado et al. 2001)
- *Albania, metal smelter:* Up to 20,300 and mean 3117 mg/kg (Shtiza et al. 2005)

The criteria for contaminated land (Dutch List 2013), following Cr concentrations in soils and groundwater, are established (in mg/kg, and μg/L) as follows: uncontaminated, 100 and 20; medium contaminated, 250 and 50; and heavily contaminated, 800 and 200. Soil quality Cr^{3+} levels are estimated, based on various criteria, within the broad range of 25–1,000,000 mg/kg (Siebielec et al. 2012).

The main purpose of remediation treatments of Cr-contaminated soils is the reduction of easily mobile Cr^{6+} to slightly mobile Cr^{3+}, or Cr adsorption by mineral or organic compounds. The phytoremediaton methods are rather limited due to a relatively low soil-to-plant transfer factor. Recently, the bioremediation of Cr-contaminated soils by actinomycetes, partly due to the bioreduction of Cr^{6+}, has been proposed (Amoroso and Abate 2012; Farmer et al. 1999).

12.3 WATERS

The worldwide median concentration of Cr in ocean water is given as 0.3 μg/L (Reimann and de Caritat 1998). In the North Pacific Ocean it is a bit lower. Its concentration in the Baltic Sea (Swedish coast) is extremely high, up to about 185,000 μg/L

(Szefer 2002). Annual Cr input to the Baltic Sea is estimated at 540 t, of which 47% is from anthropogenic sources (Matschullat 1997).

Geochemical mobility of Cr is relatively low, which resulted in its natural low concentrations in river water (Gaillardet et al. 2003). However, in general, its contents in seawater is lower than in river water, for which Cr mean content is estimated at 0.7 μg/L (Table 12.2). Some rivers, industrially polluted, may contain much more of Cr, up to 47 μg/L (range 0.4–6.9).

Chromium in water is present mainly as hydroxoanions, $Cr(OH)^{4-}$, and hydroxocations, $Cr(OH)^{2+}$, and as complexes with organic matter (Świetlik 1998). The Cr speciation in wastewater varies due to variable organic and inorganic ligands. Hexavalent Cr dominates in effluent from metallurgic and pigment industries, whereas trivalent

TABLE 12.2
Chromium in Foodstuffs (μg/kg FW)

Product	Contents
Food of Plant Origin	
Bread[a]	<11
Miscellaneous cereals[a]	30
Rice, different type[b]	8
Cabbage[c]	6
Potato[a]	31
Cucumber[c]	19
Apple[c]	11
Fresh fruits[a]	7
Strawberry[c]	32
Sugar and preserves[a]	80
Candies and chocolates[d]	1800
Food of Animal Origin	
Meat[a]	30
Meat product[a]	37
Fish[a]	40
Smoked fish[e]	125–202
Egg[a]	10
Milk[f]	16
Dairy products[a]	10

Source: [a] Rose, M. et al., *Food Addit. Contam.*, 27, 1380–1404, 2010.
[b] Jorhem, L. et al. *Food Addit. Contam.* 25, 841–850, 2008b.
[c] Ręczajska, W. et al., *Pol. J. Food Nutr. Sci.*, 55, 183–188, 2005.
[d] Iwegbue, C.M.A., *Food Addit. Contam.*, 4, 22–27, 2011.
[e] Polak-Juszczak, L., *Roczn. PZH*, 59, 187–196, 2008.
[f] Gabryszuk, M. et al., *J. Elem.*, 15, 259–267, 2010.
FW, fresh weight.

Cr is released form tannery, textile, and plating industries. All these industries release especially large amounts of Cr to surface water (ATSDR 2002b). Several methods are used for the Cr removal from wastewater, but the most common is the reduction of Cr^{6+} to Cr^{3+}, a stable form and easily precipitated, as $Cr(OH)_3$ (Chen 2013).

Aquatic plants can accumulate high amounts of Cr, as, for example, *Vallisneria spiralis* may contain this metal above 1000 mg/kg, and thus is proposed for the bioremediation of Cr-contaminated wastewater (Sinha et al. 2002). Aquatic animals, some fishes, and invertebrates (e.g., crabs) may contain elevated Cr levels. Its toxicity is variable, depending on several factors, and especially on water pH (Stoecker 2004).

Chromium content in bottom sediments is a good information on the water pollution with this metal. Its mean content in bottom sediments of San River (Poland) varies from 2 to 17 mg/kg, in sandy and mulc sediments, respectively (Bojakowska et al. 2008). Chromium content of surface-bottom sediments of harbor in Klaideda (Lithuania) depends on the granulometric composition, and is (in mg/kg, average and maximum, respectively) in sand 14.0 and 33.2, and in mud 31.5 and 67.1 (Galkus et al. 2012). Its content in stream-bottom sediments of National Park, Montgomery (Pennsylvania State) was, in 1995, within the range of 10–50 mg/kg (Reif and Sloto 1997). Sediments of Sergipe river estuary (Brazil) contain Cr from 3.25 to 74.7 mg/kg (Garcia et al. 2011). Assessment limits for Cr in sediments are established as follows (in mg/kg): effects range low, 81; effects range median, 9.6; probable effect level (PEL), 90; 110; and 100 (EPA 2000, 2013). The Environment Canada sediment-quality guidelines (USGS 2001) gave a bit similar values for Cr in lake-bottom sediments (in mg/kg) as follows: threshold effects level, 37.3; PEL, 90; and probable effect concentration, 111.

Median Cr concentration in bottled water of the EU countries is 0.123 μg/L, and is a little bit lower than those in tap water, 0.185 μg/L (Birke et al. 2010). U.S. drinking water contains Cr within the range 0.4–8.0 μg/L, with the mean value of 1.8 μg/L (ATSDR 2002b). Provisional guideline value of Cr in drinking water is established at 50 μg/L. The guideline value is designated as provisional because of uncertainties in the toxicological database (WHO 2011a).

12.4 AIR

Worldwide median Cr content of air is estimated at 0.5–0.6 ng/m³, whereas in remote areas (Antarctica) its concentration does not exceed 0.01 ng/m³ (Table 12.1). In the United States, Cr concentration in air above rural areas is <10 ng/m³, and in urban regions it ranges in 10–30 ng/m³ (ATSDR 2002b).

In 1999, atmospheric deposition of Cr on rural soils in some EU countries was as follows (in g/ha/yr): Germany, 17.8; England, 8.7; Sweden: 2.5 (Eckel et al. 2005). Data on the Cr deposition highly vary, depending on time of calculations. In general, there is a noticed decrease, with time, in amounts of Cr emission and deposition.

The main natural Cr source in the atmosphere is continental dust flux. About 70% of Cr in air is of anthropogenic origin, mainly from metal industry emissions and fuel combustion. Mean Cr contents of two moss species from mountains in Poland are a little bit lower (0.84–0.90 mg/kg) than in the same moss species from Alaska (1.02–1.14 mg/kg). This may suggest the atmospheric transfer of the metal

(Migaszewski et al. 2009). Fairly similar Cr contents of mosses (mean 2.6 mg/kg) collected in Norway, in 1990–1995 yr, to its contents (mean 1.2 mg/kg) collected in Wisconsin State (United States), in about similar period, indicated a rather stable Cr emissions to the atmosphere (Berg and Steinnes 1997; Bennett and Wetmore 2003).

Hexavalent Cr in the atmosphere is reduced to Cr^{3+}, mainly by various compounds of V^{2+}, Fe^{2+}, and As^{2+}. However, oxidation processes are also observed (ATSDR 2002b).

12.5 PLANTS

Until now there are not any clear evidences that Cr is essential in plant metabolism. Some descriptions of positive effects of Cr on plant growth have not been confirmed, whereas its phytotoxicity has been often observed and described (Fendorf et al. 2004; Laborga et al. 2007; Sharma et al. 2005; Singh et al. 2007; Stoecker 2004).

Chromium is slightly available to plants, as well as slightly translocated within plants (Vernay et al. 2007). Contents of Cr in plants are controlled by the mobile Cr contents of soils, and thus its uptake by plants is highly limited (Zayed and Terry 2003). Hexavalent Cr is easier taken up by plants than trivalent Cr. However, it is readily reduced and as Cr^{3+} is easily bind to cell walls (Zayed et al. 1998). Therefore, Cr is accumulated mainly in roots. The ratio shoot/roots of Cr contents varies highly, between 0.005 and 0.0027. The highest Cr content (in both oxidation states) was found in roots of plants of Brassicaceae family, where it was present as ionic complexes in xylem fluid. The high Cr accumulation in roots, up to 160 mg/kg, was noticed in fodder radish grown in Cr polluted soil (247 mg Cr/kg), whereas its content of shoots did not exceed 10 mg/kg (Siebielec et al. 2012). All these data clearly indicate that Cr is concentrated mainly in roots (Becquer et al. 2003).

Contents of Cr in plants have recently received much attention due to its importance as an essential micronutrient in human metabolic processes, and also because of its carcinogenic impact. Thus, an adequate Cr supply in the diet, especially plant diet, has been recently broadly investigated.

Mean Cr contents of cereal grains do not exit 90 mg/kg, and is the highest in oats grains, up to 600 mg/kg (Eriksson 2001a). Very high Cr amounts are present in some seeds and root vegetables. Grass and clover contain up to 4200 mg Cr/kg.

Toxic effects of excess Cr in plants resulted mainly in the poor protein formation, due to the disruption of N metabolisms, decreases of photosynthesis processes, and a lower S uptake. Toxic effects of excess Cr on spinach plants depend on the soil texture: 40 mg Cr/kg is toxic in sandy soils, whereas 320 mg Cr/kg is toxic in clay loamy soils (Sharma et al. 2005). The phytotoxic Cr concentrations in tops of plants are reported as follows (in mg/kg): tobacco, 18–24; corn, 4–8; and barley seedlings, 10–100. In each case, Cr^{6+} was more toxic than Cr^{3+}. Some plants grown on Cr-contaminated soils may develop Cr-tolerance mechanism, connected with increased activities of some enzymes, superoxide dismutase, and peroxidase (Dong et al. 2007; Pacha and Galimska-Stypa 1988).

The range of average Cr contents of mushroom bay bolete (*Xerocomus badius*) sampled in 1993–1998, from the northern part of Poland, were similar in caps and stalks, 0.22–0.75 and 0.22–0.71 mg/kg, respectively (Malinowska et al. 2004). Common chanterelles (*Cantharellus cibarius*) grown in mountains contained a bit higher amounts of Cr, mean 0.39 mg/kg, than those grown in the Baltic Sea coast, mean 0.20 mg/kg (Falandysz et al. 2012).

12.6 HUMANS

The total amount of Cr in the human body (70 kg) is estimated at 1.2 mg, but can be as much as 12 mg, of which in the blood 6–10 µg/L, in the bone 100–300 µg/kg, and in the tissue 25–800 µg/kg (Emsley 2011). Its concentration is relatively higher in newborn children than in adults. At old age, Cr amounts in serum decrease by 42%. Trivalent Cr tends to accumulate in epidermal tissues (hair, etc.), bones, liver, kidney, spleen, lungs, and in the large intestine. Accumulation in other tissues, especially muscles, seems to be strictly limited or nonexistent (Pechova and Pavlata 2007).

Humans can be exposed to Cr by breathing air, drinking water, eating food Cr, or through skin contact with Cr compounds. The most common health problem in workers exposed to Cr involves the respiratory tract. These health effects include irritation of the lining of the nose, runny nose, and breathing problems (asthma, cough, shortness of breath, and wheezing). Workers have also developed allergies to Cr compounds, which can cause breathing difficulties and skin rashes. Concentrations of Cr in air that can cause these effects may be different for different types of Cr compounds, with effects occurring at much lower concentrations of Cr^{6+} compared to Cr^{3+}. However, the concentrations causing respiratory problems in workers are at least 60 times higher than levels normally found in the environment.

Rates of Cr uptake from the gastrointestinal tract are relatively low and depend on a number of factors, including valence state (Cr^{6+} is more readily absorbed than Cr^{3+}), the chemical form (organic Cr is more readily absorbed than inorganic Cr), the water solubility of the compound, and gastrointestinal transit time. Once absorbed into the bloodstream, Cr^{6+} is rapidly taken up by erythrocytes, after absorption and reduced to Cr^{3+} inside the red blood cells. This reduction occurs by the action of glutathione. In contrast, Cr^{3+} does not readily cross red blood cell membranes, but binds directly to transferrin, an Fe-transporting protein in the plasma.

Hexavalent Cr crosses biological membranes easily, reacting with protein components and nucleic acids inside the cell, while being deoxygenated to Cr^{3+}. In general, Cr^{6+} is more toxic than Cr^{3+}. Trivalent Cr is the most stable oxidation state, in which Cr is found in living organisms and does not have the capacity to cross cell membranes easily (Mertz 1993). Regardless of the source, Cr^{3+} is widely distributed in the body and accounts for most of the Cr in plasma or tissues. The greatest uptake of Cr^{3+} as a protein complex is by bone marrow, lungs, lymph nodes, spleen, kidney, and liver. Cr levels in the lungs are consistently higher than in other organs.

Cr^{3+} is an essential dietary nutrient, whereas Cr^{6+} poses a significant risk of lung cancer. Cr^{6+} compounds can produce effects on skin and mucous membranes. These include irritation, burns, ulcers, and an allergic type of dermatitis. Reduction of Cr^{6+} to Cr^{3+} inside of cells may be an important mechanism for the Cr toxicity, whereas this process, outside of cells, is a major mechanism of protection (ATSDR 2012a).

Chromium is regarded as essential in humans and animals, which takes part in various metabolic processes. Especially, stable Cr^{3+} is known as *biological active chromium*, or glucose tolerance factor, which is essential to glucose, protein, and fat metabolism (Anderson 1981).

Without Cr^{3+} in the diet, the body loses its ability to use sugars, proteins, and fat properly, which may result in weight loss or decreased growth, improper function of the nervous system, and a diabetic-like condition. Therefore, Cr^{3+} compounds have been used as dietary supplements and are beneficial if taken in recommended dosages.

Chromium is a component of enzymes that controls glucose metabolism and synthesis of fatty acids and cholesterol. Thus, its deficiency leads to severe impairment in glucose tolerance, which finally leads to diabetes and atherosclerotic disease (ATSDR 2012a).

Tissue levels of Cr tend to decrease with age, which may be a factor in the increase of adult-onset diabetes disease whose incidence rose more than sixfold during the second half of the twentieth century. This increase may also mirror the loss of Cr from diets, because of soil deficiency and the refinement of foods. Higher fat intake also may inhibit Cr absorption. Even mild deficiencies of Cr can produce symptoms other than problems in blood sugar metabolism, such as anxiety or fatigue. Abnormal cholesterol metabolism and increased progress of atherosclerosis are associated with Cr deficiency.

It is estimated that 25%–50% of the U.S. population being deficient in Cr because of very low soil levels of Cr and the loss of this element from refined foods, especially sugar and flours. There are about 6 mg of Cr stored in the bodies of people who live in the United States; tissue levels of people in other countries are usually higher, and those higher levels tend to be associated with a lower incidence of diabetes and atherosclerosis. There is less hardening of the arteries in Asians, who have five times higher Cr tissue levels than Americans (Cefalu and Hu 2004).

Cr^{6+} has been reported to be toxic and carcinogenic to humans owing to its oxidizing potential and easy permeation of biological membranes. International Agency for Research on Cancer (IARC) has classified Cr^{6+} in group 1 (human carcinogen) and Cr metallic and Cr^{3+} in group 3 (not classifiable as to its carcinogenicity to humans) (IARC 2012a). However, Cr^{3+} is an essential nutrient required for normal energy metabolism. The Institute of Medicine of the National Research Council determined an adequate intake (e.g., level typically consumed by healthy individuals) of 20–45 μg Cr^{3+}/day for adolescents and adults (ATSDR 2012a). Estimated safe and adequate daily dietary intakes for Cr is 50 ± 200 μg (WHO 2004). Data on daily Cr intake in European countries are ranged from 22 to 146 μg, but most of them are under 100 μg, whereas in the United States, the reported intake is between 23 and 62 μg (Van Cauwenbergh et al. 1996).

Chromium is present in the foodstuffs (Table 12.2), both as the inorganic form and organic complexes. Most of Cr present in food is in the trivalent form, which is an essential nutrient, when the toxic hexavalent form of Cr is not normally found in food (Noël et al. 2003).

Controversy surrounding Cr supplementation is due, in part, to the substantial variability in the results of studies that have evaluated the effects of Cr in patients with or without diabetes. Results from some trials have indicated that Cr supplementation increases muscle gain and fat loss, associated with exercise and improves glucose metabolism and the serum lipid profile in patients with or without diabetes. In contrast, results of other studies have indicated little or no benefit of Cr on any of these variables (Cefalu and Hu 2004).

12.7 ANIMALS

The essentiality of Cr, as a required element, necessary for animals is well accepted. In cattle, it has been observed a favorable response to Cr supplementation, especially if the animals are under some physiologic stress. Cr supplementation is important for dairy cattle. Positive Cr effects include improved humoral and cell-mediated immune response, energy status (reduced liver triglyceride accumulation), and increased milk production in primiparous cows, but not multiparous cows. It is evident that cattle require Cr in their diets. For livestock, the maximum tolerable concentration of Cr in the diet is set at 3000 mg/kg for the oxide form, and 1000 mg/kg for the chloride form of the trivalent Cr (NRC 2001).

13 Cobalt [Co, 27]

13.1 INTRODUCTION

Cobalt (Co), a metal of the group 9 in the periodic table of elements, reveals both chalcophillic and siderophillic tendencies. Its content on the Earth's crust is within the range of 9–12 mg/kg. The highest Co contents are in mafic rocks, up to 200 mg/kg, whereas in acidic igneous rocks, its contents range from 1 to 15 mg/kg. It is also likely to be concentrated in black shales. Its distribution in sedimentary rocks is as follows (in mg/kg): argillaceous (shales), 14–20; sandstones, 0.3–10; and limestones, 0.1–3.0. Coal contains Co within the range 5–40 mg/kg. In fly ash, it may be accumulated up to about 30 mg/kg.

Cobalt occurs mainly at +2 oxidation state, but also at +3 and +4. Its geochemical behavior is similar to Fe and Mn. Due to its siderophillic character, it is likely to form minerals with S and As, such as linnelite, Co_3S_4; smaltite, $Co(Ni)As_3$; cobaltite, CoAsS; arsenosulfide, CoAsS; and safflorite, $(Co,Fe)As_2$. In hypergenic zones and in soils, it may occur as erythrite, $Co_2(AsO_4)_2 \cdot 8H_2O$. Oxides of Mn, both abiogenic and biogenic origins, reveal a great capacity to absorb Co.

World mine Co production (without the United States) in 2010 was 88 Mt, of which Congo, 45; Zambia, 11; and Russia, 6.1 (USGS 2011). However, China is considered to be the world's leading producer of refined Co, and as an important supplier of Co to the United States. Recovery of Co is mainly from Cu–Co and Ni–Co sulfide concentrates. It is also obtained from laterite and arsenide ores. In the United States up to 25% of Co is obtained from recycling. About 50% of Co is used in superalloys (strong, resistant to heat and corrosion), mainly for aircraft, and up to 20% in various other metallic applications. It is also used (~30%) in a variety of chemicals.

13.2 SOILS

The worldwide mean value of Co content in soils has been estimated at 10 mg/kg. The Co content of soils is inherited mainly from parent materials. Higher levels of Co, mean 12 mg/kg, is often noticed in heavy loamy soils (Table 13.1). In other soils, its mean contents do not exceed 7 mg/kg. However, in some specific locations, especially where parent materials are enriched in Co, some maximum, elevated Co contents are noticed (in mg/kg) in cambisols, 60; calcisols, 70; and histosols, 50. Soil-forming processes, which differ in various climatic zones, have a great impact on Co distribution in soil profiles. Higher Co contents of surface layer of soils are in arid and semiarid climatic zones, whereas very low Co contents are in light soils, in humid climatic zones.

Soils over serpentine rocks contain Co up to 520 mg/kg. Its elevated contents (116–122 mg/kg) are also observed in ferralsols of various countries (Kabata-Pendias 2011). Soils in arid and semiarid regions (e.g., Egypt) contain Co within the range 17–27 mg/kg (Nassem and Abdalla 2003). Cobalt contents of soils of South Africa range from

TABLE 13.1
Cobalt Contents in Soils, Water, and Air

Environmental Compartment	Mean/Range
Soil (mg/kg)	
Light sandy	5.5
Medium loamy	7.0
Heavy loamy	12.0
Calcerous (calcisols)	7.0
Organic (histosols)	4.5
Water (μg/L)	
Rain	0.3–1.7
River	0.15
Sea, ocean	0.001–0.02
Air (ng/m³)	
Industrial region	48
Urban areas	0.13–3.0
Greenland	0.07–0.15
Antarctica	0.0001–0.12

Source: Data are given for uncontaminated environments, from Kabata-Pendias, A. and Mukherjee, A.B., *Trace Elements from Soil to Human*, Springer, Berlin, Germany, 2007; Reimann, C. and de Caritat, P., *Chemical Elements in the Environment*, Springer, Berlin, Germany, 1998.

1.5 to 68.5 mg/kg. Light soils, developed from glacial deposits, under temperate humid climate contain usually much smaller amounts of Co. Thus, arable soils of Sweden contain Co in the range of 0.4–14 mg/kg (Eriksson 2001a). Mean Co content in organic surface soils of Norway is 1.35 mg/kg, and is lower than mean value (1.91 mg/kg) for all investigated soils (Nygard et al. 2012). Cobalt contents of soils in Spitsbergen range from 0.06 to 0.35 mg/kg (Gulińska et al. 2003). In reference soil samples of the United States, Co ranges from 5.5 to 29.9 mg/kg, and in Chinese soils it is within the range 5.5–97 mg/kg (Govindaraju 1994).

Hydrous oxides of Fe and Mn reveal a great affinity for the adsorption of Co. Thus, Co distribution in soil horizons is similar to that of Fe and Mn. Concretions of Fe–Mn accumulate Co within the range of 20–390 mg/kg, and in some specific soils may concentrate up to about 20,000 mg/kg (Kabata-Pendias 2011). Mn oxides influence the oxidation of Co^{2+} to Co^{3+}, and the formation of hydroxyl species, $Co(OH)_2$, that easily precipitate at the oxide surface. The sorption of Co by Mn oxides increases with pH and is controlled by the redox potential. Besides Mn oxides, birnessite (an oxide mineral of Mn) and goethite (an oxyhydroxide mineral of Fe) play an important role in the Co adsorption.

Cobalt speciation in soils depends on several factors, of which Eh and pH play the most significant role. Reduction of soil Eh and decreased soil pH elevate the Co mobility. Various redox mechanisms may affect the Co sorption by Mn oxides: (1) oxidation of Co^{2+} to Co^{3+}, by Mn oxides; (2) formation of hydroxyl species, $Co(OH)_2$, that easily precipitate at Mn oxides surface; (3) reduction of Mn^{4+} to Mn^{3+} in the oxides crystal lattice; and (4) replacement of Mn^{4+} or Mn^{3+} by Co^{3+}. Soil texture, clay minerals, and soluble organic matter (SOM) play an important role in the behavior of Co in soils. Cobalt is easily fixed by clay minerals, especially by montmorillonite and illite, and amorphous Fe–Al oxides. Some Co-organic ligands may stimulate Co sorption by Fe hydroxides, and also increase Co phytoavailability, at the absence of Fe hydroxides. In most soils, Co is rather slightly mobile, and thus its concentration in soil solution ranges from 0.3 to 87 μg/L. Its ionic species in soil solution are Co^{2+}, Co^{3+}, $CoOH^+$, and $Co(OH)_3^-$ (Kabata-Pendias and Sadurski 2004).

Cobalt contents of soils have been for years a great challenge for animal (especially ruminants) farms. Low Co contents of soils, or its low phytoavailability have affected its deficiency in animal nutrition. Adequate Co supply to animals was when the ammonium acetic acid–ethylenediaminetetraacetic acid extractable Co contents of soils are within the range 3.6–4.7 mg/kg (Nassem and Abdalla 2003). According to these authors, low extractable Co contents in pasture soils may cause its deficiency in animals. Its deficiency in grazing animals is associated mainly with alkaline and calcareous soils, light leached soils, and high SOM content. The application of Co sulfate compounds, or EDTA chelates, to soils are common practices in controlling the Co deficiency in ruminants.

Cobalt contaminations of soils have been noticed in several regions. The highest Co accumulation (up to about 520 mg/kg) is reported for soils of some Co-mining areas. Also, soils around metal-processing industries contain elevated Co levels (Table 13.2). Roadside and urban street dusts contain Co up to 14 mg/kg, and may be of its significant source in nearby soils (Kabata-Pendias 2011). Soils within the city

TABLE 13.2
Cobalt Enrichment and Contamination of Surface Soils (mg/kg)

Site and Pollution Source	Country	Range/Mean
Mining areas	New Zealand	10–520
	United States	13–85
Metal-processing industries	United States	42–154
	Canada	10–127
	Norway	20–70
	Great Britain	67
Sludged farmland	Netherlands	3.3–12.4

Source: From various sources, as compiled by Kabata-Pendias, A., *Trace Elements in Soils and Plants*, 4th ed., CRC Press, Boca Raton, FL, 2011.

of Łódź (Poland) contain Co up to 100 mg/kg, as effects of contaminations from the power plant and motor traffic (Jankiewicz and Adamczyk 2007).

The criteria for contaminated land (Dutch List 2013), following Co concentrations in soils and groundwater, are established (in mg/kg, and µg/L) as follows: uncontaminated, 20 and 20; medium contaminated, 50 and 50; heavily contaminated, 300 and 200.

13.3 WATERS

Cobalt is slightly mobile in terrestrial systems and therefore its C_w/C_c ratio is very low, about 0.08 (Gaillardet et al. 2003). In addition, while Co is an in soluble species, it is easily adsorbed by organic complexes and Fe–Mn hydroxides.

The worldwide concentration of Co in ocean water ranges from 0.001 to 0.02 µg/L (Table 13.1). In Baltic Sea, its concentration is fairly similar, within the range 0.001–0.07 µg/L (Szefer 2002). Its world river flux to seawater is estimated as 5.5 kt/yr (Gaillardet et al. 2003). Yearly flux of Co to the Baltic Sea is calculated for 20 t from anthropogenic sources, and 3 t from natural aerial deposition (Matschullat 1997).

The worldwide mean Co concentration in river water is 0.12 µg/L, within the range of 0.02–0.43 µg/L. River water of Western Siberia contain Co from 0.2 to 2 µg/L, and its input to the Baikal Lake is 3 t/yr (Vietrov 2002). Concentration of Co in water of the Kola River (Russia) is (in µg/L) 0.03–0.83 in dissolved phase and 0.003–0.19 in suspended phase (Pekka et al. 2004). Mean cobalt concentrations in Nordic lakes are (in µg/L) as follows: 0.96 in Norway, 1.4 in Sweden, and 2.4 in Finland (Skjelkvåle et al. 2001). Surface water around Cu-mine and smelter areas contain increased Co levels, up to about 6000 µg/L (ATSDR 2002b).

Cobalt in water is present in several species: Co^{2+}, Co^{3+}, $CoCl^{2+}$, and $CoHCO_3^+$, and is coprecipitated as $CoSO_4$ and $CoCO_3$. All Co species, but in particulate Co^{3+}, are adsorbed by mineral and organic particles and are deposited in bottom sediments. Especially high adsorption capacity have Fe–Mn hydroxide particles. Cobalt content in bottom sediments is a good information on the water pollution. Average Co content in river-bottom sediments is estimated at 13 mg/kg; however, it varies highly and in industrially polluted rivers, it may be up to 50 mg/kg. Its mean content in bottom sediments of San River (Poland) varies from 1 to 8 mg/kg, in sandy and mule sediments, respectively (Bojakowska et al. 2007). Soluble species of radiocobalt may be released from some nuclear reactors to water, and therefore its contents may increase in the future (ATSDR 2004b).

Median Co concentrations in both bottled and tap water of the EU countries is 0.023 µg/L, (Birke et al. 2010). Drinking water in the United States contain (at an average) Co <2 µg/L (ATSDR 2004b).

Aquatic organisms may accumulate elevated amounts of Co. Mussels may contain, in soft tissue, Co up to 900 µg/kg FW, whereas its contents in fish range from 1 to 10 µg/kg FW. Especially high Co levels (<100–2100 µg/kg) were found in planktons (Szefer 2002).

13.4 AIR

Cobalt concentration in the atmosphere of the remote region (Antarctica) is <0.12 ng/m^3, whereas air of urban areas contain it up to 3 ng/m^3 (Table 13.1). Very high Co concentrations may be in the atmosphere of metal industrial regions. Atmospheric Co deposition in some EU countries is reported to be (in g/ha/yr) as follows: 1.5–6 in the United Kingdom; 5.6–27 in Germany; and 0.19 in Sweden (Kabata-Pendias and Pendias 1999).

Mean Co contents of two moss species from mountains in Poland are lower (160–180 μg/kg) than in the same moss species form Alaska (410–420 μg/kg). This may suggest that atmospheric Co input is higher in Alaska than in Poland (Migaszewski et al. 2009).

13.5 PLANTS

In general, Co is taken up and transported in higher plants by the transpiration flow. However, several soil factors and ability of plants control its phytoavailability. Relatively low Co^{2+} mobility in plants restricts its transport from stems to leaves. Mobile Co in soils is easily available to plants. Also other Co forms, like bound to Fe and Mn oxides, are easily phytoavailable, whereas Co fixed by carbonates and SOM are hardly available to plants (Bhatacharya et al. 2008). Silanpää and Jansson (1992), after investigating Co in wheat and corn from 30 countries, have stated that soil texture is the most significant parameter controlling its levels in plants.

Mean Co contents in cereal grains vary within the range of 5–270 μg/kg, being the lowest in Norway and the highest in Egypt. In Sweden, the Co content in cereal grains is (in μg/kg) as follows: wheat, 1.1–18; barley, 4.4–40; and oats, 10–300 (Eriksson 2001a). Range of Co amounts in grass is 60–270 and in clover 100–570 μg/kg. Liming and different mineral fertilizers reduce the Co phytoavailability and create a risk of a low Co content in fodder plants, which resulted in Co deficiency, mainly in ruminants. Foliar applications of Co in solutions are effective in the correction of the Co deficiency.

Among various native plants grown in the polluted area of the Northern Europe (Kola Peninsula), crowberry (*Empetrum nigrum*) has the highest capacity to accumulate Co, up to 10,510 μg/kg (Reimann et al. 2001). Excess of Co may inhibit some biological processes (e.g., photosynthesis and respiration) in some plant organisms. In higher plants, however, there are some evidences that it stimulates chlorophyll formation. Favorable effects of Co on the plant growth, and especially on N-fixation processes are evident. Cobalt is a component of the vitamin B_{12} and cobamide coenzyme, which are involved in the fixation of N_2 in root nodules of legumes. Co coenzymes have been detected in nonlegumes, but it is unknown whether these compounds are originated from microorganisms associated with plants. Thus, beneficial effects of Co on plant metabolisms are not yet understood, and most possible it is cross-linked with several interactions with other chemical elements (Kabata-Pendias 2011).

Cobalt deficiency inhibits plant growth, mainly of various legumes. Its deficiency in fodder plants is considered mainly from the viewpoint of ruminants requirement, as it is the main precursor of vitamin B_{12}. Critical Co levels in ruminant diets vary from 80 to 100 μg/kg. Deficiency of Co may be controlled by the application of Co salt to soils, and in most cases, the effect of that treatment may last for several years. However, there are various opinions, and often it is advised to apply Co salts directly to livestock.

Excess Co in soils is transported mainly to plant leaves. Co toxicities are white leaves margins and tips, especially of young leaves; several enzymatic processes of plants may be inhibited. Some plants grown in soils with 25–50 mg Co/kg, and/or with 140 mg Co/L in soil solution, suffer from excess Co. Plant sensitivity to increased Co contents is various. Commonly reported critical Co levels in plants are 30,000–40,000 μg/kg. Cereals are known to be the most sensitive plants to excess Co and may tolerate it at 10,000–20,000 μg/kg. Some plants may develop a mechanism of Co tolerance, which is observed mainly in metalliferous plant species. Especially *Nyssa sylvatica* grown in Co-contaminated soil may accumulate this metal up to 800,000 μg/kg. It is recommended for the phythoextraction of Co and ^{60}Co, as well as for the remediation of contaminated soils (Malik et al. 2000). Among various native plants grown in the polluted area of Northern Europe (Kola Peninsula), crowberry (*Empetrum nigrum*) has the highest capacity to accumulate Co, at the pollution-background ratio of 206 (Reimann et al. 2001).

The mushroom, common chanterelles (*Cantharellus cibarius*) grown in mountains contained lower amounts of Co, mean 390 μg/kg, than those grown in the Baltic Sea coast, mean 100 μg/kg (Falandysz et al. 2012).

13.6 HUMANS

Cobalt has been identified in most tissues of the body, with the highest concentrations in the liver. Total body burden is estimated at 1.1–1.5 mg, of which 85% are in the form of vitamin B_{12}, with 0.11 mg in the liver (WHO 2006b). Concentrations in body fluids are well below the μg/L level; mean concentrations reported in the serum range from 0.1 to 0.3 μg/L. Considerable differences have been found in the levels of Co in hair, ranging from 0.4 to 500 μg/kg; 0.05–2.7 μg/L in the blood; 0.1–0.6 μg/L in the plasma; and 0.1–1.5 μg/L in the urine (Catalani et al. 2011).

Approximately 50% of the Co that enters the gastrointestinal tract will be absorbed. Cobalt absorption is increased among individuals who are Fe deficient. Its water-soluble forms are better absorbed than insoluble forms (WHO 2006b).

The only known essential role of Co in animals and humans is being a component of vitamin B_{12} as Co^{3+}. Absorbed Co^{2+} is not known to have any biological function (EFSA 2012b). In animals with the capacity to synthesize cyanocobalamin, Co from orally administered Co^{2+} is deposited in tissues, in the form of vitamin B_{12}. As a component of vitamin B_{12}, Co is essential in the body; therefore, it is found in the most tissues. Vitamin B_{12} is a water-soluble vitamin, with a key role in the normal functioning of the brain and nervous system, and for the formation of blood. Vitamin B_{12} occurs in foods of animal origin and represents only a small fraction of Co intake. The recommended dietary allowance (RDA) of vitamin B_{12} is 2.4 μg/day, which contains 0.1 μg of Co (WHO 2004).

Cobalt has both beneficial and harmful effects on human health. Cobalt is beneficial for humans because it is part of vitamin B_{12}, which is essential to human health, and is used for the treatment of anemia, also for pregnant women. At very high exposure levels, it increases red blood cell production. Exposure of humans and animals to levels of Co normally found in the environment is not harmful. Occupational, increased exposure to Co occurs in several industries.

Harmful health effects of Co can also occur. Workers who breathed air containing 0.038 mg Co/m^3 (about 100,000 times more than normal concentration in ambient air), for 6 hours, had a trouble breathing. Serious effects on the lungs, including asthma, pneumonia, and wheezing, have been found in people exposed to 0.005 mg Co/m^3, while working with hard metal, a Co–W carbide alloy. People exposed to 0.007 mg Co/m^3, at work have also developed allergies to Co, which resulted in asthma and skin rashes (ATSDR 2004b).

Increased lung cancer risk is related to exposure to hard-metal dust containing Co and W carbide. International Agency for Research on Cancer (IARC) established that Co metal with tungsten carbide is "probably carcinogenic to humans" (Group 2A), Co metal without W carbide is possibly "carcinogenic to humans" (Group 2B), and Co sulfate and other soluble Co(II) salts are possibly "carcinogenic to humans" (Group 2B) (IARC 2006).

A radioactive isotope, ^{60}Co is sometimes used in the radiation therapy, for cancer treatments. Co treatment in radiation therapy has been found to be more effective in shrinking tumors and destroying cancer cells in the lungs than other isotopes. However, exposition to radioactive Co may be very dangerous to health. Radiation from radioactive Co can also damage cells in body during eating, drinking, breathing, or touching anything that contains radioactive Co. Exposure to lower levels of radiation might cause nausea, and higher levels can cause vomiting, diarrhea, bleeding, coma, and even death (ATSDR 2004b).

Human dietary intake of Co is highly variable. Most of the Co ingested is inorganic. Vitamin B_{12}, which occurs almost entirely in food of animal origin, accounts for only a very small Co fraction. Vegetables contain inorganic Co but little or no vitamin B_{12}. Richest sources of Co are some products of animal origin, nuts, chocolates, and rice (Table 13.3). No specific RDA is suggested for Co.

The estimated population average intake of Co is reported to be (in mg/day): 0.005–0.04 in the United States; 0.011 in Canada; 0.012 in the United Kingdom; and 0.029 in France. These figures are most likely included already in the contribution of products from animals routinely supplemented with Co(II) compounds (EFSA 2012b).

13.7 ANIMALS

Cobalt is also essential for the health of animals, especially of cattle and sheep, which can synthesize vitamin B_{12} in the digestive tract by microbial action. Efficiency in the incorporation of Co in vitamin B_{12} in ruminants is low and inversely related to the Co intake. This incorporation rate may range from 3% to 15%. Besides covering the requirements for vitamin B_{12} synthesis, Co may play a role in rumen

TABLE 13.3
Cobalt in Foodstuffs (mg/kg FW)

Product	Content
Food of Plant Origin	
Bread[a]	0.02
Bread, rusk[b]	0.006
Wheat grain[c]	0.044
Miscellaneous cereals[a]	0.01
Rice, white[d]	0.033
Rice, brown[d]	0.069
Vegetables[b]	0.006
Fruits[b]	0.009
Nuts and oilseeds[b]	0.041
Milk chocolate[c]	0.093
Cocoa[c]	0.92
Food of Animal Origin	
Meat[a]	0.004
Meat[b]	0.008
Offal[a]	0.060
Fish[b]	0.007
Shellfish[b]	0.046
Milk[b]	0.001
Milk[e]	0.0013
Cheese[b]	0.018

Sources: [a] Ysart, G. et al., *Food Addit. Contam.*, 16, 391–403, 1999.
[b] Leblanc, J.C. et al., *Food Addit. Contam.*, 22, 624–641, 2005.
[c] Sager, M., *J. Nutr. Food Sci.*, 2, 123, 2012.
[d] Batista, B.L. et al., *Food Addit. Contam.*, 3, 253–262, 2010.
[e] Gabryszuk, M. et al., *J. Elem.*, 15, 259–267, 2010.
FW, fresh weight.

fermentation, by increasing fiber digestion from low-quality forages. Nonruminants require the intake of vitamin B_{12}, because they lack the ability to synthesize the vitamin, in significant amounts, by digestive tract microbiota. However, pigs and poultry are known to synthesize small amounts of this vitamin by hindgut bacteria.

The deficiency of B_{12} is characterized by loss of appetite, wasting, anemia, poor fertility rates, failure to thrive (particularly in young animals), and even death, in severe cases, unless the deficiency is corrected. It affects mainly sheep and cattle, with younger stocker being more sensitive, though it can occur in all ruminant animals and does affect also goats and deer.

Tolerance of ruminants to Co is very high, and greatly in excess of the requirements. Typical signs of chronic Co toxicity in most species are reduced feed intake and body weight, emaciation, anemia, hyperchromasia, debility, and increased liver Co. Signs of chronic Co toxicity are similar, in many respects, to its deficiency. Feeding supplemental Co compounds, under assessment up to the maximum total Co content in feed, set by the current EU legislation (2 mg Co/kg complete feed) is considered safe, for all animal species/categories (EFSA 2012b).

14 Copper [Cu, 29]

14.1 INTRODUCTION

Copper (Cu), a metal of the group 11 in the periodic table of elements, occurs in the Earth's crust within the range of 25–75 mg/kg. It is concentrated mainly in mafic igneous rocks, up to 120 mg/kg, whereas in acidic igneous rocks its contents range from 5 to 30 mg/kg. Its distribution in sedimentary rocks is as follows (in mg/kg): argillaceous (shales), 40–60; sandstones, 2–30; and limestones, 2–10. Coal contains Cu within the very broad range of 12–280 mg/kg. In fly ash, it may be accumulated up to about 70 mg/kg.

Copper occurs mainly at +2 oxidation state, and sometimes +1. Due to its strong affinity for S, it is likely to form minerals mainly with S, such as covellite, CuS; chalcosite, Cu_2S; chalcopyrite, $CuFeS_2$; and bornite, Cu_5FeS_4. During the weathering of Cu sulfides, Cu is incorporated in oxide and carbonate minerals, of which the most common are tenorite, CuO; cuprite, Cu_2O; malachite, $Cu_2CO_3(OH)_2$; and azurite, $Cu_2(CO_3)_2\ (H_2O)_2$. There are several host minerals of Cu, such as mica, sphaleryt, pyrite, galena, and magnetite. In some sedimentary deposits, it may occur as native Cu. It may also be very easily adsorbed by freshly precipitated Al and Mn oxides (Violante 2013).

World mine Cu production (rounded) in 2010 was 16.2 Mt, of which Chile, 5.5; Peru, 1.3; China, 1.2; and the United States 1.2 (USGS 2011). Copper in all old and new, refined or remelted scrap (160 kt) contributed to about 35% of its supply in the United States. It reveals versatile properties, and therefore has a wide range of applications, very often as Cu alloys. It is used in building construction (50%), electronic products (20%), transportation equipment (10%), consumer and general products (10%), and industrial machinery and equipment (8%). It is also used in agriculture, mainly as fertilizers and pesticides. It is also a fodder additive in livestock and poultry nutrition.

14.2 SOILS

The worldwide average Cu content is estimated at 14 mg/kg. However, it highly varies for different countries, within the range between 14 and 110 mg/kg. Copper contents are closely associated with soil textures; in most cases, the lowest is in sandy soils, 3–30 mg/kg, and the highest in loamy soils, 7–140 mg/kg. Some histosols may also contain elevated amounts of Cu, up to about 120 mg/kg (Table 14.1). Some soils, derived from parent rocks with high Cu levels, may contain much higher Cu, for example, soils in Carpathian rocks have Cu up to 22,360 mg/kg (Čurlik and Šefčik 1999).

Agricultural soils (3045 samples) of the United States contain Cu within the range <0.6–495 mg/kg (mean, 18 mg/kg) (Holmgren et al. 1993). Agricultural soils (32,000 samples) of Poland contain Cu from 0.2 to 725 mg/kg (mean, 6.5 mg/kg)

TABLE 14.1
Copper Contents in Soils, Water, and Air

Environmental Compartment	Range
Soil (mg/kg)	
Light sandy	1–7
Medium loamy	4–100
Heavy loamy	7–140
Calcerous (calcisols)	7–70
Organic (histosols)	1–120
Water (μg/L)	
Rain	0.02–0.3
River	0.27–3.55
Sea, ocean	<1–8
Air (ng/m³)	
Industrial and urban regions	5–4900
Greenland	0.03–0.06

Sources: Data are given for uncontaminated environments, from Kabata-Pendias, A. and Mukherjee, A.B., *Trace Elements from Soil to Human*, Springer, Berlin, Germany, 2007; Reimann, C. and de Caritat, P., *Chemical Elements in the Environment*, Springer, Berlin, Germany, 1998.

(Kabata-Pendias 2011). In soils of the EU countries, its mean concentrations range from 5 to 50 mg/kg. Mean Cu content in organic surface soils of Norway is 10.8 mg/kg, and is a bit lower than the mean value (11.6 mg/kg) for all investigated soils (Nygard et al. 2012). Copper contents of soils in Spitsbergen are within the range 0.1–1.2 mg/kg (Gulińska et al. 2003).

In most cases, Cu is accumulated in the upper few centimeter layer, mainly due to bioaccumulation and anthropogenic sources. However, because of its great sorption by soluble organic matter (SOM), carbonates, clay minerals, and Mn–Fe hydroxides, it may also be concentrated in deeper soil layers. Goethite reveals a strong affinity for the Cu sorption, which is controlled by organic acids (Perelomov et al. 2011). Its sorption on Al–Fe oxides may be increased by the presence of sulfate and phosphate compounds (Violante 2013).

Copper occurs in most soils as the cation $Cu(H_2O)_6^{2+}$, which is easily adsorbed by organic and mineral soil components. Mobility depends highly on Cu species and soil pH. Some species (e.g., CuO, Cu^{2+}, $Cu[OH]^+$, $Cu_2[OH]_2^{2+}$) are mobile in acidic soils (pH < 7), and other species (e.g., $Cu[OH]_3^-$, $Cu[OH]_4^{2-}$, $Cu(CO_3)_2^{2+}$) are mobile in alkaline soils (pH > 7). All these Cu species, as well as Cu-organic complexes, which are considered to be the most common Cu forms, occur in soil solutions (Kabata-Pendias and Sadurski 2004). Main variables affecting the Cu mobility are SOM, dissolved organic carbon, dissolved organic matter (DOM), Fe minerals, and pH. The overall solubility of both cationic and anionic forms of Cu decreases at pH 7–8 (Ponizovsky et al. 2006). However, soil clay fraction has the highest impact on Cu behavior.

Precipitation of $CuCO_3$, especially in calcareous soils, is a main process affecting the Cu activity in most soils. Attenuation of Cu mobility may also result from the Cu–Ca substitution in calcites present in calcareous soils, or from the precipitation of Cu–carbonate compounds in other soils (Ma et al. 2006). Organic Cu complexes also have a crucial impact on its bioavailability, and on migration within soil profile. Great impact on Cu behavior in soils have all species and minerals of Fe (Cornell and Schwertmann 2003; Contin et al. *vide* Dąbkowska-Naskręt 2009).

Phytoavailability of Cu depends on its amounts in soils, and on the molecular weight of Cu complexes. There are some opinions, however, that its availability to plants is influenced by Cu species and not by the total content (Allen 1993). Mobility and phytoavailability of Cu is reduced by the presence of Fe–Al–oxyhydroxide coating colloids; oxyhydroxide particles of Fe, Mn, and Al; and by DOM. However, the impact of DOM from sewage sludge is reported to mobilize Cu in soils (Ashowrth and Alloway 2004). Organic compounds of a low molecular weight, from the decay of organic residues, and those added to soils, mainly with sewage sludge, highly increase its mobility and availability to plants (Bahaminyakamwe et al. 2006). Forest litter (especially its compound—tannin) also increases the Cu solubility, but with time, it leads to the reduction of mobility (Karczewska et al. 2013). Chemical speciation of Cu in sewage sludge controls its bioavailability and toxicity to plants (Fjällborg and Dave 2003). Copper added to soils with sludge is easily phytoavailable and very mobile, and therefore may cause ground and surface water pollutions. Average Cu concentration in urban wastewater (UWW) is 0.2 mg/L, and its presence is very common, up to 75% of investigated UWW from the United Kingdom (ICON 2001). Its content in sewage sludge applied to agricultural land in Germany is 305 mg/kg (average) and in the United Kingdom is 373 mg/kg (median) (ICON 2001).

Soil contamination by Cu has been studied for several decades and a large database has been already presented. There are several significant sources, such as fertilizers, sewage sludge, manure, agrochemicals, polluted irrigation waters, industrial wastes and emissions, and urban pollution. Highly increased Cu levels are in soils surrounding Cu mines and smelters (Table 14.2). Mean Cu content of soil, along the high traffic road in Poland, is 65.23 mg/kg (the highest 195.76), and at the 10 km distance, it is 12.55 mg/kg (the highest 21.30) (Niesiobędzka 2012).

Agricultural input of Cu to soils is significant in some countries. In the EU countries, it varies (in g/ha/yr) from 29 in the Czech Republic to 2771 in Italy. Minus load charge to farmland soils was noticed only in Switzerland and Norway. The highest loads of Cu were observed in sludged and vineyard soils as 3,905 and 13,923 g/ha/yr, respectively (Eckel et al. 2005). Also, soil amended with poultry litter accumulated much Cu, up to 1400 mg/kg (Nachtigall et al. 2007). Fernandez-Calvino et al. (2008) reported that increased Cu levels, 246 mg/kg, in vineyard soils increased its contents, up to 209 mg/kg, in river-bottom sediments. The maximum

TABLE 14.2
Copper Contamination of Surface Soils (mg/kg)

Site and Pollution Source	Country	Maximum Content
Old mining area	United Kingdom	2000
Metal-processing industries	Belgium	1089
	Bulgaria	2015
	Canada	3700
	Romania	1387
	Russia[a]	4622
Sludged, irrigated, fertilized farmlands	United Kingdom	800
	Netherlands	265
	Poland	1600

Source: After compiled by Kabata-Pendias, A., *Trace Elements in Soils and Plants*, 4th ed., CRC Press, Boca Raton, FL, 2011.
[a] Smelter on Kola Peninsula.

allowable Cu loading to arable soils has been established in the EU countries for 12 kg/ha/yr. Acceptable Cu contents in farmland soils of Germany was established, accordingly to soil texture, as follows (in mg/kg): heavy loams, 100; medium loams, 60; and sand, 30 (Eckel et al. 2005).

The criteria for contaminated land (Dutch List 2013), following Cu concentrations in soils and groundwater, are established (in mg/kg and μg/L) as follows: uncontaminated, 200 and 50; medium contaminated, 500 and 200; and heavily contaminated, 3000 and 800. Soil quality Cu levels are estimated, based on various criteria, within the range of 100–3100 mg/kg.

Remediation of Cu-contaminated soils has been widely investigated. Commonly applied techniques are based mainly on the addition of materials of a great capability for Cu adsorption, such as (1) organic matter; (2) Fe hydroxides; (3) carbonates; (4) phosphates; and (5) clay minerals, mainly bentonite and vermiculite. Especially nanoscale Fe particles have a high fixation capacity (Zhang 2003). However, all cited remediation methods, do not limit completely the Cu phytoavailability. The best remediation effects, as indicated by yield and chemical composition of lettuce, was the application of high rate of peat and superphosphate (Wróbel 2012). Also, peat applied together with liming significantly decreases the content of mobile Cu species (Nowak-Winiarska et al. 2012).

Under microorganisms activities, Cu may be available to plants, even from the remediated soils. Some bacteria, especially in the rhizosphere, may accumulate Cu, and thus control its availability (Wang et al. 2008). Bioremediation of Cu-contaminated soils by actinomycetes, due to their very high resistance and great bioaccumulation abilities is proposed (Amoroso and Abate 2012). Under the impact of some endomycorrhizal fungi, Cu nanoparticles are formed at the soil–root interface and influence its availability (Manceau et al. 2008).

14.3 WATERS

Median Cu concentration in world ocean water is estimated at 0.25 μg/L and in river water at 1.48 μg/L (Reimann and de Caritat 1998). However, its concentration in water varies highly, depending on pollution source and water pH (Table 14.1). Its world river flux in seawater is estimated at 55 kt/yr (Gaillardet et al. 2003). Yearly input of Cu to the Baltic Sea is calculated as 1300 t for river water, and as 1200 t for atmospheric deposition. Up to about 80% of Cu input is from anthropogenic sources (Matschullat 1997).

The speciation of Cu in water controls biochemical processes and its bioavailability. It occurs in forms of various cations, soluble salts such as carbonates and sulfates, and complexes with organic colloids. Copper species in water depend highly on the pH. At about neutral pH value in freshwater dominates $Cu(OH)_2$, and also $CuCO_3$ and Cu^{2+} are common (EPA 1986). In seawater, the major Cu species are $Cu(OH)Cl$ and $Cu(OH)_2$ (Eisler 1998).

Concentration of Cu in water is in dynamic equilibrium with its contents in surface-bottom sediments. Copper content in bottom sediments is a good indicator on its content in water. Average Cu content in river-bottom sediments is estimated at 9 mg/kg; however, it varies highly, and its content in industrially polluted rivers may be up to about 500 mg/kg (Kabata-Pendias and Pendias 1999). Its mean content in bottom sediments of San River (Poland) varies from 2 to 48 mg/kg, in sandy and mule sediments, respectively (Bojakowska et al. 2007). Copper content of surface-bottom sediments of harbor in Klaipeda (Lithuania) also depends on the granulometric composition, and is (in mg/kg, average and maximum, respectively) as follows: in sand 5.5 and 23.4; in mud 21.9 and 29.5 (Galkus et al. 2012). Sediments of rivers and lakes of some industrial regions may contain elevated amounts of Cu (in mg/kg, after Kabata-Pendias and Mukherjee 2007):

- India, river near the city Chennai, 760–939
- Russia, lake near the Cu–Ni smelter, <905
- Sweden, lake near smelter area, <2000
- United Kingdom, southwest industrialized region, <590

Surface layer of bottom sediments of rivers and lakes of uncontaminated areas contain Cu within the range 7–107 mg/kg (in Wales and Lebanon, respectively).

Copper contents in stream-bottom sediments of National Park, Montgomery (Pennsylvania State) were, in 1995, within the range of 10–50 mg/kg (Reif and Sloto 1997). Sediments of Sergipe river estuary (Brazil) contain Cu from 4.92 to 32.7 mg/kg (Garcia et al. 2011). Cu concentrations in the wetland sediments from the effluent at the Savannah River site (Aiken, SC) have increased, during a five-year period, from 4 to 205 and 796 mg/kg, respectively, in the organic and floc sediment layers (Knox et al. 2006).

Assessment limits for Cu in sediments are established as follows (in mg/kg): effects range low, 34; effects range median, 270; probable effect level (PEL), 197; 110; and 86 (EPA 2000, 2013). The Environment Canada sediment-quality guidelines (USGS 2001) gave a bit similar values for Cu in lake-bottom sediments (in mg/kg): threshold effects level, 35.7; PEL, 197; and probable effect concentration, 149.

Copper is easily taken up from waters by phyto and zooplankton. Some aquatic weeds (e.g., gibbous duckweed, *Lemna gibba* L.) may accumulate up to 60%–80% of its concentrations in water (Khellaf and Zerdaout 2010). Some aquatic fungal strains have a great capacity to retain Cu, within the range of 1.7–7.5 g/kg (Simonescu and Ferdes 2012).

Contents of Cu in the aquatic fauna vary highly, and is usually higher in the coast water than in the middle of sea basins. Detritivores contain more Cu (up to about 100 mg/kg) than omnivores (up to about 50 mg/kg) (Cain et al. 1992). The highest contents of Cu are reported for crabs (58 mg/kg) and crayfish (56 mg/kg) (Kabata-Pendias and Mukherjee 2007). Some Cu cations, such as Cu^{2+}, $CuOH^{+}$, and $Cu_2(OH_2)^{2+}$, are more toxic to aquatic biota than other species.

Median Cu concentration in bottled water of the EU countries is 0.251 μg/L, and is much lower than those in tap water, 5.65 μg/L (Birke et al. 2010). The guideline Cu concentration in drinking water is established at 2000 μg/L by the WHO (2011a), and by the European Economic Community in bottled water at 1000 μg/L (Diduch et al. 2011).

Several techniques are proposed for cleaning water from Cu. It can be removed from water by the precipitation at high pH (9–11). Some chemical reagents are used for formations of insoluble Cu complexes (Chen 2013). The volcanic rock from Northern Rwanda reveals a great adsorption capacity for Cu, 10.87 mg/g, and is used for its removal from wastewater (Sekomo et al. 2012).

14.4 AIR

Copper concentration in the atmosphere varies greatly from 0.03 to 0.06 ng/m^3 in the remote region (Antarctica) to 4900 ng/m^3 in the industrial regions of the EU countries (Table 14.1). The background Cu content in air is proposed at 4 ng/m^3, and its calculated median in air of remote regions is 2.6 ng/m^3 (Reimann and de Caritat 1998).

Sources of Cu in air are mainly anthropogenic, but some amounts of Cu are from rock weathering, volcanic emissions, thermal springs, and blown dust from terrestrial components. Global emission of Cu has been recently reduced, and in 1995, it was calculated at 35 kt (Pacyna and Pacyna 2001). Its average deposition in the EU regions was estimated at 34 g/ha/yr, and varied from 5 to 100 g/ha/yr, with the lowest being in Finland and the highest in Austria (Nicholson et al. 2003).

Average Cu content, 7 mg/kg, of moss sampled in Norway during 1990–1995 was lower than in mosses sampled earlier, 20 mg/kg (Berg and Steinnes 1997). Mean Cu contents of two moss species from mountains in Poland are a bit higher (6.3–7.6 mg/kg) than that of the same moss species from Alaska (4.4–5.5 mg/kg). This may suggest that atmospheric Cu input is higher in Poland than in Alaska (Migaszewski et al. 2009).

14.5 PLANTS

There are several evidences for the active Cu uptake by plants; however, the passive absorption is likely to occur, especially at the toxic Cu level, in the growth media. Rate of its uptake differs widely, depending on the metal species. Copper in plant roots is present in both metallic and organic complexes. It is relatively slightly mobile, and thus its concentrations are higher in roots than in shoots. Its concentration increases with plant growth. Contents of Cu in various plants from unpolluted regions of different countries range from X to XO mg/kg. However, under both natural and human-induced conditions, several species of plants can accumulate much more Cu, especially in roots and storage tissues.

Copper is an essential metal in plants and plays a significant role in various physiological processes, such as respiration, photosynthesis, water permeability, protein metabolism, and functions of various enzymes, involved mainly in the oxidation–reduction reactions. In several enzymes (e.g., plastocyanin and phenoloxydase), Cu is linked by the metal-S-cluster. It is involved in the synthesis of DNA and RNA, and in mechanisms of disease resistance (mainly to fungal diseases) of plants (Marschner 2005).

Copper deficiency affects physiological processes, and therefore the plant production. Cu deficiency in crops is widespread, especially in Europe. Common crop plants that are highly sensitive to the Cu deficiency are wheat, oats, sunflower, alfalfa, carrot, lettuce, spinach, onion, and citrus trees. Cu-deficiency level in plants show genetic differences; however, in most cases, the Cu levels about 2 mg/kg are likely to be inadequate for most plants. Despite the general Cu tolerance of most plant species and genotypes, its excess is highly toxic. Copper toxicity to plants may occur when its contents in soils range between 25 and 40 mg/kg, and soil pH is below 5.5. The most common symptoms of its toxicity are (1) Cu-induced chlorosis, resulted in the low photosynthesis efficiency; (2) damage to DNA; (3) damage to membrane permeability; (4) disturbed protein complexes; and (5) root malformation.

Several plants and bacteria are resistance to excess Cu, mainly due to its binding by small proteins (Puig et al. 2007). Cu-tolerant plants and Cu hyperaccumulators have been broadly used in geochemical prospecting for Cu-ore deposits (Kabata-Pendias and Pendias 1999). Among various native plants grown in the polluted area of the Northern Europe (Kola Peninsula), crowberry (*Empetrum nigrum*) has the highest capacity to accumulate Cu, at the pollution-background ratio of 33 (Reimann et al. 2001).

Plants growing in Cu-polluted sites may accumulate elevated amounts of this metal. Especially, plants from the Cu-smelter regions and from the plantations using Cu fungicides may contain its extremely high levels, over 1000 mg/kg. This is of a real problem, especially in old plantations of citrus, coffee, cacao, tea, olives, and vineyards, where Cu fungicides have been applied for a long time.

There are several interactions between Cu and both major and trace elements in soils. The most serious interactions are as follows:

- Cu absorption is reduced at high P levels
- Cu deficiency in plants with high levels of N
- Cu low availability to plants at high levels of $CuCO_3$ or other Cu compounds
- Cu–Zn, competitive inhibition of root absorption
- Cu–Fe, increased levels of both metals decreases their absorption
- Cu–Mn, interactions are related to N metabolism
- Cu–Mn, interactions are both synergistic and antagonistic in the uptake processes under variable conditions

A number of studies have been carried out on Cu levels in edible plants, as about 30% of daily intake of Cu by adults in the Europe is from cereals and potatoes. Monitoring studies carried out in Poland (7000 sample sites) gave the following geometric mean Cu concentrations (in mg/kg): cereal grains, 3.7; potatoes, 4.5; and grasses, 5.5 (Kabata-Pendias 2011). These values are fairly similar to those presented for other countries. Food plants in the United States contain the following amounts of Cu (range, in mg/kg FW, after Ensminger et al. 1995):

- Vegetable: 0.1–3.2, lowest in celery roots, highest in garlic cloves
- Fruits: 0.3–4, lowest in grapes, highest in avocados
- Cereals: 0.3–13, lowest in oats, highest in rye
- Nuts: 0.2–23.8, lowest in coconut, highest in Brazil nuts

Mean Cu contents in cereal grains vary within the range of 3.6–5.3 mg/kg; the highest concentration is reported for wheat from Serbia (Škribić and Onjia 2007). Copper in wheat grains from seven EU countries ranges within 1.3–10 mg/kg (Eriksson 2001a). Range of mean Cu amounts in grasses from various countries is 2–10 mg/kg, and in clover 7–17 mg/kg. Grass growing along the high traffic roads, and at 10 km distance from the road, contains Cu within the ranges 5.3–34.4, and 1.2–7.4, respectively (Niesiobędzka 2012).

The mushroom common chanterelles (*Cantharellus cibarius*) grown in mountains of Poland contains higher amounts of Cu, mean 52 mg/kg, than those grown in the Baltic Sea coast, mean 30 mg/kg (Falandysz et al. 2012).

14.6 HUMANS

Copper is in almost every cell of the human organism. Its total content in the body is about 70 mg, of which in the blood 1 mg/L, in the bone 1–25 mg/kg, and in the tissue 2–10 mg/kg (Emsley 2011). The highest concentrations of Cu are in the brain and the liver. About 50% of its content is in bones and muscles, 15% in the skin, 15% in the bone marrow, 8%–15% in the liver, and 8% in the brain (Angelova et al. 2011).

Following the ingestion of Cu, its levels in the blood rapidly rises. It is predominantly bound to albumin. Copper in all living organisms forms an essential component of many enzymes (cuproenzymes) and proteins. The biochemical role for Cu is primarily catalytic, with many Cu metalloenzymes acting as oxidases, active in the reduction of molecular oxygen (e.g., Cytochrome c oxidase and superoxide dismutase).

In the body, Cu shifts between the Cu^{+} and Cu^{2+} forms, though the majority of the body's Cu is in the Cu^{2+} form. Ability of Cu to easily accept and donate electrons explains its important role in oxidation–reduction reactions, and in scavenging free radicals. Cytochrome c oxidase plays an essential role in cellular energy, lysyl oxidase, participates in cross-linking of collagen and elastin, which form the connective tissue. The impact of lysyl oxidase helps maintain the integrity and elasticity of connective tissue in heart and blood vessels, but also plays a role in bone formation.

Copper is necessary in human nutrition for the normal Fe metabolism and the formation of red blood cells. Anemia is a clinical sign of deficiency of both Fe and Cu.

Cu enzymes, ceruloplasmin (feroxidase I and feroxidase II) has the ability to oxidize Fe^{2+} to Fe^{3+}, which are connected to the protein transferring, for the transportation to red blood cells, and blood formation (Angelova et al. 2011).

A number of reactions essential to normal function of the brain and nervous system are catalyzed by Cu enzymes. The Cu enzyme, tyrosinase, is required for the formation of the pigment melanin. Melanin is formed in cells called melanocytes and plays a role in the pigmentation of the hair, skin, and eyes.

Deficiency or excess of Cu in the organism is observed in metabolic disturbances and in various diseases and conditions. Diseases with too low concentrations of Cu are Menkes syndrome, Parkinson's disease, protein loss, nephrosis, exudative enteropathy, and others.

There are two well-known genetic diseases affecting Cu metabolism, Menkes and Wilson diseases. Menkes kinky-hair disease is a problem with Cu transport or absorption. Wilson's disease is characterized by increased liver Cu content, leading to severe hepatic damage, followed by increased brain Cu levels and neurological problems. Menkes disease results in pathology resembling Cu deficiency, as opposed to the pathology of Wilson's disease, which resembles Cu toxicity (Stern et al. 2007).

Dietary interactions of Cu with sucrose or fructose, animal proteins, *S*-amino acids, histidine, and ferrous iron may inhibit Cu absorption to varying degrees, in animal models. Ascorbic acid supplements, Mo and other dietary factors, specifically high intakes of Ca and/or P and Cd, may also inhibit Cu absorption in diets containing high amounts of compounds. The interaction between Zn and Cu is well documented in humans. High levels of dietary Zn adversely influence the Cu absorption and bioavailability (SCF 2003).

Provisional maximum tolerable daily intake established for Cu is 0.05–0.5 mg/kg bw (WHO 1982a), and estimated safe and adequate daily dietary intakes is 1.5 ± 3.0 mg (WHO 2004). In the EU countries, it has been recommended that adult males and females should consume a dietary intake of 0.9 mg Cu/day (SCF 2003). The main sources of Cu in diets are cereals and cereal products, vegetables, and potatoes (Table 14.3).

Mean dietary Cu intakes from food of adults in various EU countries have been estimated within the range of 1.0–2.3 mg/day for males, and 0.9–1.8 mg/day for females. Vegetarian diets provided greater dietary intakes of Cu, approximately 2.1–3.9 mg/day (SCF 2003).

14.7 ANIMALS

The livestock, cattle and sheep, commonly show indications when they are Cu deficient. Swayback, a sheep disease associated with Cu deficiency, imposes enormous costs on farmers worldwide, particularly in the Europe, the United States, and many tropical countries.

Copper is responsible for many functions in the animals, of which one is hemoglobin formation. Copper also plays a role in bone cell function, pigment production, hair, hoof and horn formation, and animal growth.

Signs and symptoms of Cu deficiency include change in hair coat color, diarrhea, decreased weight gain, unthrifty appearance, anemia, fractures, lameness, decreased

TABLE 14.3
Copper in Foodstuffs (mg/kg FW)

Product	Content
Food of Plant Origin	
Bread[a]	1.66
Miscellaneous cereals[a]	2.21
Rice, white[b]	2.13–5.57
Rice, brown[b]	3.38–5.17
Green vegetables[a]	0.58
Potatoes[a]	1.12
Sugar and preserves[a]	1.80
Fruits[a]	0.786
Nuts[a]	9.15
Food of Animal Origin	
Carcass meat[a]	1.44
Meat products[a]	1.16
Offals[a]	52.8
Fish[a]	0.91
Fish, freshwater[c]	0.08–0.17
Milk[d]	0.02
Eggs[d]	0.56

Sources: [a] Rose, M. et al., *Food Addit. Contam.*, 27, 1380–1404, 2010.
[b] Batista, B.L. et al., *Food Addit. Contam.*, 3, 253–262, 2010.
[c] Lidwin-Kaźmierkiewicz, M. et al., *Pol. J. Food Nutr. Sci.*, 59, 219–234, 2009.
[d] Szkoda, J. and Żmudzki, J., *Copper and Molybdenum in the Environment*, The Polish Academy of Sciences, Warsaw, Poland 1996.
FW, fresh weight.

disease, cardiac abnormalities such as blood vessel and heart rupture, and elevated levels of serum cholesterol, triglycerides, and glucose.

Excess intake of Cu may lead to chronic Cu poisoning. The range of clinical manifestations associated with Cu toxicosis in cattle is large. Generally, cattle demonstrate anorexia, reduced body weight, dehydration, and alterations in feces, followed by icterus, hemoglobinuria, and death (Minervino et al. 2009).

Cattle diets should contain about 4–10 mg/kg of Cu, for calves and cows, respectively. Less than this amount results in a primary Cu deficiency. Secondary Cu deficiency occurs when there is an interrelationship with other elements, most commonly with S, Mo, and Fe. Diet and water sources high in S or Mo can interfere with the Cu uptake. Acceptable Cu:Mo ratios are from 5:1 to 10:1 in the diet.

TABLE 14.4
Copper in Organs of Different Species of Animals (mg/kg FW)

	Tissues		
Animal	**Muscle**	**Liver**	**Kidneys**
Cattle, agricultural area[a]	1.6	80.1	4.97
Cattle, industrial area[a]	2.2	92.7	5.31
Cattle[b]	0.55	18.47	3.42
Pigs[b]	0.61	4.87	6.12
Horses[b]	2.05	5.40	6.82
Wild animals[b]	2.13	9.60	6.29

Sources: [a] Waegeneers, N. et al., *Food Addit. Contam.*, 26, 326–332, 2009a.
[b] Szkoda, J. and Żmudzki, J., *Copper and Molybdenum in the Environment*, The Polish Academy of Sciences, Warsaw, Poland 1996.
FW, fresh weight.

A Cu:Mo ratio of 2:1 or less can cause severe interference with Cu absorption and result in the Cu deficiency. Soil ingestion, due to overgrazing, may affect an excess of Cd, Zn, and Ca, and can cause Cu deficiencies in cattle.

Concentration of Cu in animal body varies depending on the environment and tissue species (Table 14.4).

15 Fluorine [F, 9]

15.1 INTRODUCTION

Fluorine (F) is an element of the halogen group 17 in the periodic table of elements. Its total content in the upper crust of Earth is estimated at up to 625 mg/kg. In igneous rocks its concentrations vary between 300 and 1200 mg/kg, being the highest in mafic rocks. It is likely to accumulate in argillaceous sediments and in black shales (500–800 mg/kg). Its lowest contents are in calcareous rocks (50–350 mg/kg). Hard coal contains F usually up to 100 mg/kg, but higher contents, up to 200 mg/kg, especially in bituminous coals, are noticed. High levels of F (as fluorapatite and fluorspar) are associated with various phosphate deposits. Fluorine is a common constituent of magmatic and volcanic exhalations, and may also occur in some rocks as gaseous nebulae.

Fluorine, occurring mainly at −1 oxidation state, is a yellow gas, very toxic, and reactive with both organic and inorganic substances. Metallic fluorides are highly toxic, whereas organic fluorides are almost harmless. Its most common mineral, fluorite (also called fluoride or fluorspar), CaF_2 is often associated with sellaite, MgF_2, and is widely distributed in both lithosphere and hydrosphere. Other common minerals are cryolite, $Na_3(AlF_6)$; fluorapatite, $Ca_5(PO_4)_3F$; and topaz, $Al_2F_2SiO_4$, a popular gemstone, and depending on the addition of some metals may be of various colors.

World mine F production (excluding the United States) as fluorspar (rounded) in 2010 was 5400 kt, of which the People's Republic of China produced 3000 kt; Mexico produced 1000 kt; and Mongolia produced 450 kt (USGS 2011). Also, F gas is produced at about 2.4 kt/yr in some countries.

A high proportion of F goes to the aluminum industry and to steel industries, mainly as hydrogen fluoride (HF) and hydrofluoric acid. It is also greatly used in glass, ceramic, and plastic production. Most feldspar consumed by the glass industry is for the manufacture of container glass. Because of environmental initiatives, fiberglass consumption for thermal insulation is forecast to expand steadily. Domestic feldspar consumption has been shifting from ceramics toward glass markets. Another growing segment in the glass industry is solar glass. As a highly oxidizing element, F is broadly used in various chemical processes. Fluorine is a common compound of freon gases and chlofluorocarbons used for the refrigeration.

HF is a colorless gas and is the main source of F for industrial processes. It is lighter than air and easily dissolves in water. It is produced from the mineral fluorite and also obtained as a by-product of the P-fertilizer production. Common range of F in P fertilizers is 8,500–38,000 mg/kg.

Fluorine has been added, for a long period, to municipal drinking water (at the level up to 1.9 mg/L) for the prevention of tooth decay. It is still added to toothpastes.

15.2 SOILS

The average F contents in soils have been estimated to be 329 and 360 mg/kg, for European and the U.S. soils, respectively, but within the broad range of <1–1360 mg/kg (Table 15.1). Its highest concentrations are in heavy loamy soils and the lowest in organic soils. Its F content in soils is inherited form the parent material, but its distributions reflect soil-forming processes and soil texture. Soils may accumulate relatively high amounts of F, especially in B horizons, usually enriched in the amorphous Al hydroxides. Soils derived from bedrocks enriched in F usually contain its increased level (>1000 mg/kg), and are of a great environmental problem because they are associated with the endemic provinces of fluorosis (Ermakov 2004; Chen et al. 2008).

The behavior of F in soils is variable and highly controlled by pH, clay contents, concentrations of Ca and P, and Al hydroxides. The most common F compound in soils is fluoroapatite, but other fluorides are also present. Its mobile species are NaF, KF, and NH_4F, whereas AlF_3, CaF_2, and MgF_2 are very slightly mobile. In soil solutions, both cationic (AlF^{2+}, AlF^{+}) and anionic F species (F^-, AlF_4^-, AlF_6^{3-}, SiF_6^{2-}) occur. In certain soils, in which CaF_2 and mobile Al species are present, F is fixed and in a relatively insoluble aluminum fluorosilicate, $Al_2(SiF_6)_3$, form. In calcareous soils, F is very slightly mobile, whereas in sodic and some acidic soils, it is

TABLE 15.1
Fluorine Contents in Soils, Water, and Air

Environmental Compartment	Range
Soil (mg/kg)	
Light sandy	80–205
Medium loamy	175–462
Heavy loamy	470–1360
Calcerous (calcisols)	470–680
Organic (histosols)	<1–350
Water (µg/L)	
Rain	50
River	50–2700
Sea, ocean	30–1350
Air (ng/m³)	
Industrial and urban regions	1–7

Source: Data are given for uncontaminated environments, from Kabata-Pendias, A and Mukherjee, A.B., *Trace Elements from Soil to Human*, Springer, Berlin, Germany, 2007; Reimann, C. and de Caritat, P., *Chemical Elements in the Environment*, Springer-Verlag, Berlin, Germany, 1998.

quite mobile. Some clay minerals, especially illite and chlorite groups, have a great capacity for the F fixation. The greatest absorption of F by soil mineral components is within the pH 6–7. Usually F contents of surface soil horizons are lower than its concentrations in lower horizons. This indicates a low F affinity for organic matter. However, in some tropical soils, it may occur organically bound, easily phytoavailable F, which may be attributed to the synthesis by certain microorganisms (Kabata-Pendias 2011).

The input of F to arable soil with P fertilizers has become an environmental problem. A long-time application of P fertilizers, as well as sewage sludge and pesticides, may highly elevate F contents of soils, and increase its phytoaccumulation (Wu et al. 2007). Fluorine is strongly retained by soils, forming complexes with soil components. In soils, it is transported to surface water, through leaching or runoff of particulate-bound fluorides. Leaching removes only a small amount of F (up to 6% of F added) from agricultural soils, which is received up to 20 kg/yr with the use of P fertilizers (Kabata-Pendias 2011). The seepage water of these soils contains F between 50 and 200 μg/L.

Soil pollution by F, especially in some industrial regions, is also of an ecological problem (Table 15.2). The highest F sources are from Al smelters and P-fertilizer factories. The most important hazard of F-contaminated soils concerns changes in the soil properties due to the great chemical activity of hydrofluoric acid, which may be formed from both solid and gaseous F pollutants. Several destructions are observed in soils heavily contaminated by F: (1) loss of organic matter and humic–mineral complexes; (2) reduction of enzymatic activity; and (3) low N fixation by microorganisms. Excess F in agricultural soils may result in improper plant growth and its elevated contents in plant tissues. Remediation of F-polluted soils is now of a great ecological concern.

Fluorine concentrations in groundwater vary within 20–1500 μg/L, being always higher in southern regions. The criteria for contaminated land (Dutch List 2013), following F concentrations in soils and groundwater, are established (in mg/kg and μg/L) as follows: uncontaminated, 200 and 300; medium contaminated, 400 and 1200; and heavily contaminated, 2000 and 4000.

TABLE 15.2
Fluorine Contamination of Surface Soils (mg/kg)

Site and Pollution Source	Country	Maximum Content
Old mining region	England	19,900
Al-processing industry	Czech Republic	1,350
	Poland	3,200
P-fertilizer industry	Canada	2,080
	Poland	385
China-clay industry	England	3,560

Source: From Kabata-Pendias, A., *Trace Elements in Soils and Plants*, 4th ed., CRC Press, Boca Raton FL, 2011.

15.3 WATERS

The median F concentration in the world ocean water is estimated at 1.3 mg/L, within the range of 0.03–1.35 mg/L. Nozaki (2005) calculated its average concentration in the North Pacific Ocean as 2.8 mg/kg. Most of the fluorides in the oceans are received from rivers; lesser amounts come from atmospheric deposition.

River water contains F from 0.05 to 2.7 mg/L (Table 15.1), but its median value for stream water in Finland is 0.0008 mg/L (Reimann and Caritat 1998). In some spa water, F concentration can reach 12 mg/L. Its content in groundwater is associated with both geological and anthropogenic sources. Excess F is reported for calcareous bedrocks and anthropogenic factors. Increased F concentrations are in arid and semiarid regions, where it is of a great health concern (Jacks et al. 2005). Riverine flux of F for some industrial regions is of environmental risk.

Fluorine is easily evaporated from water basins to the atmosphere, but also is deposited into water. In natural water, below pH 5, F ions form strong complexes with Al, and thus concentration of free F is low. At higher pH, dominate Al–OH complexes and thus free F level increases. Both Ca carbonates and phosphates fix F, and remove from the solution.

Fluorides are accumulated in some marine aquatic organisms. Toxic effects due to fluorosis were observed in some species of mussel, mullet, crab, and shrimp in an estuary where wastes from an aluminum plant were released.

Rainwater and snow-melt water in regions of the Kola Peninsula contain about 0.05 mg F/L. Rainwater sampled (during 1998–2001) around Italian volcanoes contains F within the broad range of 0.005–450 mg/L (Bellomo *vide* Kabata-Pendias and Mukherjee 2007).

Fluorine concentration in drinking water is of a special concern, as it is the main source of this element in the nutrition. Its level in water of the most countries varies between <0.1 and 3 mg/L, but in some countries (mainly in dry and hot climates) it can reach up to >5 mg/L.

Median F concentration in bottled water of the EU countries is 0.21 mg/L, and in tap water is 0.09 mg/L (Birke et al. 2010). In the United States, bottled water and tap water contain this element within the range of 0.05–0.34 mg/L, and 0.05–1.0 mg/L, respectively (ATSDR 2002b). Threshold F concentration for drinking water is established at 1.5 mg/L, by the WHO Guideline (WHO 2011a), whereas by the European Economic Community it is established for bottled water at 5 mg/L (Diduch et al. 2011). Addition of F to drinking water, relatively common in the last century, is now definitely stopped (Limeback 2001).

15.4 AIR

Concentrations of F in the atmosphere are highly variable, due to its differentiated sources: (1) marine aerosols, (2) volcanic eruptions, and (3) industrial emissions, especially from coal burning. In the air, gaseous HF is absorbed by atmospheric water (rain, clouds, fog, and snow), forming an aerosol or fog of aqueous hydrofluoric acid. F particulates fall down on land or surface water, by wet and dry deposition.

HF is the most abundant gaseous F released into the atmosphere. Its natural sources of F are from volcanoes, emitting 0.6–6 Mt/yr. About 10% emission is from large eruptions. Its concentration is higher in air of industrial regions. The largest anthropogenic emissions of HF are the electrical utilities, and the Al industries. Also, the production of P fertilizers release some amounts of F. In gaseous phase, F occurs mainly as F_2 and HF. However, it may be emitted from industries in various other compounds, but the most abundant gaseous F released into the atmosphere is HF. Due to reactions with various materials, both in vapor and in aerosols, several compounds, including *S*-hexafluoride and fluorosilicic acid, are formed in the air. The predominant mode of degradation of inorganic fluorides in the air is hydrolysis. However, some fluorides emitted by industries in particulate matter are stable compounds that do not readily hydrolyze.

The common F concentrations in air of inhabited regions in various countries range from 1 to 7 μg/m³ (Table 15.1). Longer influence of increased F levels in the atmosphere has detrimental effects on plants, humans, and animals. Especially sensitive to elevated F concentration in air are coniferous trees. Acceptable F levels for the forest regions in Poland have been established for 0.02 μg/m³, whereas for the whole country it is 2 μg/m³ (Gramowska and Siepak 2002).

15.5 PLANTS

Phytoavailability of fluorine is relatively low. It is taken up by plants passively by roots, and is apparently easily transported in plants. There are some observations that F is bound in plants to mobile organic complexes. Its concentrations in some plants show a relationship with hot-water soluble F in soil. Often, its availability increases with decreasing soil pH. In soils with higher pH, and especially with soluble Ca compounds, F bioavailability decreases due to the precipitation of CaF_2 on the root surface, which resulted in lower F contents in shoots and higher in roots (Maćkowiak et al. 2003).

Usually, F concentrations in shoot tissues of plants do not exceed 30 mg/kg, and are higher than in roots. Although plants can uptake F easily from soils, especially from F-contaminated soils, its increased levels in plants are very often from airborne F compounds. Plants grown in F-contaminated soils and/or exposed to industrial emissions may significantly contain elevated amounts of F (Table 15.3). Especially some plants growing in acidic soil enriched in F may accumulate elevated amounts of F. Pine needles are common and good indicators for the F pollution. Pine needles of trees from vicinities of the Al smelter contain F above 1000 mg/kg, whereas its background contents is calculated at below 20 mg/kg. Needles of trees surrounding phosphate factories contain following amounts of F (in mg/kg): Scots pine, >200; Norway spruce, >100; Douglas fir, >50; and control trees, around 20 (Karolewski et al. 2000). Lichens used for the biomonitoring of pollution accumulate F up to 243 mg/kg, near the F point source (Geebelen at al. 2005).

Usually leaves of most plants absorb F deposited on leaf surfaces easily. Great accumulation of F in old tea leaves resulted from both root uptake and leaf absorption (Siemiński *vide* Kabata-Pendias and Mukherjee 2007). In most cases, F taken up by plants from air is much higher than those from soils. Several factors affect

TABLE 15.3
Excessive Levels of Fluorine in Plants Grown in Contamination Sites (mg/kg)

Pollution Source	Country	Plant/Part	Range/Maximum
Mine wastes	United Kingdom	Grass and tops	130–5450
Al industry	Australia	Shrubs and leaves	150–500
	Czech Republic	Grass and tops	1330
	Norway	Birch and leaves	230
	United Kingdom	Lichens	27–241
Phosphate rock processing	Canada	Tree and foliage	71–900
	United States	Sagebrush and tops	100–360
China-clay industry	United Kingdom	Grass and tops	788–1543
Fiberglass plant	Canada	Vegetation and foliage	71–900

Source: From Kabata-Pendias, A., *Trace Elements in Soils and Plants*, 4th ed., CRC Press, Boca Raton FL, 2011.

plants accumulation of airborne F, but the most important are F concentration in the atmosphere and the duration of exposure. Some amounts of F remain on the leaf surfaces and can be easily washed off with water.

Mean F contents in cereal grains vary within the range of 0.2–5.5 mg/kg, with the lowest in oats and the highest in barley. Range of F in forage legumes and grasses is 1.4–7.8 mg/kg and 3–6.8 mg/kg, respectively (Kabata-Pendias 2011).

Plants do not require F for metabolic processes. But it may be toxic, when is absorbed by leaves. Fluorine is considered to be the most phytotoxic among other airborne pollutants. In general, elevated F contents resulted in a lower plant growth and yield reduction. The most important harmful effects of airborne F are as follows:

- Decrease of oxygen uptake and respiratory disorder
- Decrease of assimilation
- Reduction of chlorophyll
- Inhibition of starch synthesis
- Inhibition of pyrophosphate function
- Injured cell organelles and cell membranes
- Disturbance of DNA and RNA
- Synthesis of fluorooctate, hazardous F compound

It can be generalized that susceptible plants could be injured by foliar F concentrations at the range of 20–150 mg/kg, intermediate plants may tolerate F contents at about 200 mg/kg, and highly tolerant plants did not suffer even with F content up to 500 mg/kg.

Excessive F contents in food and fodder plants are of the greatest concern because most of the F contamination can easily enter the food and fodder chains. The commonly reported ranges for F contents in fodder plants vary from 20 to 50 mg/kg.

15.6 HUMANS

Fluorine occurs naturally as the negatively charged ion fluoride (F^-). Some amounts of F are present in the body, at about 2.6 g in adults, of which about 95% is found in the bones and teeth. Concentrations of F in the blood of Americans, receiving fluoridated drinking water, ranged from 20 to 60 µg/L. Once in the body, F is absorbed into the blood through the digestive tract.

The mean F plasma level in human, with 5.03 mg F/L in their drinking water, is 106 µg/L. Concentration of F in bones varies with age, sex, and part of bones, and is believed to reflect an individual's long-term exposure to F (WHO 2002a). Fluoroapatite forming in bones, after the reaction of F with hydroxyapatite, hardens tooth enamel and stabilizes bone minerals. At usual intake levels, F does not accumulate in soft tissues. Fluorine therapy may be beneficial for the treatment of osteoporosis, but serious side effects have been associated with the high doses of F used to treat osteoporosis.

Fluoride crosses the placenta and is transferred from the mother to the fetus. It is eliminated from the body primarily in the urine. In infants, about 80%–90% of an F dose is retained; in adults, it is approximately 60% (WHO 2002a). There is a possibility of an adverse effect in case of high F exposure on children's neurodevelopment (Choi et al. 2012). Fluoride has both beneficial and detrimental effects on human health, with a narrow range between the amounts of its intake. The lowest dose that could trigger adverse symptoms is considered to be 5 mg/kg bw, with the lowest potentially fatal dose considered to be at 15 mg/kg bw.

Both Ca and Mg form insoluble complexes with F and are capable of significantly decreasing F absorption. Diet low in chloride (salt) may increase F retention by reducing its urinary excretion. In humans, the only clear effect of inadequate F intake is an increased risk of dental caries (tooth decay). Calcium supplements, as well as Ca- and Al containing antacids, can decrease the absorption of F.

Elevated F intake may involve gastrointestinal irritation, joint pain in the lower extremities, and the development of Ca deficiency, as well as stress fractures. One of the most important contributions of F chemistry to the quality of human life has been the creation of compounds that induce anesthesia. Modern inhalation anesthetics is almost entirely dependent on F chemistry. Ingestion of excess F, most commonly in drinking water, can cause fluorosis, which affects the teeth and bones. Moderate amounts of F lead to dental effects, but long-term ingestion of large amounts can lead to severe skeletal problems. The dental effects of fluorosis develop much earlier than the skeletal effects in people exposed to large amounts of F. Dentifrice products for adults generally contain F at concentrations ranging from 1000 to 1500 mg/kg; some products for children use contain lower levels, ranging from 250 to 500 mg/kg. Dental products, such as toothpaste, mouthwash, and F supplements, have been identified as significant sources of F.

In skeletal fluorosis, F accumulates in the bone progressively over many years. The early symptoms of skeletal fluorosis include stiffness and pain in the joints. In severe cases, the bone structure may change and ligaments may calcify, resulting in pain and impairment of muscles. Early stages of skeletal fluorosis are characterized by increased bone mass. If very high F intake persists over many years, joint pain and stiffness may result from the skeletal changes.

Acute high-level exposure to F causes immediate effects of abdominal pain, excessive saliva, nausea, and vomiting. Seizures and muscle spasms may also occur. Moderate-level chronic exposure (above 1.5 mg/L of water—the WHO guideline value for F in water) is more common. People affected by fluorosis are often exposed to multiple sources of F, such as in food, water, air (due to gaseous industrial waste), and excessive use of toothpaste. However, drinking water is the most significant source of F. Skeletal fluorosis and an increased risk of bone fractures occur at a total intake of 14 mg F/day, and an increased risk of bone effects is at total intakes about >6 mg F/day. In some regions, the indoor burning of F-rich coal also serves as an important source of F.

Fluorine in water is mostly of geological origin. Water with high levels of F is mostly found at the foot of high mountains. Intake of F with water and foodstuffs is the primary causative factor for the endemic skeletal fluorosis, which is a major public health problem, and has socioeconomic impact, affecting millions of people in various regions of Africa, China, and India.

Fluoridation of public drinking water in the United States was initiated more than 50 years ago. Since then, a number of adverse effects have been attributed to water fluoridation. At the level of 1 mg/L, F impairs the production of collagen in the organism. It causes decomposition of collagen and results in osteoporosis, bone tumors, fragility of bones and teeth, and also destroys connective tissues, which keep the organism in proper position.

The possible relationship between serum Pb levels in children and chemicals used to fluoridate has also been observed. Silicofluorides used to fluoridate water, increase blood Pb levels. Long-term accumulation of F is also likely to affect thyroid function, and may affect the pineal gland. Elevated F content of a body may also increase risk of various cancers (Limeback 2001). Some data appear that fluoridation of water may be a harmful procedure. With the inflow of new information the awareness of dangers associated with the overall effect of F increases, and more and more countries give up fluoridation of drinking water (Meler and Meler 2006).

The fluoride content of most foods is low (less than 0.5 mg/kg). Rich sources of F include tea, which concentrates F in its leaves, and marine fish, which are consumed with their bone (Table 15.4). Foods generally contribute only 0.3–0.6 mg of the daily intake of F. An adult male residing in a community with fluoridated water has F intake range from 1 to 3 mg/day.

Direct contact with F can result in tissue damage. At high concentrations, F can cause irritation and damage to the respiratory tract, stomach, and skin following inhalation, oral, and dermal exposure, respectively. At very high F doses, fluoride can bind with the serum calcium, resulting in hypocalcemia and possibly hypercalcemia; the severe cardiac effects (e.g., tetany, decreased myocardial contractility, cardiovascular collapse, and ventricular fibrillation) observed at or near lethal doses are probably due to the electrolyte imbalance. Metallic fluorides are very toxic. Organic fluorides are generally much less toxic and are often quite harmless (ATSDR 2003).

The major source of dietary F in the U.S. diet is drinking water. When water is fluoridated, F contents are adjusted to between 0.7 and 1.2 mg/L. The ESADDI for F is 1.5 ± 4.0 mg/day (WHO 2004).

Fluorine gas is extremely corrosive and toxic. The free element has a characteristic pungent odor, detectable in concentrations as low as 20 μg/L, which is below

TABLE 15.4
Fluorine in Foodstuffs (mg/kg FW)

Product	Content
Food of Plant Origin	
Bread	0.39
Apple	0.03
Grape	0.49
Carrot	0.03
Beer	0.45
Tea, instant, powder	897.72
Tea, instant, powder[b]	3.35
Wine, red	1.05
Wine, white	2.02
Food of Animal Origin	
Beef, cooked and raw	0.22
Pork products	0.04–0.38
Cheese	3.2–3.6
Milk	0.03
Fish, smoked, different[a]	0.32–0.61
Herring[a]	0.44
Canned fish[a]	1.39–2.78

Source: USDA, *National Fluoride Database of Selected Beverages and Foods*, Nutrient Data Laboratory Beltsville, Human Nutrition Research Center Agricultural Research Service, U.S. Department of Agriculture, Beltsville, MD, 2005.

[a] Usydus, Z. and Szlinder-Richert, J. Iodine and fluorine in fish products. *Bromat. Chem. Toksykol.* [in Polish], 42, 822–826, 2009.

[b] Tea with tap water.

the safe level. Exposure to low concentrations causes eye and lung irritation. Limited data exist on the toxicity of F gas; the two possible routes of exposure to F are inhalation or dermal contact with the gas. The primary health effects, in both humans and animals, of acute F inhalation are nasal and eye irritation (at low levels), and death, due to pulmonary edema (at high levels). In animals, renal and hepatic damage have also been observed (ATSDR 2003).

16 Gallium [Ga, 31]

16.1 INTRODUCTION

Gallium (Ga), a metal of the group 13 in the periodic table of elements, reveals chalcophillic properties and resembles the geochemistry of Al. Its content on the Earth's crust is within the range 1–25 mg/kg. Rocks contain Ga from 1 to 25 mg/kg; its lowest amounts are in ultramafic and calcareous rocks. In argillaceous sediments, its distribution is associated with clay minerals. Its concentration in both coal and fly ash may be up to about 45 mg/kg.

Gallium occurs mainly at +3 oxidation state, but also at +1 and +2. Its concentrations in ores of other metals are very small. It forms only few minerals. Those, containing about 30% of Ga are gallite, $CuGaS_2$; sohngeite, $Ga(OH)_3$; and tsungeite, GaO(OH). However, it may be associated with several minerals, such as feldspars, amphibolites, and micas. The electro-negativities and low ionic potential suggest that Ga may be amphoteric, and may form different complexes with some anions (e.g., Ga_2O, Ga_2S, Ga[OH]) easily.

World primary Ga production capacity in 2010 was estimated to be 184 t; scrap refinery capacity, 177 t; and recycling capacity, 141 t (USGS 2011). Most gallium is produced as a by-product of Al production from bauxite, and the remainder is produced from zinc-processing residues. Its concentration in bauxite varies, and the highest (0.08%) is reported to be in the bauxites from tropic regions. It is also obtained at sphalerite (ZnS) and other sulfide metals explorations.

The melting temperature of Gallium is very low, at about 30°C, and therefore it has been primarily used as an agent to make low-melting alloys. The alloy galinstan, containing 68.5% of Ga, has a melting point at –19°C, and is used instead of Hg for some thermometers. Nowadays, almost all forms of Ga are used in the field of microelectronics, due to its semiconducting properties. GaAs, GaN, and InGaN compounds are used in semiconductors, solar cells, photodetectors, and light-emitting diodes.

Gallium and ferric salts behave similarly in biological systems, and therefore it is used to mimic iron ions in medical applications. Productions of Ga-containing pharmaceuticals and radiopharmaceuticals have been developed.

16.2 SOILS

Average Ga content of soils is estimated at 15.2 mg/kg, and its abundance varies from <3 to 70 mg/kg (Table 16.1). Its concentrations in soils of various countries are reported as follows (in mg/kg): United States, San Joachim Valley, 16–35; Japan, 18–23; Russia, 6–48; and Sweden, 3.4–16 (Kabata-Pendias 2011).

Gallium, present in soils mainly as $Ga(OH)_3$, is relatively mobile. However, it is mobilized during weathering processes, and by complexing with organic matter

TABLE 16.1
Gallium Contents in Soils, Water, and Air

Environmental Compartment	Range
Soil (mg/kg)	
Light sandy	<3–30
Medium loamy	5–50
Heavy loamy	5–70
Calcerous (calcisols)	5–30
Organic (histosols)	7–50
Water (μg/L)	
Rain	0.002–0.04
River	0.001–0.022
Sea, ocean	0.0012–0.03
Air (ng/m³)	
Remote region	<0.001–0.14

Sources: Data are given for uncontaminated environments, from Kabata-Pendias, A. and Mukherjee, A.B., *Trace Elements from Soil to Human*, Springer, Berlin, Germany, 2007; Reimann, C. and de Caritat, P., *Chemical Elements in the Environment*, Springer, Berlin, Germany, 1998.

(Welch et al. 2004). Its distribution in soils reflects a positive correlation with clay fractions, and Fe and Mn hydroxides. It is also likely to be accumulated in soluble organic matter. Relatively high concentrations of Ga are reported for bioliths and organic soils. Due to its relatively stable contents in soils, it is considered as a useful marker for the characterization of soil genesis.

Gallium is emitted mainly from Al industries and during coal combustion. Its higher concentrations are noticed in some sludge and composts. Concentrations of Ga in some wastes are as follows (in mg/kg): wastewater treatment plants, 2.2–6.4; composts, 1.6–1.9; and motorway dust, 4.3–5.3. P fertilizers contain Ga within the range of 0.7–0.3 mg/kg, and solid pig manure –0.5 mg/kg. Impact of waste materials on Ga levels in soils has not been yet reported.

16.3 WATERS

Gallium concentration in worldwide rivers ranges from 0.001 to 0.022 μg/L, and its annual river flux is calculated at 1.1 kg (Gaillardet et al. 2003). Seawater contains Ga within the range 0.0012–0.03 μg/L (Table 16.1).

In fresh seawater, mobile Ga species are $Ga(OH)_3^{\circ}$ and $Ga(OH)_4^-$, which are formed in more acidic water. Residence time of mobile Ga in deep water varies between 100 and 700 years, depending on the locations, and it is 3–5 times longer than the residence time of Al (Shiller 2003).

Rainwater, sampled in Norway, contained Ga at the range of 0.03–0.04 μg/L (Reimann and de Caritat 1998), and those sampled in Sweden, in 1999, had Ga between 0.002 and 0.011 μg/L (Eriksson 2001a).

Median Ga concentration in bottled water of the EU countries is 0.002 μg/L, and is lower than those in tap water, 0.07 μg/L (Birke et al. 2010). Water containing $GaCl_3$ at the concentration 3.5 mg/L may be toxic to some animals.

16.4 AIR

The atmosphere from remote regions contains Ga at amounts below 0.14 ng/m^3 (Table 16.1).

Gallium concentrations in moss samples collected in Norway during 1999 was 1.1 mg/kg, and in samples collected during 2000 was 0.09 mg/kg, which clearly indicates its lower emission (Berg and Steinnes 1997). Mean Ga contents of two moss species from mountains in Poland (0.05–0.1 mg/kg) are quite similar to the same moss species from Alaska (0.09–0.1 mg/kg). This may suggest that atmospheric Ga concentrations in both regions are quite close (Migaszewski et al. 2009).

16.5 PLANTS

Gallium is commonly present in plants, and the higher Ga/Al ratio in plants than in soils, in which plants grown can indicate its selective uptake by plants. There is no evidence for the necessity or toxicity of Ga in plants, although there are some suggestions for its beneficial role in the growth of microorganisms, and some fungi (e.g., *Aspergillus*). Ectomycorrhizal beech roots and organic soil horizon from unpolluted regions of Sweden contain Ga at amounts 0.06 and 0.41 mg/kg, respectively (Tyler 2005).

Gallium contents of wheat and barley from Sweden average at about 0.001 mg/kg (Eriksson 2001a). Its much higher concentrations (0.02–5.5 mg/kg) are reported to be in some herbage. Also in lichens and bryophytes from various regions, Ga concentrations are relatively high, 2.2–60 and 2.7–30 mg/kg, respectively (Kabata-Pendias 2011). Moss sampled in Norway contained Ga up to 16 mg/kg (Berg and Steinnes 1997).

16.6 HUMANS

Persons with a mass of 70 kg contain less than 1 mg of Ga in the body (Emsley 2011). Data on natural Ga levels in humans are scanty. Gallium concentrations are always below the μg/kg level. Its distribution in several human brain areas, evaluated by radiochemical neutron activation analysis, was found to be dishomogeneous (Speziali et al. 1989).

Gallium is an element of increasing biological interest: it is involved in problems related to environmental pollution (Ga compounds are used in electronics industry) and to clinical treatments. Moreover, as its chemical behavior is similar to that of Al, it could play a role in the health effects attributed to this element. However, it has no proven benefit from its function in the body.

Following exposures, Ga is transported in blood, bound to transferring, and may have an impact on upregulating the transferrin receptor. The gallium–transferrin complex appears to be the primary mechanism by which the Ga ions are presented to the target cellular system (IARC 2006).

Pure Ga is not a harmful substance for humans, to touch. However, it is known to leave a stain on hands. Even the Ga radioactive compound, Ga citrate, can be injected into the body and used for Ga scanning, without harmful effects. Although it is not harmful in small amounts, Ga should not be consumed in large doses. Some Ga compounds can be very dangerous. For example, acute exposure to Ga^{3+} chloride can cause throat irritation, difficulty in breathing, chest pain, whereas its fumes can cause even very serious conditions, such as pulmonary edema and partial paralysis.

Gallium nitrate (brand name Ganite) has been used as an intravenous pharmaceutical to treat hypercalcemia, associated with tumor metastasis to bones. Gallium is thought to interfere with osteoclast function. Gallium-67 salts, such as Ga citrate and Ga nitrate, are used as radiopharmaceutical agents in a nuclear medicine imaging procedure, commonly referred to as a gallium scan. The form or salt of Ga are not important, as it is the free dissolved ion Ga^{3+}, which is the active radiotracer.

Gallium arsenide (GaAs), is extensively used in the microelectronics industry, because of its photovoltaic properties. The toxicity of GaAs, a compound extensively used in defense as a superior semiconductor material, in ground- and space-based radar and in electronic warfare, is not well known. Results from recent reports on experimental animals indicate that GaAs produces profound effects on lung, liver, immune, and hematopoietic systems. GaAs is found to be soluble in aqueous solution and forms unidentified Ga and As species, upon dissolution. Different As species, which are formed following the exposure to Ga, may lead to various toxic effects (Flora and Das Gupta 1994).

Exposure to GaAs can only be monitored by determining As concentrations. Several reports describe the assessment of exposure to As during GaAs production and use. Although the solubility of GaAs in pure water is very low, its dissolution in body fluids is greatly enhanced by endogenous chelating molecules. When incubated in artificial body fluid (Gamble's solution), GaAs progressively releases both Ga and As. Gallium arsenide is carcinogenic to humans (Group 1) (IARC 2006).

Over the past two to three decades, Ga compounds have gained importance in the fields of medicine. Although Ga has no known physiologic function in the human body, certain characteristics enable it to interact with cellular processes and biologically important proteins, especially those of the Fe metabolism. This has led to the development of a certain Ga compound, as diagnostic and therapeutic agents in medicine, especially in the areas of metabolic bone disease, cancer, and infectious disease (Chitambar 2010).

Gallium has shown efficacy in the treatment of several diverse disorders: (1) accelerated bone resorption, with or without elevated plasma Ca; (2) autoimmune disease and allograft rejection; (3) certain cancers; and (4) infectious disease. Gallium is effective in suppressing bone resorption and, when present, concomitant elevated plasma Ca. This antiresorptive activity has led to its clinical use in treating hypercalcemia of malignancy and Paget's disease of bone. Gallium has also shown clinical

TABLE 16.2
Gallium in Foodstuffs (mg/kg FW)

Product	Content
Wheat[a]	0.118
Wheat flour[a]	0.148
Cereal and cereal products[b]	0.001
Rice[a]	0.127
Orange juice (Florida)[c]	0.030–0.040
Orange juice (Brazil)[c]	0.063–0.145
Sweeteners, honey and confectionery[b]	0.002
Fat and oil[b]	0.016
Shrimps and mussels[b]	0.008

Sources: [a] Li, Y.-H. et al., *Electroanal*, 17, 343–347, 2005.
[b] Millour, S. et al., *J. Food Compos. Anal.*, 25, 108–129, 2012.
[c] McHard, J.A. et al., *J. Agric. Food Chem.*, 27, 1326–1328, 1979.
FW, fresh weight.

efficacy in suppressing osteolysis and bone pain, associated with multiple myeloma and bone metastases, and has been suggested as a treatment for osteoporosis.

In addition to antiresorptive activity on bone, Ga also has anabolic activity. It also shows specific immunomodulating activities. It is effective in suppressing adjuvant-induced arthritis, experimental encephalomyelitis, experimental autoimmune uveitis, and other diseases. Some studies have suggested possible efficacy in mouse models for asthma and type I diabetes. Several *in vitro* experiments have found Ga effective at inhibiting T-cell and macrophage activation and in suppressing the secretion of certain cytokines by these cells (Bernstein 1998).

Due to its chemical similarity to Fe, Ga can be substituted for Fe in many biologic systems and inhibit Fe-dependent processes. When Ga ions are mistakenly picked up by bacteria, such as *Pseudomonas*, the ability of the bacteria to respire is interfered with, and the bacteria die. The mechanism behind this is that Fe is redox active, which allows for the transfer of electrons during respiration, but Ga is redox inactive (Kaneko et al. 2007).

There is suggestion on the potential usefulness of Ga maltolate for the prevention and control of *Rhodococcus equi* infections. Gallium maltolate, an orally absorbable form of Ga^{3+} ion, is in clinical and preclinical trials as an important treatment for a number of cancers, infectious disease, and inflammatory disease (Martens et al. 2007).

Data on Ga concentration in foodstuffs are very scarce (Table 16.2), and there are no data on its intake with food.

17 Germanium [Ge, 32]

17.1 INTRODUCTION

Germanium (Ge) is a metalloid of the group 14 in the periodic table of elements. Its concentration in the Earth's crust averages 1.5 mg/kg. Its similar contents are present in igneous rocks. In argillaceous sediments, Ge ranges from 1 to 2.5 mg/kg, and in calcareous rocks is <0.3 mg/kg. Its geochemical behavior is similar to Si and Sn, the neighboring elements of the same group. Hard coal contains Ge within the range 1–780 mg/kg, and fly ash contains 6–250 mg/kg.

Preferable Ge oxidation state is +4, and may also be +2. It exhibits various properties such as lithophilic, siderophilic, and chalcophilic, and easily substitutes Si in silicate minerals, and forms complexes with oxygen. Its common minerals are germanite, $Cu_2(Ge,Fe)S_4$, and argyrodite, Ag_8GeS_6. It is likely to be associated with Pb–Zn–Cu sulfide ores.

During weathering, Ge is easily mobilized, mainly as $Ge(OH)_2$, and is readily absorbed from aquatic systems, by clay minerals, Fe hydroxides, and organic matter (OM). This might explain its elevated concentrations in several coals, where it is often associated with the occurrence of sphalerite, (Zn,Fe)S (Font et al. 2005).

World mine Ge production (rounded) in 2010 was 120 kt, of which China produced 80; Russia produced 5; and the United States produced 4.62 (USGS 2011). Generally, it is a by-product obtained from smelter dusts of the Zn production and from Pb–Zn–Cu sulfide ores. Now it is mined primarily from sphalerite, and also it is recovered from Ag, Au, and Cu ores. Significant amounts of Ge are contained in ash and flue dust generated in the combustion of certain coals for electric power, and its recovery by means of flotation is highly effective, at 100% of its contents. The Ge emission from coal burning is calculated globally at about 700 t/yr.

Germanium is an important commodity in several manufacturing sectors, such as fiber-optic and infrared optics. It is also used for polymerization catalysts, and in the production of nanowires and organometallic compounds. Its most important use is in the multijunction solar cells for satellites, and terrestrial-based solar concentrator systems and infrared devices. In some memory chips, an alloy of Sb, Ge, and Ti is applied. The use of GeO_2 in catalysts for the polyester (polyethylene terephthalate) production is declined. Gallium organic compounds are used in the chemotherapy, as a strong pain reliever, especially in the Asian medicine (Moskalyk 2004). Some complexed organic Ge compounds are being presently investigated as possible pharmaceuticals, al though none have yet been proven successful.

17.2 SOILS

Data on the Ge abundance in soils are scanty. Its average content is estimated at 2.0 mg/kg, within the range of 0.5–2.5 mg/kg (Table 17.1). Its concentrations in the soils of various countries are given as follows (in mg/kg): the United States, 0.1–2.1; Sweden, 0.05–0.64; China, 1.2–3.2 (Kabata-Pendias 2011). Increased Ge levels (up to 95 mg/kg) in soils of Sweden are, most probably, effects of the contamination (Eriksson 2001a). It is likely to accumulate (up to 27 mg/kg) in the surface horizon of fresh and old leaf litters of the beech forest (Tyler 2005). Mean Ge content in organic surface soils of Norway is 1.94 mg/kg, and is lower than mean value (2.27 mg/kg) for all investigated soils (Nygard et al. 2012).

During weathering, Ge is easily mobilized as $Ge(OH)_2$ and readily fixed from aquatic systems, as $Ge(OH)_4$, by clay minerals, Fe hydroxides, and OM. Species of Ge in soils are mainly divalent cations, and anionic complexes, such as $HGeO_2^-$, $HGeO_3^-$, and GeO_3^{2-}. Germanium concentrations in soils increase with the presence of Si, whereas decrease with the presence of Fe. Increased amounts of Ge in soils developed on basaltic lavas (in Hawaii) are effects of its fixation by clay minerals and some nonsilicates (Scriber et al. *vide* Kabata-Pendias 2011).

17.3 WATERS

The median Ge concentration in the worldwide ocean water is estimated at 0.05 μg/L (Reimann and de Caritat 1998). However, there are also other given values, which might be presented as Ge concentrations between 0.05 and 0.5 μg/L (Table 17.1). In river

TABLE 17.1
Germanium Contents in Soils and Water

Environmental Compartment	Range
Soil (mg/kg)	
Light sandy	0.6–2.0
Medium loamy	1.0–2.5
Heavy loamy	1.0–2.0
Calcerous (calcisols)	0.5–1.3
Water (μg/L)	
River	0.001–0.08
Sea, ocean	0.05–0.5

Sources: Data are given for uncontaminated environments, from Kabata-Pendias, A. and Mukherjee, A.B., *Trace Elements from Soil to Human*, Springer, Berlin, Germany, 2007; Reimann, C. and de Caritat, P., *Chemical Elements in the Environment*, Springer, Berlin, Germany, 1998.

water, Ge ranges from 0.001 to 0.08 μg/L, at the world mean content at 0.007 μg/L (Gaillardet et al. 2003). According to these authors, the riverine flux of Ge is 0.25 kt/yr.

In aquatic environments, Ge occurs as cationic species, but more commonly as complex anions such as $HGeO_2^-$, $HGeO_3^-$, and GeO_3^{2-}. Also, organic Ge species, mainly as methylated forms, are present in water, and apparently dominate over mineral forms. Amounts of the MM–Ge and DM–Ge forms in oceans vary depending on the water salinity. These Ge species are rare in river water. Similar to Si and Al, natural Ge compounds tend to be insoluble in water.

Median Ge concentration in the bottled water of the EU countries is 0.029 μg/L, and is a bit higher than those in tap water, 0.011 μg/L (Birke et al. 2010).

17.4 AIR

Main sources of Ge in the atmosphere are from coal-fired plants, metallurgical industries, and waste incineration. Its emission from coal burning in the United Kingdom was estimated at about 2 kt/yr (Kabata-Pendias and Mukherjee 2007). It is also discharged in air from stack gases.

The best indicator for Ge concentrations in air is its content in mosses. Moss samples collected in 1975 and 2000 in South Sweden contained Ge on average levels at 0.091 and 0.015 mg/kg, respectively (Rühling and Tyler 2004). Mosses sampled in Norway, during 1990–1995, contained Ge within the range of 0.24–11 mg/kg (Berg and Steinnes 1997).

17.5 PLANTS

Germanium is not thought to be an essential element for plants. However, it is absorbed by plants relatively easily, possibly in the form GeO_2. Its average content in various plants is estimated at 20 mg/kg. In food plants, Ge concentrations vary from 10 to 754 mg/kg, with the lowest in barley grain, and the highest in garlic. However, there are also quite different values of Ge in food plants from the Central American region, within the range of <0.01–<0.1 (Duke 1970).

Although Ge is toxic to plants, some plants (e.g., rice) may accumulate high Ge amount (up to 1% AW). Apparently, there is an interaction between Ge and Si, and plants needing Si for the growth are sensitive to Ge. There is evidence that even a low concentration of Ge inhibits the germination and growth of some plants.

Increased levels of Ge in some plants may be due to aerial pollution. This is well illustrated by increased levels of Ge in leaves of two kinds of trees growing in the mountains and near streets of the Wrocław City (Poland), from 12–14 to 100–108 mg/kg, respectively (Sarosiek et al. *vide* Kabata-Pendias 2011).

17.6 HUMANS

Germanium amount in human body is 5 mg, of which in blood 0.4 mg/L and in tissue 0.1 mg/kg (Emsley 2011). Its absorption from gastrointestinal tract is believed to be rapid and complete, followed by a rapid excretion, mainly with urine. There is no evidence that Ge has any functional role in the body. Inorganic Ge compounds are readily absorbed and excreted mainly through the kidney, with half-lives in the order of 1–4 days. When injected intravenously, Ge concentrates in the liver, the kidney, the spleen, and the gastrointestinal tract. After oral administration, high Ge concentrations were detected in the spleen. Apparently, Ge becomes widely distributed in the human body after intake or application (Rosenberg 2009).

Limited data on Ge metabolism suggests that organic Ge is thought not to accumulate as inorganic compounds. It is not necessary to human health; however, its presence in the body has been shown to stimulate metabolism. Germanium may also play a role in the function of the immune system, immuno-enhancement, oxygen enrichment, free radical scavenging, analgesia, and metal detoxification. In organic forms, it is not considered as carcinogenic. Organic Ge appears to inhibit cancer development and, in the form of the organic compound, spirogermanium, to destroy cancer cells. Some organic forms of Ge are less toxic, than inorganic ones. Excess intake of inorganic Ge may adversely affect kidney functioning. Other adverse effects of Ge are anemia, diarrhea, skin rash, muscle weakness, and peripheral neuropathy. The toxic effects of inorganic Ge compounds usually occurs at its higher doses (ICL 2011).

Certain compounds of Ge have low toxicity to mammals, but have high toxic effects against certain bacteria. However, no Ge compound has yet been demonstrated for pharmaceutical use, as either an antibacterial or a cancer chemotherapeutic agent. Some Ge compounds are quite reactive and present an immediate hazard to human health on exposure. For example, Ge chloride and germane (GeH_4) are liquid and gas, respectively, which can be very irritating to the eyes, skin, lungs, and throat.

Germanium, as an organic complex, has been touted, but not proved, as having anticancer properties in humans. Inorganic Ge toxicity results in kidney damage. Some individuals, consuming high amounts of organic Ge supplements contaminated with inorganic Ge, have died from kidney failure.

Ge, although essentially limited in Japan, is used in the production of organogermanium compounds such as Ge sesquioxide, *organic Ge* or ^{132}Ge, for medical applications. The interest in Ge sesquioxide was excited and also strongly promoted after they had screened a large number of Ge compounds for their biological effects. ^{132}Ge not only possesses antitumor activity but also increases the production of interferon, without showing detectable signs of cytotoxicity. Even when used as a food supplement, ^{132}Ge has been shown to act as an immune stimulant and to enhance the activity of natural killer cells and interferon levels. If adverse effects, and sometimes even fatal cases, were reported as consequences of treatment with Ge-containing drugs, this could often be attributed to the ingestion of inorganic Ge, or ^{132}Ge contamination.

As a consequence, many countries such as the United States and the EU countries have issued a ban on the import and sale of Ge sesquioxide, as a nutritional additive. The use of a Ge application in a person who died from acute renal failure resulted in high Ge concentrations in vertebrae, kidney, brain, and skeletal muscle. This fact has led to the ban of organogermanium compounds (particularly ^{132}Ge), as food supplements, due to the possible contamination of organic Ge compounds by GeO_2 from the manufacturing process, which is still enforced. Hair and nails may be allowed to monitor excessive Ge uptake (Rosenberg 2009).

TABLE 17.2
Germanium in Foodstuffs (mg/kg FW)

Product	Content: Highest[a]	Content: Lowest[b]
Bread	0.003	<0.002
Miscellaneous cereals	0.004	<0.002
Green vegetables	<0.002	<0.0004
Other vegetables	0.002	<0.0007
Potatoes	<0.002	<0.0004
Fruits	<0.002	<0.0003
Nuts	0.002	<0.002
Food of Animal Origin		
Carcass meat	0.002	<0.001
Offal	0.002	0.002
Fish	0.002	<0.0007
Milk	<0.002	<0.0003
Dairy products	0.002	<0.001

Sources: [a] Ysart, G. et al., *Food Addit. Contam.*, 16, 391–403, 1999.
[b] Rose, M. et al., *Food Addit. Contam.*, 27, 1380–1404, 2010.
FW, fresh weight.

In the absence of occupational exposure, food represents the most important source of Ge uptake for humans (Table 17.2). Most foods have Ge less than 0.01 mg/kg, although there are some data for higher Ge contents in some vegetables. Potatoes, garlic, carrot, and soya contain Ge (in mg/kg) as follows: 1.85, 2.79, 0.60, and 9.39, respectively. The highest Ge content is noticed for ginger tablets, 9.96 mg/kg (McMahon et al. 2006).

The population dietary exposure in the United Kingdom, reported in 1994, was 0.004 mg/day (Ysart et al. 1999), and in 2006, it was 0.0001–0.0015 mg/day (Rose et al. 2010).

18 Gold [Au, 79]

18.1 INTRODUCTION

Gold (Au), a chemical element of the group 11 in the periodic table of elements, is one of the metals of the so-called group noble metals. Its abundance in the Earth's crust averages 3–4 μg/kg. Its content in ultramafic igneous rocks is up to 5 μg/kg, and in sedimentary rocks, its content varies from 2–6 and in calcareous rocks and sandstones, its content is in the range of 3–7 μg/kg. In coal, Au may be concentrated within the range of 10–50 μg/kg.

The main oxidation state of Au is +1, but may also exist from 0 to +5. Gold reveals a tendency to occur in the free oxidation state and remains in this form, as it has a great chemical resistance. Most often, it occurs in the form of the native metal, and may be a minor constituent of some types of ores. Recently, Au became an important by-product from some nonferrous Cu and Pb ores. Relatively common Au minerals are associated with Te such as calverite, $AuTe_2$; montbrayite, Au_2Te_3; petzite, $Au,AgTe_2$; and sylvanite, $Au,AgTe_4$.

Gold was one of the first metals to be recognized by humans, and due to its chemical and physical properties, it became very important throughout the human civilization.

World mine Au production (rounded) in 2010 was 2500 kt, of which Australia produced 255; China produced 345; Russia produced 190; Uzbekistan produced 190; Peru produced 170; and Ghana produced 100 (USGS 2011). In some countries, Au production is partly under reported. There are several Au-mining methods, depending on the type of its deposits. Flotation methods are common, based on the amalgamation with Hg, and bioextraction methods, using some microorganisms, mainly *Thiobacillus ferrooxidans* (Reith and McPhail 2007). Gold is also obtained from some ore materials by extractions, using Na-cyanide and Ca-hydroxide. Both methods cause health risks, because the workers contact with toxic chemicals (Veiga and Baker 2004).

Gold is widely used in jewelry, coinage, dental alloys, electroplating electronics, and antiarthritic drugs.

18.2 SOILS

The average Au contents of soils is estimated at 3 μg/kg. In organic soils, it may be elevated up to 8 μg/kg (Table 18.1). However, some alluvial soils, especially soils with postmining Au deposits, and soils with Au-enriched sludge may contain much higher Au levels, within the range from 43 to 473 μg/kg. Sewage sludge contains Au at about 800 μg/kg. Forest soil mull fractions and clay mineral fractions, especially in soils from Au-mineralized areas, may contain Au up to 5000 μg/kg. Soil bacterial communities (e.g., *Bacillus cereus*) may accumulate Au in spores, up to 1100 μg/kg.

TABLE 18.1
Gold Contents in Soils, Water, and Air

Environmental Compartment	Range
Soil (μg/kg)	
Light sandy	1–2
Medium loamy	1–5
Organic (histosols)	1–8
Water (ng/L)	
River	0.02–119
Sea, ocean	0.02–4.5
Air (pg/m³)	
Polluted sites	300
Remote regions	0.01–1.5

Sources: Data are given for uncontaminated environments, from Kabata-Pendias, A. and Mukherjee, A.B., *Trace Elements from Soil to Human*, Springer, Berlin, Germany, 2007; Reimann, C. and de Caritat, P., *Chemical Elements in the Environment*, Springer, Berlin, Germany, 1998.

Gold forms in soils several, mainly anionic, complexes such as $AuCl_2^{2-}$, $AuBr_4^-$, $Au(CN)_2^-$, $Au(CNS)_4^-$, and $Au(Se_2O_4)_2^{3-}$, which are relatively mobile. In tropical lateritic soils, several other Au complexes are formed. No simple Au cation exists in soil solution (Kabata-Pendias 2011).

Various bacteria (cyanobacteria, in particular), including archaea, control the Au solubilization and precipitation. Cyanogenic plants are known to mobilize Au by cyanide realized into soil solution. Excreted forms of calcrete and carbonate concretions are often associated with Au precipitation. It is likely to be accumulated also in rhizosphere soils. Elevated Au contents in peat and algal mats are observed around Au-mining areas, as well as in surroundings of some other metals mining areas.

The global balance of the Au accumulation in sewage streams is estimated at 360 t/yr (Eisler 2004). Sewage sludge may be a major source of Au when applied to soils. Increased levels of Au in some sewage sludge are (in μg/kg) as follows: 500–3000 in the United States and 500–4500 in Germany.

18.3 WATERS

World median Au concentration in seawater is estimated at 4 ng/L. Its mean amounts range from 0.02 to 4.5 ng/L in the North Pacific and North Atlantic oceans, respectively (Table 18.1). Several chloro-, bromo-, iodo-, and hydroxyl complexes of Au occur in water (e.g., $AuCl_2^-$ and $AuClBr^-$).

In sea- and river-bottom sediments, Au may be present in cationic species, of different oxidation states such as Au°, Au^{+}, and Au^{3+}. Gold occurs in several types of deposits such as lode (metalliferous ores), residual, alluvial, bench, ancient rivers, and flood layers. A lode deposit is the original source of its deposits, mainly as pieces of ore. Pieces of Au ore are also likely to be deposited in cracks or crevices in the rocks, at the bottom of streams and rivers.

Some aquatic biota (e.g., brown algae) may accumulate considerable amounts of Au, especially near the mining and smelting areas. Mackerel (*Pneumatophorus japonicus*) and common mussel (*Mytilus edulis*) may concentrate Au up to 12 μg/kg and up to 38 μg/kg, respectively.

18.4 AIR

Gold concentrations in air of remote areas do not exceed 1.5 pg/m³, whereas it may be up to 300 pg/m³ in industrial regions (Table 18.1). Its wet deposition in Sweden was estimated at 0.07 g/ha/yr, and in the United Kingdom between the range of <0.07 and <0.2 g/ha/yr (Cawse, Eriksson *vide* Kabata-Pendias and Mukherjee 2007).

Gold particulates in the atmosphere are mainly from volcanic emissions in both gaseous and aerosol forms. It has been calculated that Mt. Etna, Italy, emitted 880 kg Au during the eruption, and Mt. St. Helen, United States, emitted 1100 kg of Au (Reimann and de Caritat 1998).

Gold content in lichens nearby Au mines accumulated from 0.22 to 1.45 mg Au/kg, whereas the background values for Au in this area is 0.12 mg/kg (Limbong et al. 2003).

18.5 PLANTS

Gold is easily phytoavailable, and when it enters the root vascular systems, it is easily transported to the tops. In reducing media, however, Au precipitates on a cell surface, and thus inhibits membrane permeability. Therefore, in most plants, its content in roots is higher than in above-ground parts. Ability of microorganisms to mobilize elemental Au has been often observed. Especially microorganisms in the plant rhizosphere are responsible for the Au mobility and availability (Eisler 2004).

The background Au contents of plants growing in uncontaminated soils is within the range of <0.001–<0.005 mg/kg (Girling, Greger *vide* Kabata-Pendias 2011). Plants from Au-contaminated sites contain Au commonly within the range of 0.007–0.17 mg/kg. Several Au-accumulator plants (e.g., *Artemisia persia*, *Prangos popularia*, and *Pinus laricio*) may contain Au at level of 0.1–100 mg/kg. In most plants, Au concentration in seeds is higher than in leaves and stems. Forest mushroom (*Boletus edulis*) from uncontaminated site accumulates Au up to 0.235 mg/kg (Falandysz et al. 2012).

Gold concentration in plants and shrubs has been extensively used for the biogeochemical survey for Au-mineralized areas, and some plants (e.g., *Sedum aceotatum*, *Phacelia sericea*, and *Artemisia* sp.) have been successfully used. Various plants collected from the mineralized areas of British Columbia (Canada) contained Au within the range of 0.7–6.5 mg/kg. Several plants are relatively resistant to higher

Au concentrations in tissues. Au toxicity leads to necrosis and wilting, due to the loss of turgidity in leaves.

Gold content in lichens is also a good indicator for elevated Au levels in the surroundings. Its contents in lichens from the vicinity of the Au mine in New Zealand vary between 224 and 7450 mg/kg (Williams et al. *vide* Kabata-Pendias 2011).

18.6 HUMANS

Gold content in the human body (70 kg) is estimated at less than 0.2 mg, of which in the blood, 1–40 µg/L; in the bone, 16; and in the liver, 0.4 µg/kg (Emsley 2011). It is not known to perform any essential functions in humans. Metallic gold (Au^{0}) is arguably considered to be the least corrosive and most biologically inert of all metals. Human exposure to Au^{0} is relatively common. However, it does not mean that Au^{0} has been shown to be completely inert in mammals. Metallic Au can be gradually dissolved by thiol-containing molecules, such as cysteine, penicillamine, and glutathione to yield gold complexes (Sun et al. 2011).

Gold demonstrates excellent biocompatibility within the human body (the main reason for its use as a dental alloy), hence the number of direct applications of Au, as a medical material. Gold also possesses a high degree of resistance to bacterial colonization, which makes it the material of choice for implants, which are at risk of infection.

Injectable Au thiolate drugs such as aurothiomalate, aurothioglucose, and aurothiopropanol sulfonate, and the oral drug auranofin, are widely used for the treatment of difficult cases of rheumatoid arthritis (Sadler and Guo 1998). In recent times, injectable Au has been proven to help reduce the pain and swelling of rheumatoid arthritis and tuberculosis.

It is important to draw a distinction between the properties of Au in its bulk form, and properties exhibited when it is present in the form of tiny nanoparticles (NPs). At the nanoscale, the properties of gold can be markedly different. Colloidal Au preparations (suspensions of AuNPs) in water are intensely red colored and can be made with tightly controlled particle sizes, up to a few tens of nanometers, by the reduction of Au chloride with citrate or ascorbate ions. Colloidal Au is used in research applications in medicine, biology, and materials science. The technique of immunogold labeling exploits the ability of the Au particles to adsorb protein molecules onto their surfaces.

Many new nanogold-based biomedical products are being developed for drug delivery, cancer therapy, diagnostic devices, and biosensing. AuNPs are widely used in consumer products, including cosmetics, food packaging, beverages, toothpaste, automobiles, and lubricants. With this increase in consumer products containing AuNPs, the potential for worker exposure to AuNPs will also increase. There are limited data on the *in vivo* toxicology of AuNPs, meaning that the absorption, distribution, metabolism, and excretion of AuNPs remain unclear (Sun et al. 2011a).

The high accumulation of AuNPs in the kidneys suggest that the kidneys are the major accumulation site for metal NPs. There is a difference in the pattern of nanoparticle distribution between male and female kidneys (Sun et al. 2011a).

TABLE 18.2
Gold in Foodstuffs (μg/kg FW)

Product	Content
Food of Plant Origin	
Bread	1.0
Miscellaneous cereals	2.0
Green vegetables	<0.4
Other vegetables	0.5
Potatoes	0.4
Fruits	<0.1
Beverages	0.4
Nuts	0.5
Sugar and preserves	0.4
Food of Animal Origin	
Carcass meat	1.0
Offal	1.0
Fish	1.0
Eggs	0.4
Milk	0.4
Dairy products	0.4

Source: Ysart, G. et al., *Food Addit. Contam.*, 16, 391–403, 1999.
FW, fresh weight.

Recent advances in wet chemical synthesis and biomolecular functionalization of AuNPs have led to a real expansion of their potential biomedical applications, including biosensorics, bioimaging, photothermal therapy, and targeted drug delivery. Whereas about 80 reports on the *in vivo* biodistribution and *in vitro* cell toxicity of AuNPs are available in the literature; there is a lack of correlation between both fields, and there is no clear understanding of intrinsic NPs effects (Khlebtsov and Dykman 2011).

Because Au is not particularly toxic as a metal, it is present in small Au flakes in some sweets and drinks, such as Goldwasser. Indeed, gold is approved in the EU countries as a food additive, and is assigned the number E175.

In foodstuffs, Au concentration was estimated at a level of <0.1–2 μg/kg FW (Table 18.2), and its daily intake was found to be 1 μg/day (Ysart et al. 1999).

19 Hafnium [Hf, 72]

19.1 INTRODUCTION

Hafnium (Hf), a metal of the group 4 in the periodic table of elements, has crystallographic and chemical properties very analogous to those of Zr, a metal of the same group. Its terrestrial abundance ranges between 3 and 5 mg/kg, and similar contents are in igneous rocks. A bit higher amounts, at about 6 mg/kg, may also be present in some sandstones and argillaceous sediments. Its lowest contents, <0.4 mg/kg, is in calcareous rocks. Hard coal contains Hf within the range of 0.5–1 mg/kg.

The oxidation state of Hf is +4. It is highly associated with Zr, and occurs mainly in the following Zr minerals: hafnon (called also zircon), $(Zr,Hf)SiO_4$ and baddeyite, $(Zr,Hf,Ti,Fe,Th)O_2$. It may also be absorbed by other minerals, mainly biotite and pyroxene.

World primary Hf production statistics are not available. However, some earlier estimation was for about 60 t/yr (Kabata-Pendias 2011). Hafnium is produced as a by-product during Zr processing. Resources of Hf in the United States are estimated to be about 130,000 t.

Hafnium is used mainly in filaments and electrodes, and in nuclear industries, especially in power plants. Hf and Zr are used interchangeably in certain superalloys, but in others, only Hf may be used. The alloys with Nb, Ti, and W have a special application in nuclear processes, mainly for reactor-control rods and coatings of cutting tools. Also, alloys of Hf with Pd and Pt are relatively common. Some amounts of Hf are applied in photographic flashes. Biometallic materials may contain Hf, as it has a good biocompatibility and osteoconductivity (Szilagyi 2004).

19.2 SOILS

The worldwide mean Hf content of soils is estimated at 6.4 mg/kg, within the range of 2–20 mg/kg, in minerals soils. Its abundance in organic soils is reported to be much more lower (Table 19.1). Hafnium content in soils is reported to range/average in various countries as follows (in mg/kg): Bulgaria, 1.8–18.7; Canada, 1.8–10; Russia, 20.8; Sweden, 7.6; the United Kingdom, 5 (Kabata-Pendias 2011).

Mean Hf content in surface soils, with high organic matter content, is 0.07 mg/kg, and is a bit lower than mean value (0.1 mg/kg) for all other organic soils in Norway (Nygard et al. 2012).

Increased Hf levels in coarse fractions of desert dusts resulted from its association with detrital fraction (Castillo et al. 2008).

TABLE 19.1
Hafnium Contents in Soils and Water

Environmental Compartment	Range
Soil (mg/kg)	
Light sandy	2–19
Medium loamy	5–20
Organic (histosols)	0.0X–0.X[a]
Water (ng/L)	
Rain	<1–2
River	4–110
Sea, ocean	3.4–7.0
Air (ng/m³)	
Urban	20–590
Greenland	0.04–0.06

Sources: Data are given for uncontaminated environments, from Kabata-Pendias, A. and Mukherjee, A.B., *Trace Elements from Soil to Human*, Springer, Berlin, Germany, 2007; Reimann, C. and de Caritat, P., *Chemical Elements in the Environment*, Springer, Berlin, Germany, 1998.

[a] Order of magnitude, Markert, B. and Lieth, H., *Fresen. Z. Anal. Chem.*, 327, 716–718, 1987.

19.3 WATERS

The worldwide Hf average concentration in stream and river water in calculated to be 6 ng/L; however its range is very broad, from 4 to 110 ng/L. Seawater contain Hf within the range of 3.4–7 ng/L (Table 19.1).

Median Hf concentration in bottled water of the EU countries is 0.65 ng/L, which is a bit lower than in those in tap water at 0.8 ng/L (Birke et al. 2010).

Elevated Hf contents, up to 38 mg/kg, in some river sediments in Germany are a result of contamination (Reimann and de Caritat 1998).

19.4 AIR

Hafnium content in the atmosphere from the remote areas (0.04–0.06 ng/m^3) is very much lower than in urban/industrial regions (20–590 ng/m^3) (Table 19.1). This clearly indicates its anthropogenic sources. According to Szilagyi (2004) hazardous Hf level in air is 0.5 mg/m^3.

19.5 PLANTS

The common range of Hf in plants is 0.01–0.4 mg/kg. Although Hf is slightly available to plants, plants growing in soils amended with sewage sludge contain elevated amount of Hf, up to, on an average, 3 mg/kg. This may suggest its higher uptake, or impact of soil dust with its higher concentration. Hafnium uptake from aerial sources may also explain its higher amounts, up to 11 mg/kg, in spruce bark (Reimann and de Caritat 1998).

Hafnium has no known metabolic role in plants, and has a very limited ability to travel up with the food chain. This is the reason why Hf concentrations for most food samples were below the detection limit. The highest levels of Hf were observed in red kidney beans, 0.006 mg/kg (Howe et al. 2005). According to Szefer and Grembecka (2007), Hf occurs in plant food in low levels, from 0.00001 to 0.01 mg/kg FW, in vegetables and vines, respectively.

Mean contents of Hf in cereal grains range from 0.04 to 0.06 mg/kg, in wheat and barley, respectively (Eriksson 2001a).

19.6 HUMANS

Hafnium does not normally cause any health problems; nevertheless, all its compounds should be regarded as toxic, although it has no known toxicity. No signs and symptoms of chronic exposure to Hf have been reported in humans.

The metal dust presents a fire and explosion hazard. It is completely insoluble in water, saline solutions, and body chemicals. Exposure to Hf can occur through inhalation, ingestion, and eye or skin contacts. Overexposure to Hf and its compounds may cause mild irritation of the eyes, the skin, and mucous membranes.

Hafnium oxide nanoparticles were reported to be used in the radiotherapy, as a promising anticancer agent for human patients (Maggiorella et al. 2012).

20 Indium [In, 49]

20.1 INTRODUCTION

Indium (In) is a very soft, malleable, and easily fusible metal of the group 13 in the periodic table of elements and is chemically similar to Ga and Th, elements of the same group. Its terrestrial abundance ranges between 0.11 and 0.25 mg/kg, and is likely to be more concentrated in ultramafic igneous rocks (<0.2 mg/kg) than in sedimentary rocks (<0.07 mg/kg). Hard coal contains In within the range of 0.1–0.2 mg/kg), and sewage sludge, up to about 0.15 mg/kg.

The main oxidation state of In is +3, and may be +1 and +2. It is likely to precipitate under conditions that form Fe and Mn hydroxides, and therefore it is associated with Fe-bearing minerals. Due to its chalcophilic properties, In forms sulfide minerals such as indite, $FeIn_2S_4$, and roquésite, $CuInS_2$. It is also associated with other sulfide minerals such as sphaleryt, ZnS; galena, PbS; and cahlcopyrite, $CuFeS_2$. Indium is very rarely found as grains of native metals, and its primary sources are Zn ores. Some amounts of In may also be recovered from slags, during the production of Pb and Sn.

In 2010, world mine production of In was 574 t, of which China produced 300; Korea produced 80; and Japan produced 70 (USGS 2011). A large amount of In is obtained from scraps recycling. Its current primary applications are to form transparent electrodes indium tin oxide and in touchscreens. It is applied in thin films to form lubricated layers, for alloys of a low melting point, and to some Pb-free solders. Alloys containing In are also used to replace Hg in alkaline batteries, and in absorbers for nuclear reactors.

Some tests have been made to use the radionuclide, ^{111}In, for the radioimmunoassay and as an antibody in cancer patients.

20.2 SOILS

Worldwide mean In contents in mineral soils are reported to range from 0.01 to 0.5 mg/kg, with the mean value of 0.06 mg/kg. Its higher concentrations, up to 2.6 mg/kg are found in organic soils (Table 20.1). During weathering, In oxidizes to In^{3+}, and behaves similarly to other cations, mainly to Fe^{3+}, and easily precipitates with Fe hydroxides. Because In is associated with organic matter (OM), it is likely to be concentrated in surface-soil horizons. Increased level of In in the forest litter, as compared with its contents of leaves, also indicate its sorption by OM (Rühling and Tyler 2004). Mobile In species, at acidic and neutral pH range, are $InCl^{2+}$ and $In(OH)^{2+}$. Above pH 9.5, the anion $In(OH)_4^-$ is likely to occur. Its distribution in soils indicates the association with clay fraction. Microorganisms

TABLE 20.1
Indium Contents in Soils and Air

Environmental Compartment	Range/Mean
Soil (mg/kg)	
Light sandy	<0.2–0.5
Medium loamy	0.03–0.6
Organic (histosols)	<0.1–2.6
Air (pg/m³)	
Industrial regions	20–1200
Remote regions	0.05–78
Antarctica	0.05

Sources: Data are given for uncontaminated environments, from Kabata-Pendias, A. and Mukherjee, A.B., *Trace Elements from Soil to Human*, Springer, Berlin, Germany, 2007; Reimann, C. and de Caritat, P., *Chemical Elements in the Environment*, Springer, Berlin, Germany, 1998.

reveal a greater resistance to In elevated concentration. Inhibited activities of nitrate-forming bacteria are observed in soils with In contents within the range of 5–9 mg/kg.

Indium contents in soils of different countries are reported as follows (mean or range in mg/kg): Brazil, 0.15; China, 0.03–4.1; Japan, 0.08–1.92; Sweden, 0.007–0.6; the United States, <0.2–0.5. Its elevated amounts in Chinese and Japanese soils are from the vicinities of Zn–Pb smelters.

Composts and sewage sludge contain high amounts of In, 14–20 and 34–94 mg/kg, respectively, and may be its significant sources in soils. Also, In in motorway dust may be elevated, up to 0.07 mg/kg (Kabata-Pendias 2011).

20.3 WATERS

Indium is considered to be the least soluble elements, and its concentration in ocean water is estimated to be at about 0.1 ng/L (Nozaki 2005). Its contents in the Atlantic Ocean is given as 6.25–10.71 pmol/kg (Alibo et al. *vide* Kabata-Pendias and Mukherjee 2007).

The dominant In species in water are $In(OH)_3$ and $InCl_2^+$. In highly chlorinated water, it may be also present as In^{3+}. Indium is likely to be deposited in bottom sediments.

20.4 AIR

Indium concentrations in the atmosphere of remote areas vary between 0.05 and 78 pg/m^3, with the lowest in Antarctica. Air of urban/industrial area contains much higher In amounts, up to 1200 ng/m^3 (Table 20.1). The main anthropogenic source of In in air is from waste and coal combustions. Its natural sources are mainly from volcanic emissions, which are clearly proofed by its variable contents in various layers of the ice cores (Matsumoto and Hinkley 2001).

Indium content of moss (*Pleurozium schreberi*) has decreased significantly from 110 μg/kg in 1975 to 13 μg/kg in 2000 (Rühling and Tyler 2004). This clearly indicates that the In emissions have decreased during that period.

20.5 PLANTS

Although In is readily available to plants, its accumulation is not observed. Physiological effects of increased In levels in plants is associated with In-induced toxicity in roots, which occur in plants grown in culture solution, with its contents within 1–2 mg/kg. A few data on In in plants are reported by Fergusson (*vide* Kabata-Pendias 2011) as follows (in μg/kg FW): beets, 80–300; vegetables, 30–710; fruit tree leaves, 0.64–1.8; tomato leaves, 0.64–1.8. Mean content of In in cereal grains sampled in Sweden contain <5 μg/kg (Eriksson 2001a).

Plants grown in soils amended with sewage sludge may contain much higher amounts of In, up to 300 μg/kg in beets. Also, elevated In contents (up to 2100 μg/kg FW) was reported for unwashed grass from industrial regions (Kabata-Pendias 2011).

20.6 HUMANS

Indium content in human body is about 0.4 mg (Emsley 2011). It is not known to have any biochemical functions. Ionic In is nephrotoxic, and may cause various damages. It may resemble Hg functions. At extremely high doses, ionic In causes focal necrosis in the liver (Castronovo and Wagner 1971).

Human exposure to In compounds may occur in the semiconductor industry, during the manufacture (sawing, grinding, or polishing) of semiconductor wafers, or during the maintenance of the production equipment. It may also be a result of In isotope utilization for organ scanning, and for the treatment of tumors. Radioactive ^{111}In (in very small amounts on a chemical basis) is used as a radiotracer to nuclear medicine tests.

Hydrated In oxide causes damage to those organs that contain phagocytic cells, which clear the insoluble particles from the blood after intravenous injection. Focal necrosis was found in the liver, the spleen, and the bone marrow. Damage was also found in the thymus and lymph nodes. At its extremely high doses, convoluted tubules of the kidney may be damaged. Hydrated In oxide caused extensive

TABLE 20.2
Indium in Foodstuffs (mg/kg FW)

Product	Content
Food of Plant Origin	
Bread	<0.02
Miscellaneous cereals	<0.02
Vegetables	<0.004
Canned vegetables	0.096
Fresh fruits	<0.003
Fruit products	0.031
Food of Animal Origin	
Carcass meat	<0.01
Offal	<0.01
Fish	<0.007
Milk	<0.003

Source: Rose, M. et al., *Food Addit. Contam.*, 27, 1380–1404, 2010. With permission.

FW, fresh weight.

hemorrhage and marked thrombocytopenia. Fibrin thrombi were observed in the liver (Castronovo and Wagner 1971). Indium compounds damage the heart, the kidneys, and the liver, and may be teratogenic. Hydrated In oxide is 40 times more toxic than ionic In.

All food groups contain In, in relatively small amounts (Table 20.2). After Rose et al. (2010), the concentration of In in food products was below the limit of detection, in most cases, except canned vegetables and fruit products. The population dietary exposure to In in the United Kingdom was 0.005–0.019 mg/day, and is not of a toxicological concern.

21 Iodine [I, 53]

21.1 INTRODUCTION

Iodine (I) is an element of the halogen group 17 in the periodic table of elements, and is the least reactive of the elements in this group. Its content in the continental Earth's crust is variable and is estimated between 0.15 and 1.4 mg/kg, in the upper layer, and 0.45–0.8 mg/kg, in the bulk crust. When I is heated, it sublimates, that is, it goes from being a solid to a vapor without going through the liquid state.

The highest contents of I are in sedimentary rocks; up to 38 mg/kg in argillaceous, and in some carbonate sediments (especially hardened calcium carbonate, caliche deposits). In organic-rich shales, its content is up to 17 mg/kg. Igneous acidic rocks contain I up to 0.5 mg/kg. During weathering, I is rapidly released and transported to ocean and sea basins. Most of I, nearly 70%, exists in ocean and sea sediments.

Increased I contents in some phosphates, bituminous shales, and coal are associated with organic matter (OM). Its mean content in coal is about 1 mg/kg. It is concentrated in some nitrate deposits, especially in Chile saltpeter, up to 400 mg/kg. The suggestion of its atmospheric origin in saltpeter seems to be the most reasonable, as I is very volatile from seas as CH_3I and I_2.

Iodine, occurring mainly at –1 oxidation state, is a gray-black substance, very easily sublimate to a blue-violet vapor, without going through the liquid phase. It forms only a few own minerals, such as lodargirite (AgI) and marshite (CuI). They are rare in the oxidized zones of metamorphosed Pb–Zn–Ag deposits and of porphyry Cu deposits. Other I minerals, also not very common, are lautarite, $Ca(IO_3)_2$; bellingerit, $Cu_3(IO_3)_6{\cdot}2H_2O$; dietzete, $Ca(IO_3)_2{\cdot}CrO_4$; and salesite, $Cu(IO_3OH)$. Sodium iodate, $NaIO_3$, is present in saltpeter, which is its main source. Iodine may also be obtained from some seaweeds (e.g., kelps).

In 2010, iodine world mine production, without the United States (rounded) was 29,000 t, of which (in t) Chile produced 18,000; Japan produced 9,000; and China produced 590 (USGS 2011). As in previous years, Chile is the world's leading producer of I, followed by Japan and the United States. Chile accounted for more than 50% of the world I production.

Seawater contains I at about 0.05 mg/L. Its total content in seawater is approximately 34 Mt. Seaweeds of the Laminariaceae family are able to extract and accumulate up to 0.45% of I on a dry weight basis, and has been its source (Szefer 2002). The seaweed industry represents its major source. Although not economical, the production of I is also as a by-product of gas, nitrate, and oil. Now, the most economic I production is from caliche deposits.

Among many I isotopes, ^{127}I is the only stable isotope in nature. Two radioactive isotopes ^{129}I and ^{131}I are the most common products of atomic reactors, and have been released in various proportions into the environment.

Iodine is broadly used in a number of chemicals and pharmaceutics, for both external and internal medical applications. It is also added in nutrition (for both human and animals). There is no adequate substitute for I, so other substances cannot replace its applications. Iodine is used as a disinfectant, but in some EU countries it is prohibited. Health aspects of I deficiency in the food chain is a real problem in several regions of the world.

In some countries, there is a recommendation to use iodized salt in food, as a simple way to assure adequate I amounts in the human diet. However, the Ministry of Health of China announced the reduction of I content in salt, owing to fears that iodized salt is causing a rise in thyroid disease.

Radioactive ^{131}I is used in medical diagnosis. Iodine is a catalytic component in some chemical reactions and is used in the production of some dyes, inks, and photosensitive photographic compounds.

21.2 SOILS

The worldwide mean I content in soils is estimated at 2.8 mg/kg. However, its concentrations vary highly, from <0.1 to >10 mg/kg (Table 21.1). In volcanic ash soils, its content may be very high, above 100 mg/kg. Iodine accumulation in the surface and subsurface soil layers is affected by aerial deposition and biogenic adsorption.

TABLE 21.1
Iodine Contents in Soils, Water, and Air

Environmental Compartment	Range
Soil (mg/kg)	
Light sandy	<0.1–10
Medium loamy	0.3–10
Heavy loamy	5–>10[a]
Organic (histosols)	1–10
Water (µg/L)	
River	2–15
Sea, ocean	50–70
Air (ng/m³)	
Urban/industrial regions	100–350
Greenland	3–4
Antarctica	0.08

Source: Data are given for uncontaminated environments, from Kabata-Pendias, A. and Mukherjee, A.B., *Trace Elements from Soil to Human*, Springer, Berlin, Germany, 2007; Reimann, C. and de Caritat, P., *Chemical Elements in the Environment*, Springer, Berlin, Germany, 1998.

[a] The highest contents are in soils derived from volcanic ash.

Atmosphere is considered the main source of I, which is emitted mainly from sea and ocean surface. In some soils of islands, as well as in seacoast areas, higher I accumulations, up to around 150 mg/kg, are reported. In areas where I contents in soils is <1 mg/kg (e.g., Sri Lanka) symptoms of I deficiency disorders (IDD) in the populations are observed (Chandrajith et al. *vide* Kabata-Pendias 2011). Usually, light soils in humid climate regions are poor, whereas clayed soils are enriched in I.

Iodine distribution in soils is associated with soluble organic matter (SOM), hydroxides of Fe and Al, and clay minerals, especially chloride–illite group. Some microorganisms may play a significant role in I cycling in soils, due to their great capability of I absorption. Microorganism biomass may contain up to above 3% of I present in surface soil layer. Some fungi (e.g., *Penicillium chrysogenum*) may accumulate even much higher amounts of I.

Iodine behavior in soils varies depending on its species. In general, large proportions of I in soils occurs in organically bound forms. The absorption rate differs for I species and is higher for I^- than for IO_3^-. In the aquatic phase of soil are present mainly anions I^-, I_3^-, IO_3^-, and $H_4IO_6^-$, of which the first two are the most common. The anionic form (I^-) is very mobile in soils and is easily leached out from soils, especially under anaerobic conditions. However, this I form is easily available to plants. There is an estimation that only a small fraction (<1%–25%) of the total I in soils is easily mobile, and thus phytoavailable. Iodine occurs in soils to a large extent in fixed forms, being adsorbed by humic and fresh OM, as well as on clay and crystal lattice of minerals. Usually, organoiodine species are slightly mobile in soils. Exchange of volatile I compounds between soil surface and the atmosphere is relatively a common process. Iodine levels in soils are highly dependent on the atmosphere precipitation. Therefore, the distance from seas and oceans has an influence on I status in soils.

Soil acidity favors I sorption by soil components. Alkali soils of arid and semiarid regions have elevated amounts of I, due to both salinity and a low I mobilization under alkaline conditions. Contents of I correlate positively with clay fraction, but only of noncalcareous soils. Liming soils is an important factor reducing the mobility of all I species, and thus reducing its bioavailability.

Soils in the surroundings of some industries (e.g., fossil-fuel combustion plants and kelp-burning facilities), as well as in close distance to high traffic roads, have usually elevated I contents. Additional I sources in soils may be due to some sewage sludge.

Great attention has been recently focused on the I radioisotopes, ^{129}I and ^{131}I, released into the environment during nuclear bomb testing and nuclear accidents, especially after the 1986 Chernobyl reactor accident. These are, however, depleted relatively faster from soils, due to their volatilization and slow vertical migration (Kashparov et al. *vide* Kabata-Pendias and Mukherjee 2007).

21.3 WATERS

The median I values in the worldwide seawater are estimated within the range of 50–70 µg/L (Reimann and de Caritat 1998). Much lower I concentrations are seen in river water, 2–15 µg/L (Table 21.1). The dominant I species in seawater are iodide ion (I^-), and organically bound compounds. However, iodate (IO_3^-) anion is also common in oceans and brine pools water. Some I species are unstable in water, and are transferred to iodate in alkali media, and to iodite in acid media. The photochemical, as well as biological oxidation, of iodide ions to iodate ions in surface water is a significant source of this element in the atmosphere.

The average I content in river water is estimated within the range of 2–15 μg/L (Table 21.1). However, in some stream water, it may be concentrated up to 198 μg/L (Reimann and de Caritat 1998). Also, in some lakes (e.g., in central Kazakhstan), its content can reach up to 160 μg/L (Fuge and Johnson *vide* Kabata-Pendias 2011).

Relatively high concentration of the isotope ^{129}I was measured in Swedish rivers, apparently from some nuclear reprocessing facilities in France and England (Kekli et al. 2003).

Median I concentration in bottled water of the EU countries is 4.8 μg/L, and in tap water it is 3.2 μg/L (Birke et al. 2010). Low I concentration in drinking water may have an impact on cardiovascular disorders. There are still inadequate data to establish a health guideline value, and lifetime exposure of I through water disinfection (WHO 2011a).

21.4 AIR

Median I concentration of air in remote regions is 1.1 ng/m^3, within the range of 0.08–4.4 ng/m^3, whereas in contaminated areas its content may be up to 350 ng/m^3 (Table 21.1). Atmosphere above seas and oceans contains variable I amounts; its mean value are as follows (in ng/m^3): Japanese Sea, 200; Baltic Sea, 1,090; and Black Sea, up to 48,000.

Organic I species dominate in air, and may be up to about 90% of its total concentration. Most probably these I species are methyl I, which is easily transferred to inorganic I forms. All iodine species, including its radionuclides, can be transported by air for relatively long distances. Radioactive I may be formed naturally, due to chemical reactions, high in the atmosphere, and may be changed to a stable element very quickly.

The oceans are the most important source of natural I in the air, water, and soil. Iodine in the oceans enters the air from sea spray or as gases. In the air, I can combine with water or with particles, and in such forms falls to the ground, from which it is easily emitted. Thus, its emission from the surface of seawater, volatilization from soils, and volcanic gases control its concentration in the atmosphere. Average I deposition in seacoast regions is estimated at 7 mg/m^2/yr (Fuge and Jonhson *vide* Kabata-Pendias 2011).

21.5 PLANTS

Iodine has not been shown to be essential to plants, although there are some reports on its stimulating impact on plant growth. Its toxic concentration in nutrient solution is higher than soluble I content of soils, and therefore its toxicity has not been observed under natural field conditions. However, there are some reports on I toxicity to rice plants grown in submerged paddy fields (Yuita *vide* Kabata-Pendias 2011). Also, in some polluted regions, I toxicity symptoms may be observed in plants.

Iodine contents of plants are a function of both plant species and growth condition. However, in general, the phytoavailability of I is low, as, most often, it is strongly fixed by soil particles. There are some indications that soil properties, especially parent materials, have a real impact on I contents in plants (Anke 2004). Also, distance of fields from seas and oceans is of real importance. Thus, there is a great variation in I contents of various plants. Because I in water is easily phytoavailable, marine plants may contain greater amounts of I (53–8800 mg/kg) than terrestrial plants (<0.01–10 mg/kg). Iodine concentration in algae seems to be associated with its pigmentation. On an average, its contents in various algae are (mean, in mg/kg) green, 58; red, 383; and brown, 2489 (Fuge and Johson *vide* Kabata-Pendias and Mukherjee 2007).

Iodine is easily volatized from the soil–plant systems, mainly in the methylated form (CH_3I). This is apparently due to methylation processes, activated by roots and/or microorganisms. Organically bound I in soils is scarcely available to plants, but after the decomposition of SOM, it becomes phytoavailable. The dominant I species in plants is apparently iodide, and about 65% of its total content is bound to proteins.

Plants take up I relatively easily from soil solution, and presumably, the iodide, I^-, is much more available than the iodate, IO^-. Both I species are more concentrated in roots than in shoots, which could be due to their absorption at the root surface. However, there are also some reports that its higher contents are in top plants. Apparently, several other factors have an impact on I distribution within plant tissues. Plants are capable of absorbing I directly from the atmosphere, both through the cuticle and as adhesive particles on the surface of hairy leaves. Adequate I levels in food and feed plants are required in human and animal nutrition. Therefore, I fertilization or foliar application have been investigated. The most effective in both treatments was potassium iodide (Strzetelski et al. *vide* Kabata-Pendias 2011).

Vegetables grown in some EU countries contain relatively high amounts of I. The highest amounts are reported for (in mg/kg) the vegetables: cabbage, 9–19; onion—bulbs, 8–10; potato—tubers, 3–5. In fodder plants, I contents are relatively similar, and vary (in mg/kg) as follows: in grass from <1 to 7 and in clover from <0.1 to 0.5. Relatively high I contents are reported for mushrooms, 5–10 mg/kg (Falandysz et al. 2012).

21.6 HUMANS

Content of I in human body is estimated at 10–20 mg; more than 95% of total I is accumulated in the thyroid gland (Emsley 2011). The only known roles of I in metabolism are its incorporation into the thyroid hormones, thyroxine (T4) and triiodothyronine (T3), and into the precursor iodotyrosines. Thyroid hormones, and therefore I, are essential for mammalian life. Both hormones have multiple functions in the energy metabolism of cells, in the growth, as a transmitter of nervous stimuli, and as an important factor in brain development. Iodine deficiency reduces the production of thyroid hormones in humans and animals, leading to morphological and functional changes of the thyroid gland, and reduction of the formation of thyroxin. Elemental iodine, I_2, is toxic, and its vapor irritates the eyes and lungs. The maximum allowable concentration of I in air, at work places, is 1 mg/m^{-3}. All iodides are toxic, if taken in excess.

Iodine deficiency is an important health problem throughout most parts of the world. Mountainous regions, such as the Himalayas, the Andes, and the Alps, and flooded river valleys, such as the Ganges, are among the most severely I-deficient areas in the world.

Globally, it is estimated that two billion individuals have insufficient I intake. Although goiter is the most visible sequelae of I deficiency, the major impact of hypothyroidism, due to I deficiency, is impaired neurodevelopment, particularly early in life. In the fetal brain, inadequate thyroid hormone impairs myelination, cell migration, differentiation, and maturation. Moderate-to-severe I deficiency during pregnancy increases rates of spontaneous abortion, reduces birth weight, and increases infant mortality. Offsprings of deficient mothers are at high risk for cognitive disability, with cretinism, being the most severe manifestation. Moderate-to-severe I deficiency during childhood reduces somatic growth. Correction of mild-to-moderate I deficiency in primary school-aged children improves cognitive and motor function. Iodine prophylaxis of deficient populations with periodic monitoring is an extremely cost-effective approach to reduce the substantial adverse effects of I deficiency, throughout the life cycle (Zimmermann 2011).

Iodine deficiency can lead to a wide spectrum of health problems, ranging from mild intellectual impairment to severe mental retardation, growth stunting, apathy, and impaired movement, speech, or hearing. Cretinism, in which most of these abnormalities occur, represents the extreme form of early I deficiency and is rare. Much more widespread is intellectual blunting, which may afflict as many as 50 million of the estimated 1.6 billion *at-risk* people, living in I-deficient regions.

Because of decreased production of thyroid hormones, I deficiency causes compensatory hypertrophy of the thyroid gland, as it attempts to make more thyroid hormone, resulting in a goiter—a disfiguring condition that is common in high-risk areas. Health problems arising from the lack of I are known as IDD (Zimmermann 2011). Universal salt iodization provides the most effective and affordable means to prevent IDD throughout the world.

Selenium deficiency can exacerbate the effects of I deficiency. Iodine is essential for the synthesis of thyroid hormone, but selenium-dependent enzymes (iodothyronine deiodinases) are also required for the conversion of thyroxine (T_4) to the biologically active thyroid hormone, triiodothyronine (T_3). Additionally, deficiencies of vitamin A or Fe may exacerbate the effects of I deficiency (FNB/IOM 2001).

Numerous measures have been undertaken to improve I supply to human diets, for example, using iodized salts, other vehicles for I, supplementation of foods of plant or animal origin or supplementing I to animal feed, in order to increase its content of food of animal origin (Flachowsky 2007).

Estimated dietary requirements for adult humans range from 0.08 to 0.150 mg/day. In 1988, Joint FAO/WHO Expert Committee on Food Additives set a provisional maximum tolerable daily intake (PMTDI) for I at 1 mg/day (0.017 mg/kg bw/day) from all sources, based primarily on data on the effects of iodide. However, recent data from studies in rats indicate that the effects of I in drinking water on thyroid hormone concentrations in the blood differ from those of iodide (WHO 2011a).

Iodine content in food and water primarily depends on its contents in soils. Thus, the average I content in foods shown in Table 21.2 cannot be used universally for estimating its intake by people of various regions.

TABLE 21.2
Iodine in Foodstuffs (mg/kg FW)

Product	Content
Food of Plant Origin	
Bread[a]	0.013
Cereals and grain products[a]	0.003–0.068
Bean[b]	0.013
Eggplant[b]	0.009
Potatoes[b]	0.009
Milk chocolate[c]	0.740
Cocoa[c]	0.107
Food of Animal Origin	
Veal[d]	0.026
Beef[d]	0.054
Pig, muscle[d]	0.045
Pigs, liver[d]	0.140
Fish[a]	0.004–0.830
Canned fish (various, in oil)[e]	0.047–0.390
Smoked fish, various[e]	0.108–1.135
Herring, salted[e]	0.492
Milk[d]; milk[f]; milk[g]	0.027; 0.309; (0.125–0.170)
Cheese[b]	0.158
Cheese, country[g]	0.250

Sources: [a] Chung, S. et al., *Food Addit. Contam.*, 6, 24–29, 2013.
[b] Hassanein, M. et al. *Pol. J.Food Nutr. Sci.*, 9/50, 23–25, 2000.
[c] Sager, M., *J. Nutr. Food Sci.*, 2, 123, 2012.
[d] Flachowsky, G., *Lohmann Inf.*, 42, 47–59, 2007.
[e] Usydus, Z. and Szlinder-Richert, J., *Bromat. Chem. Toksykol.*, 42, 822–826, 2009.
[f] Gabryszuk, M. et al., *J. Elem.*, 15, 259–267, 2010.
[g] Udeh, K.O. et al., *Pol. J. Food Nutr. Sci.* 10/51, 35–38, 2001.
FW, fresh weight.

The influence of cooking on I levels in food is minimal, except for boiling, as I dissolves into the water and will be withdrawn, unless it is eaten as soup. Iodine content in marine fish is higher than those in freshwater, and I content in egg yolk is higher than that in whole egg (Chung et al. 2013).

Iodine is characterized by a high risk of deficiency in human nutrition, but there is a low difference between requirements (0.15 mg/day for adults) and the upper level, PMTDI (1 mg/day). Therefore, I belongs to the trace elements of a high risk of deficiency from the global view, and the high risk of excess (Flachowsky 2007). Recommended dietary allowances for adult were set at 0.150 mg/day (WHO 2004).

21.7 ANIMALS

In animals, I intoxication may occur when toxic doses of I are very high. Severe or fatal intoxication in calves occurs after the prolonged administration of I, at a dose of 10 mg/kg/day. Iodism in cattle is manifested by a persistent cough, hyperthermia, naso-ocular discharge, in-appetency, depression, dermatitis, and alopecia. In sheep, depression, anorexia, hyperthermia, cough, changes in the respiratory system, and sometimes death were reported. No specific antidotes against iodism are available. Remission of the problem requires the elimination of I sources (Paulíková et al. 2002). I requirements of food-producing animals vary between 0.15 and 0.6 mg/kg of feed, according to various scientific committees (Flachowsky 2007).

22 Iridium [Ir, 77]

22.1 INTRODUCTION

Iridium (Ir), a noble metal of the group 9 in the periodic table of elements, is one of the rarest elements in the Earth's crust, present in the range of 0.02–0.05 µg/kg in the upper and bulk continental zones, respectively. Due to the tendency of Ir to bind with Fe, its content in the Earth's crust is lower than in deeper layers. It is concentrated mainly in the Fe–Ni core and in some meteorites. Its geochemical behavior is close to that of Co and Ni. Like other platinum group metals, it is very hard, brittle, silver-white, and the second densest metal (22.4 g/cm^3). Iridium contents of some coals may be up to 200 µg/kg.

Iridium was, presumably, emitted in higher amounts with volcanic dusts (Kracatau, in 1883), and later on, was deposited in sedimentary rocks. Increased Ir contents in sedimentary rocks, at the border layer between Cretaceous and Paleogene deposits, is apparently related to an effect of the meteorite collision, 65 million years ago (Dai et al. *vide* Kabata-Pendias 2011).

Main oxidation state of iridium is +4, but it may vary from +2 to +6, and has 34 radioisotopes. It occurs as natural alloys with Rh and Os, mainly in arsenosulfide minerals. Its typical minerals are irarsite ([Ir,Ru,Rh,Pt]AsS) and hollingworthite ([Rh,Pd,Pt,Ir]AsS). It may also be associated with other minerals, including olivine and ilmenite.

The annual production of Ir is estimated at about 3 t, and it is mined from Ni–Cu ores deposits. Various Ir salts and organometallic compounds are used in industrial catalysis. It is also used in alloys, for the corrosion and acid resistance at high temperature, and for electrodes in the chloralkali processes. Ir radioisotopes are used in some radioisotope thermoelectric generators. Additionally, ^{192}Ir is used as a source of gamma radiation for the treatment of cancer, especially when a sealed radioactive source is placed inside or next to the area requiring treatment.

22.2 ENVIRONMENT

Average content of Ir in Swedish arable soils is given as <4 µg/kg (Eriksson 2001a). Soil humus layer from the Kola Peninsula contains Ir up to 5 µg/kg, and in soils along motorways in Austria, its average content is 1.1 µg/kg (Fritsche and Meisel *vide* Kabata-Pendias 2011). Swedish arable soils contain Ir at the average value of <40 µg/kg (Eriksson 2001a). Dust collected along highways in Germany contains Ir in the range of 1.2–3.5 µg/kg, and plants contain from <0.02 to 0.4 µg/kg (Djingova et al. 2003).

Cereal grains from Sweden contain Ir at the average value of <0.1 µg/kg (Eriksson 2001a). Median Ir concentration in ocean water is 0.13 pg/L, whereas in the Baltic Sea, it ranges from 0.13 to 1.17 pg/L (Anbar et al. *vide* Kabata-Pendias and Mukherjee 2007).

In streams and lakes, its content may be up to about 1.20 pg/L (Reimann and de Caritat 1998). Rainwater collected in Sweden, during 1999, contained Ir at the mean value of <100 pg/L (Eriksson 2001a).

22.3 HUMANS

The total amount of Ir in human body has not been reported; however, the total amount in the tissues is 20 ng/kg (Emsley 2011). Its concentrations in human fluids are 7.4 ng/L in the blood and 18 ng/L in the urine (Hoppstock and Sures 2004). Ir is clinically inert, but Ir chloride is moderately toxic when ingested. It does not normally cause problems, as it is relatively unreactive, but all Ir compounds should be regarded as highly toxic. Most Ir compounds are insoluble and so, if taken, would not be absorbed by the body. The radioactive isotope ^{192}Ir is a γ-ray emitter, and is used in the radiation therapy (Emsley 2011).

Iridium concentrations is relatively high in miscellaneous cereals, 2 μg/kg fresh weight (FW). Its contents in several foods, bread, carcass meat, offal, milk, and nuts is about 1 μg/kg FW. Other products contain Ir <1 μg/kg FW. Mean dietary exposure for adults in the United Kingdom have been estimated at 2 μg/day (Ysart et al. 1999).

23 Iron [Fe, 26]

23.1 INTRODUCTION

Iron (Fe), a metal of the group 8 in the periodic table of elements, is a major constituent of the lithosphere; its global abundance is calculated to be from around 4.5% to >5%, and it is not considered as a trace element in rocks and soils. It plays a special role in the behavior of several trace elements and is in the intermediate position between macro and micronutrients in plants, animals, and humans. Iron reveals variable tendencies: siderophillic, chalcophillic, and lithophillic. It is the most important and widely used metal.

Highest Fe contents are in mafic rocks, up to 8%, whereas in acidic igneous rocks, its contents are up to 3%. Iron abundance in sedimentary rocks is about 4%, being more concentrated in argillaceous ones. In coal, it may occur up to about 2%, and in fly ash, up to 4%.

Iron occurs at several oxidation states, from +2 to +6, of which the most common is +3. Geochemistry of Fe is very complex and is largely determined by the easy change of the state of oxidation, and easy chemical reactions with other metals. Its behavior is also closely linked to the cycling of O, S, and C. Iron-ore minerals are mainly ferric oxides (hematite), hydrated ferric oxides (goethite), and various other minerals, such as siderite, pyrite, and ilmenite. *Bog iron ores* are small deposits of siderite, resulting from the precipitation of soluble $Fe(HCO_3)_2$ in lakes, swamps, and shallow shelf regions.

The fate of Fe in weathering processes depends largely on the Eh–pH system, and on the stage of the oxidation of Fe compounds. The general rules governing its behavior are that oxidation and alkaline conditions promote the precipitation of Fe, whereas reducing and acidic conditions promote the mobilization of Fe compounds (Cornell and Schwertmann 2003).

Common iron minerals, including those that are present in soils, formed pedogenically and biologically are as follows:

- Hematite, α-Fe_2O_3, occurs in soils of arid, semiarid, and tropical regions, and most often is inherited from parent rocks.
- Maghemite, γ-Fe_2O_3, is formed in highly weathered soils of tropical zones, and often occurs as concentrations associated with other Fe minerals.
- Magnetite, Fe_2O_3, is mainly inherited from parent rocks, and is often accompanied by maghemite.
- Goethite, α-FeOOH, is the most common mineral in soils, over broad climatic regions. Its crystallinity and composition may differ, depending upon conditions in which it has formed.
- Lepidocrocite, γ-FeOOH, is common in poorly drained soils of humid temperate regions.

- Ferrihydrite, $Fe_2O_3 \cdot nH_2O$, is not stable, but common mineral in soils.
- Ilmenite, $FeTiO_3$, not common in soils, is inherited from parent rocks and is resistant to weathering.
- Pyrite, FeS_2, and other ferrous sulfide minerals are widely distributed in submerged soils.
- Vivianite, $Fe_3(PO_4)_2 \cdot 2H_2O$, and hydrous ferric phosphate, $FePO_4 \cdot nH_2O$, are formed mainly under the influence of bacterial processes.
- Siderite, $FeCO_3$, occurs mainly in bog ores and sedimentary deposits.

Microbial oxidation of Fe occurs under both aerobic and anaerobic conditions, and stimulates the precipitation of poorly crystalline Fe oxides, which have a great affinity for the sorption of several cations, mainly Cu, Cd, Mn, Ni, Pb, and Zn. About 18 Fe minerals are listed as possibly biologically induced. Siderophores are Fe-chelated compounds secreted by microorganisms (also possibly by grasses), and are the strongest soluble Fe^{3+} binding agents.

Since about 3000 BC, iron has been used by humans and has played a dominant role in human civilization. It is the most commonly used metal in all civilizations.

Estimated world mine Fe production (rounded) in 2010 was (pig iron and raw steel, respectively) 1000 and 1400 Mt, of which China produced 600 and 630; Japan produced 82 and 10; Russia produced 47 and 66; the United States produced 29 and 90 (USGS 2011). A great proportion of Fe is obtained from Fe recycling and steel scrap.

Iron is the least expensive and the most widely used metal. In most applications, Fe and steel compete with other nonmetallic and/or metallic materials. However, as a very chemically reactive metal, Fe tarnishes rapidly in air and water. Alloys of Fe with other metals increase its resistance to corrosion. Nevertheless, effects of corrosion, on a global scale, are a serious source of this metal in various environmental compartments. Rusting (corrosion) is the reaction of Fe with oxygen present in air and with water, and leads to the formation of rust, red-brown hydrated Fe_2O_3.

23.2 SOILS

Iron abundance in soils averages about 3.5%, but varies highly from below 1% in light sandy soils, up to 5% in some organic soils (Table 23.1). Lateritic soils, formed in hot and wet tropical areas, are especially enriched in Fe (up to 30%), and are rusty-red because of Fe oxides. In soils, Fe occurs mainly in the forms of oxides and hydroxides, as small nanoparticles or in amorphous forms, associated with other soil particles and minerals. Iron minerals also are formed pedogenically and biologically. Free Fe minerals and compounds are used as a key characterization for soils and soil horizons.

Goethite is the most frequently occurring mineral, and is involved in sorption processes: (1) absorption of metals at external surface; (2) solid-state diffusion of metals; and (3) metal binding inside mineral particles. A strong ability of goethite to bind some metals suggests that it might be used, as well as other Fe hydroxides, for amelioration of metal-contaminated soils (Kumpiene et al. 2005). Especially Fe–Mn concretions reveal a great sorption capacity to several trace elements. They may accumulate Pb up to 12,000 mg/kg, and Zn up to 1,800 mg/kg. Relative metal

TABLE 23.1
Iron in Soils, Water, and Air

Environmental Compartment	Range
Soil (%)	
Light sandy	0.1–1.0
Medium loamy	0.8–2.8
Heavy loamy	1.5–3.0
Calcerous (calcisols)	0.1–0.5
Organic (histosols)	0.03–5.0
Lateritic	0.4–30
Water (μg/L)	
Rain	10–40
River (world median)	66
Sea (world median)	30
Air (ng/m³)	
Industrial and urban regions	130–14,000
Greenland	160–170
Antarctica	0.5–1.2

Sources: Data are given for uncontaminated environments, from Kabata-Pendias, A. and Mukherjee, A.B., *Trace Elements from Soil to Human*, Springer, Berlin, Germany, 2007; Reimann, C. and de Caritat, P., *Chemical Elements in the Environment*, Springer, Berlin, Germany, 1998.

affinities for the Fe-hydroxide surface have been presented as follows: Cu > Zn, Co, Pb > Mn > Cd. Several Fe compounds, as well as steel shots, have been proposed for the remediation of soil contaminated with trace metals (Kabata-Pendias 2011).

Amounts of metals bound to both amorphous and crystalline Fe oxides in polluted soils are fairly similar and governed mainly by soil properties. According to Karczewska (2004) relative amounts of total metal contents bound to Fe oxides (in % of total contents) are as follows:

- By amorphous Fe oxides: Cu, 3–24; Pb, 1–54; Zn, 3–25; As, 43–90
- By crystalline Fe oxides: Cu, 5–29; Pb, 3–53; Zn, 15–34; As, 5–46

Distribution of Fe minerals and compounds in soil profile is highly variable and reflects various soil processes, of which the most important are hydrolysis and the formation of complex Fe species. Especially, the formation of Fe compounds such as phosphates, sulfides, carbonates, and in particular, amorphous Fe hydroxides, greatly influence its behavior (Thomson et al. 2006).

Easily soluble Fe species are very low, from 0.01% to 0.1% of the total Fe content. The common range of Fe concentration in soils solution is 30–550 μg/L, but in very acidic soils, it may be above 2000 μg/L. In soil solution Fe is present in cationic species as follows: Fe^{2+}, Fe^{3+}, $FeCl^+$, $Fe(OH)_2^+$, $FeH_2PO_4^+$; and in anionic species, it is as follows: $Fe(OH)_3^-$, $Fe(SO_4)_2^-$, $Fe(OH)_4^{2-}$ (Kabata-Pendias and Sadurski 2004).

Although Fe in soil, under most conditions, is slightly mobile, its organic complexes and chelates are relatively mobile species. These compounds are largely responsible for Fe migration between soil horizons, and for its leaching from soil profiles. Nevertheless, in most soils, there is an observed higher Fe input than output, in surface soil layers.

Iron solubility is low in soils at the alkaline pH range, and its mobility increases with increasing soil acidity. Thus, Fe cations in acid anaerobic soils may be toxic to plants, whereas in alkaline well-aerated soils, it may not be available in enough quantity to plants. Processes of reduction and oxidation of Fe are strongly controlled by microorganisms.

Iron deficiencies in soils, for certain crops are relatively widespread, but most common are in aridic climate zones, and are related to calcareous, alkaline, and other specific soils. However, the assessment of Fe availability to plants is very difficult, due to several factors involved in these processes.

There are several sources of Fe contamination, including industrial and municipal wastes. Iron in solid municipal waste is mainly in forms of carbonates and sulfides (1.4 and 1.9 mg/kg, respectively) and has not been changed after composting (Ciba and Zołotajkin 2001).

23.3 WATERS

The global mobility index (C_w/C_c ratio) of Fe is very low (<0.01). Thus, its contents in surface water are relatively low. Iron concentrations in ocean water are estimated within the range of 25–743 μg/L (Reimann and de Caritat 1998). The most common Fe concentration is given as 10–100 μg/L, and the worldwide mean at 30 μg/L (Table 23.1). In the Baltic Sea, its contents range from 0.3 to 35 μg/L (Szefer 2002). Worldwide riverine flux of Fe in seawater is estimated at 2.47 Mt/yr (Gaillardet et al. 2003), and at 7 Mt/yr (Kitano 1992).

The worldwide average Fe concentration in river water, in dissolved load of <0.2 μm, is estimated at 66 μg/L, within the range of 11–739 μg/L (Gaillardet et al. 2003). The behavior and chemistry of Fe in water systems are very complex and controlled by several parameters, of which the redox potential is the most significant. Iron in stream water is subjected to various photochemical and microbial oxidation reactions. Also, concentration of dissolved organic matter has an important impact on Fe species in water. The predominated form of Fe are colloids, but several hydrous ions of Fe^{3+} and Fe^{2+} may also be common, mainly as species: $Fe(OH)^{2+}$, $Fe(OH)_3^0$ and $Fe(OH)_4^-$. Very stable in water are two forms Fe^{3+} and $Fe(OH)_2^-$. Most probably, however, the solubility of Fe^{3+} in seawater is controlled by organic complexation.

The colloidal Fe oxides play a dominate role in the sorption and coagulation of other colloidal substances and ions. Dissolved Fe compounds readily precipitate in the most aquatic environments, and form various multimetallic concretions in bottom sediments. Bacteriogenic, mainly amorphous, Fe hydrous oxides, together with Mn oxides and organic matter (OM), play especially important role in the fate and transport of dissolved metals in subterranean water systems. Some organic ligands, present in seawater may strongly bind Fe, and prevent its deposition in bottom sediments. Iron is considered to be the most mobile metal among others deposited in bed sediments.

Iron is an essential micronutrient for phytoplankton, as an important component in processes of photosynthesis and respiratory electron transports. In oxic seawater, however, Fe is present predominantly in the insoluble species, and thus it may be deficient for the phytoplankton. Thus, the addition of fine Fe particles may highly generate the growth of plankton biomass.

Iron concentration may be increased in some aquatic environments (e.g., mine drainage or iron pickling wastewater). Oxidation of Fe^{2+} to $Fe_2O_3 \times H_2O$ is applied for the deferrization of water.

Rainwater contains Fe within the range of 10–40 μg/L; the lowest value is for the coastal region of Norway, and the highest for the contaminated region of Kola Peninsula (Reimann and de Caritat 1998).

Median Fe concentration in bottled water of the EU countries is 1.26 μg/L, and is a bit lower than those in tap water, estimated at 3.21 μg/L (Birke et al. 2010). No guideline value for Fe in drinking water is proposed. Reason for not establishing a guideline value is a lack of health concern, at Fe levels found in drinking water (WHO 2011a).

Fe concentration in the wetland sediments from the effluent at the Savannah River site (Aiken, SC) is 18,421 mg/kg, and its highest amounts, 8,089 mg/kg, is fixed in soluble OM fraction. Iron in these sediments occurs also as crystalline oxides, 4652 mg/kg, and amorphous oxides, 3737 mg/kg (Knox et al. 2006).

Iron contents in stream-bottom sediments of National Park, Montgomery (Pennsylvania State) were, in 1995, within the range of 11,000–430,000 mg/kg. The assessment limits for Fe in these sediments are established as follows (in mg/kg): effects range low, 200,000; effects range median, 280,000; probable effect level, 250,000; and threshold effects level, 190,000.

23.4 AIR

The origin of Fe in air is from both terrestrial and industrial sources. Concentrations of Fe in the atmosphere of different cities, from various continents, range broadly from 130 to 14,000 ng/m^3, and are closely associated with industrial activities (Table 23.1). It is extremely elevated in air of various regions, when compared to its contents in air of remote regions, especially Antarctica.

Iron content in moss growing in Norway vary from 120 to 21,000 mg/kg (average 660 mg/kg), and resulted from atmospheric deposition (Berg and Steinnes 1997).

Aerial dust of urban regions is composed of 33%–38% (weight) of Fe particles, for which annual deposition is estimated at 16,800–43,200 g/ha/yr. Iron particles deposition in remote areas of Europe range from 300 to 5700 g/ha/yr (Manecki et al. *vide* Kabata-Pendias and Pendias 1999).

23.5 PLANTS

Iron mobility in soils is the main factor that governs its phytoavailability. Main features of the Fe uptake by plants, and Fe transport between plant organs may be summarized as follows:

- Various Fe species are taken up, mainly as Fe^{2+}, but also Fe^{3+} and Fe chelates.
- Plant roots may reduce Fe^{3+} to Fe^{2+}, which is the fundamental process in the Fe absorption by most plants.
- At Fe deficiency conditions, roots (especially of cereals and other Gramineae) release mugineic acid, which mobilize Fe species.
- In xylem exudates, Fe occurs mainly in unchelated forms.
- Fe transportation within plants is mediated largely by citrate chelates and by soluble ferritins (transferins).
- Pine trees, growing in most sites, uptake less Fe by root system than from aerial deposition.

At conditions of the Fe deficiency, roots of some plants can develop various mechanisms to enhance its availability through the reduction of Fe^{3+}, and/or chelation in phytosiderophore forms, that are efficient in mobilizing Fe (Marschner 2005). On root surface of some plants grown in waterlogged soils (e.g., rice) is formed Fe plaque, due to the oxidation of Fe^{2+} to Fe^{3+}, and the precipitation of Fe oxides on the root surface. This plaque is composed mainly of ferrihydrite and goethite, and highly inhibits the Fe uptake (Zhu et al. 2005). Easily available Fe species are its complexes with some humate forms.

Both Fe uptake and transportation within plant organs are strongly affected by several soil and plants factors. In general, a high degree of the oxidation of Fe compounds, Fe precipitation on carbonates and/or phosphates, and competition of trace metal actions with Fe^{2+} for the same binding sites of chelating compounds are responsible for a low Fe uptake and for the disturbance of its transport within plant organs.

Iron is considered to be the key metal in the energy transformation needed for several live processes of plant cells. Its essential roles in the plant metabolism, presented in oversimplified and generalized forms, are as follows:

- Several Fe proteins, mainly transferrins, ferritins, and siderophores, are involved in transport, storage, and binder systems.
- Fe occurs in heme and nonheme proteins and is concentrated in chloroplasts.
- Fe highly influences the formation of chlorophyll.
- Organic Fe complexes are involved in the electron transfer.
- Fe is directly implicated in the metabolism of nucleic acids.
- Both cations, Fe^{2+} and Fe^{3+}, play a catalytic role in various metabolic processes.

Iron deficiency affects several physiological processes, and therefore retards plant growth and yield. Its deficiency in several crops is now a major worldwide problem.

Fe deficiency may occur at very different levels in plants, and is dependent on various factors, such as soils, plants, and climates. The common initial sign of its deficiency is interveinal chlorosis of young leaves. Several fruit trees and cereals, oats and rice in particular, are very susceptible to Fe chlorosis. The control of the Fe deficiency is often not sufficient. There are some suggestions that this problem may be partly resolved by plant breeding and selection of genotypes of cultivated plants.

Interactions between Fe and other metals are described as the most common factors affecting its deficiency:

- Excess of Mn, Ni, and Co decreases Fe mobility in soils and plants.
- Fe–Zn interaction leads to the formation of franklinite, $ZnFe_2O_4$, which depresses the availability of both metals.
- Fe–P interaction, in both soils and plants, causes the precipitation of $FePO_4{\cdot}2H_2O$. The appropriate P/Fe ratio is fundamental to plan Fe mobility, and thus availability.
- Fe–S interaction causes a low Fe mobility and availability.
- Fe–Se interaction is associated with the immobilization of Se by Fe minerals and compounds.
- Fe interaction with several trace metals causes chlorosis due to the immobilization of Fe.

Excess iron is toxic to most plants. However, some plants are resistant to this, which is explained as follows:

- Plant resistance to excess Fe is associated with different reactions, such as oxidation, immobilization, and exclusion of mobile Fe species.
- Plants high in nutrients, especially in Ca and Si, can tolerate high levels of Fe.
- Roots of some plants are able to oxidize Fe and deposit it on the root surface.
- Mycorrhizas have a great capability to bind Fe at root surface or in root cells.
- Plants adapted to waterlogged conditions are more tolerant to high Fe levels than plants grown in well-aerated soils.

Appropriate Fe content in plants is essential for plant growth and for nutrient supply to humans and animals. Natural Fe contents of fodder plants range from 18 to about 1000 mg/kg. The nutritional requirement of grazing animals is usually met at Fe concentration within the range of 50–100 mg/kg.

The common range of Fe in cereal grains is from 30 to about 100 mg/kg. Higher Fe contents are very seldom cited. Usually higher Fe concentrations are in barley and oats grains (up to 200 mg/kg), than in wheat and rye grains (up to 50 mg/kg). Edible parts of vegetable contain Fe within mean values from 33 to 65 mg/kg, in carrot and soybean seeds, respectively. Some nuts and almonds may have higher levels of Fe, up to 67 mg/kg (Jędrzejczak *vide* Kabata-Pendias and Mukherjee 2007).

The common mushroom, chanterelles (*Cantharellus cibarius*) grown in the mountains of Poland, contains higher amounts of Fe, mean 180 mg/kg, than those grown in the Baltic Sea coast, mean 43 mg/kg (Falandysz et al. 2012).

23.6 HUMANS

Total Fe in human body is 4 g, of which in the blood, the average is 415 mg/L (range 380–450); in the bone, it varies between 3 and 380 mg/kg; and in the tissue, the average is 180 mg/kg (range 20–400) (Emsley 2011). The largest Fe fraction is present in the erythrocytes as hemoglobin, myoglobin, and heme-containing enzymes. Significant Fe amounts are as metalloprotein ferritin and hemosiderin, mainly in the spleen, liver, bone marrow, and striate muscle.

Iron has several vital functions in the body. It serves as a carrier of oxygen to the tissues, from the lungs by red blood cell hemoglobin. It is also an integrated part of important enzyme systems in various tissues. Fe-containing oxygen storage protein in the muscles, myoglobin, is similar in structure to hemoglobin, but has only one heme unit and one globin chain. Several Fe-containing enzymes, cytochromes, also have one heme group and one globin protein chain. Their role in the oxidative metabolism is to transfer energy within the cell, and specifically in the mitochondria. Other key functions for the Fe-containing enzymes (e.g., cytochrome P450) include the synthesis of steroid hormones and bile acids; detoxification of foreign substances in the liver; and signal controlling in some neurotransmitters, such as the dopamine and serotonin systems in the brain. Iron is reversibly stored within the liver as ferritin and hemosiderin, whereas it is transported between different compartments in the body by the protein transferrin (Hallberg 1981).

There are two kinds of dietary Fe: heme Fe and nonheme Fe. Heme Fe^{2+} present in hemoglobin and myoglobin is well absorbed and is relatively unaffected by diet composition. Nonheme Fe, the form of Fe^{3+} present in vegetables and in human's staples, generally is poorly absorbed, and is greatly affected by enhancing or inhibiting substances in the diet. The primary sources of heme Fe^{2+} are the hemoglobin and myoglobin from consumption of meat, poultry, and fish, whereas nonheme Fe^{3+} is obtained from cereals, pulses, legumes, fruits, and vegetables. Average absorption of heme Fe from meat-containing meals is about 25%. Absorption of heme Fe can vary from about 40% during Fe deficiency to about 10% during Fe repletion (Hallberg et al. 1997).

Heme Fe can be degraded and converted to nonheme Fe if foods are cooked at a high temperature for too long. Calcium is the only dietary factor that negatively influences the absorption of heme Fe and nonheme Fe.

Iron compounds used for the fortification of foods will only be partially available for absorption. Reducing substances must be present for Fe to be absorbed. Presence of meat, poultry, fish, and ascorbic acid (e.g., certain fruit juices, fruits, and certain vegetables) in the diet enhance Fe absorption. Other foods contain chemical entities (ligands) that strongly bind ferrous ions, and thus inhibit absorption. Examples are phytates present in cereal grains, seeds, nuts, vegetables, roots, and fruits, as well as certain Fe-binding polyphenols (e.g., tea, coffee, cocoa, certain spices, certain vegetables, and most red wines) and Ca (e.g., from milk and cheese).

Iron deficiency is the most common nutritional problem leading to anemia. Its deficiency ranges from depleted Fe stores without functional or health impairment to Fe deficiency with anemia, which affects the functioning of several organ systems. Its deficiency can delay normal infant motor function (normal activity and movement) or mental function (normal thinking and processing skills). Iron-deficiency anemia during pregnancy can increase risk for small or early (preterm) babies. Iron deficiency can cause fatigue that impairs the ability to do physical work in adults and may also affect memory or other mental function in teens.

Recommended dietary allowances for Fe depend on age, sex, and conditions; for example, Fe (mg/day) intakes for males and females aged between 25 and 50 are as follows: 10 (male) and 15 (female) (WHO 2004). As a precaution against storage in the body of excessive Fe, in 1983, Joint FAO/WHO Expert Committee on Food Additives established a provisional maximum tolerable daily intake of 0.8 mg/kg bw (WHO 2011a).

Iron overload may also happen: it is the accumulation of excess Fe in body tissues. Excess of pharmaceutical Fe may cause toxicity and therapeutic doses may cause gastrointestinal side effects. Chronic Fe excess, for example, in primary and secondary hemochromatosis, may lead to hepatic fibrosis, diabetes mellitus, and cardiac failure. Iron stores may also be related to cardiovascular risk. Fe-amplified oxidative stress may also increase DNA damage, oxidative activation of precancerogens, and support tumor cell growth. Due to these mechanisms, high Fe stores may present a health hazard (Schümann 2001). Iron content in foodstuffs is variable, ranging from <1 to about 70 mg/kg FW (Table 23.2).

TABLE 23.2
Iron Content in Foodstuff (mg/kg FW)

Product	Mean/Range
Food of Plant Origin	
Bread[a]	21
Miscellaneous cereals[a]	32
Rice, various species[b]	4.7
Green vegetables[a]	11
Carrot[c]	3.5 (1.3–12.6)
Other vegetables[a]	7.5
Potatoes[c]	3.5 (1.8–6.9)
Fruits[a]	3.2
Strawberries[c]	3.9 (1.2–15.0)
Nuts[a]	34
Curry, various[d]	24–135
Sugar and preserves[a]	9.4
Food of Animal Origin	
Carcass meat[a]	21
Offal[a]	69
Fish[a]	16
Freshwater fish[e]	0.9–3.2
Eggs[a]	20

(Continued)

TABLE 23.2
(Continued) Iron Content in Foodstuff (mg/kg FW)

Product	Mean/Range
Milk[a]	4.1
Milk[f]	0.8
Dairy products[a]	12

Sources: [a] Ysart, G. et al., *Food Addit.Contam.*, 16, 391–403, 1999.
[b] Jorhem, L. et al., *Food Addit. Contam.*, 25, 841–850, 2008b.
[c] Szteke, B. et al., *Roczn. PZH*, 55S, 21–27, 2004.
[d] Gonzalvez, A. et al., *Food Addit. Contam.*, 1, 114–121, 2008.
[e] Lidwin-Kaźmierkiewicz, M. et al., *Pol. J. Food Nutr. Sci.*, 59, 219–234, 2009.
[f] Gabryszuk et al. (2010).
FW, fresh weight.

The mean daily intake of Fe in the United Kingdom is 15 mg (Ysart et al. 1999). In Poland, during 1996–2004, the average Fe intake in women and men was 12 and 15 mg/day, respectively (Marzec et al. 2004).

23.7 ANIMALS

Iron is essential and plays an important role in animals. Animals have much higher Fe intakes than humans and do not absorb its compounds in the same way. More than 50% of the Fe in the mammalian body is present in hemoglobin, with smaller amounts present in other Fe-requiring proteins and enzymes, and in protein-bound stored Fe. The immune status of the organism and its resistance against infections depends on the Fe supply. Iron deficiency in animals is very often caused by a reduced absorption in the intestinal tract, because of components in the feed-forming complexes with Fe of very low solubility, or inhibitors reducing the absorption processes. Iron deficiency inhibits the myeloperoxidase activity and thus decreases the bacteriocide effect of the leucocytes, and may lead to anemia, defect of development, and susceptibility to diseases. In spite of this, when exposed to infections, the physiological mechanisms reduce the blood concentration of available Fe.

Fe requirement is approximately 50 mg/kg diet in beef cattle. Young calves are fed with milk diets and need 40–50 mg Fe/kg, for growth and prevention of anemia. Requirements of older cattle are probably lower than in young calves.

Iron toxicity causes diarrhea, metabolic acidosis, hypothermia, and reduced gain and feed intake. Dietary Fe concentrations as low as 250–500 mg/kg have caused Cu depletion in cattle. In areas where drinking water or forages are high in Fe, dietary

TABLE 23.3
Iron in Animal Tissues (mg/kg FW)

	Tissues		
Animal	**Muscle**	**Liver**	**Kidney**
Pigs	1–26 (6)[a]	4–144 (55)	7–218 (45)
Cattle	4–27 (17)	12–98 (38)	11–306 (45)
Wild game	5–70 (30)	4–170 (48)	52–200 (68)
Chicken	0.5–56 (4)	2–101 (52)	–
Fish	2–14 (6)	–	–

Source: Szkoda, J. et al., *Roczn. PZH*, 55S, 61–66, 2004.
[a] In parentheses are mean values.
FW, fresh weight.

Cu may need to be increased to prevent Cu deficiency. The maximum tolerable concentration of Fe for cattle and poultry diet has been estimated at 1000 mg Fe/kg, for pigs at 3000 mg/kg (Szkoda et al. 2004).

Concentrations of Fe in the animals depend on the kind of tissue and the specimen (Table 23.3).

24 Lead [Pb, 82]

24.1 INTRODUCTION

Lead (Pb), a metal of the group 14 in the periodic table of elements, occurs in the Earth's crust at the mean content of 15 mg/kg. It is concentrated in acidic magmatic rocks (up to 25 mg/kg, in granites) and in argillaceous sediments (up to 40 mg/kg). In sandstones and limestones, its contents do not exceed 10 mg/kg. Lead contents of coal vary within the broad range from 10 to 1900 mg/kg, and may be concentrated in fly ash up to about 50 mg/kg.

In the terrestrial environment, two kinds of Pb are present: primary and secondary. Primary Pb is of geogenic origin and is present in various minerals, whereas secondary Pb is of radiogenic origin from the decay of U and Th. The ratio of Pb isotopes, especially ^{210}Pb, with a half-life of 22.3 years, is used for dating the host materials, and for the identification of its anthropogenic pollution sources.

Lead occurs mainly at +2 oxidation state, and sometimes +4. It has high chalcophilic properties, and thus its most common, primary mineral is galena, PbS. Other common minerals are anglesite, $PbSO_4$; minium, Pb_3O_4; pyromorphite, $Pb_3(PO_4)_3Cl$; and minetesite, $Pb_5(AsO_4)_3Cl$. Goethite reveals a strong affinity for the Pb sorption, which is controlled by organic acids (Perelomov et al. 2011).

World mine Pb production (rounded) in 2010 was 4100 kt, of which China produced 1750; Australia produced 620; the United States produced 400; Peru produced 280; Bolivia produced 90; and Russia produced 90 (USGS 2011). The largest worldwide use of Pb is for lead–acid batteries. It is also utilized for buildings constructions, in alloys (pewters, Sn–Pb and fusibles), for solders, and anticorrosion paint (*red lead*, Pb_3O_4).

Lead content in crude oil is low, at about 4 mg/kg. For many years (since the beginning of the 1920s to the mid-1970s, in the United States), Pb was added to petrol, at amounts around 650 mg/kg, as an antiknock element, in the form of tetraethyllead. It is still used as an additive in some gasoline for some aviation and automotive racing. The major sources of Pb emissions have historically been from fuels in motor vehicles (mainly cars and trucks), and industrial sources. Although the use of tetraethyl and tetramethyl Pb has been forbidden in the most countries, there are still some markets for these gasoline additives, especially in the aviation gasoline.

Lead environmental pollution is presently of a great concern. Data from the ice cores from Greenland indicate that the large-scale Pb pollution started about 2500 years ago, from the smelting and mining of Pb and Ag, during the Greek and Roman eras. Lead concentrations in the ice cores decreased after the decline of the Roman Empire, and did not surpass Roman levels, until the start of the Industrial Revolution. In the twentieth century, Pb concentrations increase to 200 times its levels, of the pre-Greek and Roman times. Since the passing of the Clean Air Act in 1970 (U.S. EPA), Pb concentrations have begun to fall. However, there are also data that Pb amounts in the glacier of Greenland has increased during the last 100 years, from 0.05 to 0.5 µg/L (Buat-Menard *vide* Kabata-Pendias 2011).

24.2 SOILS

The worldwide mean Pb contents in soils is estimated at 27 mg/kg. Lead background average contents given for soils of different countries vary (in mg/kg) from 18 (Sweden) to 27 (China). Mean Pb content in organic surface soils of Norway is 43.6 mg/kg, and is a little bit lower than mean value (46.4 mg/kg) for all investigated soils (Nygard et al. 2012). Soils in Spitsbergen, Norway, contain Pb within the range of 0.14–9.33 mg/kg (Gulińska et al. 2003), whereas its contents of soils in Alaska range from 4 to 349 mg/kg (Gough et al. 1988).

Pb distribution in mineral soils shows a positive correlation with fine granulometric fraction. Its contents in various soil groups vary within the range of 3–90 mg/kg, being the highest in cambisols and histosols, and the lowest in arenosols (Table 24.1). Due to widespread Pb pollution, most soils are likely to be enriched in this metal, especially the top horizon.

TABLE 24.1
Lead Contents in Soils, Water, and Air

Environmental Compartment	Range
Soil (mg/kg)	
Light sandy	5–40
Medium loamy	10–50
Heavy loamy	10–90
Calcerous (calcisols)	15–65
Organic (histosols)	2–80
Water (µg/L)	
Rain	10–560
River	0.04–4.8
Sea, ocean	<0.03–3.8
Air (ng/m³)	
Urban/industrial regions	120–13,000
Remote regions	10–25
Antarctica	0.2–1.2

Sources: Data are given for uncontaminated environments, from Kabata-Pendias, A. and Mukherjee, A.B., *Trace Elements from Soil to Human*, Springer, Berlin, Germany, 2007; Reimann, C. and de Caritat, P., *Chemical Elements in the Environment*, Springer, Berlin, Germany, 1998.

During weathering, Pb sulfides are slowly oxidized, and in soils are fixed by clay minerals, hydroxides, and soluble organic matter (SOM), which increased with increasing pH (Mboringong et al. 2013). Lead reveals a strong affinity for SOM, and the formation of inner-sphere metal complexes (Basta et al. 2005). Lead adsorption on Al–Fe and Mn oxides may be increased due to the presence of sulfate and phosphate compounds (Violante 2013). The geochemical characteristics of Pb^{2+} resemble the divalent alkaline-earth group elements, thus it has an ability to replace K, Ba, Ca, and Sr, in both minerals and sorption sites (Kothe and Ajit 2012).

Pb distribution in soil profiles is not uniform and reveals a great association with hydroxides, especially of Fe and Mn. Its concentration in Fe–Mn nodules may be very high, up to 20,000 mg/kg. It may also be accumulated in calcium carbonate, phosphate, and some feldspar particles, especially at pH > 6. The formation of hydrocerussite, $Pb_3(CO_3)_2(OH)_2$, and pyromorphite, $Pb_5Cl(PO_4)_3$, in soils contaminated with Pb, has also been observed (Kabata-Pendias 2011).

Usually Pb is accumulated in surface-soil horizons, mainly due to its sorption by SOM. However, in some soils, mainly podzolic, it may be concentrated in deeper soil layers. The Pb fixation by clay minerals also plays a significant role in its distribution. Lead species in soils are slightly mobile, but some Pb–OM complexes and acidity increase its mobility.

Lead concentration in soil solution is relatively low, within the range of <1–60 μg/L, and is highly depending on methods used for obtaining of soil solution. It occurs in soil solutions as cationic species such as Pb^{2+}, $PbCl^{+}$, and $PbOH^{+}$, and as anionic species such as $PbCl_3$ and $Pb(CO_3)_2^{2-}$ (Kabata-Pendias and Sadurski 2004). Due to microbial activities, Pb adsorbed at the surface of Fe minerals may be dissolved in the soil solution (Perelomov and Kandeler 2006). According to Pampura et al. (2007), the predictability of free metal ion values for Pb in soil solution is high and may be useful for the critical load calculation.

The background Pb content of soils is inherited from parent rocks. However, due to the widespread Pb pollution, most soils, especially top horizons, are likely to be enriched in this metal. Sometimes, it is difficult to separate the data for background Pb levels in soils from those of anthropogenic sources. Low Pb levels in soils from remote regions suggest that the baseline value of this metal in the most of worldwide soils should not be much higher than lithogenic concentration, estimated at 20 mg/kg. However, its higher levels are noticed in all soils, also from uncontaminated regions (Table 24.1). Very often Pb level is elevated in rhizospheric soils (up to above 10,000 mg/kg), which resulted in its higher contents of some plants. Some microorganisms (*Rhizopus arrhizus*) accumulate 33 mg Pb/g, within the broad range of pH 3–7 (Perelomov et al. 2013).

Lead pools in mountain forest soils of the National Park in Poland range between 0.16 and 15.6 mg/kg, and is concentrated in forest litter, often over 100 mg/kg. Its significantly higher levels are in the lower altitudinal zones as compared to the higher zones (Szopka et al. 2013). Lead is likely to be concentrated in the upper soil layer, and in all soil layers, lead predominates its residual fraction over other species (Makuch 2012).

Lead contamination of soils from mining and industrial activities is an old problem, and began when our ancestors learned to use fire. The estimation, using Pb isotopes, informed that between 40% and 100% of the total Pb in contaminated soils in France come from the Medieval workshops (Baron et al. 2006).

The steadily increased amounts of Pb in surface soils, both arable and uncultivated, has been of a real health problem, as it is of a great hazard to man and animals from both, the food chain and dust inhalation. In Pb mining areas, it is dispersed due to the erosion and weathering of tailing materials. Most reports are on elevated Pb concentrations in vicinities of metal-processing industries (Table 24.2). However, its highest content (135,000 mg/kg) is reported to be in soils around battery industries. Increased Pb levels in top soils are also resulted from Pb shots, in hunting areas and in military-shooting sites. Robinson et al. (2008) calculated that over 400 t of Pb enters annually Swiss soils of such military areas. There are estimations that about 2 billion lead pellets are deposited into the environment worldwide every year (hunting and recreational shooting). It will take between 100 and 300 years to degrade

TABLE 24.2
Lead Contamination of Surface Soils (mg/kg)

Site and Pollution Source	Country	Maximum Content
Old mining area	United Kingdom	21,546
Metal-processing industries	Belgium	14,000
	Germany	3,074
	Greece	18,500
	Holland	1,334
	Japan	2,100
	Poland	8,000
	Russia	3,645
	United States	6,500
	Zambia	2,580
Battery manufactory	Chile	1,058
	Poland	3,800
	United States	135,000
Urban gardens and vicinities	Canada	888
	Jamaica	897
	United Kingdom	15,240
	United States	10,900
Sludged farmlands	Germany	>800
	Holland	253
	United Kingdom	3,916
Roadside soils	Germany	885
	Japan	397
	Poland	310
	United States	7,000

Source: After Kabata-Pendias, A., *Trace Elements in Soils and Plants*, 4th ed., CRC Press, Boca Raton, FL, 2011.

these pellets, depending on soil conditions and the climate. Lead particles become more rapidly degraded in acid soils or water, at a high concentration of dissolved oxygen (Mezhibor et al. 2011).

Peat bogs are good records of atmospheric Pb deposition due to a great capacity to accumulate and strong fixation of this metal. The anthropogenic Pb deposition, back at least to Greek and Roman times, has been traced in peat cores of several European countries. Lead concentrations in the present-time peat bogs is about 100 times higher than in peat layer of preanthropogenic times. Accumulation of Pb in peat bogs and soils are regarded as sink of deposited atmospheric Pb, which may be its significant source to the fluvial systems, due to peat erosion processes (Rothwell et al. *vide* Kabata-Pendias 2011).

The main Pb pollutants emitted from industries often occur in mineral forms: PbS, PbO, $PbSO_4$, and $PbO{\cdot}PbSO_4$. In automobile exhausts, it is mainly in the forms of halite salts: PbBr, PbBrCl, Pb(OH)Br, and $(PbO)_2PbBr_2$. These compounds in soils may be easily converted into other compounds and fixed by soil particles. Some Pb species may be methylated, due to both biological and chemical processes. The Pb contamination of roadside soils was, for quite a time, of a real environmental concern. Although Pb additives in petrol are banned in most countries, its increased levels (up to about 7000 mg/kg) in soils along high traffic roads have been still observed. Tetraethyl and tetramethyl Pb present in soils along highways are converted into water-soluble compounds; nevertheless, they are not significantly leached down soil profiles (Ou et al. *vide* Kabata-Pendias 2011). Soils within the city of Łódź (Poland) contain Pb up to 60 mg/kg, as effects of contaminations from the power plant and motor traffic (Jankiewicz and Adamczyk 2007).

Other important source of Pb are paints, commonly used in home and garden areas. The most common forms of added Pb are $PbCrO_4$ and $PbCO_3$. Lead is added to paints to speed up drying, increase durability, maintain a fresh appearance, and resist moisture that causes corrosion. Soils around pylons painted with *red lead* before 1970 contain still elevated Pb levels, up to 783 mg/kg (Brokbartold et al. 2012).

In some countries (e.g., the United States and the United Kingdom) there are regulations, since 1978, prohibiting the use of Pb paints; however, these paints may still be found in older properties painted prior to this regulation, and are still used in road markings. The content of Pb is higher in soils around old houses, than in new ones. The geometric mean Pb in 4650 garden soils of the United Kingdom is calculated as 298 mg/kg (Thornton *vide* Kabata-Pendias 2011). Elevated Pb contents of garden soils are of a special risk due to *pica soil* (hand-to-mouth) behavior of children. The limit for Pb in soils of play areas is estimated at 400 mg/kg, and in surrounding sites, up to 1200 mg/kg. However, Pb levels above 1000 mg/kg are hazardous to cattle. Soil-quality Pb levels are estimated, based on various criteria, within the range of 60–1000 mg/kg.

The criteria for contaminated land (Dutch List 2013), following Pb concentrations in soils and groundwater, are established (in mg/kg and μg/L) as follows: uncontaminated, 50 and 20; medium contaminated, 150 and 50; and heavily contaminated, 600 and 200. Groundwater in the United States contains Pb within the range of 5–30 μg/L, and in the European parts of Russia its concentration is within 0.1–8 μg/L (Kabata-Pendias and Mukherjee 2007).

The observations of Pb balance in soils of various ecosystems show that its input greatly exceeds its output. Lead added to agricultural soils with various materials vary from <3 to 96 g/ha/yr, from lime and sewage sludge, respectively. According to Eckel et al. (2005), agricultural Pb input to soils varies from 1.3 to 139 g/ha/yr, in Norway and France, respectively.

Lead, as pollutant, enters soils in various and complex compounds, thus its reactions with soil components highly differ. Due to low mobility and phytoavailability of Pb in soils, there was little concern about soil pollution previously. However, soon it was observed that Pb is easily taken up by roots, and may also be translocated to shoots. Pb limitation of enzymatic activities of soil microbiota has also been reported. Pb accumulation in surface soils has become a great ecological significance.

Remediation of Pb-contaminated soil was recently broadly investigated. Several methods have been applied. Although Pb is slightly phytoavailable, there are some hyperaccumulators (e.g., corn, sunflowers, and *Thlaspi* spp.) that are used for the phytoextraction and/or phytostabilization. The phytoremediation is more effective after the addition of chelators (ethylenediaminetetraacetic acid [EDTA], diethylenetriaminepentaacetic acid), which increase Pb mobility. Some plants may stabilize Pb, for a certain period, by its fixation in rhizosphere zones.

Other methods are based on the Pb stabilization, for which lime (especially Fe-rich lime) and P fertilizers, as well as some minerals (e.g. barite, $BaSO_4$) may be applied (Courtin-Nomade et al. *vide* Kabata-Pendias 2011). Addition of a brown-coal preparation (composed of coal, fly ash, and peat) decreased Pb phytoavailability (Kwiatkowska-Malina and Maciejewska 2013).

24.3 WATERS

The worldwide median concentration of Pb in ocean water is calculated at 0.03 μg/L, within the range of <0.03–0.27 μg/L (Reimann and de Caritat 1998). In the North Pacific Ocean, it is lower, estimated at 0.003 μg/L (Nozaki 2005). The Baltic Sea contains Pb at the range from 0.05 to 3.6 μg/L, with the highest being in coastal water (Szefer 2002). The average worldwide Pb concentration in river water is estimated at 0.08 μg/L, within the range of 0.04–3.8 μg/L (Table 24.1). River water in Pb-mineralized areas and near to Pb-polluted sources may contain this metal up to about 45 μg/L (Monbesshora et al. *vide* Kabata-Pendias 2011).

The annual global input of Pb to marine basins with river water and atmospheric precipitations is estimated at 40 and 20 kt, respectively (Kitano *vide* Kabata-Pendias and Mukherjee 2007). Annual Pb load to the Baltic Sea in 1990 was estimated at 2800 t, in about equal proportion from river flux and atmospheric deposition (Matschullat 1997). Above 90% of Pb discharged to this sea basin is of the anthropogenic origin.

Lead concentration in rainwater should be very low, at about 10 μg/L. However, rainwater is very often polluted, by roofs and tanks, and therefore rainwater collected in towns may contain Pb up to >500 μg/L

The ratio of Pb isotopes in the northeast Pacific Ocean was used for the identification of pollution sources; the surface layer of water contained Pb of the industrial dust from Japan, whereas deeper layers were polluted by the North American emissions (Flegal et al. *vide* Kabata-Pendias 2011).

Lead behavior in water depends on the pH and dissolved salts of other elements and organic matter. In seawater, its species are mainly $PbCO_3$ and $PbCl_2$. Surface water and groundwater are predominated by Pb^{2+}, $PbOH^+$, $PbHCO^{3+}$, and $PbSO_4$. However, all mobile Pb species will eventually precipitate in the bottom sediments, especially in neutral or alkaline water. Lead entered into sediments may become again resuspended, and enters as solid particles into the water column.

Lead contents in stream-bottom sediments of National Park, Montgomery (Pennsylvania State) were, in 1995, within the range of 30–60 mg/kg (Reif and Sloto 1997). Sediments of Sergipe River estuary (Brazil) contain Pb from 8.14 to 31.1 mg/kg (Garcia et al. 2011). The Pb concentration in the wetland sediments from the effluent at the Savannah River site (Aiken, SC) is 59 mg/kg; its highest amounts, 18.4 mg/kg, are amorphous oxide species, and 16.7 mg/kg are fixed in SOM fraction (Knox et al. 2006).

Bottom sediments are considered as the long-term sink for Pb, in which it is stored in unavailable forms. Thus, the best indicator of the Pb pollution in river water is its content in bottom sediments, where its background levels are estimated at 30–45 mg/kg. The highest concentrations of Pb in polluted rivers, from industrial and mine areas, range between 700 and 2600 mg/kg, in Poland and the United Kingdom, respectively (Kabata-Pendias 2011). Lead content of surface-bottom sediments of harbor in Klaipeda (Lithuania) also depends on granulometric composition, and is (in mg/kg, average and maximum, respectively) as follows: in sand 10.9 and 25.1 and in mud 7.6 and 32.8 (Galkus et al. 2012).

The assessment limits for Pb in sediments are established as follows (in mg/kg): effects range low (ERL), 47; effects range median, 220; probable effect level (PEL), 91.3; 250; and 179 (EPA 2000, 2013). The Environment Canada sediment-quality guidelines (USGS 2001) gave a bit similar values for Pb in lake-bottom sediments (in mg/kg): threshold effects level, 35; PEL, 91.3; and probable effect concentration, 128.

Lead concentrations in drinking water, where it occurs in the forms of $Pb(OH)_2$, $PbCO_3$, and Pb_2O, have been of a great concern. Several products are used for its removal, especially from tap water, where most Pb is the result of corrosion in the water distribution and home-plumbing system. The volcanic rock from Northern Rwanda is used for the Pb removal from polluted water in some countries, as it reveals a great adsorption capacity for Pb, 9.52 mg/g (Sekomo et al. 2012).

Median Pb concentration in bottled water of the EU countries is 0.023 μg/L, and is much lower than those in tap water, estimated at 0.118 μg/L (Birke et al. 2010). The maximum Pb level in drinking water in the United States is 15 μg/L; above this level, further treatment of tap water is necessary (EPA 1999). Provisional guideline value for drinking water established by WHO is 10 μg/L (WHO 2011a). In most countries, tap and well drinking water contains Pb within the levels of 2–3 μg/L.

Lead is easily taken up from water by phyto and zooplanktons, and may occur in concentrations up to 3000 mg/kg in bryophytes. It is also likely to be accumulated in invertebrates. Its concentration above 0.19 and 0.30 mg/L is harmful to *Daphnia magna* and *Cyclop* sp., respectively (Offem and Ayotunde 2008). Excess lead is toxic to fish and water birds. Some bird populations (condors in particular) are at risk of Pb poisoning. In fish, it may damage the respiratory system due to the coating of gills and may destroy antibacterial skin mucous layer. Recently, it has been observed that Pb concentration in fish of some sea basins is declining. Lead contents in muscles of herring (*Clupea harengus*) from the Baltic Sea vary broadly from 0.01 to 0.03 mg/kg FW, and in shells of mollusks (*Mytilus edulis*) range between <0.5 and 5.3 mg/kg FW (Szefer 2002).

24.4 AIR

Concentration of Pb in air varies highly, from about 1 ng/m^3 in the Antarctic regions to above 10,000 ng/m^3 in urban/industrial areas (Table 24.1). The common range of natural Pb levels in air is calculated to be within the range of 0.5–10 ng/m^3, and of polluted air, in the range of 70–8000 ng/m^3. Pb concentration in air acceptable by the U.S. EPA (2008) is 150 ng/m^3.

Lead enters into the atmosphere from both natural and anthropogenic sources, which include the combustion of coal, oil, and gasoline; incineration of wastes; roasting and smelting processes; and cement production. On a global basis, anthropogenic inputs of Pb, in the last century, predominated (at about 96%) its natural sources and was estimated at about 334 kt/yr (Nriagu and Pacyna 1988). During 1955–1995, Pb emission in European countries decreased from 62.53 to 28.39 kt/yr (Pacyna and Pacyna 2001). As a result of the regulatory efforts of the EPA to remove Pb from motor vehicle gasoline, emissions of Pb from the transportation sector declined by 95% between 1980 and 1999, and its levels in the air decreased by 94% between 1980 and 1999. There are prognoses that the Pb anthropogenic emission will continue decreasing.

The chemical species of Pb emitted into air depend on its sources. $PbCl_2$, PbO, PbS, $PbSiO_3$, $PbSO_4$, $PbCO_3$, and Pb are released from coal combustion and industry, whereas PbO is mainly released from oil combustion. The size of Pb particles depend on its sources; the smallest, in the range of 0.1–1.0 μm, are from automobile exhaust. Lead particles/aerosols in the atmosphere are finely deposited in terrestrial and aquatic systems.

Lichens and mosses are very sensitive indicators of Pb aerial pollution. It has been observed that there is a continuous decrease of Pb concentration during the recent decades. Average Pb contents of moss sampled in Norway during 1990–1995 ranged between 0.88 and 59 mg/kg (Berg and Steinnes 1997). Mosses sampled earlier in Sweden during 1969–1985 contained more Pb, from 20 to 50 mg/kg. Mean Pb contents of two moss species from mountains in Poland, sampled in 2000 were higher (9.5–10.1 mg/kg) than those from Alaska (0.63 mg/kg). This suggests that the atmospheric Pb input is higher in Poland than in Alaska (Migaszewski et al. 2009).

24.5 PLANTS

Lead occurs in all plants, but it has not yet been shown to play any essential roles in their metabolic processes. Although even a very low Pb content may inhibit some vital plant processes, its poisoning has seldom been observed in plants growing in field conditions. However, Pb in plants received much attention, as a major metallic pollutant in the environment, very toxic to plants, animals, and humans.

Plants may uptake Pb from water, soils, and air, and its contents in plants is highly influenced by environmental factors. Lead is slightly available to plants, and strongly governed its species in soils and plant factors. Several soil properties, such as SOM, cation exchangeable capacity (CEC), pH, and granulometric composition, as well as root surface areas and root exudates play a significant role in the Pb phytoavailability. Lead uptake by plants is passive, mainly by root hair, and it is stored to a considerable degree in root cell walls. Nevertheless, Pb contents of plants highly correlate with its levels in soils, and thus its plant content is very useful for geochemical prospecting (Kovalevsky *vide* Kabata-Pendias 2011). A variable Pb absorption from its atmospheric deposition by plants was observed, which was three times higher in spruce forest (90 mg/kg) than in pasture grass (30 mg/kg).

Although there is no evidence that Pb is essential to plants, there are some reports on the stimulating effects on plant growth of some Pb salts, mainly $Pb(NO_3)_2$. Nevertheless, some plant species, ecotypes, and bacterial strains are able to develop Pb-tolerance mechanism. This tolerance seems to be associated with the properties of membranes. Lead may be strongly bound to cell walls, mainly due to the sorption by pectic acid (Lane et al. *vide* Kabata-Pendias 2011). Also, the galacturonic acid reveals an affinity to fix Pb (Polec-Pawlak et al. 2007). Lead has a marked influence on the elasticity and plasticity of cell walls, resulting in their increased rigidity.

Plants can relatively easily uptake Pb from soil and nutrient solutions. Large amounts of Pb are concentrated in the roots, but some amounts, at about 3% of Pb root content, are transported to the above-ground biomass. Pb uptake by roots is mainly intracellular and it may be aggregated in vacuoles (Meyers et al. 2008). The main process responsible for Pb accumulation in root tissues is its deposition along cell walls, as Pb pyrophosphate and other Pb precipitates. Similar Pb deposits observed in roots, stems, and leaves suggest that it is transported and deposited in a similar manner in all plant tissues. Lead accumulated by plants, after the decay of this plant biomass, may be more easily available to other plants. Availability of Pb in soil to plants increases from soils with low soil pH and CEC, as well as with low contents of SOM, clay minerals, Fe hydroxides, and P compounds. Plants grown on Zn–Pb waste deposits containing Pb up to 22,265 mg/kg contain this metal up to 117 mg/kg (Wójcik et al. 2014).

There are several interactions between Pb and both major and trace elements. The most serious interactions are as follows:

- Pb–Zn, antagonism adversely affects the translocation of each element from roots to tops.
- Pb–Cd, stimulating effects of Pb on Cd uptake by roots may be a secondary effect of the disturbance of transmembrane systems.
- Pb–Ca, occurs due to Pb inhibition of some enzymes associated with Ca.
- Pb–P, low Pb phytoavailability is due to the formation of insoluble Pb–phosphate compounds
- Pb–S, Pb transfer from roots to top is higher at low S contents.

Airborne Pb, its major source of pollution, is also readily taken up by plants through foliage. Thus, its concentrations are always higher in older parts of plants than in shoots or flowers. Much controversy exists in the literature on the question of how much airborne Pb is fixed onto the hairy or waxy cuticles of leaves and how much Pb is actually taken into foliar cells. It has been suggested that most of the Pb pollution (up to 95% of the total Pb content) can be removed from the leaf surface by washing with detergents. Lead contents of leaves, needles, and sprouts of all plants correlate with its atmospheric concentrations. Elevated Pb content, up to 165 mg/kg, in edible mushrooms (*Boletus* sp.) growing in the smelting area, also originated from the atmospheric deposition (Komarek et al. *vide* Kabata-Pendias 2011). The common mushroom, chanterelles (*Cantharellus cibarius*), grown in mountains of Poland contains higher amounts of Pb, mean 0.86 mg/kg, than those grown in the Baltic Sea coast, mean 0.43 mg/kg (Falandysz et al. 2012).

Elevated Pb contents are usually in various plants growing in vicinities of metal smelters, Pb mines, Pb recycling factories, and roadsides. Also, sewage sludge may be a great source of Pb in plants (Table 24.3). The worldwide experiment carried out with young wheat and corn plants ($N = 1892$) showed that plant Pb is a function of soil Pb extracted with ammonium acetic acid–ethylenediaminetetraacetic acid solution, and that only soil pH reveals a slight effect on Pb availability (Silanpää and Jansson 1992).

Organolead compounds are easily available to plants. Alkylated Pb compounds (released from vehicle gasoline) are quickly converted to water-soluble Pb tetraakyls in soils, which are easily phytoavailable, and mobile in plant tissues. Organolead compounds in plants are more toxic than inorganic forms and disturb some fundamental biological processes such as photosynthesis, growth, mitosis, and so on.

There is still a great attention oriented toward the Pb levels in plant foodstuffs, especially in bread and potatoes, which are a significant part of human diets. The proposed levels of Pb proposed by the ATSDR (2007b) are (in mg/kg FW) 0.2 and 0.1, for cereals and potatoes, respectively. The highest bioaccumulation of Pb is, in general, reported for leafy vegetables (mainly lettuce and spinach) grown in contaminated areas.

TABLE 24.3
Lead Contents in Plants Grown in Unpolluted and Polluted Soils (mg/kg)

	Unpolluted Soil		Polluted Soil[a]	
Plant	**Maximum**	**Mean**	**Maximum**	**Mean**
Spinach leaf	0.2	0.05	22.6	14.9
Spinach root	2.5	1.4	23.6	16.1
Amaranthus leaf	1.6	0.1	14.5	12.2
Amaranthus root	2.3	1.4	68.2	38.6
Cabbage leaf (inner)	1.0	0.6	10.1	7.2
Cabbage root	3.5	2.7	15.6	12.4

Source: Srikanth, R. and Reddy, S.R.-J. *Food Chem.*, 40, 229–234, 1991.
[a] Soil with sewage sludge.

The background Pb contents in barley and wheat grains are established at 0.2 and 0.5, respectively. Its contents in cereal grains from seven EU countries ranges in 0.007–0.013 mg/kg (Eriksson 2001a). Range of mean Pb amounts in grasses from various countries is 0.4–4.6 mg/kg and in clover 1.3–8 mg/kg. Grass growing along the high traffic road may contain Pb up to about 20 mg/kg.

24.6 HUMANS

The total amount of Pb in human body varies with age, occupation, and environment. It has been estimated that a 70 kg man contains an average 120 mg of Pb, of which in the blood 0.2 mg/L, in the bone (in mg/kg) between 5 and 50, and in the tissue 0.2–3 (Emsley 2011).

The main sources of Pb exposure are paints, water, food, dust, soil, kitchen utensils, and leaded gasoline. The majority of cases of Pb poisoning are due to oral ingestion and absorption through the gut. Absorption of Pb from the gastrointestinal tract is influenced by physiological factors (e.g., age, fasting, Ca and Fe status, and pregnancy), and the physicochemical characteristics of ingested materials (e.g., particle size, mineralogy, solubility, and Pb species) (WHO 2011c).

Absorbed Pb is transferred to soft tissues, including the liver and the kidneys, and to the bone tissue, where it accumulates with age. The principal vehicle for the transportation of Pb from the intestine to the various body tissues is the red blood cells, in which Pb is bound primarily to hemoglobin. In the blood, approximately 99% of the Pb is found in the erythrocytes, leaving about 1% in the plasma and the serum. Concentration of Pb in the plasma is more significant than those in the whole blood as a means of distribution to target organs, that is, the brain, lungs, the spleen, the renal cortex, the aorta, teeth, and bones. Lead in blood has an estimated half-life of 35 days; in soft tissue, it is 40 days. Residence period of Pb in bone is up to 30 years, with Pb concentrations in bone and teeth increasing as a function of age. The biological half-life of Pb may be considerably longer in children than in adults. Lead binds to thiol groups and other ligands in proteins. Its toxicity has been attributed to the inhibition of enzymes and interference with Ca, Mg, and Zn homeostasis. Lead-induced oxidative stress contributes to Pb poisoning for disrupting the delicate pro-oxidant/antioxidant balance that exists within mammalian cells. The mechanisms for Pb-induced oxidative stress include the effect of Pb on membrane, DNA, and antioxidant defense systems of cells. Antioxidant nutrients including, vitamin E, C, B_6, β-carotene, Zn, and Se, may have a beneficial role in Pb-induced oxidative stress (Hsu and Guo 2002).

Absorption of Pb is higher in children than in adults. Children can absorb 40%–50% of an oral dose of water-soluble Pb compared to 3%–10% for adults. Children who are Fe or Ca deficient have higher blood Pb concentrations than children who are Fe or Ca replete. Absorption of Pb may increase during pregnancy. More than 95% of Pb is deposited in skeletal bone as insoluble phosphate. Autopsy studies have shown that 90%–95% of the body's burden is present in cortical bone and teeth. In adults, 80%–95% of the total body burden of Pb is found in the skeleton, compared with about 73% in children. Lead can be transferred from the mother to the fetus and to infants during breastfeeding (Concha et al. 2013; Leikin and Paloucek 2008).

In humans, the central nervous system is the main target organ for Pb toxicity. Ingestion of high amounts of Pb from the environment by children, particularly when anemic, has been associated with reduced intelligence and impaired motor function.

Infants and young children are more vulnerable than adults to the toxic effects of Pb, and they also absorb Pb more readily. Even short-term, low-level exposures of Pb by young children are considered to have an effect on neurobehavioral development. The most critical effect of low-level Pb exposure is on intellectual development in young children and, Pb crosses the placental barrier and accumulates in the fetus. For children, the weight of evidence is greatest, and evidence across studies is most consistent, for an association of blood Pb levels with impaired neurodevelopment, specifically reduction of intelligence quotient (IQ). The developing brain was identified as the most vulnerable organ for Pb exposure. Under certain conditions, such as pregnancy and osteoporosis, bone resorption can result in increased concentrations of Pb in blood. Lead readily crosses the placenta and is transferred into breast milk.

The toxic effects of Pb have been principally established in studies on people exposed to Pb in the course of their work. Short-term exposure to high levels of Pb can cause brain damage, paralysis (lead palsy), anemia, and gastrointestinal symptoms. Long-term exposure can cause damage to the kidneys, reproductive, and immune systems, in addition to effects on the nervous system.

Neurotoxicity, cardiovascular effects, and renal toxicity were identified as the critical effects. Lead toxicity results in central nervous system, hematological and cardiovascular systems, and kidney damage. Lead interferes with the activity of several enzymes involved in the biosynthesis of heme. Signs of acute Pb intoxication include dullness, restlessness, irritability, poor attention span, headaches, muscle tremor, abdominal cramps, kidney damage, hallucinations, loss of memory, and encephalopathy. Signs of chronic Pb toxicity include tiredness, sleeplessness, irritability, headaches, joint pain, and gastrointestinal symptoms.

For adults, the adverse effect, greatest and most consistent, is a Pb-associated increase in blood pressure. As with the Pb-associated reduction in IQ, the increase is small when viewed as the effect on an individual's blood pressure, but important when viewed as a shift in the distribution of blood pressure within a population.

Pocock et al. (1983) showed a linear increase in blood Pb in thousands of men in the United Kingdom exposed to various levels of Pb in drinking water. Blood Pb in adult males increased by 6.22 μg/L for every additional 1 μg Pb/day taken from drinking water. If drinking water contains 5 μg/L, then the contribution to blood Pb from drinking water is 31 μg/L

According to WHO (2011c), anemia occurs at blood Pb levels in excess of 0.4 mg/L in children, and 0.5 mg/L in adults. International Agency for Research on Cancer (IARC) classified inorganic Pb as probably carcinogenic to humans (Group 2A, evidence inadequate in humans, sufficient in animals) (IARC 2013).

Based on the dose–response analyses, Joint FAO/WHO Expert Committee on Food Additives (JECFA) estimated that the previously established provisional tolerable weekly intake (PTWI) of 25 μg/kg bw is associated with a decrease of at least 3 IQ points in children and an increase in systolic blood pressure of approximately

3 mmHg (0.4 kPa) in adults. These changes are important when viewed as a shift in the distribution of IQ or blood pressure within a population. JECFA, therefore, concluded that the PTWI could no longer be considered health protective, and it was withdrawn. Because the dose–response analyses do not provide any indication of a threshold for the key effects of Pb, JECFA concluded that it was not possible to establish a new PTWI, which would be considered to be health protective. JECFA reaffirmed that because of the neurodevelopmental effects, fetuses, infants, and children are the subgroups that are the most sensitive to Pb (WHO 2011a, 2011c). Maximum permissible levels of Pb in some foodstuffs were set by the EU (EC 2006, 2008).

Lead levels in foods have declined over time in many developed countries (Table 24.4). Programs such as those that have eliminated the use of leaded petrol

TABLE 24.4
Lead in Foodstuffs (mg/kg FW)

Product, Years	Content	Maximum Tolerable Level[a]
	Food of Plant Origin	
Rye, grain, 1995–2003[b]	0.29–0.16	0.20
Wheat, grain, 1995–2003[c]	0.34–0.14	0.20
Bread, 1994[c]–2006[d]	0.02–0.011	–
Miscellaneous cereals, 1994[c]–2006[d]	0.02–0.007	
Cabbage, 1995–2003[b]	0.03–0.02	0.30
Carrot and potatoes, 1995–2003[b]	0.07–0.05	0.10
Apple, 1995–2003[b]	0.05–0.03	0.10
Milk chocolate[e]	0.03	–
	Food of Animal Origin	
Carcass meat, 1994[c]–2006[d]	0.01–<0.003	0.10
Offal, 1994[c]–2006[d]	0.11–0.065	–
Fish, 1994[c]–2006[d]	0.02–0.004	0.30
Freshwater fish, various[f]	0.01–0.02	0.30
Eggs, 1994[c]–2006[d]	0.01–<0.003	–
Milk, 1994[c]–2006[d]	<0.01–0.001	0.02
Dairy products, 1994[c]–2006[d]	0.01–<0.003	–

Sources: [a] EC, *Off. J. Eur. Union*, L, 364/5, 2006; EC, *Off. J. Eur. Union*, L, 173/6, 2008.

[b] Szteke, B., *Pol. J. Environ. Stud.*, 15, 2a, 189–194, 2006.

[c] Ysart, G. et al., *Food Addit. Contam.*, 16, 391–403, 1999.

[d] Rose, M. et al., *Food Addit. Contam.*, 27, 1380–1404, 2010.

[e] Sager, M., *J. Nutr. Food Sci.* 2, 123, 2012.

[f] Lidwin-Kaźmierkiewicz, M. et al., *Pol. J. Food Nutr. Sci.* 59, 219–234, 2009.

FW, fresh weight.

are considered to be an important factor. This reflects the success in reducing Pb contamination in food, due to replacement of soldered cans, banning of Pb in wine bottles, and the phasing out of leaded petrol. The largest contributors to overall Pb exposure are cereals and cereal products and vegetables.

Dietary exposures of the general U.K. population have declined from 0.12 mg/day estimated in 1980 to 0.006 mg/day in 2006 (Rose et al. 2010). The United States reported declines in Pb exposure for all age groups, with the greatest decline in teenage males (from 0.07mg/day in 1976 to 0.00345 mg/day in 2000); Canada and France have also reported a 50% decline in Pb exposure over the past 10–15 years (WHO 2011c). Lead intake by people of China, in 1990–2000, were much higher than in some other countries (e.g., the United States, the United Kingdom, Australia, and Japan), and were in the range of 0.084–0.392 mg/day; it was caused by the large consumption of Pb-contaminated rice (Sun et al. 2011b).

24.7 ANIMALS

Cattle, especially young calves, are extremely susceptible to Pb toxicity. In domestic animals, Pb poisoning (plumbism) is mainly one of acute toxicity, whereas the chronic form occurs most often in humans. Lead is one of the most frequently reported causes of poisoning in farm animals, especially in cattle, as well as in other ruminants.

Susceptibility to Pb poisoning is affected by the type of its compounds, ruminal or intestinal acidity, animal species, and stage of lactating and/or pregnancy. In some instances, as little as 6 mg Pb/kg bw (300 mg/kg in the total diet) over 60 days was fatal for cattle. In all domestic species, Pb poisoning causes derangement of the central nervous system, gastrointestinal tract, muscular coordination, and red blood cell synthesis. In cattle, signs of toxicity include a depressed appearance, blindness, grinding of teeth, muscular twitching, snapping of eyelids, and convulsive seizures. With sheep, Pb toxicity results in depression, anorexia, abdominal pain, and usually diarrhea. Anemia is common during chronic Pb ingestion. Osteoporosis has been observed in young grazing lambs and abortions in ewes grazing in lead-mining areas.

Chronic oral exposure of experimental animals to inorganic Pb has effects on multiple organs, including kidney and liver, and systems, as well as the cardiovascular, hematological, immune, reproductive, and nervous systems. In horses, Pb toxicity syndrome induces anemia, depression, stupor, knuckling at the fetlocks, and laryngeal paralysis, which produces an obstruction in the air passage. Large amounts of dietary Zn decrease or prevent clinical signs of Pb toxicity in horses. Protective effect occurs despite an increase in Pb retention in several soft tissues, with the high Zn uptake. Also, lower than optimal amounts of dietary Ca (and P, to a lesser extent) increase liver Pb in the young horses.

About 90% of Pb in animal blood is associated with red blood cells. Lead inhibits the utilization of Fe and the biosynthesis of heme, thus causing anemia. Lead may displace metals from enzymes, thereby causing their inactivation. Nuclei, microsomes, and mitochondria seem particularly susceptible to Pb. Lead inhibits

TABLE 24.5
Lead in Animal Tissues (mg/kg FW)

	Tissue		
Animal	**Muscle**	**Liver**	**Kidney**
Cattle, agricultural area[a]	0.003	0.082	0.212
Cattle, industrial area[a]	0.004	0.194	0.373
Cattle[b]	0.015	0.084	–
Pigs[b]	0.013	0,053	–
Sheep	0.033	0.190	–
Hunting animals[b]	0.273	0.067	–
Chicken[b]	0.010	0.032	–
Fish[b]	0.016	–	–

Sources: [a] Waegeneers, N. et al., *Food Addit. Contam.*, 26, 326–332, 2009a.

[b] Szkoda, J. et al., *Ochr. Śr. Zasobów Nat.*, 48, 475–484, 2011.

FW, fresh weight.

lipoamide dehydrogenase in the synthesis of acetyl coenzyme A and succinyl coenzyme A from pyruvate and alpha ketoglutarate.

Lead readily passes the placental barrier and accumulates in fetal bone, with lesser amount in liver, kidney, and intestine (Neathery and Miller 1975). Concentration of Pb in animals (Table 24.5) is various, depending on environment, tissue species, and age (Rudy 2009).

25 Lithium [Li, 3]

25.1 INTRODUCTION

Lithium (Li), the lightest metal of the group 1 in the periodic table of elements, is widely and uniformly distributed throughout the Earth's crust, within the range of 20–25 mg/kg. It is likely to be concentrated in acidic igneous rocks (up to 40 mg/kg) and in argillaceous sedimentary rocks and phosphate rocks (up to 75 mg/kg). Its concentration in coal is <75 mg/kg, but in fly ash may be accumulated up to 240 mg/kg.

World mine Li production (without the United States) in 2010 was 25,300 t, of which Chile produced 8,800; Australia produced 8,500; and Argentina produced 2,900 (USGS 2011). Lithium is used in organic synthesis, plastic production, ceramic and glass applications, special alloys, and batteries. Demand for Li batteries is increasing, as they are used in calculators, cameras, computers, electronic games, watches, and other devices. Compounds of Li are used as psychiatric medications.

25.2 SOILS

The mean background Li contents in the worldwide soils range from 13 to 28 mg/kg, within the range of 2–175 mg/kg (Table 25.1). The lowest abundance of Li is in sandy soils, and the highest is in heavy loamy and calcareous soils. However, the lowest Li content (mean 1.3 mg/kg) is in organic soils. Surface soils of the Piedmont region of the southeastern United States contain Li from 11.5 to 33.3 mg/kg, and from the Coastal Plains, within the range of 3.7–5.8 mg/kg. Higher contents of Li are in deeper soil horizons, and the greatest amount was found in the mother rocks, up to 59.9 mg/kg (Anderson et al. *vide* Kabata-Pendias 2011). Mean Li contents of agricultural soils of Poland varies from 6 to 15 mg/kg, in light and heavy soils, respectively (Siebielec et al. 2012). Lithium contents of soils in Spitsbergen, Norway, are within the range of 0.23–3.18 mg/kg (Gulińska et al. 2003). Mean Li content in organic surface soils of Norway is 0.64 mg/kg, and is lower than the mean value (1.1 mg/kg) for all investigated soils (Nygard et al. 2012).

During weathering, Li is released easily from the primary minerals in oxidizing and acid media, and is then incorporated into clay minerals and Fe–Mn hydroxides, accumulated in phosphate rocks, and is also easily absorbed by organic matter. Li content in soils is controlled more by conditions of the soil formation than by its initial content in parent rocks. Li distribution in soil profiles follows the general trends of soil solution circulation; however, it may be highly irregular. In the arid climatic zones, Li follows the upward movement of the soil solution, and may precipitate at top horizon, along with easily soluble salts of chlorites, sulfates, and borates. These reactions explain a relatively higher Li content of soils such as solonchaks, solonetz,

TABLE 25.1
Lithium Contents in Soils, Water, and Air

Environmental Compartment	Range
Soil (mg/kg)	
Light sandy	5–70
Medium loamy	2–130
Heavy loamy	9–175
Calcerous (calcisols)	6–105
Organic	0.01–3
Water (μg/L)	
Rain	0.04–0.12
River	0.2–3.5
Sea, ocean	170–200
Air (ng/m^3)	
Urban areas	2–8.9
Greenland	0.2–1

Sources: Data are given for uncontaminated environments, from Kabata-Pendias, A. and Mukherjee, A.B., *Trace Elements from Soil to Human*, Springer, Berlin, Germany, 2007; Reimann, C. and de Caritat, P., *Chemical Elements in the Environment*, Springer, Berlin, Germany, 1998.

kastanozems, and prairien soils. Also, intrazonal young soils derived from alluvium reveal elevated Li concentrations. However, the texture of mineral soils is the most significant factor controlling the Li status in soils, although all other parameters, such as soluble organic matter, cation exchangeable capacity and pH, are of much less importance.

In the initial processes of soil formation, Li seems to be highly mobile, whereas later it may become more stable, due to its firm bonding by clay minerals. However, water-soluble Li species in a soil profile reach up to about 5% of its total content, and therefore Li is likely to occur in groundwater of areas having elevated Li contents in rocks and soils. Exchangeable soil Li is reported to be strongly associated with Ca and Mg. Thus, in humid climatic zones, under greater rainfall, there is a net loss of Li from soils. Although soil microorganisms are relatively sensitive to increased levels of Li, some fungi, *Penicillium* and *Aspergillus*, are known to adapt easily to such growth media.

25.3 WATERS

The worldwide median concentration of Li in ocean water is given as 180 μg/L, and is not very much differentiated, within the range of 170–200 μg/L (Table 25.1). Riverine flux of Li is estimated at 69 kt/yr (Gaillardet et al. 2003).

Lithium occurs in all kinds of water, most often sorbed by colloids, in the form of hydroxide, LiOH, and seldom as the simple ion Li^+. Surface sediments of the Baltic Sea contain Li within the range of <10–84 mg/kg (Szefer 2002).

In groundwater, Li concentrations vary from 0.05 to 150 μg/L (Matschullat 1997). In some hot climatic regions, Li concentration may reach 15,000 μg/L. Water used for irrigations, with Li contents in the range 5–100 μg/L, may be harmful to crop plants.

Rainwater collected in Sweden contains Li at an average value 0.06 μg/L (Eriksson 2001a), whereas its highest concentration is given as 0.12 μg/L (Table 25.1). The maximum Li concentration in rainwater from Kola Peninsula is 0.28 μg/L (Reimann and de Caritat 1998).

Median Li concentration in bottled water of the EU countries is 14.9 μg/L, and in tap water it is 2.07 μg/L (Birke et al. 2010). In drinking water of Germany, its contents vary from 4.2 to 60 μg/L (Anke et al. 1995).

25.4 AIR

There is a paucity of data on Li in the atmosphere. Its content in air is associated with the presence of dust particles, which are mostly fine granulometric soil particles. Nevertheless, Li concentration in air of contaminated areas is significantly higher, than in air of remote regions, where it does not exceed 1 ng/m^3 (Table 25.1). The wet deposition of Li is estimated at 0.58 g/ha/yr (Eriksson 2001a).

Lithium concentration in moss is a good indicator for its contents in air. Moss (*Hylocomium splendens*) growing in Norway, during 1990–1955, contained Li within the range from 0.0027 to 2.6 mg/kg (average 0.22), and have not changed as compared with Li contents in moss collected earlier (Berg and Steinnes 1997).

25.5 PLANTS

The soluble Li in soils is readily available to plants; therefore, the plant content of this element is believed to be a good guide to the Li status of the soil. There are considerable differences in the tolerance of various plant species to Li concentrations, as well as in the ability of plants to uptake this element. The index of the biological concentrations, based on the ratio of Li in plant and ash to Li in topsoil, varies considerably for various plants, being the highest for the Rosaceae family (0.6), and the lowest for the Polygonaceae family (0.04). The highest value of this index (0.8), however, was calculated for plants of the Solanaceae family, which are known to have the highest tolerance to Li. Some plants of this family, when grown in an aridic climatic zone, accumulate Li more than 1000 mg/kg. The highest uptake of Li was reported for plant species growing on *Natric* soils or other soils, having increased contents of alkali metals. Among various native plants grown in the polluted area of the Northern Europe (Kola Peninsula), crowberry (*Empetrum nigrum*), containing 0.07 mg Li/kg, has a bit higher capacity to accumulate Li than other plants (Reimann et al. 1999).

Lithium appears to share the K^+ transport carrier, and therefore is easily transported in plants, being located mainly in leaf tissues. Li contents of edible plant parts show that some leaves accumulated a higher proportion of Li (e.g., celery, 6.6 mg/kg) than do storage roots or bulbs. Nevertheless, a higher Li content is very often reported for roots. The ratio of root to top for Li in ryegrass is 4.4, whereas for white clover it is 20. This may suggest that a difference in plant tolerance to Li concentration is related mainly to mechanisms of biological barriers in root tissues. However, Li is concentrated in above-earth parts of plants rather straight from aerial sources. This is especially noticeable in plants growing in industrial regions; for example, the leaves:roots ratio for Li in dandelion from a rural region is 0.8, and from an industrial region is 5.0 (Kabata-Pendias 2011).

Although Li is not known to be an essential plant nutrient, there is evidence that Li can affect plant growth and development. However, stimulating effects of several Li salts, reported by various authors, have never been confirmed. The observed stimulation may also be related to the influence of other factors, including secondary effects of anions associated with Li. There have been reported antagonistic effects from Rb and possible Zn, whereas synergistic effects from Fe and Mn. Calcium inhibits Li uptake by plants, whereas the addition of lime to high-Li soil may reduce toxic effects of this element. Although Li is not known yet as an essential plant nutrient, there are some evidences that Li can affect plant growth and may play some metabolic function in halophytes.

Increased Li contents in soil can be toxic to some plants. Citrus trees are probably the most susceptible to an excess Li, and their growth in salt-enriched soils can be significantly reduced due to high Li contents. In high-Li soils, damage to root tips, injured root growth, and chlorotic and necrotic spots on leaves have been observed in corn. Threshold concentrations of Li in plants are variable; for example, moderate to severe toxic effects of 4–40 mg Li/kg in citrus leaves were reported. The most resistant to a high-Li concentration are plants of the Solanaceae family, which may accumulate Li even above 1000 mg/kg.

25.6 HUMANS

Lithium content in human body (70 kg) is estimated on 7 mg, of which in the blood 4 μg/L, in the bones and tissues 1300 and 24 μg/kg, respectively (Emsley 2011), and according to Zaichik et al. (2011) its content in the rib bones is about 70 μg/kg. Lithium is absorbed efficiently in the intestine and is excreted mainly in the urine. The biochemical mechanisms of Li action appear to be multifactorial and are intercorrelated with the functions of several enzymes, hormones, and vitamins, as well as with growth and transforming factors. Lithium is thought to stabilize serotonin transmission in the nervous system, in addition to influencing Na transportation. It may even increase lymphocytic (white blood cell) proliferation, and depress the suppressor cell activity, thus strengthening the immune system. Lithium may act by altering the distribution of electrolytes within the brain.

Lithium is extensively used in psychiatric medicine, for the prevention and treatment of manic-depressive disorders (Dimitrova et al. 2013). For the last few decades, Li has been the most effective psychopharmacological drug, in the long-term treatment of patients with recurrent unipolar and bipolar affective illness, and is recommended as a treatment for several emotional and mental disorders. One of the side effects associated with its chronic use is Li-induced nephropathy, which may finally manifest itself as end-stage renal disease, in need of either dialysis or kidney transplantation (Grunfeld and Rossier 2009).

Because Li is used as a medication to treat manic-depressive disorders, the dosage should be monitored extremely carefully, to prevent toxicity of the element. It is recommended as a nutritional supplement, but is primarily used as medicine drugs. It has also been, occasionally, used in treating alcoholism, as it helps in decreasing the taste for alcohol, and generating a more cheerful attitude in life (Fołta and Bartoń 2011).

There is tentative evidence that Li may prevent Alzheimer's disease, and may change disease progress in those with early symptoms (Forlenza et al. 2012). A study provides strong evidence that geographic regions with higher natural Li concentrations in drinking water are associated with lower suicide mortality rates. Epidemiological data concerning cardiovascular mortality and high Li content in drinking water (ca <100 μg/L) showed a reduction of intestinal and cardiovascular diseases (Masironi and Shaper 1981).

Although the effects of therapeutic doses of Li are well established, little is known about the health effects of natural Li intake. Study on evaluating the association between local Li levels in drinking water and suicide mortality at district level in Austria provides strong evidence that geographic regions with higher natural Li concentrations in drinking water are associated with lower suicide mortality rates (Kapusta et al. 2011). Also, in areas with higher concentration of Li are lower rates of depression and violent crime. A link between high levels of Li in tap water and low incidences of suicides, admissions to mental hospitals, murders, and rapes is observed. Although many people can benefit from increasing the amount of Li in diet, overexposure to Li can adversely affect health. Overexposure to Li negatively affects thyroid function. People living in the Andes discovered that exposure to Li through drinking water and other environmental sources may affect thyroid function (Broberg et al. 2011).

Excess Li produces symptoms such as nausea, diarrhea, excessive thirst, increased urination, hand and foot tremors, lethargy, mental confusion, delirium, and weakness in muscles. High doses of Li may cause interference with glucose metabolism and hypothyroidism.

The available experimental evidence now appears to be sufficient to accept Li as an essential element. The provisional recommended dietary allowance for a 70 kg adult is 1 mg/day (Schrauzer 2002). Lithium occurs in all foods, though at low levels (Table 25.2).

TABLE 25.2
Lithium in Foodstuffs (mg/kg FW)

Product	Content
Food of Plant Origin	
Bread[a]	0.01
Miscellaneous cereals[a]	0.02
Rice and semolina[b]	0.004
Potatoes[a]	0.01
Vegetables, various[a]	0.01–0.03
Fresh fruits[a]	0.005
Curry, various[c]	0.12–0.4
Alcoholic beverages[b]	0.003
Food of Animal Origin	
Carcass meat[a]	0.02
Meat[b]	0.002
Meat products[a]	0.01
Offal[a]	0.03
Eggs[a]	0.01
Freshwater fish[d]	0.01–0.13
Fish[b]	0.030
Milk[a]; milk[b]; milk[e]	0.003; 0.006; 0.059

Sources: [a] Ysart, G. et al., *Food Addit. Contam.*, 16, 391–403, 1999.
[b] Leblanc, J.C. et al., *Food Addit. Contam.*, 22, 624–641, 2005.
[c] Gonzalvez, A. et al., *Food Addit. Contam.*, 1, 114–121, 2008.
[d] Lidwin-Kaźmierkiewicz, M. et al., *Pol. J. Food Nutr. Sci.*, 59, 219–234, 2009.
[e] Gabryszuk, M. et al., *J. Elem.*, 15, 259–267, 2010.
FW, fresh weight.

The intake of dietary Li is dependent upon the location and type of foods consumed. The mean dietary exposure to Li was estimated at (mg/day) 0.017, in the United Kingdom (Ysart et al. 1999); 0.144, in France (Leblanc et al. 2005); 0.348, in Germany; 0.494, in Austria; and 1.485, in Mexico (Folta and Barton 2011). The average daily Li intake of an American adult (70 kg) ranges from 0.650 to 3.100 mg (Schrauzer 2002).

26 Manganese [Mn, 25]

26.1 INTRODUCTION

Manganese (Mn), a metal of the group 7 in the periodic table of elements, is one of the most abundant trace elements in the lithosphere, and is the second, after Fe, metal in the Earth's crust, at mean concentration of 0.085%. Its common occurrence in rocks ranges from 350 to 2000 mg/kg, and higher concentrations are in mafic rocks, as well as in Mn–Fe concretions, especially deposited in ocean-bottom sediments. In sedimentary rocks, it may be concentrated up to 1000 mg/kg. Contents of Mn in coal vary within the range of 50–100 mg/kg, and in fly ash may be about 400 mg/kg.

Manganese occurs mainly at +2 oxidation stage, but may change valences up to +7. It is a member of the Fe family, and is highly associated with Fe, in all geochemical processes. There are many Mn minerals, mainly together with other metals, especially with Fe. The most common is pyrolusite, β-MnO_2; other minerals are manganite, γ-MnOOH; hausmannite, Mn_3O_4; and rodochrozite; $MnCO_3$. The Mn oxide mineral birnessite, $(Na_{0.3}Ca_{0.1}K_{0.1})(Mn^{4+},Mn^{3+})_2O_4 \cdot 1.5\ H_2O$, of an unconfirmed composition, is formed due to precipitation in lakes, oceans, and groundwater. It is a major component of desert varnish and deep sea Mn nodules, exhibits a large adsorption capacity to several metals (Cd, Co, Cu, Pb, and Zn), and has a high oxidizing potential (Feng et al. 2007). Sorption capacity of MnO_2 for several metals is comparable to that of goethite and hematite.

During weathering, Mn in minerals is oxidized, and forms secondary minerals, often as concretions and nodules. Under weathering in tropical and subtropical conditions, Mn is concentrated in residual deposits, whereas under humid climate, it is leached by acid solutions from sediments and soils. These processes resulted in Mn accumulation, as concretions in various sediments (e.g., deep-sea polymetallic concretions), and in soils.

World mine Mn production (rounded) in 2010 (without the United States) was 13,000 kt, of which China produced 2,800; Australia produced 2,400; and South Africa produced 2,200 (USGS 2011). Manganese is used as alloying element with various metals, but especially with Fe (e.g., steel). There are great uses of Mn other than in steelmaking. The amount of domestic Mn consumption is estimated to be in the range of 5%–10% of its metallurgical uses. Mn is used chiefly as MnO_2 (e.g., for dry cell batteries), and as MnO. Manganese has been used as a fuel additive methylocyclopentadienyl manganese tricarbonyl (controversial gasoline additive; MMT). Potassium permanganate is a strong oxidizing agent used for water treatments and several other purposes.

26.2 SOILS

Manganese contents in worldwide soils are highly diverse, and range approximately from 10 to 9000 mg/kg (Table 26.1). Its highest levels occur in loamy and calcerous soils. Its elevated contents are also in soils derived from mafic rocks, and rocks rich in Fe compounds and soluble organic matter (SOM).

The world grand mean Mn content is estimated at 437 mg/kg, and for U.S. soils at 495 mg/kg. On the world scale, average Mn contents vary from 270 to 525 mg/kg, in podzols and cambisols, respectively. The highest Mn contents, up to 9200 mg/kg, is reported for soils derived from basalts and andesites in Australia (Wells *vide* Kabata-Pendias 2011). Mean Mn content in organic surface soils of Norway is 279 mg/kg, and is a bit lower than mean value (330 mg/kg) for all investigated soils (Nygard et al. 2012). The distribution of Mn in soil profiles is positively associated with clay contents. However, it is usually accumulated in topsoil, as the result of its fixation by organic matter (OM).

TABLE 26.1
Manganese Contents in Soils, Water, and Air

Environmental Compartment	Range
Soil (mg/kg)	
Light sandy	10–2000
Medium loamy	50–9200
Heavy loamy	100–9300
Calcerous (calcisols)	50–7750
Organic	10–2200
Water (μg/L)	
Rain	0.5–3.0
River	0.02–130
Sea, ocean[a]	0.2–10
Air (ng/m³)	
Urban/industrial areas	50–900
Greenland	2.8–4.5
Antarctica	0.004–0.02

Sources: Data are given for uncontaminated environments, from Kabata-Pendias, A. and Mukherjee, A.B., *Trace Elements from Soil to Human*, Springer, Berlin, Germany, 2007; Reimann, C. and de Caritat, P., *Chemical Elements in the Environment*, Springer, Berlin, Germany, 1998.

[a] Range of mean values.

During weathering processes, Mn compounds are oxidized and Mn oxides are reprecipitated and form readily secondary Mn minerals. The behavior of Mn in surface deposits is very complex and governed by different factors, of which Eh–pH conditions are the most important. Presence of Mn in soils is the key to the entire status of soil redox potential (Sparks *vide* Kabata-Pendias 2011).

Manganese compounds readily leach down from the upper soil layers under cold and humid climate. Under tropical and subtropical conditions, is may be concentrated in soils, mainly in the form of concretions and nodules. Manganese is mobile in the most soil media, and occurs in soil solution in various species (Kabata-Pendias and Sadurski 2004):

- Cationic forms: Mn^{2+}, $MnOH^{+}$, $MnCl^{+}$, $MnHCO_3^{+}$, $Mn_2(OH)_2^{2+}$, and Mn_2OH^{3+}
- Anionic forms: MnO_4^{-}, HMO_2^{-}, $Mn(OH)_3^{-}$, and $MN(OH)_4^{2-}$

Because of a low mobility of Mn compounds in oxidizing systems at pH levels near neutrality, any changes in the Eh–pH conditions are important, and have an impact on its concentration in the soil solution. Concentration of Mn in soil solutions vary highly from 25 to 2000 μg/L. Solubility of Mn always increases with the increase of soil acidity. However, the ability of Mn to form anionic complexes, and to combine with organic ligands, may contribute to increased Mn mobility in the alkaline pH range. Among abiotic and biotic soil parameters, fixation by root exudates and cross-interactions with Fe hydroxides play a crucial function in the Mn mobility, and thus in phytoavailability. Also, SOM, and especially fulvic acid, has a high impact on Mn behavior in soils.

All Mn forms, minerals, concretions, nodules, exhibit a great adsorption capacity to various metals, which usually increases with increasing pH. In the case of Pb, this increase is from 20% to 80% of the total Pb concentrations in solution, at 5 and 8 pH, respectively (Wilson et al. *vide* Kabata-Pendias 2011). Manganese oxides have a relatively high total surface (30–300 m^2/g) and cation exchangeable capacity value (150–320 cmol/kg). The most readily sorbed metals by Mn oxides are Cu, Co, and Pb, which may be unavailable to plants. However, due to the reducing and oxidizing properties, Mn oxides can increase the mobilization of some metals, under specific soil conditions. Biologically mediated processes are the most significant in the Mn–redox cycling in soils (Perelomov et al. 2013). They increase formation of Mn oxides, as well as stimulate the release of Mn from some compounds. The microbial dissolution of Mn compounds, especially due to the enzymatic reduction of Mn (+3 and +4), and due to the production of CO_2, and organic acids, is of real importance.

Some microorganisms (bacteria and fungus), on the other hand, may precipitate Mn by oxidizing Mn^{2+} to Mn^{3+} and Mn^{4+}, or stimulating the precipitation of carbonates, sulfides, and so on. Some organic acids secreted by microorganisms may release Mn from several MnO_2 compounds. Biologically mediated processes are the most significant in the Mn–redox cycling in soils. Colloidal Mn oxides reveal a great affinity for adsorption of cationic and anionic species of various elements, as well as for OM. However, due to both reducing and oxidizing properties, Mn oxides may increase the mobilization of some metals under specific soil conditions.

Processes governing Mn behavior in soils are complex, and may be presented as follows (Negra et al. *vide* Kabata-Pendias 2011):

- Reduction: $Mn^{3+} \rightarrow Mn^{2+}$, abiotic and biotic, by Fe^{2+}, Cr^{3+}, S, phenols, OM, and reducing bacteria
- Oxidation: $Mn^{2+} \rightarrow Mn^{3+}$, Mn^{4+}, can occur under both aerobic and anaerobic conditions and is biologically mediated or autocatalytic
- Mn^{3+} is the extremely reactive redox species and quickly disappears, either by accepting or donating an electron
- Mn^{2+} is either absorbed by MnO_2, or oxidized to Mn^{3+}, or Mn^{4+}
- Organic and phosphate ligands are involved in the Mn–redox cycling

Manganese budgets (input/output ratio) in soils of various ecosystems indicate a predomination of leaching processes over atmospheric input. In some forests (pine, spruce, and birch) Mn leaching from soil profiles accounts from 360 to 6100 g/ha/yr, whereas in some agriculture ecosystems the accumulation of Mn has been observed, within the range of 90–191 g/ha/yr (Eckel et al. 2005).

All Mn compounds are important, as they are essential in plant nutrition, and control the behavior of many other trace elements. They have also a considerable impact on some soil properties, and in particular on Eh and pH values. Oxidizing conditions reduce the bioavailability of trace elements, whereas reducing conditions may lead to the easy availability of several micronutrients, which may result in their toxicity to plants. Reducing impact of Mn compounds may also increase Ca mobility, which resulted in soil acidification. Increased mobility of Mn with lower soil pH resulted in its increased phytoavailability and its losses from soils (Watmough et al. 2007).

Manganese has not been considered to be a polluting metal in soils; however, the maximum allowable concentration value for Mn is estimated to range at 1500–3000 mg/kg. Its major anthropogenic sources are municipal wastewater, sewage sludge, and metal smelting processes. In some regions (e.g., Mississippi River delta), alluvial sediments concentrate Mn up to 2700 mg/kg (Mielke et al. *vide* Kabata-Pendias 2011). Also, soils irrigated with water affected by acid mine drainage contain elevated amounts of Mn. Its content in soils after sludge application increased from 242 to 555 mg/kg, during five years. After longer period of Mn addition to soils with sludge, toxic effects of Mn in some plants might be observed.

26.3 WATERS

Worldwide mean Mn concentration in seawater is calculated to range from 0.4 to 10 µg/L, with an average of about 2 µg/L (ATSDR 2002b). However, various values are given as median Mn concentrations for ocean water (in µg/L): 0.02–0.1 (Kitano 1992); 0.2 (Reimann and de Caritat 1998); and 0.02 (Nozaki 2005). In the Baltic Sea, Mn concentrations range from <1 to 3 µg/L (Szefer 2002). Manganese concentrations in water of Nordic Lakes of different countries vary as follows (mean values, in µg/L): Norway, 101; Finland, 316; and Sweden, 474 (Skjelkvale et al. 2001). Rivers of the United States contains Mn within the range of 11–51 µg/L (ATSDR 2002b).

The global riverine flux of Mn in worldwide seas is estimated at 1270 kt/yr (Gaillardet et al. 2003). According to the calculation of Kitano (1992), Mn input to worldwide seas varies from 400 to 1000 kt/yr. The majority of Mn is in suspended particulate matter, which is concentrated at deeper water layers, from 30 to over 200 m (Brügmann *vide* Kabata-Pendias and Mukherjee 2007). Suspended matters in the Baltic Sea contain variable amounts of Mn, from 20 to 73,000 mg/kg (Szefer 2002). Atmospheric input of Mn with rainwater to the oceans is an important source of its stable and soluble species in surface seawater (Willey et al. 2009). Rainwater is thought to be a main removal mechanism for the atmospheric Mn. The reductive dissolution of Mn oxides, due to microbial reductions, controls Mn mobility and its transfer to well water (Petrunic et al. 2005).

All compounds and species of Mn in water are easily transferred into colloidal forms, and precipitated in bottom sediments. These processes are highly controlled by variations in the redox conditions of water and sediments. Formation of Mn–Fe nodules has an impact on the behavior of some other metals.

Manganese content of some sediments may be very high. Mn content in rivers of some EU countries is 770 mg/kg (Odra River) and 960 mg/kg (Rhine River), respectively. Sediments of the Baltic Sea contain Mn within the range of 120–2290 mg/kg, and sediments of some gulfs may have up to 7260 mg/kg (Szefer 2002).

Manganese contents in stream-bottom sediments of National Park, Montgomery (Pennsylvania State) were, in 1995, within the range of 430–34,000 mg/kg (Reif and Sloto 1997). Sediments of Sergipe River estuary (Brazil) contain Mn from 7.2 to 251 mg/kg (Garcia et al. 2011). Mn concentration in the wetland sediments from the effluent at the Savannah River site (Aiken, SC) is 458.5 mg/kg, and its highest amounts, 365.3 mg/kg, are fixed in SOM fraction (Knox et al. 2006).

Median Mn concentration in bottled water of the EU countries is 0.81 μg/L, and is higher than those in tap water, 0.54 μg/L (Birke et al. 2010). Guideline for Mn in drinking water is not established, as it is not of health concern (WHO 2011a).

Due to reactions of Mn with some metals, especially with Cd and Pb, its presence in water may decrease metal toxicity to some water biota. However, these processes may lead to Fe deficiency in some algae.

Manganese is likely to concentrate in shells of mollusks, in the form of layers at the surface of the shell. Its contents in shells of some mollusks from the Baltic Sea is within the range of 7.5–233 mg/kg, and in some species (cockle) may be over 34,000 mg/kg (Szefer 2002).

26.4 AIR

The origin of Mn in the atmosphere is from both terrestrial and anthropogenic sources. Its concentrations in air vary from <0.02 ng/m^3 above Antarctica up to 900 ng/m^3 in industrial regions (Table 26.1). Median world Mn content in air has been estimated at 2 ng/m^3 (Reimann and de Caritat 1998). In air of some cities, not industrialized, Mn amounts average at about 10 ng/m^3 (Bankovitch *vide* Kabata-Pendias and Mukherjee 2007). One of the principal sources of inorganic Mn, as a pollutant in the urban atmosphere, is the combustion of MMT, particularly in areas of high traffic density. MMT was used as a gasoline additive in several countries, since 1990s.

The total worldwide emissions of Mn in 1983 ranged from 10.56 to 65.97 kt, with the predominant sources from coal combustion, and secondary, from nonferrous metal production, and sewage sludge incineration (Nriagu and Pacyna 1988).

Mean Mn content of mosses (*Hylocomium splendens*) from Poland was 245 mg/kg, whereas those collected in Southern Alaska was 80 mg/kg (Migaszewski et al. 2009). This indicates its anthropogenic sources. High Mn concentration, within the broad range of 28–2100 mg/kg, was reported for mosses from Sweden (Berg and Steinnes 1997).

Manganese exists mainly in two oxidation states in the atmosphere, Mn^{2+} and Mn^{4+}. Divalent Mn is a soluble oxidation state, whereas tetravalent Mn is mainly as a particulate; therefore, Mn^{2+} is in higher concentrations in rainwater (Willey et al. 2009). Concentrations of Mn^{2+} and SO_4^{2-} are significantly correlated in some rain water, as Mn stimulates the catalytic oxidation of SO_2. Manganese in air also reacts with soluble NO_2 present in rainwater.

Concentration of Mn in rainwater collected in Sweden (in 1999) varied from 0.5 to 3.1 μg/L, and was fairly similar to its contents in rainwater from the Kola Peninsula, within the range of 0.45–6.1 μg/L (Eriksson 2001a).

26.5 PLANTS

Adequate levels of available Mn are necessary for plants. Mn uptake by plants is both metabolically controlled, and due to passive absorption, especially at its high content in soils. Complex interactions between roots and microorganisms, which often resulted in the oxidation of soluble Mn^{2+} into unavailable Mn^{3+} or Mn^{4}, have presumably an impact on the Mn phytoavailability (Marschner 2005). Deficiency of Mn may occur in certain crop plants, on neutral and calcerous soils.

Manganese is easily transported within plants, which indicates that it is not bound to insoluble organic ligands (Peng et al. 2008). However, when Mn is accumulated in old leaves or sheaths, it is not translocated to young organs, under its deficiency. Generally, the most readily phytoavailable Mn is in acid, flooded, and SOM reach soils.

The most important Mn function in plants is related to the oxidation–reduction processes. It is a specific component of two enzymes, arginase and phosphotranferase, and may substitute Mg in other enzymes. It participates in the photosynthesis, being involved in O_2 and electron transport systems. It is also involved in the NO_2^- reduction process, and thus may have an impact on the N assimilation by plants. Chloroplasts are the most sensitive cells to the Mn deficiency, of which first symptoms occur as interveinal chlorosis. Mn-deficient plants have retarded growth, and lower resistance to diseased and climatic impacts. The most sensitive crop plants to Mn deficiency are oats, peas, sugar beet, and some fruit trees. The correction of Mn deficiency in crop plants may be done by both soil and foliar applications. The toxicity of Mn to some crop plants may be expected on soils with pH < 5.5, with high Mn levels, and poorly drained soils. Activities of some enzymes and hormones in plants are limited under an excess of Mn. Legumes appear to be very sensitive to excess Mn, due to affected rhizobia nodules, and thus lower the N fixation. Also, potatoes are easily affected by excess Mn. Plant tolerance to high Mn contents in soils is related to several metabolic processes, such as (1) oxidizing power of plant roots, from Mn to MnO_2; (2) complexation of Mn by low-molecular-weight compounds; (3) Mn entrapment in nonmetabolic centers; (4) Mn accumulation in roots; and (5) interactions with other elements, especially with Ca, Fe, Al, and NH_4.

The most common symptoms of Mn toxicity are Fe chlorosis and brown spots on leaves, and in some plants also browning of roots. Most plants are affected by Mn contents around 500 mg/kg, but some plants, and especially hyperaccumulators, may contain Mn within the range 1,000–10,000 mg/kg, without toxic symptoms.

Interactions between Mn and some metals may have an impact on the phytoavailability and biochemical functions. The most important is an appropriate Fe:Mn ratio, which should range from 1.5 to 2.5. Both metals have an impact on their absorption and metabolic functions. Cobalt became unavailable to plants under excess Mn. Also, other trace metals, which are especially important in the case of Cd and Pb, are likely to be fixed by Mn compounds. The interaction of Mn and P is related to both the variation in the Mn–phosphate solubility, and the Mn influence on P metabolic reactions. Depending on soil conditions, P fertilizers may either aggravate Mn deficiency in some crop plant or increase Mn uptake by other plants. These phenomena are related to soil pH and soil-sorption capacity. Elevated Si contents in plants may reduce Mn toxicity. Increased Mn levels may induce Ca and Mg deficiencies in some plants. Antagonistic interactions between Mn and K, Na, and N have also been observed. Most of these interactions, however, are very variable, depending on plant characteristic and soil properties.

There is a very wide variation in Mn contents among plant species, grown in the same soil and site, from 30 mg/kg in *Medicago truncatula* to about 500 mg/kg in *Lupinus albus*. Worldwide contents of Mn in grass range from 17 to 334 mg/kg and in clover from 25 to 119 mg/kg.

Food plants also contain variable amounts of Mn, with the highest being in beet (both sugar and red) roots (36–113 mg/kg) and the lowest in apples (1.3–1.5 mg/kg). A remarkable variation in Mn contents among plant species, stage of growth, different organs, and various ecosystems is noticed. A relatively small variation in Mn contents is seen in cereal grains, with an average from 18 to 48 mg/kg, throughout the world. Highest Mn amounts are reported for wheat grains (16–103 mg/kg) and for oats grains (17–121 mg/kg). Relatively high Mn contents (10–42 mg/kg) are in nuts and almonds (Kabata-Pendias and Mukherjee 2007).

The common mushroom, chanterelles (*Cantharellus cibarius*) grown in the mountains of Poland contains higher amounts of Mn, mean 32 mg/kg, than those grown in the Baltic Sea coast, mean 21 mg/kg (Falandysz et al. 2012).

26.6 HUMANS

The human body contains about 12 mg of Mn, which is stored mainly in the liver, kidneys, bones, and pancreas (Emsley 2011). Normal ranges of Mn levels in the body fluids are (in μg/L) 4–15 in the blood; 1–8 in the urine; and 0.4–0.85 in the serum. Average tissue concentrations are typically between 0.1 and 1 mg Mn/kg FW. Mn concentration in the liver is slightly higher, 1.2–1.7 mg/kg FW, and the lowest concentrations are in bones and fat, about 0.1 mg/kg FW (ATSDR 2012b).

Manganese is an essential dietary mineral for mammals. Its biological function acts as both a constituent of metalloenzymes and an enzyme activator. It is a component of metalloenzymes such as superoxide dismutase, arginase, and pyruvate carboxylase, and is involved in amino acid, lipid, and carbohydrate metabolism

(EFSA 2013). It is also a cofactor for a number of important enzymes, including cholinesterase, oxidoreductases, transferases, hydrolases, lyases, isomerases, ligases phosphoglucomutase, pyruvate carboxylase, mitochondrial superoxide dismutase and several phosphates, peptidases, and glycosyltransferases. In certain instances, Mn^{2+} may be replaced by Co^{2+} or Mg^{2+}.

Mn and Fe have many physicochemical similarities, and there is a possibility of competition between these elements. Excess Mn interferes with the absorption of dietary Fe. Increased Mn concentrations inhibit the metabolic function of the Fe-dependent enzyme, aconitase. Because Fe is the most prevalent nutritional deficiency in the world, there is the potential health risk associated with Fe deficiencies exacerbating the brain Mn burden. Long-term exposure to excess Mn levels may result in the Fe-deficiency anemia. Increased Mn intake impairs also the activity of Cu metalloenzymes.

Adverse effects resulting from the Mn exposure in humans are associated primarily with inhalation in occupational settings. In the workplace, exposure to Mn is most likely to occur by inhalation of Mn fumes or Mn-containing dusts. This is a concern mainly in the ferromanganese, iron and steel, dry-cell battery, and welding industries. Exposure may also occur during Mn mining and ore processing. Workers in the Mn-processing industry are at the most risk.

Oral exposure to Mn, especially from contaminated water, can also cause adverse health effects, which are similar to those observed from inhalation exposure. An actual threshold level, at which exposure to Mn produces neurological effects in humans, has not been established (EFSA 2013). Well water rich in Mn can be the cause of its excessive intake.

Chronic low-dose Mn intoxication is strongly implicated in a number of neurodegenerative disorders, including Alzheimer's disease, Parkinson's disease, and amyotrophic lateral sclerosis.

It may also play a role in the development of multiple sclerosis, restless leg syndrome, and Huntington's disease. Excess accumulation of Mn in the brain results in a neurological syndrome with cognitive, psychiatric, and movement abnormalities. Highest concentrations of Mn in the brain are at the basal ganglia, which may precipitate a form of parkinsonism, with some clinical features that are similar, and some that are different to those in Parkinson's disease (Guilarte 2010).

Manganese overexposure, by both ingestion and inhalation, is most frequently associated with manganism, a rare neurological biphasic disorder.

Manganese is a natural component of most foods (Table 26.2). Oral exposure is the primary source of absorbed Mn, although its absorption through gastrointestinal is relatively low. Dietary Mn intakes were significantly lower in nontea drinkers (3.2 mg/day) than in tea drinkers (5.5 mg/day), depending upon the value used for Mn levels of black tea. Tea drinking is a major source of dietary Mn, and Mn intakes commonly exceed proposed adequate values of 1.8–2.3 mg Mn/day, and may exceed upper limits of 10–11 mg/day (Hope et al. 2006). Nuts, chocolate, cereal-based products, pulses, fruits, and fruit products are rich sources of Mn (Table 26.2).

Formal recommended dietary allowance for Mn has not been established, but WHO has proposed the estimated safe and adequate dietary intake at 2–5 mg/day, for adults (WHO 2004).

TABLE 26.2
Manganese in Foodstuffs (mg/kg FW)

Product	Content
Food of Plant Origin	
Bread[a]	8.01
Miscellaneous cereals[a]	7.91
Rice, white[b]	13.2–27.1
Rice, brown[b]	61.6–146.4
Green vegetables[a]	2.06
Potatoes[c]	1.35
Strawberry[c]	4.21
Nuts[a]	24.9
Walnuts[d]	13.2–48.7
Cocoa[e]	38.8–53.4
Milk chocolate[e]	2.73–6.02
Food of Animal Origin	
Carcass meat[a]	0.129
Meat products[a]	2.75
Offal[a]	2.65
Fish[a]	0.72
Freshwater fish[f]	0.19–0.23
Milk[g]	0.023

Sources: [a] Rose, M. et al., *Food Addit. Contam.*, 27, 1380–1404, 2010.
[b] Batista, B.L. et al., *Food Addit. Contam.*, 3, 253–262, 2010.
[c] Szteke, B. et al., *Roczn. PZH*, 55S, 21–27, 2004.
[d] Markiewicz, K. et al., *Bromat. Chem. Toksykol.*, 39, 237–241, 2006.
[e] Sager, M., *J. Nutr. Food Sci.*, 2, 123, 2012.
[f] Lidwin-Kaźmierkiewicz, M. et al., *Pol. J. Food Nutr. Sci.* 59, 219–234, 2009.
[g] Gabryszuk, M. et al., *J. Elem.* 15, 259–267, 2010.
FW, fresh weight.

People, mainly vegetarians, ingest relatively high amounts of Mn, which may exceed its estimated daily dietary intake. However, the bioavailability of Mn from vegetable sources is substantially decreased by dietary components, such as fiber and phytates, some elements (Fe, Ca, Zn, P, and Mg), and some vitamins (B_1, B_2, and C) (ATSDR 2012b).

Mean intake of Mn by adults in some EU countries is around 3 mg/day (EFSA 2013), in the United Kingdom, 5.2 mg/day (Rose et al. 2010). The dietary Mn deficiency does not occur in the general population. The absorption of Mn has been observed to be lower in men than in women.

26.7 ANIMALS

Manganese deficiency is dangerous for animals. Mn-deficient animals exhibit impaired growth, skeletal abnormalities, reproductive deficits, ataxia of the newborn, and defects in lipid and carbohydrate metabolisms. Its nutritional deficiency, in cattle, sheep, and pigs may cause infertility and skeletal deformities, including enlarged joints, pain, knuckling at the fetlocks, and twisting of the legs. It is a rare deficiency in dogs and cats. For animals, Mn is an essential component of over 36 enzymes, which are used for the carbohydrate, protein, and fat metabolism.

For some animals, Mn lethal dose is quite low. Excess Mn can cause lung, liver and vascular disturbances, declines in blood pressure, failure in the development of animal fetuses, and brain damage. Its excessive intake may present a cardiovascular hazard (ATSDR 2012b).

27 Mercury [Hg, 80]

27.1 INTRODUCTION

Mercury (Hg), a metal of the group 12 in the periodic table of elements, is the only metal that exists in the liquid state in natural conditions. Its average content in the Earth's crust is 0.07 mg/kg, and is much lower in igneous rocks (0.004–0.008 mg/kg) than in sedimentary rocks (0.01–0.4 mg/kg). It is likely to be concentrated in argillaceous sediments. Contents of Hg in coal vary within the range of 0.2–10 mg/kg, and in fly ash, the content is around 0.01 mg/kg.

Mercury occurs mainly at +2 oxidation stage, but may also have valences from +1 to +3. It easily forms amalgams with noble metals. It may also be methylated by microorganisms in various natural environments and reveals a chalcophillic character, and thus combine readily with sulfur. Its main mineral, of commercial importance, is cinnabar (metacinnabar), HgS. Calomel (mercurous chloride), Hg_2Cl_2, is rarely found in nature. Sulfide (e.g., schuetteite, $Hg_3SO_4O_2$), selenide, telluride, and so on of Hg may occur in various ore deposits. There are several Hg host minerals such as amphiboles, sphere, sphalerite, and other sulfides. The most important geochemical features of Hg are as follows: (1) affinity to form strong bonds with S; (2) formation of organomercury compounds; (3) formation of gaseous compounds; (4) volatility of metallic form; and (5) low affinity for oxygen bound to carbon. The characteristics of Hg in the environment are its number of chemical and physical forms, such as Hg^0, $HgCl_2$, HgO, HgS, CH_3HgCl, and $(CH_3)_2Hg$. All these compounds behave differently in various ecosystems. Some of the mercuric salts are soluble in water, whereas organomercuries are neither soluble nor do they react with weak acid and bases.

World total (rounded) production (without the United States) of Hg in 2010 was 1960 t, of which (in t) China produced 1400; Kyrgyzstan produced 250; and Chile produced 150 (by-product) (USGS 2011). Its production has not changed much during the last 10 years. However, there is some information that due to recent regulations and restrictions in the primary Hg production, real data are not available (Kabata-Pendias 2011). In some countries its production was lowered or stopped, and Hg is obtained as a by-product. At the small-scale, it is obtained at the gold mining.

Major uses of Hg are still in (1) chloralkali production; (2) vinyl, chloride monometer production; (3) artisanal gold mining; (4) batteries and some control instruments; (5) pharmacology; (6) stomatology; (7) wood impregnates; and (8) paints. Its use, especially in agriculture (pesticides and seeds treatments), in electric lamps has been recently forbidden. Thimerosal (merthiolate; $C_3H_3HgNaO_2S$) is used since the 1980 as preservative in vaccines and in various cosmetics and medical preparations. Also, calomel (mercurous chloride, Hg_2Cl_2) was used in medicine as a diuretic, as well as in some cosmetics. There is a lot of disagreement about the dangers of Hg in the medical and dental professions, and some recent decisions may forbid Hg addition to medical and pharmaceutical products. However, WHO and American Academy of Pediatrics have recently accepted the addition of thimerosal to various vaccines for children.

Galistan, an alloy of Ga, In, and Sn, or alternatively, digital thermometers, now replaces the Hg used in traditional Hg thermometers. Hg-cell technology is being replaced by newer diaphragm and membrane-cell technology at chloralkali plants. Light-emitting diodes that contain In in lamps, batteries with Li, Ni–Cd, and Zn batteries, are presently substitutes for Hg instruments.

Mercury has been known since prehistrorical times, and was used by alchemists in China since 2000 BC. For over a century, it has been known as an environmental pollutant. Released Hg with industrial wastewater in Minamata (Japan) in 1959 resulted in neurological damage of many people. Effects of Hg uses are still observed.

Increased Hg load into the environment due to its use in Au mines became a real environmental concern (mainly in Amazonian region and Siberian subregion). All metals can form amalgams with Hg. The amalgam Au + Hg used in the extraction of Au from ore is of environmental risk in the Au-mine districts. There is an estimation that about 1000 t Hg/yr is used and lost by artisanal miners.

Mercury is considered a global, hazardous pollutant, which is widespread, mobile and easily bioaccumulated. Its emission from various industries, and coal combustion is still of a great environmental concern. Current, worldwide anthropogenic Hg sources is calculated at about 2909 t/yr, and its emission from fossil-fuel-fire power plants is given as 1422 t/yr (Pirrone et al. 2009).

27.2 SOILS

The worldwide average content of Hg in soils is estimated at 1.1 mg/kg, within the range of 0.01–1.5 mg/kg; however, it seldom exceeds 1 mg/kg (Randall and Chattopadhyay 2004). Higher Hg contents are in heavy loamy soils (Table 27.1). Its lowest contents are reported for soils of Sweden, 0.043 mg/kg (Kabata-Pendias 2011). The background levels of Hg in soils are not easily established, due to the widespread Hg pollution. Soil quality Hg levels are estimated, based on various criteria, within the range of 1–23 mg/kg. The contamination of soils (Dutch List 2013), following Hg concentrations in soils and groundwater, is established (in mg/kg and μg/L) as follows: uncontaminated, 0.5 and 0.2; medium contaminated, 2 and 0.5; and heavily contaminated, 10 and 2.

Mercury contents of virgin soils are inherited mainly from the parent rocks. In some regions, however, degassing and thermal activity of the Earth may be its second sources. Soils of volcanic areas contain Hg up to 7.45 mg/kg, at Mt. Etna (Italy), and up to 0.23 mg/kg at Mt. St. Helen (United States). Mercury in soils of municipal lawns in Wrocław City (Poland) vary within the range of 0.05–1.14 mg/kg, and does not exit standard content established for urban soils, at 2.0 mg/kg (Dradrach and Karczewska 2013). Agricultural top soils of Spain contain Hg within the range of 0.001–0.22 mg/kg (Rodrigues et al. 2008). The maximum Hg contents of some Brazilian soils vary from 1.6 to 29.1 mg/kg (Melo 2012).

Mercury is highly associated with soluble organic matter (SOM) and S levels in soils, and thus it is concentrated mainly in surface layers. Especially, raw humus material reveals a great capacity for binding Hg, which is the main source of atmospheric deposition. In all soils, its highest sorption is at pH range from 4 to 5. Soil sorption

TABLE 27.1
Mercury Contents in Soils, Water, and Air (mg/kg)

Environmental Compartment	Range
Soil (mg/kg)	
Light sandy	0.08–0.7
Medium loamy	0.01–1.2
Heavy loamy	0.02–1.5
Calcerous (calcisols)	0.01–0.5
Organic	0.04–1.2
Water (ng/L)	
Rain	<20
River	1–7
Sea, ocean	0.5–3.0
Air (ng/m³)	
Urban/industrial areas	0.17–38.0
Remote regions	0.01–0.06
Greenland	0.04–0.08

Sources: Data are given for uncontaminated environments, from Kabata-Pendias, A. and Mukherjee, A.B., *Trace Elements from Soil to Human*, Springer, Berlin, Germany, 2007; Reimann, C. and de Caritat, P., *Chemical Elements in the Environment*, Springer, Berlin, Germany, 1998.

capacity is positively correlated with organic matter (OM) and cation exchangeable capacity, and is greater for organic Hg compounds than for $HgCl_2$. Concentration of Hg in soil solution is very low, at about 2.5 μg/L, and preferable as cationic species, such as Hg_2^+, Hg^{2+}, $HgCl^+$, and $HgCH_3^+$. It can also form anionic species, such as $HgCl_3^-$ and HgS_2^{2-} (Kabata-Pendias and Sadurski 2004). Hg^{2+} species are very mobile in soils and easily make complexes with various anions (Cl^-, OH^-, and S^{2-}) and SOM; therefore, Hg concentration in soil solution is relatively low.

Specific properties of Hg species in soils are as follows: (1) easy volatilization, Hg^0 and $(CH_3)Hg$; (2) easy mobility, $HgCl_2$, $Hg(OH)Cl$, and $Hg(OH)_2$; (3) low mobility, CH_3Hg^+ and CH_3HgS^-; (4) nonreactive species, HgS, $Hg(CN)_2$, Fe, and Mn complexes; and (5) Hg^{2+} bound to some SOM.

Mercury, as a contaminant in soils, occurs mainly as Hg^0, but small amounts of Hg_2^{1+} and Hg^{2+} are also present (Melo 2012). Al and Mn oxides, organic C, and SOM have a great adsorption capacity for Hg^{2+}. Also, goethite reveals a strong Hg adsorption.

The transformation of Hg compounds, especially methylation of elemental Hg, plays the most important role in the Hg cycling. The methylated Hg is readily mobile and easily taken up by living organisms. Mercury adsorbed by SOM may be both easy methylated and/or converted to the elemental state (Hg^0). Methylation processes of Hg may be abiotical, but more often are stimulated by microorganisms, which are common in anaerobic, organic-rich environments (Gray et al. 2003). These processes stimulate the reduction of cationic Hg^{2+} to the elemental state, Hg^0, which resulted in the volatilization of Hg from media. It is relatively easily vaporized from surface soil, mainly due to microbial processes. This process increased after water addition to soil.

Bacteria have an ability to adjust to increased Hg level in soils. Clay-mineral complexes on the surface of bacterial cell walls are apparently involved in mechanisms of Hg resistance (Tazaki and Asada *vide* Kabata-Pendias 2011). Mercury methylated by microorganisms is more toxic to biota than its metallic forms (Perelomov and Chulin 2013).

Mercury may form various ionic species, both cationic and anionic; however, it is not very mobile during weathering processes. Thus, Hg in soils is retained mainly as slightly mobile organocomplexes. At soil, pH near neutrality predominates $Hg(OH)_2$, which is readily adsorbed by both mineral and organic compounds.

Mercury contents in top soils greatly differ, depending on the soil origin and pollution sources. Its highest concentration (up to 2695 mg/kg) is reported for soils of Hg-mining district in Spain (Table 27.2). However, elevated Hg contents also are seen in soils from vicinities of chloralkali plants, and in soils sludged or irrigated.

TABLE 27.2
Mercury Contamination of Surface Soils (mg/kg)

Site and Pollution Source	Country	Range/Maximum
Old mining area or ore deposit	United Kingdom	0.2–3.4
Mercury-mining areas	Spain[a]	0.13–2695
Chloralkali or chemical industry	Canada	0.3–5.7
	Egypt	0.1–0.5
	France	0.1–0.3
	United Kingdom	3.8
Urban garden and park	Canada	0.02–1.14
	Israel	0.04–0.08
	Japan	0.06–0.24
	United Kingdom	0.25–15.0
	United States	0.6
Sludged or irrigated farmlands	Germany	0.43–24
	Holland[b]	10.0
	Hungary	0.02–0.15
	Japan	0.29–0.71
	Poland	0.12–0.35
	Sweden	0.8
Application of fungicides	Canada	9.4–11.5

Sources: After Kabata-Pendias, A., *Trace Elements in Soils and Plants*, 4th ed., CRC Press, Boca Raton, FL, 2011.

[a] Almaden mining district. Molina et al. (2006).

[b] Soils flooded by Rhine River water.

The greatest amounts of Hg came from sewage sludge, which may contain it up to 6 mg/kg. Total amount of Hg in waste of 11 EU countries, in 1999–2000, has been estimated at 752 t; however, the authors have pointed out that several countries have not calculated this carefully (Mukherjee et al. 2004).

Besides industrial sources of Hg, it has been added to agricultural soils with fungicides and seed dressing (now prohibited, but used for over 20 years). Also, fly ash used for the land reclamation, and to limit the mobility of Cd, Cu, Pb, and Zn, may be a serious source of Hg in soils (Kumpiene et al. 2005).

Mercury, as a contaminant is soils, occurs mainly as Hg^0, but small amounts of Hg_2^{1+} and Hg^{2+} are also present (Melo 2012). Methylated Hg is present as a result of natural microbial transformation. Its complexes in soils are as follows: Hg–Fe, Mn > Hg residual > Hg–OM (Melo 2012). Soil Hg contamination is often an effect of its anthropogenic deposition over the last 100 years (Fitzgerald *vide* Kabata-Pendias and Mukherjee 2007).

27.3 WATERS

The worldwide mean Hg concentration in seawater is estimated at 30 ng/L (Reimann and de Caritat 1998). However, it widely varies depending on sampling sites; in water from open seas, it is within the range of 0.5–3.0 ng/L, and in coastal and bay water, it varies from 2 to 65 ng/L. Uncontaminated rivers and streams contain Hg within the range of 1–7 ng/L. It occurs in water in several forms such as Hg^0, Hg^+, Hg^{2+}, $HgCl_2^0$, HgS_2^-, and in various methylated species (Me–Hg). Up to 60% of Hg^{2+} and Me–Hg are particle-bound, and the rest of Hg is dissolved or dissolved organic carbon-bound phases (Nriagu *vide* Kabata-Pendias and Mukherjee 2007). Unfiltered water from Hg mine in Spain contain Hg up to 13,000 ng/L and methyl Hg up to 30 ng/L, and mine water in California contain up to 19,000 and 4.5 ng/L, respectively (Gray et al. 2004). Several techniques are proposed for cleaning wastewater from Hg, by adding Na sulfide or some coagulants (e.g., $FeCl_3$). These methods may clean up water with Hg concentration <10,000 ng/L (Chen 2013).

Mercury methylation and bioaccumulation in water depend on several parameters, such as Hg loading, microbial activity, pH and redox potential, temperature, and several other factors. Me–Hg species are relatively easily accumulated by water biota (Perelomov and Chulin 2013). A great part (>95%) of total Hg in both, fresh and saltwater fish is in the form of Me–Hg. It is taken up by fish straight from the water, as well as from their fodder. In general, the highest Hg content is in fish muscles and the lowest is in gonads. In muscles of fish from the Baltic Sea, its contents vary between 0.2 and 0.33 mg/kg, whereas in female gonads its range is 0.02–0.05 mg/kg (Voigt *vide* Kabata-Pendias and Mukherjee 2007). The highest means Hg levels in commercial fish in the United States are reported to be 0.73 and 1.45 mg/kg FW, in mackerel king and tilefish, respectively; and the lowest contents are 0.01 and 0.11 mg/kg FW, in salmon and cod, respectively.

Mercury contamination of fish from gold-mining areas in America, Asia, and Africa has been investigated. Hg levels in fish from Brazilian Amazon, near mining areas, varied from 2.21 to 6.11 mg/kg in Caras and Trairas fish, respectively.

Especially important factor is Hg deposition in bottom sediments. Its content of surface-bottom sediments of harbor in Klaideda (Lithuania) depends not only on its concentration in water but also on the granulometric composition of sediments, and is (in mg/kg, average and maximum, respectively) in sand 0.02 and 0.07 and in mud 0.04 and 0.15 (Galkus et al. 2012). Mercury contents in stream-bottom sediments of National Park, Montgomery (Pennsylvania State) were, in 1995, within the range of 0.04–0.12 mg/kg (Reif and Sloto 1997).

Assessment limits for Hg in sediments are established as follows (in mg/kg): effects range low, 0.15; effects range median, 0.71; probable effect level (PEL), 0.49; 2; 4 (EPA 2000, 2013). The Environment Canada sediment-quality guidelines (USGS 2001) gave the following values for Hg in the lake-bottom sediments (in mg/kg): threshold effects level, 0.17; PEL, 0.49; and probable effect concentration, 1.06.

Median Hg concentration in both bottled and tap water of the EU countries is <5 ng/L (Birke et al. 2010). Guideline value for inorganic Hg in drinking water is established at 6 μg/L (WHO 2011a).

27.4 AIR

The worldwide mean Hg concentration in the atmosphere of remote regions varies within the range of 0.01–0.06 ng/m^3, whereas in urban/industrial areas it is between 0.17 and 38.0 ng/m^3. Mercury content of air above Greenland varies from 0.04 to 0.08 mg/m^3 (Table 27.1).

Mainly, vapor-phase elemental Hg^0 (total gaseous mercury—TGM) is present in the atmosphere. There is also $HgCl_2$, but represents less than 5% of the total Hg concentration. It may occur in both forms, as reactive gaseous mercury and total particulate mercury. Metallic Hg^{2+} present in air is relatively quickly deposited within the distance of about 100 km from its sources, whereas Hg^0 is insoluble in water and thus is transported, as particles, for long distances. Mercury adsorbed by small air particles (10 μm) is more mobile in the atmosphere and easily inhaled by people than that fixed by coarse particles (>10 μm). Mean Hg content of rainwater is estimated at <0.002 μg/L.

Significant Hg sources in air are coal combustion and the inputs of various industrial emissions. Global anthropogenic Hg fraction constitutes 40%–50% of its total emission (Horvart et al. *vide* Kabata-Pendias and Mukherjee 2007). There are also natural sources of Hg emission, which are from (1) the earth surface; (2) previously deposited Hg waste; (3) volcanoes; and (4) geothermal activities. Mercury is also emitted from the boreal forest and wetlands. The highest emissions (up to 3.5 ng/m^2/h) were observed at the forest floor with the moss and grass cover. However, there was also noticeable uptake of Hg by the forest floor, depending on weather conditions.

There is the lack of exact quantification of the Hg emission from natural sources. However, there are suggestions that natural sources account for about 10%, estimated at 5.5–8.9 kt of Hg currently being emitted and re-emitted to the atmosphere from all sources. Nevertheless, there are estimations that the Hg deposition to the Greenland ice sheet is significantly influenced by anthropogenic inputs from North America, Asia, and Europe, linked especially to coal burning and solid waste incineration. There are some suggestions that Hg concentrations were higher in snow dated from the late 1940s to the mid-1960s, than in more recent snow. Luckily, various forecasts for the Hg emission in Europe are lower for 2020 than those for 2010. However, these estimations highly vary and are 64, 125, and 180 t/yr, depending on the authors. The general tendency in reducing Hg emission from the anthropogenic sources is clearly shown by its significant decrease in the 1990s (by 60%) compared to its emission in the 1970s (Pacyna and Pacyna 2001).

Mosses are very sensitive indicators for Hg concentration in air. Its average content was 0.08 mg/kg of moss sampled in Norway during 1990–1995 (Berg and Steinnes 1997). Mean Hg content of mosses (*Hylocomium splendens*) from Poland was 0.084 mg/kg, whereas of those collected in Southern Alaska was 0.131 mg/kg (Migaszewski et al. 2009). The authors concluded that this variation in Hg contents of mosses is linked to both the location of emission point sources and underlying geology.

27.5 PLANTS

Mercury is easily phytoavailable; therefore, in most cases, its concentrations in plants increase with its elevated contents in soils. Plants differ in their ability to uptake Hg, and can develop a tolerance to its high contents in growth media. Plants uptake Hg mainly by roots, where it is accumulated. Its translocation to shoots is relatively small. This is well illustrated by *Asparagus acutifolius* from the Almaden district (Hg mine in Spain), which contains Hg (in mg/kg) 7.7 in stems and 298.2 in roots. Mercury is also taken up from the atmosphere. Some plants, such as lettuce, spinach, and mushrooms are likely to take up more Hg than other plants, most probably also from the air (Table 27.3). Some plants (e.g., Indian mustard, *Brassica juncea*) reveal the ability to uptake great amounts of Hg from contaminated soil. The highest contents of Hg in this plant grown in soils with Hg at 1000 mg/kg, was as follows (in mg/kg): 264–325 in shoots and 1775–2089 in roots (Su et al. *vide* Kabata-Pendias 2011).

All forest berries grown in the polluted area of the Northern Europe (Kola Peninsula) contain similar amounts of Hg, <0.04 mg/kg (Reimann et al. 2001). Common mushroom, chanterelles (*Cantharellus cibarius*) grown in mountains of Poland and in the Baltic Sea coast contain similar amounts of Hg, about 0.037 mg/kg (Falandysz et al. 2012). Another mushroom, red aspen boletes (*Leccinum aurantiacum*) contain Hg within the range of 0.27–1.3 mg/kg. Bioconcentration factor was much higher for mushroom caps than for stipes (Falandysz et al. 2012).

Mercury content of plants has recently received much attention because of its pathway into the food chain. The background levels of Hg in vegetables and fruits vary from about 0.003 to 0.09 mg/kg DW, and from 0.006 to 0.07 mg/kg FW. However, plants cultivated in industrial regions may contain its higher concentrations. The Hg contents of cereal grains seem to be fairly similar for the same kind of plants grown in various countries, and vary from <0.0001 to 0.034 mg/kg. In countries where

TABLE 27.3
Mercury Content of Plants Grown in Contaminated Sites (mg/kg)

Site and Pollution Source	Country	Plant and Part	Range/Maximum
Mining area and metal-processing industry	Yugoslavia	Edible mushroom	37.6
	Yugoslavia	Carrot, roots[a]	0.5–0.8
Chloralkali or chemical industry	Egypt	Lettuce, leaves	0.09–0.35
	Egypt	Radish, roots	0.03–0.29
	Finland	Edible mushroom	1.10–4.70
	Switzerland	Spinach, leaves	0.11–0.59
	Switzerland	Corn, grain	0.07–0.14
	Finland	Edible mushroom	72–200
Sludged or irrigated farmland	Canada	Bromegrass[b]	0.09–2.01
	Japan	Brown rice, grain	4.9
Application of fungicides	Canada	Potato, leaves	1.1–6.8
	Canada	Lettuce, leaves	0.1–0.3
	Sweden	Oat, grain[b]	0.63

Source: After Kabata-Pendias, A., *Trace Elements in Soils and Plants*, 4th ed., CRC Press, Boca Raton, FL, 2011.

[a] Fresh weight basis.
[b] Pot experiment.

dressing of seeds with mercuric compounds was used in the last century, Hg content in crop seeds was elevated up to 0.17 mg/kg.

Mercury has a strong affinity for sulfhydryl/thiol groups that are involved in enzymatic reactions. Thus, it inhibits synthesis of proteins in plant leaves, and disrupts some metabolic processes (Chen 2013). Especially inhibited are photosynthesis, chlorophyll formation, exchange of gases, and respiration. Also, limitations of seedling growth and root development are observed.

Although Hg is strongly bound to amino acids of several enzymes and proteins, especially S-rich proteins, it seems to be relatively easily transported within plants. In roots it is associated with high-weight molecules. Toxic effects in young barley were observed at the Hg level of 3 mg/kg. However, the critical Hg levels vary, depending on plants from 1 to 8 mg/kg. The toxicity of volatilized elemental Hg, and of some methylated compounds is the most serious for plants, especially young ones.

The tolerance to increased levels of Hg in higher plants also was reported, which may be due to its inactivation at the membrane sites. Particularly resistant to high Hg concentrations is the transgenic plants, *Arabidopsis thaliana*, with the genes for mercuric ion reductase (reduction of toxic Hg^{2+} to the relatively inert Hg^0). This plant, as well as others with such a gene, may be useful as a phytoextractive plant to grow on Hg-contaminated sites.

27.6 HUMANS

Mercury content in human body has been estimated at 6 mg, of which in the blood 8 µg/L, in the bone 0.5, and in the tissue 0.2–0.7 mg/kg (Emsley 2011). It accumulates in the thyroid and pituitary glands, in the brain, kidneys, liver, pancreas, testes, ovaries, and the prostate gland. It is concentrated in the kidneys at 50%–90% of the body burden. Its levels (mean and range, in µg/kg) are as follows: 6.7; 2.4–12.2 in the occipital cortex; 49, 21–105 in kidneys; 6.7, 1.9–22.1 in the gray brain matter; and 3.8, 1.4–7.1 in the white brain matter (WHO 1991a).

Exposure to Hg is commonly evaluated using Hg concentrations in blood and urine, although hair may also be used as a biomarker of Hg exposure. The concentration of total Hg in blood is accepted as a reasonable biomeasure of MeHg exposure. The WHO has estimated that the average total blood Hg concentration, for the general population is approximately 8 µg/L; however, individuals with high fish consumption can have concentrations in blood as high as 200 µg/L. Typical total Hg concentrations in urine have been reported to be in the 4–5 µg/L range (Health Canada 2010b).

The health effects of Hg are diverse and depend on its chemical form, the dose, the duration of exposure and the exposure route, and also dietary content of interacting elements, especially Se. There are three forms of Hg from a toxicological point of view: metallic Hg, inorganic Hg salts, and organic Hg compounds. The most common organic Hg compounds in nature are methylmercury (CH_3Hg^+, MeHg) and dimethylmercury ($[CH_3]_2Hg$). Metabolism of Hg compounds to other forms of Hg can occur within the tissues of the body.

Amounts of Hg added to some pharmaceuticals, such as thiomersal (ethyl mercury), which is used as a preservative in some vaccines, are very small by comparison with its other sources. Claims have been made that thimerosal in vaccines may be a cause of autism and related disorders; however, this theory has yet to be confirmed.

All forms of Hg have adverse effects on health at high doses. However, the evidence that exposure to very low doses of Hg from fish consumption, the receipt of dental amalgams, or thimerosal in vaccines has adverse effects is open to wide interpretation (Clarkson et al. 2003).

There are substantial differences in the toxicity of elemental Hg metal, inorganic Hg salts, and organometallic Hg compounds. Less than 0.01% of Hg^0, approximately 10% of inorganic Hg and 95% of organic Hg, is absorbed from the gastrointestinal tract following oral ingestion (Langford and Ferner 1999).

Methylmercury is the most toxic form. It affects the immune system, alters genetic and enzyme systems, and damages the nervous system, including coordination and the senses of touch, taste, and sight. Methylmercury is particularly damaging to developing embryos, which are five to ten times more sensitive than adults. Dimethylmercury can be extremely toxic. However, the exposure to MeHg is usually by ingestion, and it is absorbed more readily and excreted more slowly than other forms of Hg. Elemental Hg(0), the form released from broken thermometers, causes tremors, gingivitis, and excitability when vapors are inhaled over a long period of time.

The toxicity of Hg salts varies with their solubility. Inorganic Hg salts are water soluble, irritate the gut, and cause severe kidney damage. Following oral exposure, inorganic Hg salts show limited absorption, which is related to their water solubility. Inorganic Hg compounds are not lipid soluble and do not readily cross the blood–brain barrier or placenta membranes. Ionic species of inorganic Hg readily bind to sulfhydryl groups of various thiol-containing compounds, such as glutathione, cysteine, and metallothionein. Kidneys exhibit the greatest concentration of Hg following exposure to inorganic Hg compounds (WHO 2011b).

Ingestion of other common forms of Hg, such as the salt $HgCl_2$, which damages the gastrointestinal tract and causes kidney failure, is unlikely from environmental sources.

Exposure to elemental Hg is hazardous because of its potential release of toxic Hg vapor, which is readily absorbed into the body through inhalation. If elemental Hg is ingested, it is absorbed relatively slowly and may pass through the digestive system without causing damage. The vapor pressure at room temperature is about 100 times the safe amount, so poisoning can occur if Hg metal is spilled into crevices or cracks in the floorboards, so that heating metallic Hg greatly increases the associated risks, as inhalation is the usual route of toxicity. Dentists are occasionally poisoned this way. Inhalation of Hg vapor may cause respiratory, cardiovascular, kidney, and neurological effects (Health Canada 2010).

Inhaled Hg vapor accumulates in the body, in particular the central nervous system, which is the site of its major toxic actions. Approximately 70%–80% of Hg vapor is absorbed in humans. Once absorbed, a proportion of Hg is taken up by the red blood cell, while some remains in the bloodstream, allowing its rapid distribution around the body, including the central nervous system. Within the red blood cells, liver, and central nervous system, the metal is oxidized through the catalase–peroxide pathway to Hg^{2+} oxide. Mercury excretion from the body starts almost immediately after absorption, following a variety of routes, though principally by the kidneys. Chronic poisoning markedly affects the central nervous system and kidney.

The organic Hg compounds are very lipid soluble: 90%–100% of an oral dose is absorbed. Organic Hg compounds, which are fat soluble, can cross the blood–brain barrier and cause neurological damage, but inorganic Hg salts are not lipid soluble, and do not cross this barrier in significant amounts. Usually Hg^{1+} compounds are of low solubility and significantly less toxic than Hg^{2+} compounds. Inorganic Hg salts present a far greater hazard than elemental Hg if ingested orally, owing to their greater water solubility. For Hg^{2+} chloride, the lethal dose may be as small as 0.5 g, compared with 100 g Hg metal. Mercury salts are usually nonvolatile solids, so poisoning by inhalation is rare, though toxicity may arise if aerosols are deposited in the lungs. Once adsorbed, the Hg^{1+} form will readily react with the thiol groups of amino acids such as cysteine. The majority of the dose of an ingested inorganic Hg salt accumulates either in the liver, where it is excreted in the bile, or in the kidney, where it is excreted in the urine (Langford and Ferner 1999).

Following absorption, organic Hg is distributed to all tissues, including hair, with highest accumulation in the kidneys. Organic Hg is demethylated in the body to inorganic Hg, which accumulates primarily in the liver and kidneys. MeHg is estimated to have a biological half-life of approximately 50 days. The majority of Hg in the body is excreted through the feces, with a small amount excreted as inorganic Hg in the urine.

The primary effects associated with oral exposure to organic Hg compounds are neurological effects and developmental neurotoxicity; effects are similar for both acute and chronic exposure. Because MeHg is hydrophobic, it easily penetrates the cell membrane and inactivates the enzymes that participate in protein synthesis within the central nervous system. Especially, it disrupts cell division, formation of microtubule and movement of neuroaxis. Symptoms of organic Hg toxicity include a tingling sensation in the extremities; impaired peripheral vision, hearing, taste, and smell; slurred speech; muscle weakness and an unsteady gait; irritability; memory loss; depression; and sleeping difficulties. Exposure of a fetus or young child to organic Hg can result in effects on the development of the nervous system, affecting fine motor function, attention, verbal learning, and memory. The gastrointestinal effects include nausea, vomiting, and abdominal pain; higher doses can cause diarrhea and an exposure-related colitis. Other symptoms include the discoloration of the gums (similar to poisoning with inorganic compounds), sialorrhoea, and perioral paraesthaesia. Central nervous effects after slight exposure include numbness in the limbs; as the level of exposure increases, there are tremors, ataxia, dysarthria, and visual field constrictions (Langford and Ferner 1999).

The chronic toxicity of MeHg is best exemplified by the epidemic poisonings in Japan and Iraq. In Minamata and Niigata, Japan, MeHg poisoning resulted from the ingestion of fish that had accumulated MeHg and other Hg compounds that were released from industrial sources into surface water. By a process of biomagnification, the MeHg accumulated to sufficient amounts in predatory fish to poison over 800 people, who relied on the fish for their nutrition. Many of them were infants *in utero*.

In Iraq, over 6000 individuals were hospitalized and 459 individuals died, as a result of consuming bread prepared with flour made from wheat and barley treated with a methylmercurial fungicide. Methyl Hg concentration in the wheat flour ranged from 4.8 to 14.6 mg/kg. The clinical symptoms included paresthesia, visual disorders, dysarthria, and deafness.

International Agency for Research on Cancer (IARC) in 1993 determined that MeHg compounds are possible human carcinogens (Group 2B), based on inadequate human data but sufficient animal data showing a link to certain cancers, particularly renal cancer. Metallic Hg and inorganic Hg compounds were determined to be not classifiable as to their carcinogenicity (IARC 1993, 2013).

WHO established a provisional tolerable weekly intake (PTWI) for inorganic Hg of 4 μg/kg bw. The previous PTWI of 5 μg/kg bw for total Hg was withdrawn. The new PTWI for inorganic Hg was considered applicable to dietary exposure to total Hg from foods other than fish and shellfish (WHO 2011b). European legislation set a maximum level for total Hg in fish and fishery products of 0.5 mg/kg FW, except for some predatory species with limit 1.00 mg/kg FW (EC 2006).

Total Hg levels in foods other than fish products are generally low (range 0.0001–0.050 mg/kg). The highest levels are found in fungi (Table 27.4).

TABLE 27.4
Total Hg in Foodstuffs (mg/kg FW)

Product	Content
Food of Plant Origin	
Bread[a]	<0.003
Miscellaneous cereals[a]	<0.003
Wheat[b]	0.0028
Rye[b]	0.0020
Green vegetables[a]	<0.0005
Cabbage[b]	0.0005
Potatoes[b]	0.0006
Mushroom fresh, various[c]	0.028
Apple[b]	0.0004
Food of Animal Origin	
Meat and meat products[a]	<0.002
Offal[a]	0.004
Freshwater fish[d]	0.01–0.19
Fish[a]	0.056
Shellfish[e]	0.017
Fish[f]	0.006–1.5
Seafood other than fish[f]	0.004–0.067

Sources: [a] Rose, M. et al., *Food Addit. Contam.*, 27, 1380–1404, 2010.
[b] Jędrzejczak, R., *Food Addit. Contam.*, 19, 996–1002, 2002.
[c] Wojciechowska-Mazurek, M. et al. *Bromat. Chem. Toksykol.*, 44, 143–149, 2011.
[d] Lidwin-Kaźmierkiewicz, M. et al., *Pol. J. Food Nutr. Sci.*, 59, 219–234, 2009.
[e] Leblanc, J.C. et al., *Food Addit. Contam.*, 22, 624–641, 2005.
[f] Chung, S.W.C. et al., *Food Addit. Contam.*, 25, 831–840, 2008.
FW, fresh weight.

The mean intake of total Hg depends on structure of consumption. In the United Kingdom, it was estimated at 1.0–3.0 μg/day (Rose et al. 2010). Students in Hong Kong intake 7.9 μg Hg daily, and fish (59%), shellfish (14%), and cereals (11%) made the greatest contribution to the dietary exposure (Chung et al. 2008). The mean dietary intake of Hg in various countries ranges from 2 to 20 μg/day per person (WHO 2011b). Average dietary exposure to total Hg (μg/kg bw, per day) is estimated in range: the United States 0.01–0.02 (96%–100% from fish and shellfish) to China (Zhoushan Island) 0.47–0.92 (87%–95% from fish and shellfish) (WHO 2011b).

Fish is a dominant source of human exposure to MeHg (Table 27.5). According to the WHO (2011b), total Hg levels in shellfish samples ranged from 0.002 to 0.86 mg/kg. No shellfish species contained MeHg at concentrations > 0.5 mg/kg

TABLE 27.5
Mercury and Methylmercury in Some Fish (mg/kg FW)

Fish	Total Hg	MeHg
Canned tuna[a]	0.001–2.581	0.003–1.0307
Fish (cod, salmon)[a]	0.012–2.529	0.021–0.507
Sword fish[b]	–	0.16–1.53

Sources: [a] Torres-Escribano, S. et al., *Food Addit. Contam.*, 27, 327–337, 2010.
[b] Park, J.-S. et al., *Food Addit. Contam.*, 4, 268–274, 2011.
FW, fresh weight.

(range 0.002–0.451 mg/kg), with the maximum concentration found in edible crab. Total Hg levels in fish samples ranged from 0.001 to 11.4 mg/kg, with the maximum concentration in marlin.

The proportion of total Hg contributed by MeHg generally ranged between 30% and 100%, depending on the species of fish, size, age, and diet. Furthermore, in about 80% of these data, MeHg accounted for more than 80% of total Hg. However, a few submitted data showed proportions of methylmercury of about 10% or less (WHO 2011b).

27.7 ANIMALS

Mercury concentration of animal tissues is various and highest is in fish muscles (Table 27.6). Content of Hg of cattle muscle and liver does not change with age (Rudy 2009).

TABLE 27.6
Mercury in Animal Tissues (mg/kg FW)

	Tissue	
Animal	**Muscle**	**Liver**
Cattle	<0.001	0.002
Pigs	0.001	0.001
Hunting animals	0.004	0.010
Poultry	<0.001	0.001
Sheeps	<0.001	0.003
Rabbits	<0.001	0.001
Fish	0.036	–

Source: Szkoda, J. et al., *Ochr. Śr. Zasobów Nat.*, 48, 475–484, 2011.
FW, fresh weight.

28 Molybdenum [Mo, 42]

28.1 INTRODUCTION

Molybdenum (Mo), a metal of the group 6 in the periodic table of elements, occurs in the Earth's crust at the mean content of 1.5 mg/kg, within the range of 1–2 mg/kg. In acidic magmatic rocks and granites, its content is up to 2 mg/kg, and in argillaceous sediments, up to 2.5 mg/kg. In coal, its contents vary between 3 and 6 mg/kg, and in fly ash, it may be concentrated up to 15 mg/kg.

Molybdenum reveals lithophilic properties, and may occur at various degrees of oxidation, from +2 to +6, but most often has the highest degree. Its behavior is unusual, as in the most environmental compartments it is likely to form oxyanions. The most common Mo mineral is molybdenite, MoS_2; less frequent minerals are molibdite, MoO_3; wulfenite, $PbMoO_4$; and powellite, $CaMoO_4$. Several other Mo minerals, mainly oxidates, are present in various sediments and are associated mainly with Fe and Ti minerals. Especially Mn–Fe nodules reveal a great capability to concentrate Mo, up to 2000 mg/kg (Fairbridge *vide* Kabata-Pendias 2011). Under anoxic and S-rich environments, which dominate in sea-bottom sediments, Mo may be accumulated in Mn–Fe nodules, up to about 500 mg/kg (Szefer 2002).

Molibdenite, a primary Mo ore deposit, is a major source of this metal. It is also a by-product of some metal ores, mainly of Cu and W. World mine Mo production (rounded) in 2010 was 234 kt, of which China produced 94; the United States produced 56; Chile produced 39; and Peru produced 12 (USGS 2011).

Molybdenum is an alloying element in steel and cast-iron production. It is also used for refractory materials in high-temperature electric furnaces, and as catalyst in chemistry. It is used in some red and orange pigments.

28.2 SOILS

The worldwide average soil content of Mo is established at 1.1 mg/kg, within the range of <1–2 mg/kg. However, its higher concentrations, up to about 7 mg/kg, are also noticed in soils (Table 28.1). Mean Mo content in organic surface soils of Norway is 0.56 mg/kg, and is a bit lower than mean value (0.67 mg/kg) for all investigated soils (Nygard et al. 2012).

It behaves like both a chalcophille and a lithophille element, and in aerobic condition, mainly as molybdate oxyanion, MoO_4^{2-}. Molybdenum is likely to form various complex compounds and minerals in soils, for example, wulfenite, $PbMoO_4$; ferrimolibdite, $Fe_2(MoO_4)_3 \cdot 8H_2O$, as well as other semicrystalline Fe–Mo forms. Under reducing conditions, Mo easily form soluble thiomolybdates (e.g., MoS_4^{2-}).

During weathering processes, Mo in minerals is oxidized to oxyanions, which occurs mainly at lower pH values. These Mo anions are readily coprecipitated by organic matter (OM), $CaCO_3$, and by several cations. Strong adsorption of Mo reveal

TABLE 28.1
Molybdenum Contents in Soils, Water, and Air (mg/kg)

Environmental Compartment	Range/Maximum
Soil (mg/kg)	
Light sandy	0.1–3.7
Medium loamy	0.4–6.4
Heavy loamy	0.7–7.2
Calcerous (calcisols)	0.3–7.4
Organic (histosols)	0.3–7.4
Water (μg/L)	
Rain	0.01–2.6
River	0.04–1.3
Sea	1.6–6.3
Air (ng/m³)	
Industrial regions	1.0–10.0
Remote regions	<0.2

Sources: Data are given for uncontaminated environments, from various sources, mainly: Kabata-Pendias, A. and Mukherjee, A.B., *Trace Elements from Soil to Human*, Springer, Berlin, Germany, 2007; Reimann, C. and de Caritat, P., *Chemical Elements in the Environment*, Springer, Berlin, Germany, 1998.

hydroxides of Al, Fe, and Mn. Especially Mn concretions have a great capacity to fix Mo, up to above 400 mg/kg (Kabata-Pendias and Sadurski 2004). Behavior of Mo in soils is highly dependent on pH and Eh conditions, and thus is very variable. The mobility and availability of Mo to plants is highly governed by soil pH and drainage conditions. In acidic soils (pH < 5.5), and especially in those with a high Fe oxide levels, Mo is hardly available to plants, whereas from wet alkaline soils, it is most easily taken up by plants. The low phytoavailability of Mo, which sometimes occurs in peat soils, is affected by the strong fixation of Mo by humic acid, and possibly by other OM. The highest proportion of Mo, up to about 50% of its content, is fixed by various OM compounds. Water soluble fraction of Mo is about 4%, and residual fraction, about 15% of its total contents.

Liming of acid soils is a common practice to increase Mo availability to plants. However, at higher rate of liming, Mo mobility may decrease due to its adsorption by $CaCO_3$. Increased Mo phytoavailability should be controlled, as its elevated contents in fodder are not good for animals. Soils with high amounts of Mo, mainly ferrasols, and other soils in arid and semiarid regions, should be ameliorated to decrease Mo contents in plants. The application of S compounds is most effective, whereas increased P level in soils may stimulate Mo uptake by plants. Some Mo compounds (e.g., Na_2MoO_4) may be used as addition in fertilizers.

Various types of industrial pollution may be responsible for elevated Mo contents in soils. Some sewage sludge contain high levels of Mo, up to about 50 mg/kg (Lahamann *vide* Kabata-Pendias 2011). Also, fly ash from some coal-fired power plants may be a potential Mo source.

Contamination of soils (Dutch List 2013), following Mo concentrations in soils and groundwater, is established (in mg/kg and µg/L) as follows: uncontaminated, 10 and 5; medium contaminated, 40 and 20; and heavily contaminated, 200 and 100.

28.3 WATERS

The concentrations of Mo in seawater range from <0.1 to 10 µg/L. Water of the Baltic Sea contain Mo within the range of 1.6–6.3 µg/L (Szefer 2002). Contents of Mo in terrestrial water are also highly variable, within the range of 0.04–1.3 µg/L (Table 28.1).

Molybdenum in seawater is relatively unreactive; in oxygenated conditions its concentration is rather stable, whereas in anoxic water, it readily precipitates in bottom sediments. The pH–Eh water regime control both Mo precipitation and speciation. Various anionic Mo species predominate in water of different pH as follows: pH < 5: $HMoO_4^-$; pH > 5: MoO_4^{2-}; pH < 6: $Mo_7O_{24}^{6-}$; and $Mo_8O_{26}^{4-}$ (Anbar 2004). Molybdenum easily forms various compounds with Cl and F.

Rainwater sampled in Sweden, in 1999, contained Mo within the range of 0.03–0.06 µg/L (Eriksson 2001a), whereas in rain collected above the Baltic Sea, in about the same period, its contents varied between 0.08 and 0.27 µg/L (Szefer 2002).

Median Mo concentration in bottled water of the EU countries is 0.275 µg/L, and is a bit higher than in tap water, estimated as 0.233 µg/L (Birke et al. 2010). As molybdenum occurs at very low concentrations in drinking water, it is not considered necessary to set a formal guideline value. Reason for not establishing a guideline value is that Mo occurs in drinking water at concentrations well below that of health concern (WHO 2011a).

28.4 AIR

Molybdenum concentration in the atmosphere highly vary, from <0.2 to 10 ng/m³, in remote and industrial regions, respectively (Table 28.1). Its highest contents are in heavy industrial, and especially in steel works, vicinities. Relatively high Mo concentrations are present in airborne ash expelled usually during the combustion of fossil fuels.

Its bulk deposition in Kola Peninsula is estimated (in kg/km²/yr) at 0.043 in remote areas, and 0.923 in polluted areas (Reimann and de Caritat 1998).

Range of Mo contents of moss (*Hylocomium splendens*) from the mountains in Poland (0.1–0.3 mg/kg) are fairly similar to those from Alaska (<0.05–0.3 mg/kg). This may suggest similar Mo atmospheric deposits in both regions (Migaszewski et al. 2009).

28.5 PLANTS

Molybdenum is essential to plants, but physiological requirement for this element is relatively low. Its phytoavailability increases with soil pH, and is proportional to its concentration in the soil solution. Molybdenum is moderately mobile in plants, and there is an assumption that Mo–S amino acid complexes are involved in this process. Also, some proteins are identified as transporters of Mo within plant cells (Fitri et al. *vide* Kabata-Pendias 2011).

Molybdenum is the essential component of several enzymes, such as nitrate reductase, xanthine dehydrogenase, sulfite oxidase, and several other enzymes. It is also involved in enzymatic activities of DNA and RNA. The basic enzymatic role of Mo is its functions a redox carrier, and is apparently reflected in its valence change. Plant requirement for Mo is related to N supply; plants supplied with NH_4–N have less need for Mo, than those supplied with NH_3–N. It is essential to microorganisms. Especially *Rhizobium* bacteria, as well as other N-fixing microorganisms have a large requirement for Mo, which may be concentrated up to 100 mg/kg. Number and weight of root nodules of clover significantly increase with elevated mobile Mo species in soils.

The most important Mo function in plants is NO_3 reduction, thus its deficiency in plants is similar to those of N deficiency. The correction of Mo deficiency may be accomplished by Mo application in soil and foliar or seed treatments. However, the preferable control of Mo deficiency is liming the soil to a pH around 6.5, as uncontrolled increase of Mo, especially in fodder plants, may be toxic to animals.

Several complex interactions between Mo and other elements have an impact on both its availability and physiological function.

Mo–Cu antagonism is strongly related to N and S metabolisms, and has variable impact on Mo uptake by plants. Soil factors that increase Mo availability usually have inhibitory effects on the Cu uptake. Also, increased levels of SO_4^{2-} reduce the Cu absorption, which may have an impact on undesirable Mo–Cu ratio in fodder plants. There are several interactions between Mo and other elements:

- Mo–Mn antagonism is associated with soil acidity; liming of soil for the correction of Mo deficiency may increase Mn toxicity.
- Mo–Fe interaction is relatively complex, but is associated mainly with Fe–Mo precipitates within root tissues that may limit Mo mobility.
- Mo–V and Mo–W interactions are due to possible substitution between these elements in several biochemical processes.
- Mo–P interaction is contradictory, as P fertilizers on Mo mobility is variable, depending on some other soil factors.
- Mo–Ca interaction is complex and highly cross-linked with the soil pH.

In general, Mo contents of plants, within the range of 0.1–0.5 mg/kg, are sufficient for their metabolism, whereas concentrations in the range of 10–50 mg/kg are toxic to most plant species. Only some legumes and cruciferous plants require more Mo.

Molybdenum concentrations in plants closely reflect its mobile pool in soils, as it is readily taken up by plants, when present in soluble species. Molybdenum is always more available from neutral and alkaline soils than from acidic ones. Some plants, particularly legumes, may accumulate very high amounts of Mo, up to 350 mg/kg, without toxicity symptoms.

Plant foodstuffs contain variable amounts of Mo, within the range of 0.07–1.75 mg/kg, with legume vegetables being in the upper range, and fruits being in the lower range. In cereal grains, Mo average contents are reported within the range of 0.5–1.0 (Eriksson 2001a). Mean Mo contents of fodder plants in various countries vary between 0.3 and 1.4 mg/kg in grasses, and 0.2–2.5 in legumes. In forage plants from areas where Mo toxicity in grazing animals was observed, mean contents of this element range (in mg/kg) from 1.5 to 5.0 in grasses and from 5.2 to 26.6 in legumes.

Vegetables grown in Mo-polluted soils, near an Mo-processing plant, accumulated this element from 124 to 1061 mg/kg in lettuce and cabbage, respectively (Hornick et al. *vide* Kabata-Pendias 2011). Vegetables grown in soil amended with municipal sludge ashes contain Mo within the range of 18–19 mg/kg, contrasted to the control values of 0.4–0.8 mg/kg (Furr et al. *vide* Kabata-Pendias 2011).

28.6 HUMANS

The human body contains about 5 mg of Mo (Emsley 2011). It occurs in higher concentrations in the liver, kidneys, and bones. Its contents in human tissue are (in mg/kg DW) in the liver, 1.3–2.9; in the kidneys, 1.6; in the lungs, 0.15; in the brain and muscles, 0.14; and in hairs, 0.07–0.16 mg/kg (EFSA 2009a). Normal Mo blood levels are 2–6 μg/L in the whole blood and 0.55 μg/L in the serum. Raised levels of Mo appear in adrenals and fat. Biological half-life of Mo may be up to several weeks in humans.

Molybdenum is considered to be an essential trace element for organisms. It functions as a cofactor for a number of enzymes that catalyze important chemical transformations in the global C, N, and S cycles. Thus, Mo-dependent enzymes are required for human health, as well as for plants.

Molybdenum functions as a cofactor for some enzymes, such as sulfite oxidase, xanthine oxidase (XO), and aldehyde oxidase, which are involved in sulfur amino acid metabolism, and purine metabolism. Of these enzymes, sulfite oxidase is known to be crucial for human health. The activity of XO is directly proportional to the amount of Mo in the body. However, an extremely high concentration of Mo reverses the trend, and can act as an inhibitor in purine catabolism and other processes. Its concentrations also affect protein synthesis, metabolism, and growth.

There is no Mo bioaccumulation, as its tissue levels are rapidly returning to normal contents, once the exposure stops. Increased exposure at the work place, or through drinking water, is balanced by increased urinary excretion. Serum levels of Mo rise at liver functional defects, hepatitis, hepatic tumors, and after certain drugs. Raised blood levels are seen in uremia, rheumatic disorders, and cardiovascular disease (EFSA 2009a).

Ingestion of food is a route of Mo exposure of the general population. In humans, 30%–70% of dietary Mo is absorbed from the gastrointestinal tract. Following gastrointestinal absorption, 25% of absorbed Mo rapidly appears in the blood and other organs. It easily crosses the placental barrier. There is no apparent bioaccumulation of Mo in human tissues (WHO 2011d).

Observations of Mo deficiency have been limited to genetic defects, that interfere with the ability of the Mo cofactor to activate molybdoenzymes, and to the one case of feeding Mo-free total parenteral nutrition (TPN). Human dietary deficiency of Mo has not been reported (EFSA 2010).

A congenital Mo cofactor deficiency disease, seen in infants, results in interference with the ability of the body to use Mo in enzymes. It causes high levels of sulfite and urate, and neurological damage. A Crohn's disease patient, receiving TPN without Mo added to the TPN solution, developed tachycardia, tachypnoea, severe headache, night blindness, nausea, vomiting, central scotomas, generalized edema, lethargy, disorientation, and coma (WHO 2011d).

In humans, tetrathiomolybdate therapy has been developed for Wilson's disease, a genetic disease in which the accumulation of Cu in tissues leads to liver and brain damage. More recently, the use of tetrathiomolybdate has been explored for the treatment of cancer and inflammatory diseases.

The rate of gastrointestinal absorption of Mo is influenced by its chemical forms. Tetravalent Mo is not readily absorbed. Sodium tungstate is a competitive inhibitor of Mo. Dietary W reduces the concentration of Mo in tissues. High levels of Mo can interfere with the body's uptake of Cu, producing Cu deficiency. Molybdenum prevents plasma proteins from binding to Cu, and it also increases the amount of Cu that is excreted in urine.

An epidemiological study in India indicated that a form of lower-limb osteoporosis may be associated with the high Mo content of the cereals consumed by the population. The results from a cross-sectional study in two settlements of a Mo-rich province of the former Soviet Union suggested that the high incidence (18%–31%) of a gout-like disease was associated with high intake of Mo (10–15 mg/day). The disease was characterized by joint pains of the legs and hands, enlargement of the liver, disorders of the gastrointestinal tract, liver, and kidney, increased blood levels of Mo and uric acid, increased XO activity, decreased blood levels of Cu, and increased urinary Cu.

The toxicity of Mo compounds appears to be relatively low in humans. But dusts and fumes, which are generated by mining or metal working, can be toxic, especially if ingested. Prolonged exposure to Mo can cause irritation to the eyes and skin, joint pains, back pains, headache, or hair changes. The maximum permissible Mo exposure, in 8 hours a day, is 5 mg/m^3. Chronic exposure to 60–600 mg/m^3 can cause symptoms including fatigue, headache, and joint pains.

Molybdenum content in food exceeds 1 mg/kg FW rarely (Table 28.2), the highest concentrations are in nuts, pulses, and offal. The estimated safe and adequate daily dietary intake for Mo is 0.075–0.250 mg. The recommended dietary allowance for Mo is different for various groups, by age and gender; for adult male and female, it was set at 0.045 mg/day (FNB/IOM 2001).

Estimated daily intake varies widely regionally, depending on the soil type. Its intakes in the United States range from 0.240 mg/day for adult men to 0.100 mg/day for women (WHO 2011d). For adults, the representative range of mean estimates of Mo intakes in different countries is 0.080–0.250 mg/day. Intakes of Mo in EU countries range from 0.096 mg/day, in the Netherlands, to 0.500 mg/day, in Germany. Its intake by children (1–17 years old) varies between 0.0064 and 0.0144 mg/day. However, in France, it is estimated at 0.106–0.119 mg/day (EFSA 2009a).

TABLE 28.2
Molybdenum in Foodstuffs (mg/kg FW)

Product	Content
Food of Plant Origin	
Bread	0.22
Miscellaneous cereals	0.32
Rice and semolina[a]	0.110
Green vegetables	0.143
Pulses[a]	0.727
Potatoes	0.068
Fruits[a]	0.010
Nuts	1.26
Nuts and oilseeds[a]	1.031
Food of Animal Origin	
Meat[a]	0.141
Offal	1.10
Meat products	0.085
Fish[a]	0.065
Shellfish[a]	0.129
Eggs	0.124
Milk[a]	0.039
Milk[b]	0.085
Milk products	0.065

Sources: Rose, M. et al., *Food Addit. Contam.*, 27, 1380–1404, 2010. Unless otherwise stated.

[a] Leblanc, J.C. et al., *Food Addit. Contam.*, 22, 624–641, 2005.

[b] Gabryszuk, M. et al., *J. Elem.*, 15, 259–267, 2010.

FW, fresh weight.

28.7 ANIMALS

In ruminants, the dietary intake of excessive Mo causes, in part, a secondary hypocuprosis. Toxicosis due to massive doses of Mo is rare. Domestic ruminants are much more susceptible to Mo toxicity than nonruminants. Resistance of other species is at least 10 times that of cattle and sheep. Hexavalent Mo is readily absorbed, following oral administration, the amount absorbed being higher in nonruminants than in ruminants (WHO 2011d).

Excess dietary Mo has been found to result in the Cu deficiency in grazing animals, especially ruminants. In the digestive tract of ruminants, the formation of compounds containing S and Mo, known as thiomolybdates, prevents the absorption of Cu, and can cause fatal Cu-dependent disorders (Suttle 2012). Dietary Mo increased the content of Cu in the kidney and brain, but decreased in the liver.

Ruminants consuming high amounts of Mo develop symptoms including diarrhea, stunted growth, anemia, and achromotrichia (loss of hair pigment).

Water-soluble molybdates, thiomolybdates, and oxothiomolybdates and Mo in herbage and green vegetables are absorbed at 75%–97% by laboratory animals. Silicates inhibit the absorption of dietary molybdates (EFSA 2009a).

29 Nickel [Ni, 28]

29.1 INTRODUCTION

Nickel (Ni) is a metal of group 10 in the periodic table of elements, which also contains Pd and Pt (platinum group metals). Its mean abundance in the Earth's crust is around 20 mg/kg. The contents of Ni in ultramafic rocks are highly elevated and range from 1400 to 2000 mg/kg. Its concentrations decrease with increasing acidity of rocks, down to the range of 5–20 mg/kg in granites. Sedimentary rocks contain Ni from 5 to 90 mg/kg in sandy and argillaceous sediments, respectively. The lowest Ni contents are in limestones, around 5 mg/kg. In coal its amounts highly vary, up to about 300 mg/kg, and in fly ash they are lower, around 90 mg/kg.

Nickel occurs mainly at +2 oxidation state, which may change up to +4. It reveals both chalcophilic and siderophilic affinity, and readily combines with Fe. Therefore, Ni–Fe compounds are common in both the Earth's core and meteorites. The Ni–Fe alloy of the Earth's core (called barysphere or NIFE) is composed of Fe/Ni within the ratio of 11:1. The great affinity of Ni for S results in their association with various compounds and minerals. The main ores of Ni are composed mainly of pentlandite, $(Ni,Fe)_9S_8$, and pyrrhotite, $Fe_{1-x}Ni$. In rocks, Ni occurs primarily as sulfides (millerite, NiS), arsenides (nickeline, NiAs), and antimonides (ullmannite, NiSbS). Nickel is also associated with several Fe minerals and forms sulfides and sulfarsenides with Fe and Co. During weathering, it is coprecipitated with Fe and Mn oxides, and becomes included in their minerals (e.g., goethite, limonite). Nickel may also be associated with carbonates, phosphates, and silicates. Organic matter (OM) and clay minerals exhibit a strong ability to absorb Ni; thus, it may be concentrated in some coals and oil, as well as in bottom sediments rich in OM.

World (excluding the United States) mine production of Ni (rounded) in 2010 was 1550 Mt, of which 265 Mt was mined in Russia, 232 Mt in Indonesia, 155 Mt in Canada, 156 Mt in Philippines, and 139 Mt in Canada (USGS 2011). Nickel has been broadly used in several industries, especially for the production of stainless steels. Most of Ni is added to various alloys resistant to oxidation and corrosion. Ni alloys are used for different tools and vessels used in medicine, food technology, and kitchen equipments. Its compounds are used as dyes in ceramic and glass manufactures. Other uses of Ni have been recently limited, partly due to its relatively high price.

29.2 SOILS

Soils throughout the world contain Ni in a very broad range; however, its mean contents reported for various countries are within the relatively low range of 13–37 mg/kg. The lowest Ni contents are in sandy and organic soils, and the highest in heavy loamy and calcareous soils (Table 29.1). However, it may be elevated in some organic soils,

TABLE 29.1
Nickel Contents of Soils, Water, and Air

Environmental Compartment	Range Mean
Soil (mg/kg)	
Light sandy	7–33
Medium loamy	11–25
Heavy loamy	20–50
Calcareous (Calcisols)	20–95
Organic	4–20
Water (µg/L)	
Rain	0.3–1.0
River	0.2–27
Ocean	0.6–1.7
Baltic Sea	0.09–1.1[a]
Air (ng/m³)	
Urban/industrial areas	4–120
Remote regions	0.9

Source: Data are given for uncontaminated environments, from various sources, mainly Kabata-Pendias, A. and Mukherjee, A.B., *Trace Elements from Soil to Human*, Springer, Berlin, Germany, 2007; Reimann, C. and de Caritat, P., *Chemical Elements in the Environment*, Springer, Berlin, Germany, 1998.

[a] Szefer, P., *Metals, Metalloids and Radionuclides in the Baltic Sea Ecosystem*, Elsevier, Amsterdam, The Netherlands, 2002.

especially peaty soils. Higher Ni contents are also in soils of arid and semiarid regions, as well as in serpentine soils, which contain Ni (in Serbia) from 0.08 to 6.22 mg/kg (Tumi et al. 2012). The highest Ni contents in such soils are in Rhodesia, up to about 7000 mg/kg (Table 29.2). High levels of Ni in uncontaminated soils, as reported from some countries (in mg/kg), are as follows: Canada, 119; China, 450; Japan, 660; and Italy, 3240 (Kabata-Pendias 2011). Soils of the remote region in Spitzbergen contain Ni within the range of 0.15–0.96 mg/kg (Gulińska et al. 2003). Its geometric mean content in soils from major agricultural areas of the United States is 16.5 mg/kg, ranging from 0.7 to 269 mg/kg (Holmgren et al. *vide* Kabata-Pendias 2011). The range of Ni contents in the surface layer of agricultural soils of Poland is 1.4–71 mg/kg (Siebielec et al. 2012). The mean Ni content in organic surface soils of Norway is 7.5 mg/kg (Nygard et al. 2012), and that in agricultural soils of Sweden is 13 mg/kg (Eriksson 2001a). Agricultural soils usually contain more Ni than not cultivated ones.

The maximum permissible Ni level in agricultural soils, established in 1997, was 50 mg/kg (Herselman et al. *vide* Kabata-Pendias 2011). However, the recent estimation for accepted Ni levels in soils, based on various criteria, is

TABLE 29.2
Nickel Enrichment and Contamination of Surface Soils (mg/kg)

Site and Pollution Source	Country	Range/Mean
Soils over serpentine rocks	Australia	770
	New Zealand	1,700–5,000
	Rhodesia	3563–7,357
Metal-processing industry	Albania	1243
	Canada	206–26,000
	The United Kingdom	500–600[a]
	Russia	304–9,288
Sludged farmland	Germany	50–84
	Holland	31–101
	The United Kingdom	23–846

Source: After Kabata-Pendias, A., *Trace Elements in Soils and Plants*, 4th ed., CRC Press, Boca Raton, FL, 2011.

[a] Soluble in HCl.

within the broad range of 50–1600 mg/kg. The criteria for contaminated land (Dutch List 2013), following Ni concentrations in soils and groundwaters, are established (in mg/kg and μg/L, respectively) as follows: uncontaminated, 50 and 20; medium contaminated, 100 and 50; and heavily contaminated, 500 and 200.

Nickel is easily mobilized during weathering processes and may easily migrate with water down soil profiles. Its species in soil solution are cationic (Ni^{2+}, $NiOH^{+}$, $NiHCO_3^{+}$) and anionic ($HNiO_2^{-}$, $Ni(OH)_3^{-}$), as well as complex compounds ($Ni(OH)_2^{\circ}$, and $NiSO_4^{\circ}$) (Kabata-Pendias and Sadurski 2004). Its concentrations in soil solution highly vary (3–150 μg/L) and may be influenced by some agricultural treatments.

Nickel in soils is slightly mobile and about 50% of its contents is associated with the residual fraction. However, in surface soil horizons, where Ni occurs in bound organic forms, it is easily chelated and mobile. However, the remobilization of Ni from solid phases is possible in the presence of humic acids (HA): fulvic acid (FA) and HA. The impact of FA and HA on the behavior of Ni is very complex, because they may also strongly adsorb this metal, as well as form mobile species. Thus, Ni species in soils are various and range from highly mobile to ones that have no reactivity. Several soil properties, but especially soluble OM (SOM), clay fractions, and pH control Ni behavior, and its phytoavailability. However, almost all Ni mobile forms are relatively easily transformed into residual fractions. Complexing ligands, such as SO_4^{2-}, Fe–Mn hydroxides, and organic acids, reduce Ni sorption in soils. Some metals may decrease their sorption, as, for example, increased Cd concentration (Selim 2012). Microorganisms may also affect Ni behavior in soils due to its accumulation: *Rhizopus arrhizus* contains 16 mg Ni/g at the pH range of 5–7.7 (Perelomov et al. 2013).

Nickel is considered as a serious pollutant that is released from metal-processing plant and from the combustion of coal and oil. Sewage sludge and some P fertilizers are also important sources of Ni in agricultural soils (Table 29.2). Some municipal sludges of America cities contain Ni up to 800 mg/kg, with a median value of 195 mg/kg (ATSDR 2002b). Application of sludge to soils usually increases the Ni mobility, due to its complexation with dissolved OM. In soils contaminated with Ni, processes of OM mineralization are disturbed and the activity of some enzymes, such as dehydrogenases, urease, and phosphatases, is decreased (Wyszkowska et al. 2008).

The Ni balance in soils of crop farms in the EU countries was estimated (in g/ha/yr) from –3.3 in Denmark to 33 in France (Eckel et al. 2005). Emission from the Ni industry in Russia has affected the elevated Ni levels in soils of Norway, up to about 1400 mg/kg of HNO_3-extractable Ni (Alamås et al. *vide* Kabata-Pendias 2011). The phytoremediation of Ni-contaminated soils has been recently broadly studied, with promising effects (Chaney et al. 2005).

29.3 WATERS

The mean nickel concentration in ocean waters is given within the range of 0.6–1.7 μg/L (Table 29.1). There is also other estimation, such as 0.1–3.0 μg/L (ATSDR 2002b). The world riverine flux of Ni is given as 305 kt/yr (Gaillardet et al. 2003). Its flux to the Baltic Sea is estimated to be 400 t/yr (Matschullat 1997). Although Ni is slightly mobile in water, its amount in rivers may vary from 0.2 to 27 μg/L, what reflex its variable sources, mainly anthropogenic. Water of Nordic lakes has fairly similar Ni concentrations, which are as follows (mean in μg/L): 7.1 in Finland, 5.7 in Sweden, and 3.2 in Norway. In the lakes of the Kola Peninsula, Ni contents of water average at 50.4 μg/L (Skjelvåle et al. *vide* Kabata-Pendias and Mukherjee 2007).

Nickel does not remain long in aquatic environments as soluble species, because it is easily adsorbed by the suspended matter and Fe–Mn hydroxides, and is deposited in bottom sediments. Nickel solubility decreases at pH between 9 and 11.5. Thus, the addition of some hydroxide compounds will stimulate its precipitation (Chen 2013). Also the stage of oxygen saturation has significant impact on the Ni concentration in water (Vignati *vide* Kabata-Pendias and Mukherjee 2007).

The mean Ni concentration in rainwater is about 1 μg/L, whereas in contaminated regions, it may be up to 60 μg/L (Reimann and de Caritat 1998). In atmospheric precipitation over the Baltic Sea, Ni concentrations vary between 0.9 and 19.7 μg/L, whereas rainwater over the Kola Peninsula contains Ni within the range of 0.1–57 μg/L, in remote and polluted regions, respectively. The average Ni content of rainwater collected in Sweden is 0.34 μg/L, and the maximum is 0.56 μg/L (Eriksson 2001a). Rainwater collected in Canada contains Ni at the concentration of <1.5 μg/L (ATSDR 2002b). Several Ni species have been identified in rainwater, mainly in complexes with anions, such as OH^-, SO_4^-, Cl^-, and HCO_3^-. Most often they occur as $Ni(H_2O)_6^{2+}$.

The median Ni concentrations in both surface water and groundwater of uncontaminated regions are in the range of 0.5–6 μg/L (ATSDR 2002b), whereas in the industrial region of Uzbekistan the following amounts of Ni are reported: surface water –30, and well water at 8 m depth –120 μg/L, respectively (Galiulina et al. *vide* Kabata-Pendias and Pendias 1999).

The median Ni concentration in bottled waters of the EU countries is 0.204 μg/L and is lower than that in tap water, estimated to be 0.381 μg/L (Birke et al. 2010). Various limits for maximum allowable concentration (MAC) have been set for Ni in drinking water in different countries, from 20 to 100 μg/L. The highest proportion of analyzed water samples (43%) in the United States contains Ni from 1 to 5 μg/L (ATSDR 2002b). Elevated Ni levels in tap water may be a result of the corrosion of Ni-containing alloys used in some water distribution systems. The guideline value for Ni in drinking water is 70 μg/L (WHO 2011a).

Water pollution with Ni is usually reflected in its accumulation in bottom sediments. Its background contents in bottom sediments of *Rhein* and Vistula Rivers are 46 mg/kg and 40 mg/kg, respectively (Kabata-Pendias and Pendias 1999). Fine granulometric fraction of sediments in polluted rivers may contain highly elevated amounts of Ni, up to around 300 mg/kg (Bojakowska et al. 2007). The nickel contents of surface-bottom sediments of harbor in Klaipeda (Lithuania) depend also on granulometric composition and are as follows (in mg/kg, average and maximum, respectively): 5.3 and 12.9 in sand and 4.2 and 18.4 in mud (Galkus et al. 2012). Sediments of Sergipe River estuary (Brazil) contain Ni from 0.3 to 28.2 mg/kg (Garcia et al. 2011).

The assessment limits for Ni in sediments are established as follows (in mg/kg): effects range low, 21; effects range median, 52; probable effect level, 36; severe effect level, 75; and toxic effect threshold, 61 (EPA 2000, 2013). The Environment Canada's sediment-quality guidelines (USGS 2001) gave the following values for Ni in the lake-bottom sediments (in mg/kg): threshold effect concentration, 22.7, and probable effect concentration, 48.6.

Nickel may be accumulated in aquatic biota. Plankton of the Baltic Sea contains Ni within the range of 0.16–16.6 mg/kg FW (Szefer 2002). Muscles of flounders have Ni from 0.2 to 0.6 mg/kg, from clean and polluted waters of Finland seacoast, respectively (Kabata-Pendias and Mukherjee 2007).

29.4 AIR

The concentrations of Ni in air vary highly from 0.9 to 120 ng/m^3, in remote and urban regions, respectively (Table 29.1). Air samples collected over the United States in the last century contain Ni up to 68 ng/m^3, and it has been suggested its decline (ATSDR 2002b).

Nickel emission to the atmosphere is estimated to be 34 kt/yr, of which 29 kt/yr is of natural origin (Nriagu and Pacyna 1988). Due to other calculations, Ni yearly emission from natural sources is 8.5 kt, whereas from anthropogenic, it is 55.6 kt (ATSDR 2002b). The main natural Ni sources are windblown dust and volcanic eruptions, and anthropogenic ones are mainly fuel oil and coal combustions, and Ni metal refining plants.

Nickel average deposition in EU countries is 10 g/ha/yr, and varies from 1.5 in Finland to 36 in Italy (Kabata-Pendias and Pendias 1999). The natural aerial Ni deposition on the Baltic Sea is estimated to be 5 t/yr, whereas from anthropogenic sources, it is 100 t/yr (Matschullat 1997).

Nickel contents of forest moss (*Pleurozium schreberi*) sampled in 1975 and 2000 have decreased to 2.8-fold, which clearly indicate its smaller global emission

(Rühling and Tyler 2004). Its contents of moss (*Hylocomium splendens*) from the mountains in Poland (0.7–1.4 mg/kg) are similar to those of the same moss species from Alaska (0.8–4.7 mg/kg). This may also suggest similar Ni atmospheric deposition in both regions (Migaszewski et al. 2009).

29.5 PLANTS

There is no evidence of an essential role of Ni in plant metabolism, although there are several suggestions that it might be needed for plants. Its essentiality for some bacteria has been proven, mainly bacteria involved in processes of nitrification and mineralization of OM. Thus, Ni is considered as essential for legumes, in which it plays a role in urease metabolism. It is involved in N transport from roots to tops and in H_2 metabolism.

Studies on phytoavailability and behavior of Ni in plants are related mainly to its toxicity and implications with respect to human and animal health. Its toxicity and biological effects are highly related to its species. The cationic form, Ni^{2+}, is more readily absorbed and more toxic than its complexed species. However, all species of Ni are easily available to plants and controlled by both soil and plant factors. Soil pH has the most pronounced impact; increased pH from 4.5 to 6.5 decreases Ni in oats by a factor of about 8. Plants uptake Ni added to soil easier than from its lithogenic sources. High cation-exchange capacity values of soils may limit its availability to plants.

The phytoavailability of Ni depends on its origin and soil properties, as well as on plants' abilities to absorb Ni, (e.g., accumulators and hyperaccumulators). About 200 plant species have been identified as Ni hyperaccumulators, which may contain more than 0.1% of this metal (Mesjarz-Przybyłowicz et al. 1994). The processes of Ni uptake are still little known; however, its availability is positively correlated with its concentrations in soil solutions. There are observations indicating an extracellular, metabolically independent Ni adsorption and a possible displacement of Mg^{+} from cell membranes. All Ni^{2+} retained by bacteria was accumulated externally to cells (Kabata-Pendias 2011). The transport and storage of Ni in plants is metabolically controlled. It is mobile in plants and is likely to be accumulated in leaves and seeds. The Ni speciation in the plant extracts indicates that regardless of chemical forms added to soils, it is found only in neutral and negative complexes (Khellat and Zerdauori 2010).

In plants under Ni stress, the absorption of nutrients, some metabolic processes, and root development are strongly retarded. Photosynthesis and transpiration processes, as well as N_2 fixation by legume plants, are also inhibited. The most common symptom of Ni phytotoxicity is chlorosis, which may be associated with Fe-induced chlorosis. Under natural conditions, Ni toxicities are associated with serpentine or other Ni-rich soils. Elevated Ni concentration in nutrient solution decreases the activity of superoxide dismutase and catalase in wheat shoots, while increasing the activity of peroxidases and glutathione *S*-transferase (Gajewska and Skłodowska 2008). Addition of Ni, at the concentration of 200 mg/kg, to two soils, heavy loamy sand and light silty loam, decreases oat yields by 65% and 40%, respectively (Wyszkowska et al. 2007).

The phytotoxic Ni concentrations in plants widely range from about 40 to over 300 mg/kg, depending mainly on plant properties. Generally, the range of excessive or toxic Ni contents in the most plant species varies from 10 to 100 mg/kg. However, native vegetation of serpentine soils (as well as Ni hyperaccumulators) may contain very high Ni levels, up to above 7000 mg/kg in leaves, without any symptoms of toxicity. The great ability of some plants to extract Ni from soils over Ni ore body may be useful for biogeochemical prospecting.

The interaction between Ni and other trace metals, Fe in particular, is apparently a common mechanism involved in its toxicity. In the presence of some cations, Cu^{2+}, Zn^{2+}, and Fe^{2+}, the absorption of Ni by plants may be inhibited. However, in some conditions, synergistic interactions may also occur. Variable effects of Ni–Cd, Ni–Pb, and Ni–Cr interactions were observed in soil bacteria.

The most common Ni accumulators are *Alyssum* spp., which are suggested for the phytoextraction of Ni from contaminated soils (Chaney et al. 2005). There are calculations that about 25 kg Ni/ha may be extracted by these plant species. The mechanism of the Ni hyperaccumulations is not well understood, but it is evidently associated with the formation of organometallic complexes. The main ligands for Ni in these plants are citrate and malate, which are involved in the metal transfer within plant tissues (Someya et al. *vide* Kabata-Pendias 2011). However, several other organic ligands (e.g., histidine) are also involved in these processes.

The easy phytoavailability of Ni is a real environmental concern. Especially elevated Ni contents in vegetable and fodder plants may be more of a health risk. The background Ni mean contents in barley and wheat grains are established at 0.41 and 0.34 mg/kg, respectively. Oat grains contain a bit higher Ni levels, between 0.2 and 8.0 mg/kg (Eriksson 2001a). The highest Ni contents are reported for plants from contaminated soils (in mg/kg): lettuce (leaves), 40–80; onion (bulbs), 47; celery (stalks), 29; soybean (seeds), 7–26; clover (tops), 3–15; and alfalfa, 44.

The range of mean Ni amounts in grasses from various countries is 0.1–1.7 mg/kg, whereas in clover it is 1.2–2.7 mg/kg. Much higher Ni contents have been reported for meadow grass (13–75 mg/kg) and forest grass (10–100 mg/kg) from the taiga zone of Western Siberia (Niechayeva *vide* Kabata-Pendias 2011). In ecosystems where Ni is airborne pollutant, the tops of plants are likely to concentrate its great amounts.

The mushrooms, common chanterelles (*Cantharellus cibarius*) grown in mountains of Poland, contain lower amounts of Ni, mean 1.4 mg/kg, than those grown in the Baltic Sea coast, mean 2.0 mg/kg (Falandysz et al. 2012).

29.6 HUMANS

The estimated average body burden of Ni in adults (70 kg) is 15 mg (Emsley 2011). Nickel concentrations are (in μg/kg FW) as follows: the lung, 7–137; the bone and kidney, 9–14; the heart 6–8; the liver 8–10; the spleen 7; whole blood 0.34–2.3 μg/L, and urine, 0.9–2.0 μg/L; and serum, 1.2 μg/L (IARC 1990).

The essentiality of Ni in humans has not been established, and Ni dietary recommendations have not been recommended for humans. Nickel deficiency has not been observed in humans, although there may be benefits from small doses of Ni, whereas exposure to its high levels may result in adverse health effects. These effects are dependent on the route of exposure and, in the case of the inhalation, on Ni species.

In postmortem tissue samples from adults, with no occupational or iatrogenic exposure to Ni compounds, the highest Ni concentrations were found in the lung, bone, thyroid, and adrenals, followed by the kidney, heart, liver, brain, spleen, and pancreas, in diminishing order. Nickel is bound to albumin, histidine, and macroglobulin, and is distributed widely throughout the tissues. It is also found in breast milk, saliva, nails, and hair. Transplacental transfer of Ni has been demonstrated in rodents.

Nickel is present in dozens of products; thus, it might be easy, for sensitive people, to develop Ni allergy. The main source of Ni exposure to the general population is food (Table 29.3). Other sources of Ni exposure include air, drinking water, soil, and household dust. Ni exposure can also occur through dermal contact with alloys containing Ni (e.g., jewelry) and with Ni-containing products, such as cosmetics

TABLE 29.3
Nickel in Foodstuffs (mg/kg FW)

Product	Content
Food of Plant Origin	
Bread[a]	0.07
Miscellaneous cereals[a]	0.16
Rice, different type[b]	0.130–0.179
Green vegetables[a]	0.09
Pulse[c]	0.33
Nuts[a]	3.02
Sugar and preserves[a]	0.31
Milk chocolate[d]	0.98
Dark chocolate[d]	3.07
Cocoa[d]	15.33
Margarine[b]	0.02
Nuts[a]	3.02
Food of Animal Origin	
Meat[c]	0.02
Meat products[a]	0.07
Offal[a]	0.02
Fish[c]	0.05
Freshwater fish[e]	0.03–0.113
Milk[a]	<0.007
Eggs[a]	<0.02

Sources: [a] Rose, M. et al., *Food Addit. Contam.*, 27, 1380–1404, 2010.
[b] Poletti, J. et al., *Food Addit. Contam.*, 7, 63–69, 2014.
[c] Leblanc, J.C. et al., *Food Addit. Contam.*, 22, 624–641, 2005.
[d] Sager, M., *J. Nutr. Food Sci.*, 2, 123, 2012.
[e] Lidwin-Kaźmierkiewicz, M. et al., *Pol. J. Food Nutr. Sci.*, 59, 219–234, 2009.
FW, fresh weight.

(generally present as an impurity), household cleaning and bleaching agents, and medical products (joint implants, intrauterine devices, and acupuncture needles). Nickel exposure can also be from inhalation of cigarette smoke (ATSDR 2005a).

Nickel and its compounds are absorbed from the respiratory tract and, to a lesser extent, from the gastrointestinal tract and skin. About 20%–35% of the inhaled Ni that is retained in the lung is absorbed into the blood, whereas only 1%–10% of ingested Ni is absorbed, depending largely on the composition of the diet. Absorption of Ni following oral exposure varies between 3% and 40%, being higher from drinking water. Most of the absorbed Ni is excreted in the urine, regardless of the route of exposure (Health Canada 2010b).

Oral Ni absorption is greater from more soluble compounds (in % of Ni content): 34, $NiNO_3$; 11, $NiSO_4$; 10, $NiCl_2$; 0.5, Ni_3S_2; 0.09, metallic Ni; 0.04, black Ni oxide; and 0.01, green Ni oxide. Its absorption was reduced by some ligands, such as cysteine and histidine, and, to a lesser extent, by proteins. Oral absorption of $NiSO_4$, given in drinking water, was 27%; and it was 0.7%, when the $NiSO_4$ was given in food (WHO 2006a).

At current levels of exposure, Ni and its compounds are not of a concern for human health, according to Health Canada and Environment Canada. However, several Ni compounds are entering the environment in a quantity or concentration and amounts and forms, which may constitute a danger to human life or health (Health Canada 2010).

Acute Ni exposure is associated with a variety of clinical symptoms and signs, which include gastrointestinal disturbances (nausea, vomiting, abdominal discomfort, and diarrhea), visual disturbance (temporary left homonymous hemianopia), headache, giddiness, wheezing, and cough. The most acutely toxic Ni compound is Ni carbonyl, $Ni(CO)_4$. This colorless, highly volatile liquid is extremely toxic and induces systemic poisoning, with the lungs and brain being especially susceptible targets. Chronic inhalation of Ni and its compounds is associated with an increased risk of lung cancer. Cigarette smoking can cause a daily absorption of Ni of 1 μg per pack, due to the Ni content of tobacco. Its contents vary from 2.2 to 2.3 μg per cigarette. About 10%–20% of Ni in cigarettes is released in mainstream smoke, and most of Ni is in the gaseous phase (IARC 1990).

Inhalation is an important route of exposure to Ni and its salts, in relation to health risks. The International Agency for Research on Cancer (IARC) concluded that inhaled Ni compounds are carcinogenic to humans (Group 1) and that metallic Ni is possibly carcinogenic (Group 2B). However, there is a lack of evidence of a carcinogenic risk from oral exposure to Ni (IARC 1990, 2012a).

Nickel is one of the most common contact allergens on the skin, not only in industrial settings but also in the general population exposed, via direct contact of the skin with Ni-containing items (WHO 2006a).

Tolerable daily intake of Ni was estimated to be 12 μg/kg bw (WHO 2011a). Nickel is present in most food groups (Table 29.3).

Data from studies in the United States give estimates of daily dietary intakes in the range of 0.101–0.162 mg/day for adults, 136–140 μg/day for males, and 107–109 μg/day for females. Estimates for pregnant and lactating women are higher with average daily intakes of 121 and 162 μg/day, respectively (ATSDR 2005a). Daily dietary intakes of

Ni were 0.082 mg in Sweden and 0.16 mg in Denmark in 1987. The dietary intake of Ni in a Canadian study ranged from 0.19 mg/day for 1- to 4-year-old children to 0.406 mg/day for 20- to 39-year-old males; the Ni intake for 20- to 39-year-old women was, on average, 0.275 mg/day (WHO 2005a). The population dietary exposure to Ni in the United Kingdom was estimated to be 0.13 mg/day in 2006, and this is the same as previously reported in 1997 and 2000 (Rose et al. 2010).

29.7 ANIMALS

Nickel is an essential element in a number of laboratory animal species, although its functional importance has not been clearly demonstrated. It is considered essential, based on the reported Ni deficiency in several animal species (e.g., rats, chicks, cows, goats). Nickel deficiency is manifested primarily in the liver functions, including abnormal cellular morphology, oxidative metabolism, and increases and decreases in lipid levels. In animals, Ni deficiency is associated with disturbances and reduced growth, impaired reproductive function, and reduced hematopoiesis.

30 Niobium [Nb, 41]

30.1 INTRODUCTION

Niobium (Nb) is a soft, gray transition metal of group 4 in the periodic table of elements, with physical and chemical properties similar to those of Ta, the metal of the same group. It was also formerly called columbium (Cb). Its abundance in the Earth's crust varies from 20 to 26 mg/kg. Its highest contents are in acidic magmatic rocks (15–60 mg/kg) and argillaceous sediments (15–20 mg/kg). The lowest Nb contents are in limestones (0.05–0.1 mg/kg). In coal, its amounts vary within the range of 5–10 mg/kg, but may also reach the level of 70 mg/kg (Finkelman 1999).

Niobium occurs mainly at +5 oxidation state, and may also have a lower oxidation. In the nature, Nb and Ta occur together, due to their great similarities in atomic radii and several properties. Both elements exhibit an affinity to associate with Fe, Ti, and Zr. It is scattered in several minerals of complex formula, for example, columbite/niobite/tantalite, $(Fe,Mn)(Nb,Ta)_2O_6$, and pyrochlore, $(Na,Ca)_2(Nb,Ta)_2O_6(OH,F)$. Niobium is also included in some Ti minerals and Zr compounds. The free Nb element is not found in nature. Columbite–tantalite minerals occur in pegmatite intrusions. Its deposits are also associated with carbonate–silicate igneous rocks.

World mine Nb production (rounded), excluding the United States, in 2010 was 63,000 t, of which 58,000 t was produced in Brazil and 4,400 t was produced in Canada (USGS 2011). Niobium is used mainly for steel alloys. These alloys are resistant to strength and temperature impacts, and thus are widely used in pipeline construction and in aerial transportation industries. It is used as a precious metal (e.g., in commemorative coins), often with Ag and Au. Niobium is also added to alloys used for surgical implants and in the stomatology.

30.2 SOILS

The worldwide Nb content of soils is established at 12.5 mg/kg. Its highest concentrations are in heavy loamy soils (Table 30.1). Its amounts in soils depend mainly on its contents in mother rocks. Thus, in soils derived from different rocks in the United Kingdom, the concentrations of Nb vary from 31 to 300 mg/kg (Ure et al. *vide* Kabata-Pendias 2011). The range of Nb contents of soils in various countries are given as follows (in mg/kg, for sandy and loamy soils, respectively): Alaska, <4 and 44; China, 9.3 and 37.6; Japan, 6.3 and 14; and Sweden, 5 and 17.

The behavior of Nb during weathering highly depends on host minerals; therefore, it may be released from some minerals (e.g., amphibolite) or may remain within resistant minerals (e.g., sphene/titanite, zircon). Thus, the accumulation of Nb in certain residual sediments has often been reported. Its elevated concentrations, >15 mg/kg, also occur across areas of aeolian sediments (e.g., loess of central Europe) and in residual soils developed on some carbonate rocks.

TABLE 30.1
Niobium Contents of Soils, Water, and Air

Environmental Compartment	Range/Mean
Soil (mg/kg)	
Light sandy	<4.0–8.0
Heavy loamy	14.0
Water (ng/L)	
Rain	0.8–10
River	2–11
Sea and ocean	<5–10
Air (ng/m^3)	6–10

Source: Data are given for uncontaminated environments, from various sources, mainly Kabata-Pendias, A. and Mukherjee, A.B., *Trace Elements from Soil to Human*, Springer, Berlin, Germany, 2007; Reimann, C. and de Caritat, P., *Chemical Elements in the Environment*, Springer, Berlin, Germany, 1998.

Most of Nb compounds are slightly mobile in both acidic and alkaline media. However, they may be mobilized by complexing with organic matter.

30.3 WATERS

The world average Nb concentration in river waters is established at 1.7 ng/L, within the range of 2–11 ng/L (Table 30.1). In ocean waters, its mean value is estimated to be 10 ng/L, whereas Nozaki (2005) gave the mean Nb level <5 ng/L in the North Pacific. In aquatic media, Nb is most often in forms of $Nb(OH)_5$ and as anion $Nb(OH)_6^-$. In rainwater, its concentrations do not exceed 10 ng/L. Its deposition with rainwater in Sweden was calculated as 24 mg/ha/yr (Eriksson 2001a).

The median Nb concentration in bottled water of the EU countries is 2.3 ng/L and is similar to its contents in tap water, estimated to be 2.5 ng/L (Birke et al. 2010).

Some river bottom sediments in Europe contain Nb up to 245 mg/kg, which is an evidence for both anthropogenic source and its easy leaching from soils (Reimann and de Caritat 1998).

30.4 AIR

Niobium concentration in air is roughly estimated within the range of 6–10 ng/m^3 (Table 30.1). It may occur as compounds with oxygen and nitrate, which are more common.

The contents of Nb in moss sampled in Norway, in 1990–1995, ranged between <0.002 and 0.8 mg/kg, and averaged 0.11 mg/kg (Berg and Steinnes 1997). Its contents have not been changed much during the recent decades.

30.5 PLANTS

Niobium is relatively mobile under humid conditions, and therefore may be available to plants. Its biochemical function is unknown. The common Nb amounts in food plants are very broad, within the range of 0.02–1.1 mg/kg. Cereal grains of Sweden contain Nb from 0.5 to 1.7 mg/kg (Eriksson 2001a). Its very low concentration (<0.1 μg/L) is reported for wines (Szefer and Grembecka 2007).

In some herbs from the Ural Mountains, the mean contents of Nb are 0.4 and 1.0 mg/kg, in tops and roots, respectively (Grankina et al. *vide* Kabata-Pendias 2011). Several native plants have the capability to extract Nb from soils, for example, *Rubus* species, grown in the Nb-mineralized area of Komi (Republic in Russian Federation). These plants are proposed as indicators for the Nb ore deposits.

30.6 HUMANS

According to Emsley (2011), Nb content of human soft tissues is 0.14 mg/kg, with a total amount of 1.5 mg in the whole body. Niobium contents (mg/kg) are present in the blood (0.004), the kidney (0.01), the lung (0.02), the muscle (0.03), and the testis (0.009) (Nielsen 1986).

Metallic Nb has a low order of toxicity due to a poor absorption from stomach and intestines. Niobium and its compounds may be toxic. Apart from measuring its concentration, no research on Nb in humans has been undertaken. Once inhaled, it resides primarily in the lungs and can be absorbed into bones. It interferes with Ca as an activator of enzyme systems. No side effects have been reported because it does not react with human tissues. The English diet supplied only 0.020 ± 0.004 mg Nb daily (Nielsen 1986).

Niobium and some of its alloys are physiologically inert and thus hypoallergenic. For this reason, Nb is found in many medical devices such as pacemakers (Mallela et al. 2004). It is frequently used in jewelry and has been tested for use in some medical implants. The Nb nitride coatings show a better cell adhesion than the uncoated stainless steel (Ramírez et al. 2011).

The Nb-based coatings do not have any toxic effects on cells and have a similar biocompatibility to Ti. Niobium and its compounds cause skin and eye irritation and can be toxic, but human poisoning has not been reported. When inhaled, the element is retained in the bones and lungs, and interferes with Ca.

The inhalation of Nb pentoxide and nitride results in scarring of the lungs in laboratory animals. In rats, the Nb supplement elevated Cu, Mn, and especially Zn, in variety of organs, with most marked changes occurring in the heart and liver (Nielsen 1986).

31 Osmium [Os, 76]

31.1 INTRODUCTION

Osmium (Os) is a hard, brittle, transition metal of group 8 in the periodic table of elements and is a member of the platinum group metals (PGMs). It is considered the least abundance stable element in the Earth's crust, within the range of 0.5–1.0 μg/kg. The highest concentrations of Os are in some igneous rocks and volcanic deposits. Like other PGMs, Os occurs mainly in native alloys of variable composition, and also in Pt ores.

Osmium forms compounds within the broad range of oxidation state, from −1 to +8. The most common states, however, are above +2. There are no Os minerals, but this metal is likely to be associated (at about 3%) with laurite, (Ru,Os) S_2. Some other minerals of PGMs exhibit an affinity for elevated amounts of Os, up to 30%, for example, iridosmine and osmiridium—native alloys of variable composition.

Osmium is obtained commercially as a by-product from Ni and Cu mining and processing. There are not available data on the Os production. However, there is an estimation that in 2012, the U.S. production of Os was 75 kg.

Osmium is mainly used for the production of hard-wearing alloys and for various electronic devices (e.g., fountain pen tips, electrical contacts). It is very resistance to all acids and alkalis. During metal processing, various Os compounds are formed, of which OsF_6, $OsCl_4$, and volatile OsO_4 are toxic to humans. There are also some suggestions indicating anticancer effects of Os compounds.

31.2 ENVIRONMENT

The mean Os content of soils is estimated to be 0.05 μg/kg, fairly similar to its content on the Earth's crust. Although Os is known to be released from metal-processing plants, mainly as volatile tetraoxide (OsO_4), there are no data on Os contaminations. Osmium emitted from automobile catalytic converters is accumulated along motorways, where its contents vary between 0.03 and 2.36 μg/kg (Fritsche and Meisel *vide* Kabata-Pendias 2011).

Its mean concentration in ocean waters is established at 2 pg/L (Nozaki 2005). Its concentrations in river waters range from 3.3 to 42.1 pg/L and average 9 pg/L (Gaillardet et al. 2003).

Mobile Os species are easily available to plants and are relatively toxic. Although Os is considered as biologically unreactive metal, its increased levels in food and fodder plants might be of a great health risk.

31.3 HUMANS

The total amount of Os in the human body is not known and has no significant biological role. Although the metal itself is not toxic, its volatile oxide is (Emsley 2011). Osmium tetroxide (OsO_4) is extremely poisonous and highly volatile; penetrates the skin readily; and is very toxic by inhalation, ingestion, and skin contact. Airborne low concentrations of OsO_4 vapor can cause lung congestion and skin or eye damage, and should therefore be used in a fume hood (Luttrell and Giles 2007).

Its particles can cause damage to the mucus membranes, because they act as an irritant. Prolonged exposure to the dust can damage the lungs. Concentrations of OsO_4 in air, as low as 10^{-7} g/m^{-3}, can cause lung congestion, skin damage, and severe eye damage.

The toxicity of OsO_4 in food is rapidly reduced to relatively inert compounds, by polyunsaturated vegetable oils, such as corn oil.

Osmium carbohydrate polymers, called osmarins, have been synthesized and proposed for use as active agents to treat arthritis. This proposal was based on the use of OsO_4 for about 50 years to treat human arthritis and cancer, mostly in Europe. However, this treatment is controversial because of the high toxicity of OsO_4, which is injected (in solution) into the synovial space of diseased joints. Osmium-containing material remained in the joints for as long as 5 years, suggesting a long-term biological effect. The use of the Os carbohydrate polymers for this purpose was proposed to avoid trauma associated by unselective OsO_4 oxidations. Unlike the highly toxic OsO_4, the osmarin polymers have low acute toxicities (Carraher and Pittman 2004).

32 Palladium [Pd, 46]

32.1 INTRODUCTION

Palladium (Pd) is a rare, lustrous silvery-white metal of group 10 in the periodic table of elements and is a member of the platinum group metals (PGMs). It is the least dense of them and has the lowest melting point. Its average abundance in the Earth's crust is estimated within the range of 4–10 μg/kg, but in the upper layer, it is much less, about 1 μg/kg. Relatively high Pd contents may be in ultramafic rocks, some coals, black shales, and phosphorites. Its mean content of coal in the United States is <1 μg/kg (Finkelman 1999). Basalts and shale schists contain Pd up to 0.6 μg/kg. The highest Pd contents are associated with Mn ores and Mn concretions, where it may be accumulated up to about 12 μg/kg.

Palladium has various oxidation states, from +1 to +4, and most often it is +2. It exhibits relatively strong chalcophilic character and chemical reactivity, and thus, it forms more mineral compounds than other PGMs. Its better-known minerals are as follows: potarite, PdHg; arsenopalladinite, Pd_3As; stibiopalladinite, Pd_3Sb; cooperite, (Pt,Pd)S; and braggite, (Pt,Pd,Ni)S. It may be associated with several sulfide minerals, at the approximate concentration from 0.7 to 10 mg/kg, and may also be accumulated in ilmenite, zircon, and chromite.

The global Pd production (rounded) in 2010 was estimated to be 197 t, of which 87 t was produced in Russia, 73 t in South Africa, and 11 t in the United States (USGS 2011). Recycling is also a source of Pd, mostly from scrapped catalytic converters. Above half of Pd supply is used for the automotive catalytic converters, and recently, it has been highly increased (Schäfer and Puchelt 1998). It is used to reduce emissions of various carbon compounds and nitrous oxides. However, Pd emission may also be of potential risk to human and environment. It is used in electronics, jewelry production, as well as chemical and medical instruments. Its compounds are used in photography and electroplatings, and as a catalyst, mainly in organic synthesis. Radioactive ^{103}Pd is proving to be a successful form of treatment for some types of cancer. Several Pd compounds have antiviral, antibacterial, and fungicide properties. Due to the numerous applications and limited supply sources of Pd, it is a metal of a great value.

32.2 SOILS

The mean Pd contents of soils, as reported by various authors, range from 1 to 40 μg/kg and average 20 μg/kg. According to the data of the World Health Organization (WHO 2002a), its mean content is 40 μg/kg. Agricultural soils of Sweden contain Pd within the range of <40–150 μg/kg, at the mean value of 40 μg/kg (Eriksson 2001a). Recently, Pd levels in soils and dust along roads, especially high traffic, have been elevated, within the range of 6.6–280 μg/kg (Ravindra et al. 2004). The highest Pd content, above 440 μg/kg, is in roadside soil samples from Perth, Australia (Whiteley 2005). For the most soils along heavy traffic roads of various countries, the maximum Pd contents are reported as follows (in μg/kg): Austria, 50; Brazil, 58; Germany, 47; Greece, 126; Hawaii, 105; Italy, 110; Poland, 42; and the United States (California), 115. High Pd contents are also reported for road tunnel dust, from 20 to 207 μg/kg (Helmers et al. *vide* Kabata-Pendias 2011).

Due to the use of Pd in car converters, its steady increased levels in environments along roads, as well as in sewage sludge, have been observed. Its content in sewage sludge may vary from about 10 to 20 in rural and industrial areas, respectively. Sewage sludge ashes contain Pd within the range of <100–600 μg/kg (Pyrzyńska *vide* Kabata-Pendias and Mukherjee 2007).

32.3 WATERS

Palladium concentrations in the Pacific Ocean vary from 0.04 to 0.06 ng/L (Nozaki 2005). Its world average contents in river water are from 1 to 30 ng/L, at the average of 2.8 ng/L (Gaillardet et al. 2003). Rainwater collected in Sweden, during 1999, contains Pd at the average value of <1 ng/L, and its wet deposition is calculated for <11 mg/ha/yr (Eriksson 2001a). In water, Pd may occur in both species, hydroxide, $Pd(OH)_2$ and chloride, $PdCl_2$.

According to the data of the WHO (2002a), the common ranges of its concentration are as follows (in ng/L): freshwater, 0.4–22; salt water, 19–70; and drinking water, <24.

In water contaminated with road dust, an increased Pd concentration is noticed, and its elevated mean amount in eels, up to 0.18 μg/kg fresh weight (FW), is observed (Sures et al. 2001).

Sediments in some water reservoirs along German highway contain Pd up to 35 μg/kg (Gower and Zereini *vide* Sures et al. 2001).

32.4 AIR

Palladium concentrations in air vary commonly from <270 to 280 pg/m^3. However, air above some cities and industrial regions may contain much higher Pd amounts, for example, 12,700 and 56,600 pg/m^3, in Chicago, Illinois, and Chernivtsi, Ukraine, respectively (Ravindra et al. 2004). The average Pd concentration in air of the United States is <1 pg/m^3 and varies from <0.06 to 110 pg/m^3 (WHO 2002a).

32.5 PLANTS

Palladium is most available to plants, among other PGMs. Its contents in various shrubs and trees are in the range of 30–400 μg/kg ash weight (AW). Twigs of birch trees grown close to the mineral deposits of PGMs in Canada accumulate Pd up to 4014 μg/kg AW (WHO 2002a). In contaminated regions (e.g., PGMs ore deposits in Canada), twigs of birch trees contain Pd within the range of 30–400 μg/kg AW. A great variation has been observed in Pd contents in various trees, within the range of 30–400 μg/kg AW, due to variable climatic conditions and stage of growth.

Plants exposed to automobile exhausts contain Pd within from 0.4 to 2.4 μg/kg, with the highest value for moss and the lowest for plantain (Dijingova et al. *vide* Kabata-Pendias 2011). The average Pd content in grass from roadsides in Poland is 3.2 μg/kg. Palladium contents in grass from roadside areas decrease with the distance from road 129 to 11 μg/kg, at 0 and 5 km, respectively. However, no relationship with its amounts in soils was observed (Hooda et al. *vide* Kabata-Pendias 2011).

The mean Pd content in cereal grains is calculated as 0.9 μg/kg FW, whereas in food plants it varies from 0.4 to 3.0 μg/kg, being lowest in apples and highest in nuts (WHO 2002a).

Stress effects of Pd in plants have been observed at a lower concentration, compared with other posttranslational modifications, apparently due to its relatively high biochemical activity. It is likely to be bound to high-molecular-weight proteins and is able to replace Mn in some metalloenzymes, due to similar ionic radii. Phytotoxic effects in terrestrial plants are observed at Pd concentrations from about 2 to 60 mg/L of the nutrient solution (WHO 2002a), but there are also reported toxic effect at its lower concentration, from 1 to 3 mg/L (Kabata-Pendias 2011).

32.6 HUMANS

Concentrations of Pd in human *wet* tissue samples (liver, kidney, spleen, lung, muscle, fat) were below the limit after automotive catalytic converters (ACC) were not used, since 1974. It was less than 0.6–6.7 μg/kg (Johnson et al. 1976 *vide* WHO 2002a). Direct exposure to traffic has no verifiable influence on the background Pd burden of the population. It is probable because the concentrations of Pd in the body fluids of unexposed people are <0.1 μg/L in blood and <0.3 μg/L in urine (WHO 2002a).

Exposure of the general population is through Pd in air, food, and water, and through release of Pd from dental restorations. Exposures to Pd and its compounds (in dusts or solutions) may also occur in workers of the Pd mining, smelting, refining, or recycling industries; the chemical industry, particularly in catalyst manufacture, the electronics industry, and jewelry production.

Palladium ions can be taken up by the skin and by oral and inhalative routes. Although its absorption and retention are poor, there may be a risk for sensitive persons. Under certain conditions, Pd ions (and possibly microparticles) appear to be released from metallic Pd (e.g., in dental alloys). There are also indications that very finely dispersed element Pd particles become bioavailable, when dissolved in biological media. However, precise quantitative data are not available.

Absorbed Pd is in almost all organs, tissue fluids of a body, with maximum in the kidney, liver, spleen, lymph nodes, adrenal gland, lung, and bones (of experimental animals). Palladium and its compounds are of very low to moderate acute toxicity if swallowed (depending mainly on their solubility). Several Pd salts may cause severe primary skin and eye irritations. The biological half-life of Pd in rats has been estimated to be 12 days (WHO 2002a).

A major source of health concern is the sensitization risk of Pd as very low doses are sufficient to cause allergic reactions in susceptible individuals. Persons with known Ni allergy may be especially susceptible. Workers occupationally exposed to

Pd include miners, dental technicians, and chemical workers. The latter are exposed mainly to Pd salts, several of which may have primary skin and eye irritations. The general population may come into contact with Pd mainly through mucosal contact with dental restorations and jewelry containing Pd, and possibly via emissions from Pd catalysts. Protection of the public from Pd adverse effects may be achieved by the use of alloys with high corrosion stability, and thus minimal release of Pd. Patients with an allergy to Ni may be sensitive to Pd-containing dental materials, due to Pd allergy, although this risk appears to be low (Kielhorn et al. 2002).

Palladium ions are among the most frequent sensitizers within metals (second rank after Ni). Owing to the ability of Pd ions to form complexes, they bind to amino acids (e.g., L-cysteine, L-cystine, L-methionine), proteins (e.g., casein, silk fibroin, many enzymes), and DNA or other macromolecules (e.g., vitamin B_6). Several Pd compounds have been found to have antiviral, antibacterial, and/or fungicidal properties (WHO 2002a).

Consumers are exposed mainly to Pd from jewelry and dental restorations. Allergy to Pd is nearly always together with Ni allergy, because Pd and Ni tend to cross-react.

TABLE 32.1
Palladium in Foodstuffs (μg/kg FW)

Product	Content [a]	Content [b]
Food of Plant Origin		
Bread	2	0.8
Miscellaneous cereals	0.9	0.7
Green vegetables	0.6	0.23
Other vegetables	0.5	0.3
Potatoes	0.5	0.7
Fresh fruits	0.4	0.04
Fruit products	0.5	0.2
Nuts	3.0	1.9
Beverages	0.4	0.1
Food of Animal Origin		
Carcass meat	0.4	0.5
Offal	2.0	2.2
Meat products	0.6	0.6
Fish	2.0	0.5
Milk	<0.3	0.03
Dairy products	0.4	1.8

Sources: [a] Ysart, G. et al., *Food Addit. Contam.*, 16, 391–403, 1999.
[b] Rose, M. et al., *Food Addit. Contam.*, 27, 1380–1404, 2010.
FW, fresh weight.

Its salts should be included in dental screening patch test series. In jewelry, Pd should be limited, until we know more about the risk of sensitization (Faurschou et al. 2011).

The allergenic potential seems to be associated with the ionic Pd species, which form complexes able to react with endogenous proteins. A case of occupational asthma has been reported in workers exposed to Pd in a galvanoplasty plants (WHO 2002a).

Palladium complexes are used efficiently in anticancer drugs. Its most important properties are good solubility in water, the ability to transport (through the membranes), and fortitude in cells (Abu-Surrah et al. 2008).

Palladium contents in food are low, in general below 1 μg/kg (Table 32.1). The population dietary exposure to Pd in the United Kingdom in 1994 was from 0.7 to 1 μg/day, in 2006. The greatest contribution to Pd exposure is from beverages (20%), cereals (13%), bread (12%), and potatoes (11%) (Rose et al. 2010).

33 Platinum [Pt, 78]

33.1 INTRODUCTION

Platinum (Pt) is a dense, malleable, precious, transition metal of group 10 in the periodic table of elements. It is one of the rarest in the Earth's crust, at an average abundance of approximately 5 μg/kg. It is considered a noble metal and exhibits a remarkable resistance to corrosion in various conditions. Its highest contents, up to 3 μg/kg, are in ultramafic rocks and the lowest, 0.05 μg/kg, in limestones. Its mean content of coal in the United States is <1 μg/kg (Finkelman 1999). However, there are reported much higher Pt concentrations in coal, up to 230 μg/kg (Li *vide* Kabata-Pendias 2011). It exists in higher abundances on the moon and in meteorites.

As a very nonreactive metal, Pt occurs mainly in alloys with other platinum group metals (PGMs). Its more common minerals are as follows: sperryllite, $PtAs_2$; cooperite, (Pt,Pd)S; braggite, (Pt,Pd,Ni)S; telluride, PtBiTe; arsenite, $PtAs_2$; and nigglite, PtSn. Several sulfide minerals, and ilmenite, zircon, and gadoline may be hosts for Pt, within the range of its concentrations from 0.2 to 2 mg/kg. Platinum may be combined with other PGMs and deposited in alluvial sediments, from which it was mined.

Platinum has various oxidation states, from +2 to +5, and the most common is +4 state. It occurs in some Ni and Cu ores, as well as in some native deposits, mostly in South Africa, which account for about 80% of the world Pt production. Platinum, as well as other PGMs, is obtained commercially as a by-product from Ni and Cu mining and processing.

The global Pt production (rounded) in 2010 was estimated to be 183 t, of which 138 t was produced in South Africa, 24 t in Russia, and 8 t in Zimbabwe (USGS 2011). Platinum is still used, in great quantities, for catalytic converters, laboratory equipment, electrical contacts, electrodes, thermometers, medical equipments (especially in some implantations), and jewelry. The common use of Pt is as a catalyst in some chemical reactions. Pt compounds, mainly cisplatin, *cis*-$[Pt(NH_3)_2Cl_2]$, are applied in the chemotherapy against certain types of cancer. Effluents of some hospitals are considered as a main Pt source in sewage sludge. Its contents in sewage sludge of some cities in Germany varied (in the last century) from 86 to 266 μg/kg. Platinum concentrations in ash from sewage sludge is about 300 μg/kg (Pyrzyńska *vide* Kabata-Pendias 2011).

33.2 SOILS

The common Pt contents in surface soils vary from 2 to about 40 μg/kg. Its contents in uncontaminated soils are highly dependent on its concentrations in mother rocks, and relatively high Pt contents, up to 75 μg/kg, are in soils over weathered norite. The magnetic fraction of these soils contain Pt within the range of 860 to

<3000 µg/kg (Fuchs and Rose *vide* Kabata-Pendias 2011). Various Pt amounts in soils are reported for different countries (in µg/kg): Australia, <5–70; Italy, 1.2–52; the United Kingdom, 2.3–21; and Germany, 0.35–330 (Ravindra et al. 2004).

The maximum levels of Pt, reported for roadside soils of various countries, are as follows (in µg/kg): California, 680; Hawaii, 506; Western Australia, 440; Germany, 330; Mexico, 300; Italy, 278; Greece, 141; and Austria, 134. In some road dusts in the EU countries, elevated Pt concentrations may be up to 1000 µg/kg. Its highest contents are always reported for soils along high traffic roads and in road dusts (Farago et al. 1998).

Platinum distribution in surface soils shows a positive correlation with traffic flow. The estimation of its emission from motor vehicles varies, and recent calculations showed that it might be between 0.5 and 0.8 µg Pt/km (Pyrzyńska *vide* Kabata-Pendias 2011). Such a calculation for the vicinity of Frankfurt (Germany) is at mean value of 0.27 µg/km (Zereini et al. 2001). Significant Pt sources are hospital effluents, which contain it within the range of 0.01–0.66 µg/L (Kümmerer 2013). One hospital may release about 330 g/yr. Extrapolation on a national basis amounted to an upper limit of 141 kg/yr for the Pt input in Germany. Up to 70% of Pt used in drugs is excreted and go to hospital effluents.

Platinum emissions in Germany are estimated as follows (in kg/yr): catalytic converters, 15; hospital effluents, 28–60; and sewage sludge, 100–400. About one-third of Pt concentrations in sewage sludge originates from automobile converters (Schäfer and Puchelt 1998). Important, local Pt emissions are from Ni ore mining and smelting industries. Nickel ores contain Pt within the range of 600–13,700 µg/kg, and its emission from one Ni smelter (Monchegorsk, Russia) is about 2.2 t/yr.

Platinum emitted from motor vehicles is mainly in the form of nanocrystals. Some oxidized Pt species are also emitted, and they may be easily mobilized by complexation and biomethylation, especially when they are associated with organic matter (OM) compounds. It is also likely to form complexes with inorganic ions (Cl, N, S), which are of a great mobility and bioavailability. There are estimations, however, that only about 5% of Pt in street dust is in the mobile fraction (Ljubomirova et al. *vide* Kabata-Pendias 2011).

33.3 WATERS

Platinum concentrations in seawater vary, depending on the location and water depth (increasing with depth), from 0.004 to 0.332 ng/L. The median value for the North Pacific is reported to be 0.05 ng/L (Nozaki 2005). In water from the Baltic Sea, the mean Pt concentration is 2.2 ng/L (Szefer 2002). Much higher Pt contents, up to 20 ng/L, are in the waters of some lakes in Norway (Reimann and de Caritat 1998). Rainwater collected during 1999 in Sweden contains Pt, at an average of 22 ng/L. Its wet deposition is calculated as 250 mg/ha/yr (Eriksson 2001a).

The main sources of Pt in surface water are its contents in sewage sludge from industrial areas and in effluents from hospitals. However, it was finely concluded that the main input of Pt into municipal sewage was urban and road run-off from traffic and other Pt emitting sources, and not from hospital sewage. Also road dusts fall down to water may be a significant Pt source.

Platinum levels in drinking water have been estimated to be 0.1 ng/L. In some tap water, however, it is reported to be very high, up to 60 ng/L (WHO 1991b).

In marine algae, Pt contents vary from 0.08 to 0.32 μg/kg. It is relatively easily bioaccumulated, and in some crustaceans, it may be concentrated up to 38 μg/kg (WHO 1991b).

33.4 AIR

There are only a few data on Pt concentration in air, prior to the introduction of motor vehicle catalytic converters. Data from that time may reflect Pt background levels and are <0.05 pg/m^3 in air samples taken near freeways in California and <0.6–1.8 pg/m^3 in rural areas of Germany. Data taken in Germany for the twentieth century (in pg/m^3) are as follows: rural areas, <0.6–1.8; traffic roads, <1–13; and city road dust, on average, 730 (WHO 1991b). The average Pt content in airborne particles of roadsides of some EU countries is approximately 15 pg/m^3 (Gómez et al. 2002).

Maximum Pt concentration in air recommended in Germany for the occupational exposure is 2 pg/m^3 (Hoppstock and Sures *vide* Kabata-Pendias and Mukherjee 2007).

Platinum contents of moss sampled in Finland are up to 12.2 μg/kg, whereas in moss sampled from areas of Ni industries, it may be above 6000 μg/kg (Reimann and Niskavaara 2006).

33.5 PLANTS

Apparently, plants easily take up mobile Pt species, which are mainly its complexes with soluble OM. It is accumulated in roots. Platinum may be bound by some proteins and is relatively highly toxic, especially Pt^{2+}, which is more toxic than Pt^{4+}.

The contents of Pt in crop plants vary highly between the kind of plants and growth sites. Grasses from rural regions contain Pt within the range of 0.1–0.3 μg/kg, whereas from areas close to traffic roads, they contain 0.8–96.0 μg/kg (Kabata-Pendias 2011). Some plants from the roadsides may contain Pt up to above 800 μg/kg (Ravindra et al. 2004). Platinum emitted from motor vehicles (its main pollution sources) is deposited mainly as suspended particles on the plant surface.

Cereal grains from Sweden contain Pt from 0. 2 to 0.9 μg/kg (Eriksson 2001a), whereas corn grains have Pt from 80 μg/kg in Germany to 642 μg/kg in Austria. Relatively high Pt contents are in radish roots (530 μg/kg) and the highest tobacco leaves (23,300 μg/kg) (Ravindra et al. 2004). The authors reported that barks of trees grown in remote and urban areas contain Pt from 0.07 to 38 μg/kg, respectively, and in pine needles from a town, its content is up to 102 μg/kg.

33.6 HUMANS

The total amount of platinum in the human body is not known, but low, and it has no known biological role (Emsley 2011). The natural levels of Pt significantly fluctuate in the human body fluids of various types. The values of 0.8–6.9 ng/L and 0.56 μg/L for the blood and 0.5–15 ng/L and 0.18 μg/L for urine have been reported (Vaughan and Florence 1992). Pt levels of 0.1 to 2.8 μg/L in the blood are in the general population. In serum, of occupationally exposed workers, Pt levels range from 150 to 440 μg/L. The principal Pt deposition sites are the kidneys, liver, spleen, and adrenals. The high amount of ^{191}Pt found in the kidney shows that once Pt is absorbed, most of it is accumulated in the kidney and is excreted in the urine. The lower level in the brain suggests that Pt ions cross the blood–brain barrier, but to a limited extent (WHO 1991b).

Metallic Pt is considered to be biologically inert and nonallergenic, and because the emitted Pt is probably in metallic, or oxide, form, the sensitizing potential is presumably very low. Platinum from the road dust, however, can be solubilized and enters into water, sediments, soils, and the food chain (Ravindra et al. 2004).

Platinum compounds provide an excellent illustration of the need to differentiate the chemical species of an element, when evaluating its allergenic potential. Allergic symptoms including rhinitis, asthma, and urticaria have been reported after World War II in workers employed in Pt refineries and in secondary users, mainly from the manufacture or recycling of catalysts plants (WHO 2006a).

Platinum compounds, especially soluble salts, are toxic, and chronic industrial exposure to these compounds is responsible for the development of a syndrome known as platinosis, which is characterized by respiratory and cutaneous hypersensitivity. The acute toxicity of Pt depends mainly on the Pt species. Soluble Pt compounds are much more toxic than insoluble ones. Exposure to Pt salts is mainly confined to occupational environments, primarily to Pt metal refineries and catalyst manufacture plants.

Platinum nanoparticles, with different sizes, showed the different bacterio-toxic or compatible properties with the clinical pathogen (Gopal et al. 2013).

cis-Diaminedichloroplatinum(II), *cis*-$(NH_3)_2PtCl_2$, clinically called *cisplatin* is one of the most successful anticancer compounds. After the discovery of its activity, thousands of Pt complexes have been synthesized and evaluated for their anticancer activity. Research in the field of Pt-based cancer chemotherapy showed that cisplatin and its analogous compounds exhibit very similar patterns of antitumor sensitivity and susceptibility to resistance, which means that most of them produce identical adducts with DNA (Abu-Surrah et al. 2008).

The concentration of Pt in the sewage of hospitals is originating from excreted antineoplastic drugs. About 70% of Pt, administered in the form of either *cisplatin* or *carboplatin*, is excreted, and therefore, it ends up in hospital effluents. Pt concentrations, measured in the total effluents of hospitals, ranged widely from <10 ng/L to approximately 3500 ng/L. Nursing staffs are evidently concerned about the risk of hazardous exposure of Pt, due to increasing use and contact with antineoplastic drugs (Ravindra et al. 2004).

The International Agency for Research on Cancer considered the evidence for carcinogenicity of *cisplatin* for animals to be sufficient, but that for humans inadequate. Cisplatin is classified in Group 2A, that is, probably carcinogenic to humans (IARC 1987, 2013).

TABLE 33.1
Platinum in Foodstuffs (mg/kg FW)

Product	Content	
Food of Plant Origin		
	a	b
Bread	0.0001	<0.003
Miscellaneous cereals	0.0001	<0.003
Rice (different species)	–	<0.0003[c]
Vegetables	0.0001	<0.0006
Potatoes	<0.0001	<0.001
Fruits	<0.0001	<0.0005
Nuts	0.0001	<0.003
Beverages	<0.0001	<0.0005
Food of Animal Origin		
Carcass meat	0.0001	<0.0015
Meat products	<0.0001	<0.0015
Offal	<0.0001	<0.0015
Fish	<0.0001	<0.003
Milk	<0.0001	<0.0005
Dairy products	0.0001	<0.0015

Sources: [a] Ysart, G. et al., *Food Addit. Contam.*, 16, 391–403, 1999.
[b] Rose, M. et al., *Food Addit. Contam.*, 27, 1380–1404, 2010.
[c] Jorhem, L. et al., *Food Addit. Contam.*, 25, 841–850, 2008b.
FW, fresh weight.

Concentrations of Pt are below the levels of detection in most food groups (Table 33.1). Taking into account uncertainty related to samples in which Pt could not be detected, dietary exposure of Pt in the United Kingdom was estimated to be 0.0002 mg/day in 1994 (Ysart et al. 1999) and 0–0.023 mg/day in 2006 (Rose et al. 2010). The highest contribution to dietary exposure was introduced by beverages (27%), miscellaneous cereals (17%), and bread (12%) (Rose et al. 2010). In Australia, dietary intake of 0.00144 mg Pt per day for adults (male: 0.00173 mg/day; female: 0.00115 mg/day) was estimated. Platinum in materials of plant and animal origins appears to be bioavailable (Vaughan and Florence 1992).

34 Polonium [Po, 84]

34.1 INTRODUCTION

Polonium (Po) is a highly radioactive element of group 16 in the periodic table of elements, chemically similar to Bi and Te, and may follow the biochemical pathway of Se and Te. Based on its properties and behavior, it is sometimes classified as a metalloid. Its abundance in the Earth's crust is estimated to be 1 fg/kg (10^{-16}%) or 2 pg/kg (10^{-13}%). It may be accumulated in some coals and in phosphate rocks. Its highest concentrations are in uranium (U-pitchblende) ores, up to about 0.1 μg/kg, and it has been its source for quite a time. Nowadays, Po is obtained by irradiating Bi, with high-energy neutrons or protons. There are 25 known Po isotopes, with atomic masses from 194 to 218, of which 7 isotopes occur naturally. The only one occurring to any extend is ^{210}Po, with the half-life of 138.4 days.

Applications of Po are few and include heaters in space probes, antistatic devices, and sources of neutrons and alpha particles. It is highly dangerous and has no biological role. The main hazard is its intense radioactivity (as α-radiation emitter), when it is ingested, inhaled, or absorbed, and it is very difficult to handle it safely.

Uses of ^{210}Po include nuclear batteries, neutron sources, and film cleaners. It is a lightweight nuclear heat source for thermoelectric power in space satellites. However, Po needs to be replaced in these devices nearly every year, because of its short half-life. Due to its dangerous properties, it has been replaced by other sources of beta particles.

Alexander Litvinenko, a Russian dissident, was probably the first person who died of the acute α-radiation effects of ^{210}Po. It has also been suggested that Irène Joliot-Curie died from leukemia, which resulted from the exposure to ^{210}Po.

34.2 SOILS

The natural abundance of ^{210}Po in soils range from 10 to 220 Bq/kg. This is the result of ^{222}Rn decay, in the top layer of soils, as well as the fall down of ^{210}Po suspended in the air. This radionuclide is strongly, almost irreversibly fixed by soil particles, mainly by metal hydroxides and/or sulfides. It is also likely to be bound by soluble organic matter (SOM), and a close correlation between levels of ^{210}Po and SOM contents was observed (Korenkov et al. *vide* Kabata-Pendias 2011). It is not very mobile in soils. There are suggestions that it may be methylated, similar to Hg, Se, and Te.

Soils around U ore deposits in India contain ^{210}Po, on average, 125 Bq/kg (Marbagniang et al. *vide* Kabata-Pendias 2011). Its much higher concentrations (15,000–22,000 Bq/kg) are in various U mine tailings. Also environments (soils, plants, and wild animals) of phosphate rock processing are enriched in this radionuclide.

34.3 WATERS

The median Po concentration in ocean waters is roughly estimated to be 1 fg/L (Reimann and Caritat 1988). It may be accumulated by some aquatic biota, as well as in phytobenthos.

The mean concentration of ^{210}Po in drinking water in Poland, in 1998, was 0.48 mBq/L, and its annual intake per person was calculated as 0.24 Bq (Skwarzec et al. 2001). Guidance level of ^{210}Po in drinking water is 0.1 Bq/L (WHO 2011a).

34.4 PLANTS

The radioactivity of ^{210}Po in terrestrial plants is within the range of 8–12 Bq/kg. However, ^{210}Po activity in plants from the area of U deposits in India is relatively low and varies from 0.02 to 6.69 Bq/kg (Marbaniang et al. *vide* Kabata-Pendias 2011).

Radioactive ^{210}Po contained in phosphate fertilizers is absorbed by the roots of plants and stored in its tissues. Phosphate fertilizers are considered to be the main source of this radionuclide. However, ^{210}Po, as a very volatile element, is emitted from the soil surface and absorbed by broad leaves of plants. Increased ^{210}Po levels in large leaves of tobacco suggest that its source is also the atmospheric deposition. Its maximum allowable concentration in air is estimated to be about 10^{11} μCi/cm^3.

Polonium has been found in tobacco leaves, especially from plants grown on soils with phosphate fertilizers. The presence of radioactive Po in tobacco smoke has been known since the early 1960s, but this information was not spread. Tobacco plants fertilized by phosphates contain ^{210}Po, which emits alpha radiation, which is estimated to cause about 11,700 lung cancer deaths annually worldwide. Its mean activity in cigarette tobacco of various countries is estimated to be 15.1 mBq per cigarette. Cigarettes produced in Poland have ^{210}Po radiation within the range of 4.23–24.12 mBq per cigarette (Skwarzec et al. *vide* Kabata-Pendias 2011).

The range of ^{210}Po radioactivity in mushrooms varies from 20.1 to 76.5 in caps and from 9.9 to 47.9 Bq/kg in stipes. It is higher in *Boletus edulis* than in other wild mushrooms (Skwarzec and Jakusik *vide* Kabata-Pendias 2011).

34.5 HUMANS

Polonium can enter the human body via two principle ways: ingestion and inhalation of radon gas originated from ^{238}U. The main route of ^{210}Po intake in the human body is ingestion with foodstuffs. The absorption coefficient of ^{210}Po into the blood from the digestive tract was estimated to be 35%. The highest concentration of ^{210}Po in the body was observed in the skeleton and hair. The body burden of humans is 513 pCi of ^{210}Po. Excretion of these radionuclides from the body in urine is 14–15 times less than in feces. The calculated annual radiation doses of ^{710}Po in the bones, liver, and kidneys were 2.1–5.0 mrad/yr, and in other soft tissues, 0.3–0.9 mrad/yr (Ladinskaya et al. 1973).

Results showed that the activity concentrations of naturally occurring radionuclides, especially those of ^{210}Po and ^{40}K, were much higher than those of anthropogenic radionuclides. Calculations of effective dose, due to ingestion of a given radionuclide in the food, depend on the concentration of the radionuclide, the amount of food eaten, and the dose coefficient, or the effective dose per unit intake of the radionuclide. The dose coefficient is essentially dependent on the body uptake and the distribution and retention by body organs (Hunt and Rumney 2007).

^{210}Polonium is a high-energy α-emitter (radioactive half-life of 138 days) that presents a radiation hazard only if taken into the body, for example, by ingestion, because of the low range of α-particles in biological tissues. As a result, external contamination does not cause radiation sickness. Ingested ^{210}Po is concentrated initially in red blood cells, and then the liver, kidneys, spleen, bone marrow, gastrointestinal tract, and gonads. ^{210}Po is excreted in urine, bile, sweat, and (possibly) breath, and is also deposited in the hair. After ingestion, unabsorbed ^{210}Po is present in the feces. The elimination half-life time in humans is approximately 30–50 days. In the absence of medical treatment, the fatal oral amount is probably in the order of 10–30 μg. Ingested ^{210}Po can cause after radiation sickness symptoms (ARS). This is characterized by a prodromal phase, in which nausea, vomiting, anorexia, lymphopenia, and sometimes diarrhea develop after exposure. Higher radiation doses cause a more rapid onset of symptoms and a more rapid reduction in lymphocytes. The triad of early emesis followed by hair loss and bone marrow failure is typical of ARS (Jefferson et al. 2009).

^{210}Po and ^{210}Pb are the most important representatives of the natural marine radioactivity, and they can be considered the major contributors to radiation doses received by humans, among natural radionuclides in the marine environment (Table 34.1).

TABLE 34.1
^{210}Po Activity Concentration in Fish from Different Areas in the World (Bq/kg FW)

Country	Concentration
Lebanon	0.22–47.82
Syria	0.27–27.48
Japan	0.6–26.0
Denmark	0.35–0.9
Portugal	0.2–11
Australia	0.09–44.1
South Africa	2.2–20.3
Brazil	0.5–5.3
America	0.4–153.3
Poland	0.9–5

Source: El Samad, O. et al., *Leban. Sci. J.*, 11, 39–45, 2010.
FW, fresh weight.

^{210}Po, a naturally occurring radioisotope, is known to be enriched by various kinds of marine organisms. It can, however, have a toxic effect, and therefore play an important role in humans. Among the natural radioactivity in human foodstuff, ^{210}Po is the major contributor. Seaweeds rich in protein may have a high activity of ^{210}Po, which is higher in storage organs than in their hard parts. Its concentrations may vary, according to sampling sites, as well as weeds species and season (Alam and Mohamed 2011).

35 Radium [Ra, 88]

35.1 INTRODUCTION

Radium (Ra) is an almost pure white alkaline metal of group 2 in the periodic table of elements, but may be oxidized readily and become black. It is the heaviest of all the alkaline earth metals. It exists mainly as a divalent radioactive cation, composed of several isotopes, which are highly radioactive. They are common products of the U and Th decay chain. The most stable and relatively frequent radionuclide in the biosphere is ^{226}Ra, which has a half-life of 1601 years and decays into radon gas. Its chemical properties are similar to those of Ba, Sr, and Ca. It is a lithophilic element and is associated with minerals of zeolite group. Its typical mineral is radiobarite, $(Ba,Ra)SO_4$.

The average content of ^{226}Ra in the Earth's crust is 0.9 ng/kg (range 0.1–1.1 ng/kg) and may be concentrated in coal. It forms several compounds, such as $RaOH^+$, $RaCl^+$, $RaCO_3$, and $RaSO_4$. It is a natural component of the environment, but is mainly of the anthropogenic origin. Its main sources are U mining and processing, P and K fertilizers, radioactive and luminizing wastes, coal combustion, and cement factories. The mean concentration of ^{226}Ra in coal is on the order of 1 pCi/g, and in soils surrounding coal-fired power plants, its level may be up to 8 pCi/g (ATSDR 2002b). Also residues from oil and gas industries often contain Ra and its daughters. Water from oil well, when discharged into the sea and mixed with seawater, may affect the precipitation of Ra and its accumulation in bottom sediments.

Radium is not necessary for living organisms, and its adverse health effects are likely, when it is incorporated into biochemical processes, because of its radioactivity and chemical reactivity. It was formerly used in self-luminous paints for watches, nuclear panels, aircraft switches, clocks, and instrument dials. It has also been used in silk, leather, and wood industries for stains and dyes.

Radium (usually in the compound $RaCl^+$) is used to produce radon gas, which has been used for cancer treatment. Currently ^{223}Ra is still under investigation for use in the treatment of bone cancer.

35.2 SOILS

Radiation of Ra in soils range broadly from 4.4 to about 150 Bq/kg in Canada and Turkey, respectively. Increased ^{226}Ra levels are in soils around some mining and industrial plants, especially of U and P processing. Also some P fertilizers may contain its higher levels at about 75 Bq/kg. The estimated annual deposition of ^{226}Ra and ^{228}Ra in agricultural soils is 4.5 and 0.5 Bq/m^2, respectively.

Radium sorption in soils highly varies and the lowest is in sandy soils (Smith and Amonette *vide* Kabata-Pendias 2011). The influence of soil biota on its mobility is observed. Especially sulfate-reducing bacteria can influence a rapid dissolution of $RaSO_4$, mainly under reducing conditions in sludge. Thus, Ra mobility is enhanced under anaerobic conditions and in acid soils. In moist soils, its mobility and bioavailability are highly increased.

Some soils derived from granite and limestone, argillaceous limestones in particular, contain elevated amounts of ^{226}Ra. Its distribution is associated with Ca, Mg, Al, and SO_4 (Megumi and Mamuro *vide* Kabata-Pendias 2011).

Also soluble organic matter (SOM) and Fe–Mn hydroxides, as well as clay minerals, have a strong affinity for Ra and significantly decrease its exchangeable fraction, especially in neutral and alkaline pH range of soils. Apparently, ^{226}Ra may form complexes on the clay surface and also within the inner space. Its mobile species in soils amount to 1%–10% of the total contents.

35.3 WATERS

Worldwide average Ra concentration in river waters is estimated to be 0.024 pg/L, within the range of 0.002–0.09 pg/L, with the highest values for Asian rivers (Gaillardet et al. 2003). Its mean concentration in seawaters is around 1 pg/L. In general, concentrations of two radioisotopes of Ra in surface water and groundwaters are low. However, in some geographic regions, it may be elevated due to the impact of bed rocks. Its concentrations may also be higher in waters from deep wells, located near radioactive waste disposal sites.

The common ^{226}Ra content in drinking water of the United States ranges from 0.9 to 10 pCi/L (ATSDR 2002b). The maximum allowable concentration value for Ra in drinking water of Russia was established for 0.1 pg/L (Reimann and de Caritat 1998). Its levels are usually around 1 pCi/L; higher levels, >5 pCi/L, are dangerous for human. Guidance level for ^{226}Ra in drinking water is 1 Bq/L (WHO 2011a).

Radium in water is relatively easily available to aquatic biota. Marine plants and animals may contain Ra on the order of 100 times of that in the water media. Especially predatory fish, from water basins with contaminated bottom sediments (e.g., from industrial or cities effluents), have highly elevated Ra contents (ATSDR 2002b).

Significant sources of Ra are U mine tailings and processing effluents. Also water discharged from some coal mines may contain elevated Ra levels (Pluta *vide* Kabata-Pendias and Mukherjee 2007). Residues from oil and gas industries often contain ^{226}Ra and its daughters. Water from oil well, when discharged into the sea and mixed with seawater, may affect the precipitation of Ra and its accumulation in bottom sediments.

35.4 AIR

The combustion of coal is the most common and important source of Ra in the atmosphere. Radium may concentrate in fly ash, which is within the range of 1–10 pCi/g (1 pCi equal to 0.04 Bq/g). It is estimated that a single 1000-MW coal-fired power plant will discharge about 28 mCi of total radium per year.

Dust samples collected from the air of New York City contained ^{226}Ra at 8×10^{-5} and ^{228}Ra at 1.5×10^{-4} pCi/m^3 (ATSDR 2002b).

People who work at factories that process U ores or work with P fertilizers are exposed to higher levels of Ra radionuclides in ambient air.

35.5 PLANTS

The content of Ra in plants is roughly estimated to be 0.03–1.6 ng/kg. It is relatively easily available to plants and is also easily moved from roots to shoots. Soil properties, especially cation-exchange capacity and SOM, control its bioavailability. The order of plants' capability to take up Ra from soils is as follows: leafy vegetables > root vegetables > grasses > corn stems > grains. The mean value of the transfer factor of Ra (ratio of plant content to soil content) is estimated to be 6.4×10^{-3}, being lowest for potato and highest for green leek (Uchida and Tagami *vide* Kabata-Pendias 2011). The same authors reported that ^{226}Ra contents in food plants vary from <0.02 Bq/kg in barley grains to 1.43 Bq/kg in carrot roots. Vegetables collected in Poland (during 1991–1996) contain this radionuclide from 0.087 in cucumber to 0.594 Bq/kg in carrot, respectively (Rosiak and Pietrzak-Flis *vide* Kabata-Pendias 2011).

35.6 HUMANS

Radium can be taken into the human body by eating food, drinking water, or breathing air. It is not known if it can be taken in through the skin. Most of Ra is taken in by ingestion (about 80%) and will promptly leave the body in feces. Its remaining 20% enters the bloodstream and is carried to all parts of the body. Inhaled Ra can remain in the lungs for several months and will gradually enter the bloodstream and be carried to all parts of the body, especially the bones.

The metabolic behavior of Ra in the body is similar to that of Ca. For this reason, an appreciable fraction is preferentially deposited in the bones and teethes. Its amount in the bones decreases with time from the exposure, generally dropping below 10% in a few months, to 1% and less in few years. Release from the bone is slow, so a portion of inhaled and ingested Ra will remain in the bones throughout a person's lifetime. Although ingestion is the most common means of Ra entry into the body, risk coefficients for that exposure route are lower than for inhalation (Argonne 2007).

Radium has been shown to cause adverse health effects, such as anemia, cataracts, fractured teeth, cancer, and death. The relationship between the amount of Ra that the man is exposed to and the amount of time necessary to produce these effects is not known. Although there is some uncertainty as to how much exposure to Ra increases chances of developing a harmful health effect, the greater the total amount of exposure to Ra, the more likely to develop one of these diseases (ATSDR 1990a).

Radium is currently used in brachytherapy to treat various types of cancer. Brachytherapy is a method of radiation treatment, in which sealed sources are used to deliver a radiation dose, at a distance of up to a few centimeters, by surface, intracavitary, or interstitial application (Argonne 2007).

There is no evidence that exposure to naturally present levels of Ra has harmful effects on human health. However, exposure to higher levels of Ra may result in health effects, such as teeth fracture, anemia, and cataract. When the exposure lasts for a long period of time, Ra may even cause cancer and the exposure can eventually lead to death. These effects may take years to develop. They are usually caused by γ radiation of Ra, which is able to travel fairly long distances through air.

Radium undergoes radioactive decay. During this process, α, β, and γ radiations are released. Alpha (α) particles can travel only a short distance and cannot go through the skin. Beta (β) particles can penetrate through the skin, but they cannot go all the way through the human body, whereas γ radiation can go all the way through the body.

Levels of occupational exposure to Ra are difficult to assess. Workers who are occupationally exposed to Ra, through the mining and processing of U, are also probably exposed simultaneously to U, Th, and Rn by inhalation and probably dermal exposure (ATSDR 1990a).

The International Agency for Research on Cancer (IARC 2012b, 2013) has classified ^{224}Ra, ^{226}Ra, ^{228}Ra, and their decay products in Group 1, human carcinogens.

^{226}Radium may occur in many different foods, and its activities have varied considerably. The mean ^{226}Ra contents of diets in 11 cities in the United States were estimated to be from 0.52 to 0.73 pCi/kg of food consumed (0.019–0.027 Bq/kg). Estimates of the mean concentrations of ^{226}Ra in milk and beef of the United States are 0.23 pCi/L (0.009 Bq/L) and 0.22 pCi/kg (0.008 Bq/kg) fresh weight.

There is a potential for human exposure to Ra by the consumption of beef and milk derived from cattle that graze on forage grown in soils containing Ra. No information is available on the occurrence of ^{228}Ra in food (ATSDR 1990a).

Average dietary exposure to ^{226}Ra (mBq/day) in different countries is estimated as follows: Japan, 23; Poland, 42; Italy, 52; and the United States (New York), 52. The greatest source of radionuclides in diet are cereals, potatoes, milk, and dairy products (Pietrzak-Flis et al. 2001).

36 Radon [Rn, 88]

36.1 INTRODUCTION

Radon (Rn) is a radioactive, colorless, and odorless noble gas of group 18 in the periodic table of elements, occurring naturally, as an indirect decay product of the U and Th chains. Its concentration in surface layer of the Earth's crust is about 2 mg/kg. The gamma rays are produced by radon and the first short-lived elements of its decay chain are isotopes of Po, Pb, and Bi. It may form some compounds, such as RnF_2 and $RnCl_4$. It is associated with a few minerals, mainly uranite, UO_2, and zircon, $ZrSiO_4$. It has about 30 isotopes and the most stable is ^{222}Rn, with a half-life of 3.8 days. It is often the important contributor to an individual's local background radiations.

36.2 ENVIRONMENT

Radon is one of the densest substances that remain as gas under normal conditions, and as easily inhaled, it is considered a health hazard due to its radioactivity. It exhausts naturally from the ground, particularly in certain regions, especially, but not only, in regions with granitic soils. It may also be emitted directly from granites. Also soils derived from gneisses, schists, and limestones may contain its elevated amounts. Radon is easily released from these soils to the atmosphere, which is estimated, on a global scale, to be 2400 mCi (90 TBq) annually.

Despite short lifetime of ^{222}Rn, its gas from natural sources can accumulate to far higher than normal concentrations in buildings, especially in low areas, such as basements. In some locations, U tailing materials have been used for landfills and were subsequently built on, resulting in possible increased exposure to Rn. In some buildings, Rn concentration may be >200 Bq/m^3 (Talbot et al. *vide* Kabata-Pendias and Pendias 1999).

Owing to its very short half-life (4 days for ^{222}Rn), its concentration in air decreases very quickly, when the distance from its production areas increases. Its concentration ranges from <10 Bq/m^3 to over 100 Bq/m^3 in some European countries and varies greatly with seasons and atmospheric conditions.

Radon may migrate freely through faults and fragmented soils, and may accumulate in caves or water. Thus, its higher concentrations can be found in some springwaters and hot springs. Water of hot springs in health spa of Poland contains ^{222}Rn within the range of 0.9–193.1 Bq/L, at an average of 14.5 Bq/L (Pachocki et al. 2009).

The major Rn exposure is through inhalation, with background levels in ambient air of approximately 0.1–0.4 pCi/L. Its higher levels are in indoor locations, where its mean level is estimated to be 1.6 pCi/L. In some family homes, it may exceed 8 pCi/L. It is a known pollutant emitted from geothermal power stations, because it is present in the material pumped from deep underground. Its content may also be elevated in the air of coal mines. However, its environmental impact is minimal.

Radon as a noble gas should not be bioaccumulated. However, the accumulation has been reported for Rn progeny, such as ^{210}Pb in cephalopods and ^{210}Po in marine birds, mushrooms, cephalopods, and coastal sand dune wild legumes (ATSDR 2012c).

Radon concentrations in groundwaters vary from 1 to 50 Bq/L for rock aquifers in sedimentary rocks, from 10 to 300 Bq/L for wells dug in soil, and from 100 to 50,000 Bq/L in crystalline rocks.

Surveys in the EU countries have shown that Rn concentrations in surface waters are very low, usually well below 1 Bq/L. The highest contents are usually associated with high U concentrations in the bedrock. A characteristic of Rn concentrations in rock aquifers is their variability; within a region with fairly uniform rock types, some wells exhibit concentrations far above the average (EC 2001). Significant seasonal variations in concentrations have also been observed (EC 2001).

36.3 HUMANS

Raised lung cancer rates have been reported from a number of cohort and case–control studies of underground miners exposed to Rn and its decay products. These include particularly U miners, but also groups of Fe ore and other metal miners, and one group of fluorspar miners. Several small case–control studies of lung cancer have suggested a higher risk among individuals living in houses known or presumed to have higher levels of Rn and its decay products than among individuals with lower presumed exposure in houses (IARC 1988).

^{222}Radon and its decay products are carcinogenic to humans (Group 1) (IARC 1988, 2013). When Rn gas is inhaled, densely ionizing α particles emitted by deposited short-lived decay products of Rn (^{218}Po and ^{214}Po) can interact with biological tissue in the lungs, leading to DNA damage. Because even a single α particle can cause major genetic damage to a cell, it is possible that Rn-related DNA damage can occur at any level of exposure. Therefore, it is unlikely that there is a threshold concentration, below which Rn does not have the potential to cause lung cancer (ATSDR 2012c).

The primary pathway for human exposure to Rn is inhalation, both indoors and outdoors. Ambient outdoor levels are the result of Rn emanating from soil or released from coal, oil, and gas power plants. Outdoor Rn levels are typically much lower than indoor Rn levels. Soil gas intrusion into buildings accounts for the majority of indoor Rn levels. However, indoor Rn levels can also originate from water used for domestic purposes, outdoor air, and building materials.

Exposure to high concentrations can occur in any location, with geologic Rn sources. Relatively high-level occupational exposure can occur among workers in underground mines (U, P, Sn, Ag, Au, V, hard rock) and sites contaminated with Rn precursors (Ra, U, or Th), as well as in any underground places. Also air of hospitals that used Ra needles for therapy may contain elevated Rn amounts.

Lung cancer risk in cigarette smokers and recent ex-smokers is associated with Rn, as demonstrated by increased lung cancer risk, with increasing cumulative exposure. The risk to nonsmokers is 25-fold lower (ATSDR 2012c).

In view of the latest scientific data, the World Health Organization (WHO) proposes a reference level of 100 Bq/m^3 to minimize health hazards due to indoor Rn exposure. However, if this level cannot be reached, under the prevailing country-specific conditions, the reference level should not exceed 300 Bq/m^3, which represents approximately 10 mSv per year, according to recent calculations by the International Commission on Radiological Protection (WHO 2009).

Radon is used in the medical field. Archeological discoveries in the Central and Southern Europe suggest that therapeutic properties of certain waters were well recognized in ancient times, but only within the last century high concentrations of Rn were discovered to be present in these waters. Radon spas are used extensively in the Southern and Central Europe and Japan, to treat several health conditions (WHO 2009).

The therapeutic use of Rn involves the intake of Rn gas either through inhalation or by transcutaneous resorption of Rn dissolved in water. Mineral Rn water is used in the treatment and rehabilitation of chronic inflammatory and degenerative rheumatic diseases, neurological diseases, cardiovascular diseases, allergic disorders, gynecological disorders, and endocrine and vegetative disorders.

The exact mechanism of the effect of Rn on the human body is not completely understood. The hormesis theory (a metal or ionizing radiation that produces harmful biological effects at moderate to high doses may produce beneficial effects at low doses) is the best explanation of the advantageous biological effect of ionizing radiation in low doses (Zdrojewicz and Strzelczyk 2006). Radon therapy is not an approved biomedical therapy in the United States, primarily due to continuing controversy over how much Rn exposure is safe (Erickson 2007).

37 Rhenium [Re, 75]

37.1 INTRODUCTION

Rhenium (Re) is a silvery metal of group 7 in the periodic table of elements. It may be readily oxidized and become black. Its content in the Earth's crust is within the range of 0.4–0.6 μg/kg. Sometimes, it is accumulated in carbon shales and some coal, up to about 100 μg/kg.

Rhenium oxidation states vary from −1 to +7, and the most common is +7. It does not occur as free metal, and its mineral, dzhekazganite ($CuReS_4$), is very rare. It reveals geochemical similarity to Mo and may be associated with some minerals: molibdenite, zircon, and gadoline. Higher concentrations of Re are also found in some minerals of LAs.

The global Re production (rounded) in 2010 was estimated to be 48 t, of which 25 t was produced in Chile, 6 t in the United States, 5 t in Peru, and 4.5 t in Poland (USGS 2011). Rhenium is very hard and resistant to corrosion. Its main use is in superalloys for blades in turbine engines. It is an ideal metal for use at very high temperatures, which makes it suitable for rocket motors. In alloy forms with Mo, Pt, and W, it is used in various thermocouples, X-ray tube, and electrical and catalytic devices. Occasionally, it has been used for plating jewelry. Some Re compounds have recently been used in the treatment of cancer patients (Collery et al. *vide* Kabata-Pendias and Mukherjee 2007).

37.2 ENVIRONMENT

Rhenium is readily soluble, mainly as the anion, ReO_4^-, during weathering processes, especially in oxidizing and acidic media. It is precipitated and/or sorbed by various sediments, especially by shales and some coals, where it may be accumulated up to 10 mg/kg. Arable soils of Sweden contain Re at an average value of <40 μg/kg (Eriksson 2001a), whereas in soils along motorways in Australia, its content is up to 9.8 μg/kg (Frische and Meisel *vide* Kabata-Pendias 2011).

Mobile Re species are relatively easily available to plants. Some plants, mainly from U ore regions, may accumulate high amounts of Re. Its highest concentrations, up to 31 mg/kg ash weight, were found in some plants from vicinities of U mines (Bushkov et al. *vide* Kabata-Pendias 2011). Cereal grains from Sweden contain Re at an average value of <0.1 μg/kg (Eriksson 2001a).

37.3 HUMANS

Rhenium is not an essential element, and the toxic effect occurs only at its high concentrations. Because its uptake by plants is limited, the animal and human tissues contain it in very small amounts. Therefore, negative consequences because of Re content in agricultural production and human nutrition are not a reality, so its toxic effects are not important (Szabó 2009).

Data on safety dose of Re and its compounds can vary widely, depending on its forms. Soluble salts, such as the Re halides or perrhenates, could be hazardous due to both Re and associated elements. It has been described as *relatively inert* in the body and produces transient changes in blood pressure (both hypo- and hypertensive), tachycardia, sedation, and ataxia. In one comparative study, the lethal oral dose of Re was about 8 times higher than that of Mo. However, one report suggests that it could be more potent as an inhalation toxicant.

Much information is devoted to medical applications of the radioactive isotopes of Re. In contemporary nuclear medicine, α, pure β, or β/γ emitters are used for targeted therapy. Use of pure α and β and combined α/β emitters in oncology, endocrinology, rheumatology, and, a short time ago, interventional cardiology has been refined as an important alternative to more common therapeutic regimens. Its two radioisotopes, ^{186}Re and ^{188}Re, are of particular interest (Argyrou et al. 2013; Hsu et al. 2012).

38 Rhodium [Rh, 45]

38.1 INTRODUCTION

Rhodium (Rh) is a silvery-white, hard metal of group 9 in the periodic table of elements, is the least frequent element of the platinum group metals (PGMs), and is chemically inert. It reveals both siderophilic and chalcophilic properties. Its average content in the Earth's crust is estimated to be 0.06 µg/kg, and its contents in rocks vary within the range of 0.01–20 µg/kg. It may be accumulated in coal, up to about 100 µg/kg. It tends to occur along with deposits of PGMs and is primarily obtained as a by-product of mining and refining Pt. It is also mined from the Cu–Ni deposits, mainly in Monchegorsk area, the Kola Peninsula (Russia). The main minerals of these ores are pyrrhotite, magnetite, pentlandite, chalcopyrite, Ti magnetite, and pyrite, in which Rh contents vary from 4 to 2780 µg/kg (Gregurek et al. 1999).

Rhodium oxidation states vary from +2 to +6, and the most common is +3. It may be associated with some Pt minerals. Its typical arsenosulfide minerals are hollingworthite, (Rh,Pd,Pt)AsS, and irarsite, (Ir,Ru,Rh,Pt)AsS. Several other minerals, such as ilmenite, zircon, and chromite may accumulate Rh. Higher Rh amounts are accumulated in siderites and pyrrhotites, up to >100 µg/kg, and therefore, its elevated amounts around some Fe processing industries may be expected.

Rhodium is one of the most rarest and valuable precious metals. Its worldwide production is about 25 t/yr, mainly in South Africa, Russia, and Canada. About 85% of Rh is used for autocatalysts. It is also used frequently, as an alloying agent with other metals, mainly Pt and Pd. These alloys are used to make such things as catalytic converters, electrodes for aircraft spark plugs, laboratory crucibles, electrical contacts, chirurgical tools, and jewelry.

38.2 ENVIRONMENT

Rhodium contents of soils in Sweden are given as follows (in µg/kg): background soil, 0.07–0.13; urban soil, 0.14–20.5; arable soil, mean <40, and sewage sludge, 0.26 (Eriksson 2001a). Soils along motorways in Australia contain Rh, at the mean value, 13 µg/kg. The highest Rh content, above 91 µg/kg, is in roadside soil samples from Perth, Australia (Whiteley and Murray 2003).

Its content in dust samples collected (in 1999) along highways in Germany ranges from 30 to 42 µg/kg (Djingova *vide* Kabata-Pendias 2011). Its sources are mainly catalytic converters of cars. There is an estimation that all Rh in sewage sludge are from automobile emission (Schäfer and Puchelt 1998).

The median Rh concentration in worldwide ocean waters, as well as in the North Pacific, is estimated to be 0.08 ng/L (Nozaki 2005). Its similar concentration, at about 0.1 ng/L, is reported by. These authors emphasized that unlike the other PGMs, Rh is enriched in pelagic sediments (0.05–3 µg/kg), hydrothermal S deposits, and phosphorites. In ferromanganese minerals, it may be concentrated up to 44 μg/kg, and in some pebbles up to about 22 μg/kg. It seems to be relatively easily bioaccumulated by crustaceans, up to 12 μg/kg.

Rhodium concentration in some stream waters is about 100 ng/L (Reimann and Caritat 1998). Rainwater collected in Sweden during 1999 contains Rh at an average value of 1 ng/L (Eriksson 2001a).

Rhodium content of cereal grains is <1 μg/kg, whereas plants such as plantain and moss grown along highways contain Rh within the range of 1.3–5.4 μg/kg, respectively (Djingova *vide* Kabata-Pendias 2011). There are suggestions that >50% of Rh in plants is from road soil dust attached to plant surfaces.

38.3 HUMANS

Rhodium concentration in the human body is unknown; its content in human bones is under detection limits (Zaichik et al. 2011). Like the other noble metals, all of which are too inert to occur as chemical compounds in nature, Rh has not been found to play, or suspected to play, any biological role. If used in elemental form rather than as compounds, the metal is harmless (Leikin et al. 2008).

TABLE 38.1
Rhodium in Foodstuffs (μg/kg FW)

Product	Content a	Content b
Food of Plant Origin		
Bread	0.2	<3
Miscellaneous cereals	0.1	<3
Vegetables	<0.1	<0.6
Potatoes	<0.1	1
Fruits	<0.1	<0.5
Nuts	4	<3
Beverages	0.1	<0.5
Food of Animal Origin		
Carcass meat	<0.1	<1.5
Meat products	<0.1	<1.5
Offal	0.5	<1.5
Fish	0.2	<3
Milk	0.3	<0.5
Dairy products	0.1	<1.5

Sources: [a] Ysart, G. et al., *Food Addit. Contam.*, 16, 391–403, 1999.
[b] Rose, M. et al., *Food Addit. Contam.*, 27, 1380–1404, 2010.
FW, fresh weight.

The environmental concentration of Rh is slowly increasing as a result of using PGMs in automobile catalytic converters. Because finely dispersed Rh particles from exhaust converters become bioavailable by some alimentary chains or metabolic pathways, and toxic effects are expected. The main pathways by which Rh can enter the food chain is aerosol deposition, caused by Rh emission from motor vehicles and industrial sources. The primary sources from which Rh can be incorporated into the human body are therefore plants or agricultural products. This explains the increasing interest in quantifying Rh in real matrices of alimentary concern (Sánchez Rojas et al. 2007). Rhodium content in foodstuffs is very low, in general not exceeding 1 μg/kg (Table 38.1).

The dietary exposure of population to Rh in the United Kingdom has been estimated to be 0.3 μg/day in 1999 (Ysart et al. 1999) and <2.3 μg/day in 2006 (Rose et al. 2010).

39 Rubidium [Rb, 37]

39.1 INTRODUCTION

Rubidium (Rb) is a silvery-white soft metal of group 1 in the periodic table of elements, with properties similar to alkali metals. The Earth's crust contains Rb within the range of 90–110 mg/kg. Its higher contents are in acidic igneous rocks, <100–200 mg/kg, and in sedimentary argillaceous rocks, <120–200 mg/kg. The mean Rb content in coal is 25 mg/kg, but it may be concentrated up to 140 mg/kg.

The common oxidation state of Rb is +1, but may also be from +2 to +6. It is highly reactive and rapidly oxidized in air, and forms monoxide compounds: Rb_2O, Rb_6O, Rb_9O_2, and RbO_2, in excess of oxygen in media. It also easily forms salts with halides (e.g., RbBr, RbF, RbI). There are no minerals in which Rb is the predominant metal. It is associated with K minerals and may be concentrated in pegmatites. Lepidolite (mineral of mica group) is considered the principal ore mineral of Rb and contains up to 3.5% Rb oxides.

World mining of Rb has been roughly estimated to be about 1 t/yr (USGS 2001). It is obtained as a by-product from P, Li, and Cs processing, and ferrocyanide products. The most important use of Rb is in various chemical and electronic applications. Its use in superconductors is increasing. Rb-rich feldspars are used in ceramics. Rb salts are used in biomedical treatments of some patients.

39.2 SOILS

Rubidium contents in soils are largely inherited from the parent materials, and therefore, its highest contents, 100–120 mg/kg, are in soils derived from granites and gneisses. Also soils of heavy texture contain its higher amounts (Table 39.1). The lowest Rb contents are in sandy and organic soils. The significant, positive correlation between Rb contents and clay fraction in soils clearly indicates its sorption by fine fractions. Especially micaceous clays reveal a great capacity for its sorption. The range of Rb contents in soils of various countries are given as follows (in mg/kg): Bulgaria, 63–420; China, 15–140; Japan, 63–100; Poland, mean, 66; Russia, mean, 96; and the United States, 15–140 (Kabata-Pendias 2011).

Sewage sludge may contain Rb up to about 100 mg/kg, whereas P fertilizers have it at an average value of 5 mg/kg and manure at about 0.006 mg/kg. Groundwaters contain Rb from 0.05 to 2.5 μg/L.

39.3 WATERS

Rubidium concentrations in seawaters are fairly stable and range from 100 to 200 μg/L (Table 39.1). In river waters, its amount is much lower, within the range of 0.2–6.5 μg/L. The global riverine flux of Rb is estimated to be 61 kt/yr (Gaillardet et al. 2003).

TABLE 39.1
Rubidium Contents of Soils, Water, and Air

Environmental Compartment	Range/Mean
Soil (mg/kg)	
Light sandy	30–50
Medium loamy	55–140
Organic	<10–20
Water (μg/L)	
Rain	0.08–0.2
River	0.2–6.5
Sea and ocean[a]	100–200
Air (ng/m³)	
Urban areas	<1.0–<6.0
Greenland	<1.0

Sources: Data are given for uncontaminated environments, from various sources, mainly Kabata-Pendias, A. and Mukherjee, A.B., *Trace Elements from Soil to Human*, Springer, Berlin, Germany, 2007; Reimann, C. and de Caritat, P., *Chemical Elements in the Environment*, Springer, Berlin, Germany, 1998.

[a] Gaillardet, J. et al., *Treatise on Geochemistry*, Elsevier, Oxford, 2003.

Rainwaters of Europe contain Rb within the range of 0.08–0.15 μg/L, fairly similar to its worldwide values (Table 39.1). A bit higher Rb level, up to 0.25 μg/L, is reported for rain of Sweden (Eriksson 2001a).

Rubidium is relatively easily bioavailable and is accumulated up to 14 mg/kg in mussels soft tissues and up to 5.7 mg/kg in fish muscles. In the surface layer of bottom sediments of the Baltic Sea, its contents are within the range of about 20 to over 100 mg/kg (Szefer 2002).

39.4 AIR

Rubidium is very active in air and forms easily monoxide and hydroxide compounds. Its concentrations in air of remote regions is <1 ng/m^3, and in urban areas, it may be up to around 6 ng/m^3 (Table 39.1). Its deposition in the inhabited regions of the United Kingdom, during 1972–1981, was estimated to be <10 to <40 g/ha (Cawse *vide* Kabata-Pendias and Mukherjee 2007).

Mosses from Nordic countries contain Rb from 14 to 37 mg/kg (Rühling and Tyler 2004).

39.5 PLANTS

Rubidium is easily taken up by plants, similarly as other monovalent cations. The contents of available K control the Rb uptake. Increased bioavailability of Rb in acidic soils (pH 3.6–5.0) is a secondary effect of leaching losses of K. There is a suggestion that leaf Rb contents may be used as a method to assess early stage K deficiency in plants on acidic soils (Drobner and Tyler *vide* Kabata-Pendias and Mukherjee 2007).

Rubidium may partly substitute for K sites in plants, but cannot substitute for K metabolic roles. Thus, at higher concentrations, it is toxic to plants. However, in field conditions, Rb toxicity is practically unknown.

Most of the higher plant species contain Rb within the range of 20–70 mg/kg. Its contents in food plants vary highly from 3 to 15 mg/kg in cereal grains and lettuce leaves, respectively. Higher Rb amounts are in fruits (mean in mg/kg): avocado (20) and apple (50). Its highest content, at a mean value of 220 mg/kg, is reported for soybean seeds. Plants from industrial regions contain always higher amounts of Rb. Some fungi may accumulate increased amounts of Rb.

39.6 HUMANS

Rubidium is the most abundant element in the body (680 mg), which has no known biological role (Emsley 2011). Concentrations of Rb in the blood (2.8 mg/L) and serum (0.24 mg/L) are at the same range as concentrations of Zn and Cu, and it shows a high Rb bioavailability from food (Barány et al. 2005). Rb mean level of brain tissues is 5.49 mg/kg (Canavese et al. 2001). Its concentration in tissues reflects Rb intake, and Rb depletion affects mineral (Na, K, P, Ca, Mg, Fe, Zn, and Cu) status (Yokoi et al. 1996). Rubidium is rapidly and highly absorbed and excreted by the digestive tracts of mammals. It does not accumulate in any particular organ or tissue, and normally is relatively low in bones.

Rubidium is not toxic and is removed relatively quickly in perspiration and urine. However, taken in excess, it can be dangerous. The metabolism of Rb is closely related to that of K, and they show interchangeability in a variety of biological systems, with little evidence for any toxicity. Homeostatic levels of Rb in the blood of children (12 ± 3 mg/L) may suggest its essentiality (Lombeck et al. 1980). Studies in animals have shown that Rb can be administered orally.

Biological interest in Rb has been stimulated by its close physicochemical relationship to K, and its presence in living tissues, at higher concentrations relatively to those of K. It is higher than in the terrestrial environment. Relationships between K and Rb have been found in a variety of physiological processes.

There is an evidence suggesting that Rb has a role involving some neurophysiological mechanism. In the heart, Rb is lower in conductive tissue than in adjacent muscle tissue. Its content of brain differs significantly between defined functional regions, and also decreases with age.

Some neurophysiological, neurochemical, and behavioral actions of Rb are opposite from those of Li. Rubidium has a disinhibiting action and has antidepressant properties. It only becomes toxic if the concentration in the erythrocytes or muscle cells reaches 30% of the K contents. It has a long biological half-life.

Rubidium was used, during the nineteenth century, by European physicians in the treatment of cardiac conditions, syphilis, and epilepsy. Clinical studies of the effects of Rb ions on the course of manic-depressive illness, and it seems that Rb tends to increase the length of manic phases, and possibly reduces the extremes of mood. Rubidium did not seem to produce any severe side effects, at the dose

TABLE 39.2
Rubidium in Foodstuffs (mg/kg FW)

Product	Content
Food of Plant Origin	
Bread (white wheat)	9.2
Bread (rye)	4.3
Rice	2
Rice[a]	0.33–12.0
Rice[b]	2.08–13.39
Maize	2.6
Legumes (beans, lentil)	13
Cabbage	2.1
Potatoes and carrots	1.2–2.2
Citrus plants	0.7
Bananas	3.3
Apples	0.1
Coffee	88
Tea	38
Chocolate[c]	4.51–7.55
Food of Animal Origin	
Meat	1.2
Freshwater fish	0.74
Seawater fish	0.29
Shrimps	0.4
Eggs	2.5
Dairy products	2.1

Sources: Gorbunov, A.V. et al., Assessment of human organism's intake of trace elements from staple foodstuffs in central region of Russia. Preprint of the Joint Institute for Nuclear Research, Dubna, Moscow, Russia, 2004, http://wwwinfo.jinr.ru/publish/Preprints/2004/089(D14-2004-89)_e.pdf.

[a] Jorhem, L. et al., *Food Addit. Contam.*, 25, 841–850, 2008b.

[b] Batista et al. (2010).

[c] Sager, M., *Ecol. Chem. Engin.*, 17, 289–295, 2010.

FW, fresh weight.

administered, but it has a long biological half-life and caution is still required (Paschalis et al. 1978).

Rubidium is a putative anticancer agent and the urinary levels of Rb were significantly and inversely associated with risk of breast cancer and had potential to be a biomarker for breast cancer risk assessment (Su et al. 2011).

Salts of Rb are generally inert, and their toxicity is almost always a consequence of the anion, not of the Rb^{+}. Excessive exposure may trigger weakness, hypotension, muscle twitching, and other symptoms of K deficiency. It may also interfere with thyroid uptake of iodine.

It has no much information on Rb content in foodstuffs. However, published data indicate relatively high Rb content in such products as cereals, rice, some vegetables, tea, coffee, and chocolate (Table 39.2).

According to Hendrix et al. (1997), the worldwide Rb intake by adults varies (in mg/day) between 1.2 and 7: in the United States, 2.8; in Belgium, 1.87–2.45; in Japan, 2.34 (Shiraishi 2005); and in Russia, 2.2 mg/day (Gorbunov et al. 2004).

40 Ruthenium [Ru, 44]

40.1 INTRODUCTION

Ruthenium (Ru) is a hard, white metal of group 8 in the periodic table of elements and is a member of the platinum group metals (PGMs). It is resistant to acids, but reacts with alkalis, especially under oxidizing conditions. Its mean concentration in the Earth's crust is about 1 µg/kg, whereas its contents in rocks vary from 0.01 to 60 µg/kg. It usually occurs as a minor component of Pt ores. The PGMs mined in South Africa contain, on average, 11% Ru, whereas the PGMs mined in the former USSR contain only 2% Ru. It may also be associated with ores of some base metals, such as Fe, Ni, and Cu. Fission products of ^{235}U contain significant amounts of Ru, and therefore, used nuclear fuel might be its possible source.

The annual Ru production is 12–20 t, as given by various sources (USGS 2001). Most Ru is used for wear-resistant electrical contacts and the production of thick-film resistors. Its minor use is in Pt and Pd alloys, which are applied in jewelry and stomatology. It is also used for some catalytic processes and in medicine, mainly as antimalaria and immunosuppressive drugs.

Ruthenium oxidation states range from −2 to +8, and the most common are +2, +3, and +4. Chemically, it is similar to Os. Its known compounds are dipotassium ruthenate, K_2RuO_4, and potassium perruthenate, $KRuO_4$.

40.2 ENVIRONMENT

There are not many data on Ru in soils. Its average content in soils of Sweden is about 40 µg/kg, and its deposition was calculated as <11 mg/ha/yr (Eriksson 2001a). In soils along motorways, Ru amounts vary from 0.01 to 5.77 µg/kg (Fritsche and Meisel *vide* Kabata-Pendias 2011).

Ruthenium concentrations in water vary from 0.7 to 20 ng/L, in seawaters and lakes, respectively. Its highest content, about 0.1 µg/L, is in stream waters of India (Reimann and de Caritat 1998). Its relatively high accumulations are in pelagic sediments, within the range of <0.2–7 µg/kg, and in ferromanganese nodules, up to 46 µg/kg (Bertine et al. 1993). Elevated ^{106}Ru contents were found in bottom sediments of the Danube River, below the nuclear reactor (Pavlotskaya *vide* Kabata-Pendias 2011). Rainwater in Sweden contains Ru <1 ng/L (Eriksson 2001a).

Plants uptake Ru relatively easily, and its elevated contents, within the range of 0.3–0.9 µg/kg, are in plants grown along motorways (Djingova et al. *vide* Kabata-Pendias 2011). The average Ru concentration in plants is estimated to be 5 µg/kg. However, it may highly vary in various environments. In tropical forest region of Central America, its contents in plants may be up to about 2000 µg/kg (Duke *vide* Kabata-Pendias 2011).

40.3 RADIONUCLIDES

Two radionuclides, ^{103}Ru and ^{106}Ru, are released during nuclear reactions. Both radionuclides, and particularly ^{106}Ru, were deposited on soils by the fallout after the Chernobyl accident. This radionuclide is relatively mobile and migrates down soil profiles. It is also easily bioavailable and its highest proportion is concentrated in roots. A significant phytogenetic effect on the variable uptake and accumulation of both radionuclides by plants is observed.

40.4 HUMANS

There are no data on measurable content of Ru in the human body. Ruthenium anticancer drugs have attracted an increasing interest in the past 20 years and two of them have entered clinical trials. Recent research shows that Ru complexes have anticancer properties *in vivo*, and they might be a good alternative to Pt-based drugs for anticancer therapy (Antonarakis and Emadi 2010). Compared to Pt drugs, the complexes based on Ru are often identified as less toxic and capable of overcoming the resistance induced by Pt drugs in cancer cells (Bergamo and Sava 2011).

Although Pt, Rh, and Ru are used as catalysts of petrol-engined vehicles, which have elevated their concentration in roadside dust, there is little information about the biological effects of PGM metals in food. There is not any evidence for adverse effects of these metals, introduced into environments with exhaust emissions

TABLE 40.1
Ruthenium in Foodstuffs (µg/kg FW)

Product	Content a	Content b
Food of Vegetable Origin		
Bread and miscellaneous cereals	<2	<1
Vegetables and green	<2	<0.2
Vegetables and other canned vegetables	<2	0.2
Potatoes	<2	<0.3
Fruits and fresh fruit products	<2	<0.2
Nuts	<2	<1
Food of Animal Origin		
Carcass meat and offal	<2	<0.5
Fish	<2	<0.3
Milk	2	<0.2
Dairy products	2	<0.5

Sources: [a] Ysart, G. et al., *Food Addit. Contam.*, 16, 391–403, 1999.
[b] Rose, M. et al., *Food Addit. Contam.*, 27, 1380–1404, 2010.
FW, fresh weight.

(Rose et al. 2010). According to studies in the United Kingdom, Ru occurrence in food is generally below the limit of detection (Table 40.1). Taking into account uncertainty related to samples in which these elements could not be detected, the population dietary exposure to Ru has been estimated to be 0.03–0.8 μg/day in the United Kingdom (Rose et al. 2010).

41 Scandium [Sc, 21]

41.1 INTRODUCTION

Scandium (Sc) is a silvery-white transition metal of group 3 in the periodic table of elements, with properties similar to Al and Y. Its mean concentration in the Earth's crust is estimated to be 11 mg/kg, but also the range of 16–30 mg/kg is given. In higher amounts, it may be accumulated in mafic igneous rocks and argillaceous sediments. Its lowest contents, <5 mg/kg, are in calcareous rocks. Some organic raw materials, such as coal, peat, and crude oil, may accumulate elevated Sc amounts, up to 100 mg/kg ash weight (AW). Its mean content in the U.S. coal is 4.2 mg/kg, but may be elevated up to 100 mg/kg (Finkelman 1999). Its level in fly ash is up to about 5 mg/kg, mainly after burning lignite coal.

Scandium oxidation state is +3, and it may easily substitute for Al^{3+}, Fe^{3+}, and Ti^{3+}. Thus, it is likely to be associated with ferromagnesian minerals, biotite, and some phosphates, up to about 40 mg/kg. Its minerals, which are its primary sources, are thorveitite, $Sc_2(Si_2O_7)$, and kolbeckite, $ScPO_4{\cdot}2H_2O$.

The worldwide Sc mining is about 2 t/yr of Sc_2O_3, mainly in three mines: China, Ukraine, and Russia (USGS 2011). Some other sources gave Sc production at 10–50 kg/yr of Sc. The mineral thortveitite, $Sc_2(Si_2O_7)$, contains 35%–40% Sc_2O_3 and is used, as well as kolbeckite, $ScPO_4{\cdot}2H_2O$, for the Sc production. Its another important source (at about 2.5 t/yr) is a by-product from U ore processing, even though it only contains 0.02% Sc. Scandium is also associated with several rare earth element minerals.

Scandium is applied for light Al–Sc alloys used mainly in the aerospace industries. It is also used for the production of various lamps, lasers, and fluorescent materials, and as a catalyst in organic chemistry.

41.2 SOILS

The worldwide mean Sc content in soils is estimated to be 12 mg/kg, within the range of 0.8–28 mg/kg. Higher amounts are in heavy loamy soils, especially those that are derived from granitic and volcanic rocks. It may also be accumulated in some organic soils (Table 41.1). Scandium contents of soils of different countries are given as follows (in mg/kg): China, 5–28; Japan, 13–28; Poland, 0.5–7.9; Sweden, 3.5–16; the United States, 5–18; and Russia, 5 (mean value for loess deposits). Soils in remote regions of Spitsbergen, Norway, contain Sc within the range of 0.02–0.14 mg/kg (Gulińska et al. 2003).

Scandium in soils may occur as oxide, Se_2O_3, and hydroxide, $Se(OH)_3$, and is likely to form complexes such as $Sc(H_2O)_6^{3+}$ and $Sc(H_2O)_6^{2+}$. It shows an affinity for the association with PO_4, SO_4, and CO_3, as well as with some organic compounds. Elevated Sc amounts in P fertilizers, up to about 40 mg/kg, may be its source in agricultural soils.

TABLE 41.1
Scandium Contents of Soils, Water, and Air

Environmental Compartment	Range
Soil (mg/kg)	
Light sandy	0.8–1.5
Medium loamy	3–8
Heavy loamy	2.5–28
Calcerous (calcisols)	6–11
Organic	2–20
Water (ng/L)	
Rain	2–1.2
River	60–1800
Sea and ocean[a]	0.6–1.0
Air (pg/m³)	
Urban/industrial areas	30–3000
Antarctica	0.06–0.21

Sources: Data are given for uncontaminated environments from Kabata-Pendias, A. and Mukherjee, A.B., *Trace Elements from Soil to Human*, Springer, Berlin, Germany, 2007; Reimann, C. and de Caritat, P., *Chemical Elements in the Environment*, Springer, Berlin, Germany, 1998.

[a] Range of mean values.

41.3 WATERS

Scandium concentration in worldwide ocean and seawaters is within the range of 0.6–1.0 ng/L (Table 41.1). In river waters, it is in much higher amounts, up to 1800 ng/L, especially in rivers of South America and Africa (Reimann and de Caritat 1998).

The median Sc concentration in bottled waters of the EU countries is 72 ng/L, and is slightly higher than in tap waters, 61 ng/L (Birke et al. 2010). The mean Sc content in drinking water of Poland is 70 ng/L (Kabata-Pendias and Pendias 1999).

41.4 AIR

Air from remote areas of Antarctica contain Sc within the range of 0.06–0.21 pg/m^3, whereas its much higher concentrations, up to 3000 pg/m^3, are in air of urban and industrial regions (Table 41.1). Its world median values are given from 19 to 30 pg/m^3 (Reimann and de Caritat 1998). Atmospheric deposition of Cs in the United Kingdom is estimated to be 0.3–2.0 g/ha/yr (Cawse *vide* Kabata-Pendias and Mukherjee 2007). In Sweden, Cs deposition is given as 0.04 g/ha/yr (Eriksson 2001a).

Moss samples collected in Norway, in the period 1990–1995, contained Sc within the range of 0.02–3.1 mg/kg (Berg and Steinnes 1997).

41.5 PLANTS

The common Sc contents in plants usually vary within 0.005–0.07 mg/kg, but in some herb plants, they may be much higher, up to 2 mg/kg. Similar range of Sc, 0.002–0.1 mg/kg, is in food plants (Kabata-Pendias 2011). The broad range of Cs is reported for fungi, from <0.002 to 0.3 mg/kg (Bowen *vide* Kabata-Pendias 2011).

Rice straw contains Sc about 0.2 mg/kg, and its amounts increase up to 1 mg/kg in manure produced from that straw (Goto *vide* Kabata-Pendias 2011). Scandium is accumulated in roots and old leaves. Shtangeeva et al. (2004) observed that added Sc is likely to be concentrated in seedling and roots. Some interactions between Sc and Na, K, and Ca were observed by these authors.

41.6 HUMANS

The mean Sc content in the human body is estimated to be about 0.2 mg in the blood 8 μg/L and 1 μg/kg in the bone and tissue (Emsley 2011). Scandium has no biological role and is not toxic, although there have been suggestions that some of its compounds might be cancerogenic. It is mostly dangerous in the working environment, due to the fact that damps and gasses can be inhaled with air. This can cause lung embolisms, especially during long-term exposure. Scandium can be a threat to the liver when it accumulates in the human body (Horovitz 2000).

Only trace amounts of Sc are the food chain, so the average person's daily intake is <0.1 μg. Scandium content is high in some spices, but less in fresh fruits and vegetables (Table 41.2).

TABLE 41.2
Scandium in Foodstuffs (μg/kg FW)

Product	Content	
	Range	Mean
Fruit, fresh[a]	0.1–1.2	0.4
Legume[a]	0.3–1.75	0.91
Root vegetables[a]	0.2–21.0	3.4
Onion[b]	0.2–5.0	2
Peas[b]	1–10	2
Chocolate[c]	<2.6–20.0	5.5
Cheese[c]	<1.5–94	26
Curry, powder[d]	2800–3410	3100
Curry, leaf[d]	1070–1220	1140

Sources: [a] Howe, A. et al., *Environ. Geochem. Health*, 27, 19–30, 2005.
[b] Gundersen, V. et al., *J. Agric. Food Chem.*, 48, 6094–6102, 2000.
[c] Sager, M., *Ecol. Chem. Engin.*, 17, 289–295, 2010.
[d] Gonzalvez, A. et al., *Food Addit. Contam.*, 1, 114–121, 2008.
FW, fresh weight.

42 Selenium [Se, 34]

42.1 INTRODUCTION

Selenium (Se) is a nonmetal element of group 16 in the periodic table of elements, and its properties are similar to Te, an element of the same group. Its mean abundance in the Earth's crust is estimated to be 0.05 mg/kg, but sometimes higher values, up to 0.5 mg/kg, are cited. It may be a bit concentrated in mafic rocks. In sedimentary rocks Se is associated with clay fraction, and thus its highest contents are in argillaceous sediments. Its enriched concentrations in Cretaceous rocks (above 100 mg/kg) are derived from volcanic gases and dust, deposited with rain into the Cretaceous sea. Coal contains Se within the range of 0.2–10.7 mg/kg, and in fly ash it is accumulated up to about 1.5 mg/kg. Some coal of China contain Se at about 6500 mg/kg (Plant et al. *vide* Kabata-Pendias 2011). Increased Se concentrations, up to 0.1%, were observed in some U deposits, and in some metal sulfide ores. Its elevated amounts are also reported for phosphate rocks (Bech et al. 2010).

Selenium has variable oxidation state, from –2 to +6. The most common is +4, whereas in organic compounds it is mainly –2. It rarely occurs in its elemental state in nature, or as pure ore compounds. It exhibits variable geochemical properties, chalcophilic and siderophilic, that affects its easy change of oxidation states, and complex behavior in geochemical processes. Common Se species, selenites (Se^{3+}) and selenates (Se^{6+}), do not form stable compounds, and are preferably absorbed by minerals, mainly clay minerals, and Fe and Mn oxyhydroxides, which are involved in several exchange reactions. Selenium is associated also with P and S compounds.

Approximately 50 Se minerals are known, of which the relatively common are klockmanite (CuSe), berzelianite ($Cu_{2-x}Se$), clausthalite (PbSe), tiemannite (HgSe), ferroselenite ($FeSe_2$), crookesite [$(Cu,Tl,Ag)_2Se$], and cobaltomenite [$Co(SeO_3)$ $2H_2O$]. Commonly Se occurs in association with some host minerals, such as pyrite, chalcopyrite, and sphalerite.

Global Se production (rounded) in 2010 was 2260 t, of which 780 t was produced in Japan, 680 t in Germany, 200 t in Belgium, 170 t in Canada, and 140 t in Russia (USGS 2011). Impure selenium is found in metal sulfide ores, and, commercially, it is produced as a by-product in the refining of these ores, most often during Cu production.

Various compounds of Se are used in photoelectric cells, some batteries, and in solar cell technologies. Selenides (e.g., HgSe, PbSe, ZnSe) are good semiconductors. It is used as a pigment (maroon and orange colors), mainly in the production of glass and plastic, is also added to some steel alloys, and is used in lubricants for metals. Some Se organic compounds are applied in organic synthesis.

A high proportion of Se (around 20% of the total production) is used as dietary supplement for humans and livestock. It is added, mainly as sodium selenite (Na_2SeO_3), to fertilizers, insecticides, and as supplement in the plant–animal–human food chain. Although small amounts of Se are considered beneficial, it can be hazardous in larger quantities. Trace Se amounts are necessary for cellular function in humans and animals. It is a component of the antioxidant enzymes glutathione peroxidase and thioredoxin reductase, and in three deiodinase enzymes is involved in thyroid hormones.

42.2 SOILS

Selenium background contents in various soil groups range from 0.05 to 1.50 mg/kg (Table 42.1). Its lowest amounts are in light sandy soils. Soil Se is inherited mainly from parent materials, but in some regions atmospheric deposition may also have an impact. The average Se levels in sandy soils of various countries are reported as follows (in mg/kg): Finland, 0.21; Lithuania, 0.14; Poland, 0.14; Russia, 0.18; Slovakia, 0.10; Sweden, 0.23; the United States, 0.5; and Japan, <0.8–>3.0 (in various soils) (Kabata-Pendias 2011). In the remote region in Spitzbergen, soils contain Se within the range of 0.004–0.039 mg/kg (Gulińska et al. 2003).

Most soils of the temperate humid climate zones, and derived from sedimentary rocks containing low Se amounts, that are not sufficient to produce food and fodder plants with its adequate contents. Applications of farmyard manure, containing Se at about 1 mg/kg, may be a good means to increase its level in soils. Also sewage sludge, with mean Se level at 1.3 mg/kg, may be its source in Se-deficient soils.

TABLE 42.1
Selenium Contents of Soils, Waters, and Air (mg/kg)

Environmental Compartment	Range/Mean
Soil (mg/kg)	
Light sandy	0.25–0.50
Medium loamy	0.15–1.50
Heavy loamy	0.20–1.50
Calcerous (calcisols)	0.20–1.40
Organic (histosols)	0.30–1.50
Water (μg/L)	
Rain	0.04–1.7
River	0.05–22.0
Sea	0.1–0.35
Air (ng/m³)	
Industrial and urban areas	0.08–30.0
Antarctic	0.06

Sources: Data are given for uncontaminated environments from Kabata-Pendias, A. and Mukherjee, A.B., *Trace Elements from Soil to Human*, Springer, Berlin, Germany, 2007; Reimann, C. and de Caritat, P., *Chemical Elements in the Environment*, Springer, Berlin, Germany, 1998.

The Se contents of soils have received much attention, mainly in countries where its role in human and animal health has been recognized. The program for increasing Se levels in soils of Finland began in 1984 (Eurola et al. *vide* Kabata-Pendias 2011) and was based on the addition of Se (as sodium selenate) to inorganic fertilizers, at levels 6–10 mg/kg. The mean hot-water extractable Se in soil before this program was 0.006 mg/L, and after the program (in 1998) was 0.01 mg/L. However, Se that was easily available to plants was present only in the first summer after Se fertilizing, and quickly turned into unavailable species to plants species. The fixation of Se in soils was correlated positively with clay content, Fe content, and negatively with sulfuric acid-extractable P. However, nonsoluble Se compounds were, later on, changed into mobile species. Secondary effects, after this program, carried out during the period 1983–1992, resulted in increased Se levels in river water (up to 180 mg/L), and especially in bottom sediments (up to 4 mg/kg). Results of this experiment showed that Se phytoavailability varies significantly, depending on soil parameters (Hécho et al. 2012).

Elevated Se amounts in soils (called seleniferous soils) are in some ferrasols, organic soils, and other soils derived from Se-rich parent materials. They occur mainly in some central regions of the United States, and in Asia. Also, salt-affected soils may contain higher Se levels.

Selenium behavior in soils is controlled by pH–Eh soil system. Usually it is easily mobile and also oxidized in soils. Selenite ions resulting from oxidation processes are stable and readily adsorbed on minerals and organic particles. Clay minerals (particularly montmorillonite) and Fe oxides (especially goethite) reveal a great adsorption capacity. In acid soils, Se is likely to occur as Se^{4+}, which is adsorbed by Fe oxide and form ferric selenite [$Fe_2(OH)_4SeO_3$] and iron selenite (FeSe). Its maximum adsorption occurs at the pH range of 3–5, and decreased as pH increases. In alkaline Se-rich soils, predominated species is Se^{6+}, which is very weakly adsorbed. Thus, Se is very mobile in soils of arid and semiarid regions. The biological methylation of Se compounds affects its easy volatilization, which plays a significant role in its geochemical cycling.

Due to variable oxidation states of Se, its behavior in soils is very variable. The impact of pH–Eh is as follows:

- pH 7, Eh >400, oxidation Se^{6+}, major species: SeO_4^{2-}
- pH >7, Eh 200–400, oxidation Se^{6+}, major species: SeO_3^{2-}
- pH <7, Eh 200, oxidation state Se^{4+}, major species: $HSeO_4^-$
- pH <3.2, Eh <200, oxidation state Se^{2-}, major species: HSe^-, H_2Se°

Inorganic species of Se reveal variable properties, depending on its oxidation state, which are as follows:

- Selenates (Se^{6+}) are mobile as inorganic forms, especially in neutral and alkaline soils, and are not adsorbed on hydrous sesquioxides (mainly $Fe_2O_3{\cdot}H_2O$)
- Selenites (Se^{4+}) are slightly mobile in neutral and acid soils, and are easily absorbed hydrous sesquioxides and organic matter
- Selenides (Se^{2-}) are very slightly mobile in acid soils, due to the formation of stable minerals and organic compounds

The transformation between these species, as well as the formation of elemental Se take place in all soils, however, under most conditions there are very slow processes. Mobile anionic species that occur in soil solution are SeO_3^{2-}, SeO_4^{2-}, $HSeO_4^-$, $HSeO_3^-$, $HSeO_4^-$, (Kabata-Pendias and Sadurski 2004). Some of these anions, mainly SeO_3^{2-}, are adsorbed by Fe oxides, and precipitate as $Fe(SeO_3)_3$. Very weakly SeO_4^{2-} is adsorbed, especially at high soil pH. In acid soils Se is likely to occur as Se^{4+}, strongly fixed by Fe compounds, and thus become slightly mobile. In general, mobile and easy phytoavailable Se occurs in alkaline and well-aerated soils, thus in soils of arid and semiarid regions.

The stability of Se–Fe compounds and Se adsorption by various metal hydroxides and clay minerals control, to a high degree, Se behavior. Addition of PO_4 to soils may increase Se mobility and thus its bioavailalbility. The phytoavailability of different Se species in soils decreases in the following order:

Selenate > selenomethionine > selenocysteine > selenite > elemental Se > selenide

A close relationship between Se and organic C is common in most soils. Organic matter has a tendency to form organometallic complexes, which remove Se from soil solutions. Microbial processes in both, the formation and mineralization of organic Se compounds (e.g., selenomethionine, selenocysteine) play a crucial role in the Se cycling, and especially in its volatilization from Se-contaminated soils. These processes are mainly the reduction of Se^{4+} and Se^{6+}, as well as methylation of Se. Microbiota, reveal variable sensitivity to elevated Se contents, and its concentration at 5 mM/L in soil solution may inhibit activity of soil enzymes (Nowak et al. *vide* Kabata-Pendias and Mukherjee 2007). Therefore, organic amendments of soils may significantly increase the rate of Se volatilization from soils. Microorganisms, especially *Rhizobium* sp. play an important role in Se volatilization. However, not all Se species are available to microbial volatilization.

Selenium contents of soils and its behavior have received much attention, especially in countries where its deficiency in humans and animals has been recognized. However, elevated Se contents of soils in some regions, due to both geochemical and anthropological factors, are also of a great concern. Soils developed on seleniferous parent material (e.g., Cretaceous limestone) and soil contaminated from irrigation water or from coal power plants has also been a real environmental problem.

The water-soluble Se species are considered to be the fraction that is available to plants, and usually there is an observed close relationship between Se content of plants and its concentration in the soil solution, and in the diethylenetriaminepenta-acetic acid extraction, which contain about 6% of its total contents (Borowska et al. 2012). However, some other Se species may also be phytoavailable, and there is a calculation that about 45% of the total Se in soils may be taken up by plants (Kabata-Pendias 2011).

The phytoremediation of Se-enriched soils has become recently a hot topic of several experiments. Soils of regions with seleniferous parent materials and Se-contaminated soils require alleviation. There are several publications presented on "green technology of Se phytoremediation" (e.g., Banuelos 2001; Dhillon and Dhillon 2009). The Se phytoremediation methods are based on the plant potential to take up elevated amounts of Se and/or on stimulation of its volatilization. There are calculations that up to about 20% of Se contents of soils can be removed by harvested plants (Kabata-Pendias 2011). Activities of rhizobacteria, as well as other microbial activities, have a significant impact on the Se phytoremediation. *Brassica* plants reveal great capability to remove Se from soils, which may take up 740–950 g Se/ha/yr (Dhillon and Dhillon 2009). These practices are applied mainly in two regions, northwest India and west-central California.

42.3 WATERS

Mean Se concentration in seawaters is given within the range of 0.1–0.35 μg/L (Table 42.1), and its median value is estimated to be 0.2 μg/L (Reimann and de Caritat 1998). According to other sources, Se mean level in ocean waters is 0.09 μg/L (ATSDR 2002b), and in the North Pacific is 0.1 μg/L (Nozaki 2005).

The global Se average concentration in river waters is given at the value of 0.07 μg/L, with the range of 0.05–22.0 μg/L (Table 42.1), and its world average riverine flux is estimated to be 2.6 kt/yr (Gaillardet et al. 2003). Colorado River in the United States contain Se from 1 to 4 μg/L, however, in some places much higher amounts, up to 400 μg/L, are reported (ATSDR 2002b). Much of this Se is derived from industrial sources, for example, oil refineries contribute up to 75% of its load to the San Francisco Bay (Plant et al. *vide* Kabata-Pendias and Mukherjee 2007). Waters of coal mines may also contain elevated Se amounts, mainly as dimethyl selenide, $(CH_3)_2Se$.

The common range of Se in rainwaters is 0.04–1.7 μg/L, and in polar ice its content averages 0.02 μg/L. Median Se concentration in rainwater from the remote region of the Kola Peninsula is 0.5 μg/L, and increases up to 0.9 μg/L in the polluted area (Reimann and de Caritat 1998). The average Se content of rainwater of Sweden, in 1999 was 0.15 μg/L, with the maximum value of 0.26 μg/L (Eriksson 2001a). The main source of Se in rainwater is fossil-fuel combustion, but in some regions it is also volcanic eruption. Also sewage sludge, containing Se up to 280 μg/L, may be its source (ATSDR 2002b).

Groundwaters contain usually more Se than surface waters. Selenium contents of groundwater, in the region close to chemical plant, vary from 0.25 to 1.8 μg/L, and the proportion of Se^{4+} and Se^{6+} is also varied (Siepak et al. 2003). Especially elevated Se concentrations, up to 1000 μg/L, are in groundwaters in areas with seleniferous bedrocks. In some arid regions (e.g., the United States, China, Pakistan, Venezuela), Se concentrations in waters are elevated to about 2000 μg/L (Plant et al. *vide* Kabata-Pendias and Mukherjee 2007). High Se levels in soils resulted in its higher concentration in stream and river waters and bottom sediments. After Se fertilization of soils in Finland, its concentration in river waters increased up to 180 μg/L, and in bottom sediments up to about 4 mg/kg (Haygarth *vide* Kabata-Pendias and Mukherjee 2007).

Aquatic biota is relatively sensitive to Se excess in waters. Acute hazard to these organisms are established for Se at 20 μg/L in freshwater, and up to 300 μg/L in marine water (ATSDR 2002b). In general, Se is toxic to fish. The most resistant to elevated Se concentrations is salmon, but its content of about 1000 μg/kg is toxic.

Median Se concentrations in bottled water of the EU countries is 0.036 μg/L, and is lower than in tap water, estimated to be 0.115 μg/L (Birke et al. 2010). The existing guideline on Se in drinking water, last updated in 1992, established its maximum allowable concentration (MAC) of 10 μg/L, based on an adequate and safe range of its intake of 0.05–0.2 mg/day (EPA 2013). Previously proposed MAC, of 50 μg/L, resulted in chronic selenosis. Provisional guideline value for Se in drinking water established by the World Health Organization (WHO) is 40 μg/L. The guideline value is designated as provisional because of the uncertainties inherent in the scientific database (WHO 2011a).

42.4 AIR

Concentration of Se in the atmosphere is highly variable due to its differentiated sources: (1) evaporation from ocean and seawater surface, (2) volcanic eruptions, and (3) industrial emissions. Their amount in air above the South Pole is about 0.06 ng/m^3, and in urban and industrial regions it varies within the range of 0.06–30 ng/m^3 (Table 42.1). Its median content of air from polluted regions is estimated to be 4.0 ng/m^3 (Reimann and de Caritat 1998). Always its higher aerial abundance is in industrial regions. The lowest harmful Se concentration in air is estimated to be 1,000,000 ng/m^3 (ATSDR 2002b).

Enrichment of Se in marine aerosols resulted from the formation of volatile Se-organic compounds, mainly dimethyl selenide [$(CH_3)_2Se$]. Also dimethyl diselenide is an easily volatile species. Increased Se levels in mosses (>1 mg/kg) and peat (>2 mg/kg) in marine regions indicate clearly the impact of Se from the surface of seawaters (Berg and Steinnes 1997). Substantial amounts of Se may also be volatized from Se-contaminated soils through bacterial and phytovolatilization processes. There is the estimation that about 45% of Se content in coal is emitted as $(CH_3)_2Se$, which stay relatively long, about 45 days, in air. Its global emission is estimated to be >6 kt/yr, in both small particles and volatile compounds, which contribute to around 40% of its aerial abundance (Schrauzer *vide* Kabata-Pendias 2011).

Atmospheric Se deposition is clearly shown by its high accumulation in mosses and mushrooms. Mosses from the Scandinavian countries contain Se within the average values of 390–2900 μg/kg (Berg and Steinnes 1997). High Se accumulation in rootless mushrooms (up to 20,000 μg/kg) indicates also its easy absorption from the atmosphere. Several plants uptake Se from air via the leaf surface, and then accumulate it mainly in roots, as various inorganic selenite compounds. There is the estimation that up to about 80% of Se content in plants is from aerial sources (Haygarth *vide* Kabata-Pendias 2011).

42.5 PLANTS

Selenium requirements by plants differ by species, with some plants requiring relatively large amounts, and others apparently requiring none, what is still in need of further clarification. Most plants contain rather low Se levels, around 25 μg/kg, and rarely exceed 100 μg/kg. However, some plants reveal a great capability to accumulate Se, up to levels about 1000 μg/kg, that may be toxic to humans and animals.

Although Se is not an essential element for plants, with some exceptions, it is being added to soils to ensure that both food and fodder plants contain its adequate amounts for dietary needs. It should be emphasized, however, that the margin of safe Se concentration is rather narrow. Contents of Se in crop plants received recently much attention, because of its importance in the food chain. Thus, most data are for Se in food and fodder plants. Especially cereal grains, as the most common Se source in diets, have been broadly analyzed. In particular, the anticarcinogenic effectiveness of various Se compounds in plants have been recently investigated (Fordyce 2005).

The Se uptake by plants depends on several factors, however, it is always easily taken up by plants, when is present in mobile species in soils. In most cases there is a positive linear relationship between Se in plant tissues and Se content of soils, and soil pH. Soil properties have a great impact on the Se phytoavailability. It is easily taken up by plants from akaline sandy soils, in arid climate, than from acidic clay soils, with elevated content of soluble organic matter, in humid climate.

The greatest Se uptake is from alkaline soils, at pH about 6.5. However, Se uptake depends, to a great degree, on plant capability. Thus, plants have been divided into three categories: (1) plants that are accumulators and contain high amounts of Se, >1000 μg/kg, and presumably require this element; (2) plants that absorb medium Se quantities, around 100 μg/kg; and (3) plants that are nonaccumulators, with low Se contents, <30 μg/kg. Plants that hyperaccumulate Se, up to about 5000 μg/kg, are mainly of *Astragalus* family, and also of several other herbaceous species.

The kind of Se added to soils (contamination or supplementation), and various plant factors have an impact on the Se speciation, and thus its uptake by plants. Especially the formation of SeMeth by both, rhizosphere microbes and roots of some plants, have an impact on Se phytoavailability and phytovolatilization. In some specific root–soil interactions, *Rhizobium* bacteria and root exudates stimulate the oxidation of SeO_3 to SeO_4, what increases Se availability to plants. Thus, plants inoculated with bacteria reveal a higher capability to both Se uptake and Se volatilization. On the other hand, the reduction of SeO_4 to SeO_3 appears to be a rate-limiting step in the production of volatile Se compounds by plants. Inhibitory effects of S on the uptake and volatilization of Se may be reduced substantially if Se is supplied as, or converted to, SeO_3 and/or SeMeth. Interactions between Se and S, at the root–soil surface, may also control Se phytoavailability, whereas some Se compounds stimulate S adsorption by plants (Rios et al. *vide* Kabata-Pendias 2011). The phytovolatilization of Se during the growing season of some plants may be significant (e.g., about 35 mg Se/m^2 by *Salicornia* species). It seems to be a promising method for the removal of excess Se from contaminated soils.

All mobile Se species are easily taken up by plants with the water flow. Only selenates and organic Se are driven metabolically. In general, organic Se compounds are more readily available to plants than inorganic species. On the other hand, SeMeth compounds are easily transported within plants tissues. However, the mechanism of Se absorption and metabolization by plants is very similar to those of S.

The Se function in plants has been broadly investigated in recent times, but there is still little evidence that Se is essential for all plants. Recent studies on some grasses and vegetables have indicated that at proper Se addition plant growth is stimulated (Hartikainen *vide* Kabata-Pendias 2011). Although the essentiality of Se organic compounds in plants has not yet been documented, synthesis of Se proteins in some plants are reported (Terry et al. *vide* Kabata-Pendias 2011). Several Se organic compounds were found in plants, but their metabolic functions have not yet been established. Plants may synthesize various Se organic compounds. Several selenoamino acids, often in association with glutathione peroxidases, were found in both bacteria and higher plants. Predominated Se forms in plants are SeMeth (selenomethionine), in cereal grains, and legumes seeds; and Se-methyloselenocysteine (SeMSC) in vegetables (Djujić *vide* Kabata-Pendias 2011). SeMSC (the most-effective anticarcinogenic Se compound) is present mainly in garlic, broccoli, and brussels (Lyi et al. *vide* Kabata-Pendias 2011).

Selenium taken up by plants is incorporated mainly (up to 75%) in insoluble proteins. Impact of Se on metabolic process implants is variable. It may increase glutathione peroxidase activity and decrease dismutase functions. Excesses of both selenate (SeO_4^-) and selenite (SeO_3^-) are toxic to most plants, and can be attributed to a combination of three factors: (1) selenate and selenite are readily absorbed from the soil by roots and translocated to other parts of the plant; (2) metabolic reactions convert these anions into organic forms of Se; and (3) the organic Se metabolites, which act as analogues of essential S compounds, interfere with cellular biochemical reactions.

Selenium contents of wheat grains highly differ among various countries (average values, in μg/kg): Algeria, 920; Australia, 23; Canada, 21; Egypt, 340; Finland, 142; France, 36; Poland, 42; Norway, 33; Sweden, 14; and the United States, 297. In general, mean Se contents of cereal grains are higher in countries with arid climate than in countries with humid climate. There is also variation in Se content among vegetables, depending on both crop types and growth media. Mean Se contents in various plants of the United States vary as follows (in μg/kg): roots and bulbs, 407; leafy vegetables, 110; seed vegetables, 66; vegetable fruits, 54; and tree fruits, 15 (Fordyce 2005). The highest Se amounts contain brazil nuts, 200–253,000 μg/kg fresh weight (FW) (mean 14,700 μg/kg FW), which grows on calcareous soils derived from the limestone enriched in Se from volcanic eruption. Also coconuts contain elevated Se amounts, up to about 700 μg/kg FW, presumable from similar sites.

Selenium content of forage plants has also been of a great concern. Leguminous plants contain usually more Se than grasses. The average Se contents of grass from various countries vary from 13 to 350 μg/kg, whereas its mean contents of clover and alfalfa are between 15 and 672 μg/kg. However, this is not a rule, and sometimes grass may contain several times more Se than clover, from the same sites (Oldfield *vide* Kabata-Pendias 2011).

In areas with low soil Se, application of Na selenites to soils or as foliage spray are proposed for the correction of Se nutritional deficiency. However, because of toxic properties of Se salts, these practices should be carefully controlled. The addition of Se to soils (at 10 g/ha) affects its elevated contents in barley and oats grains from 19 to 260 μg/kg, and from 32 to 440 μg/kg, respectively. Much higher effect is after Se application on potatoes, which resulted in its increase in tubers from 0.47 to 1068 μg/kg (Hlušet et al. *vide* Kabata-Pendias 2011). Industrial pollution also impact Se level in plants. This illustrates well-increased Se content in forest crowberries of the Kola Peninsula, from <0.5 to 11.4 mg/kg in crowberries grown close to the Monchegorsk plant (Reimann et al. 2001).

42.6 HUMANS

Selenium content of normal adult humans can vary widely. Values from 3 mg in New Zealanders to 14 mg in some Americans reflect the profound influence of the natural environment on the Se contents of soils, crops, and human tissues. Approximately 30% of tissue Se is contained in the liver, 15% in the kidney, 30% in the muscle, and 10% in the blood plasma (WHO 2004). The average Se contents in the blood are within the range of 40–400 μg/L (Gać and Pawlas 2011). Selenium can also be found in significant amounts in the nails and hair. Approximately 50%–80% of absorbed Se is eliminated in the urine.

Following ingestion, most water-soluble inorganic and organic Se compounds in foods are relatively efficiently absorbed across the gastrointestinal tract (80%–95%), although elemental Se and Se sulfide are poorly absorbed. A number of other factors besides chemical forms may also influence the bioavailability and distribution of Se (Thomson 1998). Organic forms (e.g., selenoamino acids such as selenomethionine and selenocysteine) are absorbed more readily than inorganic forms, as well as on the overall exposure level, with absorption increasing, when Se levels in the body are low.

The nutritionally essential functions of Se appear to be discharged by some 25 selenoproteins. The specific selenoproteins include glutathione peroxidases (it is a constituent element of the entire defense system that protects the living organism from the harmful action of free radicals), thioredoxin reductases, 5-iodothyronine deiodinases, selenoprotein P, and others. Many forms of Se (including selenite, selenate, selenocysteine, and selenomethionine) are metabolized to hydrogen selenide. While the latter metabolite is the obligate precursor to the formation of selenocysteine in the specific selenoproteins, it can also be serially methylated (to methyl selenol, dimethylselenide, and trimethylselenonium ion) or converted to a selenosugar, and excreted.

Selenium aids in the defense of oxidative stress, the regulation of thyroid hormone action, and the regulation of the redox status of vitamin C and other molecules. Its activity appears to be closely related to the antioxidative properties of α-tocopherol (vitamin E) and coenzyme Q (ubiquinone). Organic and inorganic Se compounds function is in preventing certain disease that have been, in the past, associated with vitamin E deficiency. Selenium protects the organism from oxidative damage to cell membranes by destroying H_2O_2, whereas vitamin E protects against damage by preventing the formation of the lipid hydroperoxides.

Selenium role in aging processes and prevention of age-related diseases depends upon a complex combination of Se status, Se bioavailability, genetic variations in selenoprotein genes, and the regulation of downstream metabolic pathways involved in the aging process.

Selenium inadequacy is common in older people. In particular, it was reported that Se intake is reduced in elderly people and that Se status declines in an age-dependent manner. The Epidemiology of Vascular Ageing study established a relationship between Se status and longevity. Significantly higher mortality in people with low plasma Se, and an association between plasma Se concentration and mortality by cancer, suggests that inadequate Se intake may increase the vulnerability to diseases (Méplan 2011).

Blood Se levels are inversely associated with the prevalence of several types of cancer. Reduced cancer risk is associated with Se treatment, although the Se supplementation was not related to reduce in prostate cancer risk.

Very low dietary intakes of Se by humans have been associated with incidence of Keshan and Kashin–Beck diseases. Keshan disease is a cardiomyopathy that affects young women and children in the Se-deficient region of China. The acute form of the disease is characterized by the sudden onset of cardiac insufficiency, while the chronic form results in moderate to severe heart enlargement, with varying degrees of cardiac insufficiency. Kashin–Beck disease is characterized by the degeneration of articular cartilage between joints (osteoarthritis) and is associated with poor Se status in areas of northern China, North Korea, and eastern Siberia. The disease affects children between the ages 5 and 13 years (WHO 2004).

At low Se levels in persons virulence in viral pathogens may decrease. A decline in Se status in humans with immunodeficiency virus (HIV) has been observed. There is an evidence that Se supplementation can have a beneficial effect on HIV and AIDS patients (Reilly 2002).

High dietary intakes of Se have been identified in parts of Venezuela, China, and South Dakota (the United States). Excessive Se intakes can lead to chronic toxicity (selenosis), with health damages, such as loss of hair and nails, skin lesions, hepatomegaly, polyneuritis, and gastrointestinal disturbances. Different chemical forms of Se can have vastly different toxic potentials, and for elemental Se seems to be less toxic.

Occupational exposure to Se usually occurs through direct contact and/or through inhalation. Exposure to Se and its compounds occurs in both Se industries, primary and secondary. For acute occupational Se exposures, the effects vary according to the chemical form of Se, of which the elemental Se appears to be nontoxic.

Various national and international organizations have established recommended daily intakes of Se. A joint Food and Agriculture Organization (FAO)/WHO consultation recommended its intake of 6–21 μg/day for infants and children, according to age, and 26 and 35 μg/day for adult females and males, respectively. Because of concern about the adverse effects resulting from exposure to excessive levels of Se, various national and international organizations have established upper limits of exposure for Se. The FAO/WHO established an upper tolerable limit for Se of 400 μg/day (WHO 2011e).

In plant and animal tissues, Se is found mostly bound to proteins. Therefore, the most important food sources of Se are meats and seafood, because of their high protein contents; and cereals, due to their large consumption. Foods with relatively low protein levels, such as vegetables and fruits, have relatively low Se contents (Table 42.2). Global Se intakes vary significantly in the range from about 0.03 to 0.5 mg/day. Its average dietary intakes in various countries are as follows (in mg/day): Europe, 0.04–0.09; China, 0.01–0.20; and Switzerland, 0.066. In most countries, Se intake is estimated at the level of 0.06/day (Jenny-Burri et al. 2010). In the United Kingdom, the miscellaneous cereals (in 16%) and the meat products group (in 15%) made the greatest Se contribution to the population dietary exposure (Rose et al. 2010). In Switzerland, pasta made of North American durum

TABLE 42.2
Selenium in Foodstuffs (mg/kg FW)

Product	Content
Food of Plant Origin	
Bread	0.06
Miscellaneous cereals	0.07
Rice, white[a]	0.023–0.045
Rice, different type[b]	<0.1–0.3
Potatoes	<0.01
Green vegetables	0.007
Other vegetables	0.018
Fresh fruit	<0.005
Nuts	0.30
Food of Animal Origin	
Carcass meat	0.14
Meat products	0.14
Offal	0.77
Poultry	0.17
Fish	0.42
Fish smoked, various[c]	0.122–0.218
Egg	0.19
Milk	0.014
Dairy products	0.03

Sources: Rose, M. et al., *Food Addit. Contam.*, 27, 1380–1404, 2010.
[a] Batista, B.L. et al., *Food Addit. Contam.*, 3, 253–262, 2010.
[b] Jorhem, L. et al., *Food Addit. Contam.*, 25, 841–850, 2008b.
[c] Polak-Juszczak, L., *Roczn. PZH*, 59, 187–196, 2008.
FW, fresh weight.

wheat was the food with the highest Se contribution to the dietary intake, followed by meat (Jenny-Burri et al. 2010).

42.7 ANIMALS

Selenium contents of different animal tissues are concentrated mainly in the kidneys, less in the livers, and the least in the muscles (Table 42.3). Its contents in animal tissues are often in relation to their surrounding habitat.

Selenium deficiency is a global problem, related to an increased susceptibility to various diseases of animals, and in decreased productive and reproductive performance of farm animals. Optimization of the Se nutrition of poultry and farm animals resulted in increased efficiency of egg, meat, and milk production, and improved quality of products (Lyons et al. 2007).

TABLE 42.3
Selenium in Animal Tissues (mg/kg FW)

	Mean Content		
Animal	**Muscle**	**Liver**	**Kidney**
Pigs	0.039	0.112	1.389
Cattle	0.018	0.075	0.448
Hens	0.070	0.263	–
Bison	0.020	0.044	0.912
Hares from forest[a]	0.069	0.104	–
Hares from field[a]	0.119	0.342	1.155

Source: Żmudzki and Szkoda (1994).
[a] Cieśla, W. and Borowska, K., *Arsenic and Selenium in Environment*, The Polish Academy of Sciences, Warsaw, Poland, 1994.
FW, fresh weight.

Selenium deficiency is related to several nutritional disease conditions in animals. The pathological changes in animals include growth retardation, skin lesions and hair loss, visual defects, reproductive disorders, pancreas atrophy, liver necrosis, and dystrophy of the skeletal muscle and of the heart muscle. The occurrence, in animals of Se-responsive endemic deficiency diseases in various parts of the world is an excellent example of the interrelation between the geochemical environment and geographic pathology of the Se nutritional inadequacy. One of the most common diseases in sheep and cattle is muscular dystrophy caused by the Se deficiency (Khanal and Knight 2010).

Because of Se's vital role, its deficiency in animals may result in a wide variety of clinical signs. Selenium in the form of selenoproteins is critical in the formation of thyroid hormones and other endocrine systems. Adequate Se levels are also necessary for normal spermatogenesis. In severe deficiency states, myodegeneration occurs, resulting in cardiomyopathy, muscle weakness, white muscle disease (nutritional muscular dystrophy) in ruminants, and other species. Selenium is necessary for growth and fertility in animals, neutrophil and lymphocyte functions. Its deficiency in animals is very common and widespread around the globe, affecting much of South America, North America, Africa, Europe, Asia, Australia, and New Zealand (Khan et al. 2010).

Supplementation of broilers diet with Se–yeast not only increased Se concentration, but also reduced Cd concentration in the tissues. Selenium is negatively correlated with Cd, and positively correlated with Zn, Cu, and Fe (Pappas et al. 2011).

The toxic effects of Se were first discovered in 1930, when livestock ate certain plants of some wild vetches of the genus *Astragalus*, which accumulated toxic amounts of Se from soils. Lambs had visible evidence of reduced feed intake, depression, reluctance to move, and tachypnea, following minimal exercise. Major histopathological findings in animals of the high dose groups included multifocal myocardial necrosis and pulmonary alveolar vasculitis, with pulmonary edema and hemorrhage. Toxicity of Se is due to errors in its dosage in swine feed, which resulted in an initial episode of diarrhea, followed by dermatological and neurological signs. Cutaneous lesions consisted of diffuse alopecia, multifocal skin necrosis, and coronary band necrosis of the hooves. Central nervous system lesions comprised of a severe bilateral polioencephalomalacia of the ventral horns. In general, elemental Se is relatively nontoxic, whereas organic Se, in plants and grains, is more toxic to livestock.

Selenium poisoning should be generally suspected, based upon a variety of clinical signs including weight loss, poor growth rates, lameness, defective hoof growth, horizontal ridges or cracks in the hoof wall, hair loss, infertility, and acute deaths, especially when errors are made in mixing of Se into animal feeds or overdosing injectable Se products. A garlicky odor of the animals breath may be detected at overdosed Se (Khan et al. 2010).

43 Silicon [Si, 14]

43.1 INTRODUCTION

Silicon (Si) is a chemical element of group 14 in the periodic table of elements and is classified as a metalloid or a nonmetal element. It is the eighth most common element in the universe, by mass, but very rarely occurs as the pure element in nature. It is the second most abundant element in the Earth's crust (about 28% by mass). It is also the most stable element. However, under specific conditions, its compounds can be dissolved and transported, mainly in colloidal phase. Its content in magmatic rocks is up to about 35%, more in acid than in alkaline rocks. In sedimentary rocks, especially in sandstones, its content is over 40%, whereas in calcareous rocks, it is below 30%. Its content in coal may be up to about 30%. Thus, Si should not be included in the group of trace elements. However, it is considered as a trace element, with respect to its biochemical functions.

The oxidation state of Si is +4, but sometimes it may vary from −4 to +2. Its most common mineral, quartz (SiO_2), is also the most common compound in all terrestrial compartments. It is also the main component of various silicate minerals. Nonsilicate mineral containing Si is silicon carbide (carborundum), SiC.

Global Si production (rounded) in 2010 was 6900 kt, of which 4600 kt was produced in China, 610 kt in Russia, 330 kt in Norway, and 240 kt in Canada (USGS 2011). Demand for Si comes primarily from the aluminum and chemical industries. It is used in several different sectors of manufactures of steel and aluminum alloys, glass and ceramic productions, and various refractory materials. Due to semiconductor properties, Si is broadly used in various electronic devices. It is also applied in productions of cement, papers, textiles, cosmetics, and pharmaceutics.

43.2 SOILS

Silicon is the most abundant element in soils, averaging about 55%. Its mean contents in various soil groups differ and vary from 43% to 63% (Table 43.1).

Quartz (SiO_2) is the most resistant mineral in soils. It also occurs in a noncrystalline form, opal, which is presumably of a biological origin. Amorphous silicates contribute to anion adsorption processes and compete with other anions, for example, phosphates, for sites on mineral soil particles. In acidic soils, silicate and phosphate ions form insoluble precipitates that may fix other cations, for example, Fe and Al. Increased amounts of soluble organic matter in flooded soils induce a higher Si mobility, apparently due to the reduction of Fe hydrous oxides, which release adsorbed monosilicic acid.

A part of Si is released from minerals into the soil solution, and this process is controlled by both soil parameters and climatic factors. Soil pH has an especially marked impact on Si contents of soil solution, although the mobility of Si in soils

TABLE 43.1
Silicon Contents of Soils, Water, and Air (mg/kg)

Environmental Compartment	Range
Soil (%)	
Light sandy	60–80
Medium loamy	40–65
Heavy loamy	40–60
Calcerous (calcisols)	30–50
Organic (histosols)	20–35
Water (mg/L)	
Rain	0.5–7.5
River	2–25
Sea	0.5–<3
Air (mg/m³)	
PEL value	10

Sources: Data are given for uncontaminated environments from Kabata-Pendias, A. and Mukherjee, A.B., *Trace Elements from Soil to Human*, Springer, Berlin, Germany, 2007; Reimann, C. and de Caritat, P., *Chemical Elements in the Environment*, Springer, Berlin, Germany, 1998; Takeda, A. et al., *Geoderma*, 119, 291–307, 2004.

PEL, probable effect level.

cannot be predicted accurately from the pH only. Its concentrations in soil solution, mainly H_2SiO_4, range from 1 to 200 mg/L. Usually, Si is more mobile in alkaline soils. The main source of mobile Si is weathering of aluminosilicate minerals.

43.3 WATERS

The worldwide mean concentrations of Si in surface waters vary from 2 to 6 mg/L, being the highest in stream water (Reimann and de Caritat 1998). Global river input of Si to seawaters is estimated to be 145.45 Mt/yr, and the input from other sources is 30.9 Mt/yr. It is estimated that about 80% of Si in seawaters delivered with rivers is in colloidal forms and/or biochelates (Treguet et al. *vide* Kabata-Pendias and Mukherjee 2007). Its concentrations in natural waters, in the form of Si, range from 5 up to 25 mg/L.

Silicon in waters occurs mainly in the form of silica acid, H_2SiO_4, which resulted from both inorganic and biochemical processes. These processes also lead to the formation of amorphous Si, which is accumulated in bottom sediments. There is a competition between P and Si sorption on minerals particles, and thus, Si may depress P retention, which resulted in increased P mobility in various water basins. Colloidal Si is of a great importance for several organisms that exist in water and bottom sediments.

Rainwater contains Si within the range of 0.5–7.5 mg/L, but its average contents do not exceed 2.6 mg/L (Table 43.1).

The median Si concentration in bottled waters of the EU countries is 6.64 mg/L, and in tap waters, it is 4.3 mg/L (Birke et al. 2010).

43.4 AIR

Predominated Si compounds in the atmosphere are oxides and silicates, which are stemming from both natural and industrial sources. Windblown dust from deserts, especially from the Sahara, may carry huge amounts of Si into the atmosphere. High concentrations of Si in moss, up to 470 mg/kg, collected in Sweden in 1975 illustrate its increased concentrations in air (Rühling and Tyler 2004).

Elevated Si contents in air is of a health concern; thus, estimated probable effect level (PEL) value for Si is 10 mg/m^3 (Table 43.1).

43.5 PLANTS

Silicon is a common mineral constituent of plants, but it is not defined to be essential. It is readily absorbed from the growth media, mainly in the form of H_4SiO_4, from the soil solution. It is uptaken by both mechanisms, active and passive. Its biological function is slightly known. It is believed, however, that Si is necessary for the growth and development of plants, as well as for the resistance to toxicities of some chemicals, as well as to some diseases, especially fungal diseases. The amorphous Si impregnates the walls of epidermal and vascular tissues. Thus, Si strengthens plant tissues and reduces water losses.

Silicon contents vary greatly among plant species: the mean amounts in grasses range between 0.3% and 1.2%, whereas in leguminous, they range between 0.05% and 0.2%. Some plants may accumulate much higher Si amounts, up to >10%, for example, sedges, nettles, horsetails, and diatoms. The residues of these plants contribute to the formation of amorphous Si, which partly can be formed as opal.

Rice plants accumulate especially much Si, up to 10% in hulls and up to 4% in grains. The deficiency of phytoavailable Si in soils of rice plantations has deterioration impact on the crop. The addition of phytoavailable Si in the form of fertilization improved rice growth and crops.

43.6 HUMANS

The silicon quantity in a man weighing 70 kg is 1 g (Emsley 2011), which is the third most abundant trace element in the human body after Zn and Fe. Silicon is present in all body tissues, but its highest concentrations are in bone and other connective tissues, including the skin, hair, arteries, and nails (Jugdaohsingh 2007). In the last three decades, it has been suggested that Si may be an essential trace element. Silicon may have a role in a number of areas of human physiology and metabolism, especially bone and connective tissue formation, but possibly also gene expression and cardiovascular health (Powell et al. 2005). Silicon breast implants may cause autoimmune disorders and even cancer.

High Si levels are present in the bone, kidney, tendons, and walls of the aorta, with the nails containing its highest levels (up to 1500 mg/kg). Lower levels of silicic acid are present in red blood cells or serum, approximately 44 mg/kg in red cells and 20 mg/kg and in plasma. It is also accumulated in the liver, spleen, and lung, as well as in breast milk. It interacts with a number of elements, including Cu, Zn, and Ge. It promotes the Ca absorption and metabolism. Silicon may significantly reduce the absorption of Al, and thus reduce the risk of developing Al-induced Alzheimer's disease (Edwardson et al. 1993). Water-soluble forms of Si are absorbed in the intestinal tract, with excess amounts eliminated by the kidneys within 4–8 hours following ingestion (Jugdaohsingh 2007). Thus, it is unlikely for Si to accumulate in excessive amounts in healthy individuals. Patients on dialysis may accumulate Si because renal failure prevents the excretion of Si.

Silicon is an essential mineral for the bone formation. It improves bone matrix quality and facilitates bone mineralization. Increased intake of bioavailable Si is associated with increased bone mineral density. Its supplementation in humans and animals increases bone mineral density, and improves bone strength (Price et al. 2013). All tissues contain large amounts of Si, when they are perfectly healthy, but its amount decreases with age (approximately 0.1 mg/yr), and then the tissues undergo gradual degradation (Jugdaohsingh 2007). Depleted levels of Si are also in some pathological states, such as atherosclerosis or neoplastic diseases.

Silicon is nontoxic as an element and in all its natural forms, mainly silica and silicates, which are the most abundant elements, but some of its species may cause major damage to the lungs. Long-term exposure to silicates such as asbestos is a severe health problem. Diatoms and some protozoa, sponges, and plants use silicon dioxide (SiO_2) as a structural material.

Silicon is an important essential trace element in human nutrition. The recommended daily intake is about 10–25 mg (Powell et al. 2005). Its deficiency is mostly associated with losses of connective tissue components, such as glycosaminoglycanes, collagen, and elastin. The most readily absorbable form of silicon is orthosilicic acid (H_4SiO_4) (Sripanyakorn et al. 2005).

The role of Si in human biology is poorly understood, although its beneficial influences are observed in some diseases, such as osteoporosis, atherosclerosis, progress of diabetes, propagation of neoplastic process, and occurrence of heart diseases. Silicon was also found to reduce negative effects of some processes such as skin, hair, and nail aging (Jugdaohsingh 2007).

The bioavailability of Si for intestinal absorption depends on the solubility of its compounds. Foods derived from plants rather than from animals provide the highest sources of dietary Si. Unrefined cereals and grains have high Si content, especially oats and oat bran. The principal sources of dietary Si are whole grains, fruits, beverages, and vegetables (Table 43.2). In particular, high levels of bioavailable Si are in beer, which is made from barley malt (Cejnar et al. 2013).

Silicon is also available as a food supplement in tablet and solution form for bone and connective tissue health, although there are no formal data on its *in vivo* utilization or safety following sustained dosing (Jugdaohsingh et al. 2013).

TABLE 43.2
Silicon in Foodstuffs (mg/kg FW)

Product	Content
Food of Plant Origin	
Flour wheat, various	30.4–42.9
Bread, various	17.6–47.8
Breakfast cereals	13.4–233.6
Rice	8.8–37.6
Fruit, fresh	<0.2–31.5
Vegetable, fresh and boiled	1.9–51.2
Been, fresh and boiled	52.5–87.3
Vegetable, fresh and raw	<7.1
Lettuce, fresh and raw	11.1–27.8
Beers, different category[a]	19.5–63.1
Food of Animal Origin	
Meat raw[b]	5–20
Liver raw[b]	20–110
Kidney raw[b]	17–20
Fish fresh[b]	<5–10
Eggs[b]	3
Milk	0.7
Cheese	2–60

Sources: Powell, J.J. et al., *Br. J. Nutr.*, 94, 804–812, 2005. Unless otherwise indicated.

[a] Cejnar, R. et al., *Czech J. Food Sci.*, 13, 166–171, 2013.

[b] Pennington, J.A.T., *Food Addit. Contam.*, 8, 97–118, 1991.

FW, fresh weight.

According to Jugdaohsingh (2007), the average daily dietary intake of Si is 20–50 mg for European and North American populations, with higher intakes for men than for women, which is due to the higher intake of beer in males. Daily intake of Si is higher in China and India (140–200 mg/day) where grains, fruits, and vegetables form a larger part of the diet. China and India also have the lowest prevalence of hip fractures, compared to all other regions of the world (Price et al. 2013).

44 Silver [Ag, 47]

44.1 INTRODUCTION

Silver (Ag) is a soft, lustrous transition metal of group 11 in the periodic table of elements, called noble metals. Its average abundance in the Earth's crust is estimated to be about 0.08 mg/kg. In some sandstones and calcareous rocks, its contents may be enriched up to about 0.2 mg/kg. It reveals chalcophilic properties and, after rock's weathering, easily precipitates in alkaline media. Silver may concentrate in some biolits and may be accumulated in coal, up to 20 mg/kg, and in crude oil, up to 0.3 mg/kg.

The common oxidation state of Ag is +1, but may also be +2 and +3. Typical minerals of silver are argentite, Ag_2S; cerargite, AgCl; arsenide, Ag_3As; and proustite, Ag_3AsS_3. It forms several other minerals, and its host minerals are mainly sulfides. It may also occur as an alloy with other metals.

Global Ag production (rounded), excluding the United States, in 2010 was 22,200 t, of which 4,700 t was produced in Peru, 3,500 t in Mexico, 3,000 t in China, and 1,700 t in Australia (USGS 2011). Most Ag is produced as a by-product of Au, Cu, Pb, and Zn mining and refining.

Silver was discovered about 4000 BC, when it was used in jewelry and a medium of exchange in bullion. It has very high electrical and thermal conductivity, and antibacterial properties, especially in forms of nanoparticles. It is used in various electrical equipments, mirrors, and catalytic chemistry. Among all metals, Ag is unique in its behavior with oxygen and has a great catalytic power for the oxidation. In sanitation processes, it reveals catalytic oxidation of cell surface. Currently, the major part of the Ag production is used in manufacturing and photographic industries. It is also applied in some surgical equipments. Colloidal Ag is used as an antibacterial agent in ceramic filters for household waters. Silver may also be used to replace more costly metals in catalytic converters for off-road vehicles.

44.2 SOILS

The average Ag content in the worldwide soils is estimated to be 0.13 mg/kg, within the range in soils of various countries from 0.05 to 0.4 mg/kg. Its higher contents are noticed mainly in organic soils, which may contain up to 5 mg/kg. Also soils from mineralized areas contain usually elevated Ag amounts; particularly soils from the regions of old Pb mines (e.g., in Wales) contain increased Ag level, up to 9 mg/kg (Jones et al. *vide* Kabata-Pendias 2011). Elevated Ag amounts are also reported for garden soils (0.4 mg/kg) and street dust (0.2 mg/kg) (Rasmussen et al. 2001). However, garden soils, as well as vineyard soils, may contain much higher Ag levels, up to about 1 mg/kg.

The common Ag species in soils are simple cations, Ag^+, Ag^{2+}, and AgO^+. Complex Ag anions predominate in soil solutions: $AgCl_2^-$, $AgCl_3^{2-}$, and $Ag(SO_4)_2^{3-}$ (Kabata-Pendias and Sadurski 2004). Despite several mobile species, Ag is slightly mobile, especially in soils with pH > 4, and with higher organic matter (OM) contents.

Silver in soil is primarily in the form of sulfide and chloride compounds, associated with Fe, Pb, and Mn. Its behavior is controlled mainly by pH–Eh conditions and OM, which may absorb relatively high Ag amounts. The maximum binding capacity of humic acid and fulvic acid ranges from 0.8 to 1.9 mM/g, at pH 6.5. In general, the sorption of Ag by soil components is very strong, and nearly half of the total Ag is in immobile species. The contents of OM in soils may control the phytoavailability of Ag. Thus, the excess of Ag in soils of low OM contents is more phytotoxic than in soils of high OM contents. The elevated Ag contents of soils are toxic, especially to meso- and microbiota.

44.3 WATERS

Data for mean Ag concentrations in ocean waters are variable. According to Andren and Bober (2002), these values vary from <0.024 to 0.56 ng/L, whereas Reimann and de Caritat (1998) gave its concentration within the range of 2–40 ng/L, and in stream waters, it may be elevated up to 190 ng/L. Other data on Ag in waters are as follows (in ng/L): <0.01–140 in river waters and 0.2–0.5 in lake waters (Andren and Bober 2002). After the program of *clean technology* (after 1980), concentrations of Ag, especially in surface waters, have significantly decreased.

Silver in water occurs in several ionic species, associated mainly with Cl, S, and NO_3. The common species that control its mobility are Ag^+, $AgCl_2^-$, and $AgCl_3^-$. Also several organic and inorganic complex compounds of Ag are present in water. Almost all Ag species and compounds are readily scavenged by suspended minerals and deposited in bottom sediments. The highest Ag contents, 25–>30 mg/kg, are reported for sediments of river estuaries. Fine granulometric fractions of lake sediments contain Ag up to 3.4 mg/kg, whereas in sandy sediments its amounts do not exit 0.1 mg/kg (Bojakowska *vide* Kabata-Pendias and Mukherjee 2007). Bottom sediments are principal sinks for Ag, which may cause an environmental risk. Depending on several properties of sediments, such as salinity, pH, hydroxides of Fe and Mn, and OM, Ag may become slightly mobile and remains deposited for over 100 years.

Silver is a highly reactive element in aquatic environments, and only the free ionic species, Ag^+, is highly toxic, even at low levels. Luckily, this Ag cation exists in water for quite a short time, due to reactions with organic and inorganic ligands. The toxicity of Ag in seawater is still not well understood. It depends on reactions of fish gills, which under elevated Ag concentration failure the functions, and fish appear to die. However, the toxicity of Ag concentrations is relatively high and estimated for freshwater organisms between 0.85 and 1542 μg/L, and for seawater organisms, 13.3–2700 μg/L (Andren and Bober 2002). Several other water properties have also significant impact on the Ag toxicity to aquatic biota. Silver occurs naturally, mainly in the form of its very insoluble and immobile oxides, sulfides, and some salts. It has occasionally been found in groundwater, surface water, and drinking water at concentrations about 5000 μg/L.

Silver accumulation by algae plays a crucial role in its cycling in aquatic environments, and in its transfer to food chain. Zooplanktons contain Ag within the order of 10–20 mg/kg ash weight (AW), and in phytoplanktons, it averages 4 mg/kg AW. Mussels accumulate more Ag at low salinity than at high salinity. Mollusks contain Ag within the range of <0.01–4.0 mg/kg fresh weight (FW) from the Baltic Sea and 0.002–0.09 mg/kg FW from the Atlantic coast of the United States (Szefer 2002). Fish from rivers in the vicinities of towns may contain much higher amounts of Ag. Apparently, it is the most easily available to trout (Ratie *vide* Kabata-Pendias and Mukherjee 2007).

The median Ag concentration in bottled and tap waters of the EU countries is fairly similar and is <1.0 ng/L (Birke et al. 2010). Drinking water in Finland contains Ag at levels <30 ng/L (Kabata-Pendias and Mukherjee 2007). Its acceptable concentration is much higher, up to about 1000 ng/L. Its increased level, up to about 50 μg/L, after using Ag filters for water cleaning, is not harmful to humans. Due to inadequate data, health-based guideline value for Ag has not been established.

However, there is a recommendation that Ag concentrations in drinking water should not exceed 100 μg/L (WHO 2011a).

44.4 AIR

The world median Ag concentration in air is estimated to be 0.044 ng/m^3 (Reimann and Caritat 1998). The background value of Ag contents of the atmosphere of Finland is 0.014, 0.010, and 0.011 ng/m^3, during summer, autumn, and winter, respectively (Jalkanen and Häsänen *vide* Kabata-Pendias and Mukherjee 2007).

Silver in air occurs mainly in the forms of halides, sulfides, bromides, and oxides. The wet Ag deposition in Sweden (in 1999) was 0.4 g/ha/yr, whereas in the United Kingdom was up to 1 g/ha/yr (Eriksson 2001a).

44.5 PLANTS

Silver contents in plants vary greatly; the lowest are in cereals grains, <0.003 mg/kg, and the highest in grass, <5 mg/kg, and in fungi, <150 mg/kg (Eriksson 2001a; Falandysz et al. 2012). The common Ag concentrations in food plants range within the order of 0.003–0.05 mg/kg.

The amounts of Ag in most plants are related to its contents of soils. Thus, Ag can be concentrated to toxic levels in plants growing in Ag-contaminated or Ag-mineralized areas. In plants, Ag is apparently deposited in root tissues, and therefore is excluded from metabolic processes. However, when Ag is mobile (e.g., $AgNO_3$), it may be taken up to toxic levels. Most sensitive to excess Ag levels are lettuce, radish, and corn. Elevated Ag amounts reduce plant yields, often without toxic symptoms. Interactions of Ag with some cations (e.g., K, Co, Cu) may result in the inhibition of their uptake by plants.

Silver is very toxic to various bacteria, and thus, it is widely used as an ascetic substance. It may precipitate some proteins in bacteria and form insoluble complexes with ribonucleic acids.

44.6 HUMANS

Normal concentrations of Ag in human tissues are low: in human body (70 kg) it is approximately 2 mg (Emsley 2011), of which in the blood it is <0.0023 mg/L, in the liver and kidney 0.05 mg/kg (Wan et al. 1991), and in the rib bone ≤0.011 mg/kg (Zaichick et al. 2011).

Silver is not known to perform any essential function in humans. It may enter the human body through the gastrointestinal tract, lungs, mucous membranes, and skin lesions, when medicines containing Ag are also taken or applied to the skin or gums. Generally, much less Ag enters the body through the skin than through the lungs or stomach. Ingested Ag compounds are absorbed by the body at a level of about 10%, with only 2%–4% being retained in tissues. Most of the Ag transported in the blood is bound to globulins. In tissues, it is present in the cytosolic fraction, bound to metallothionein. Silver is stored mainly in the liver and skin, and in smaller amounts in other organs.

Metallic Ag is not soluble in aqueous solutions nor is it readily solubilized by any physiological mechanisms. Therefore, it is poorly absorbed after exposure and is more likely to be excreted by the body than its soluble compounds. Finally, accumulated Ag can be oxidized to Ag sulfide or Ag selenide, resulting in blue-gray pigmentation.

Soluble Ag compounds are absorbed by the body more readily as a result of their ability to bind to proteins, DNA, and RNA. The majority of occupational exposures are related to soluble Ag compounds, which seem to cause toxic effects at lower concentrations than metallic Ag and insoluble Ag compounds (Drake et al. 2005). Silver may be metabolized to all tissues other than the brain and the central nervous system (Lansdown 2010).

Colloidal Ag protein has been used as an allergy and cold medication, in eye drops to alleviate soreness, and for the treatment of various ailments. The adverse effects from extended use of colloidal Ag protein may be discolored fingernails, ocular argyrosis, and generalized argyria. If there is overexposure, Ag can accumulate in the skin, liver, kidneys, corneas, gingiva, mucous membranes, nails, and spleen. Due to Ag affinity for the thiol groups in the liver, it binds to reduced glutathione and is transported into the bile, thus depleting the amount of reduced glutathione available for biochemical pathways (Drake et al. 2005).

Most foods contain traces of Ag within the range of 10–100 μg/kg (Table 44.1). Its low levels in drinking water, generally below 5 μg/L, are not relevant to human health with respect to argyria. By contrast, Ag salts used to maintain the bacteriological quality of drinking water may be of some risk. Higher levels of Ag, up to 100 μg/L, could be tolerated, without risk to health (WHO 2011a). The mean intake of Ag in Russia is <50 μg/day and is similar to levels found for most other countries, 70–100 μg/day (Gorbunov et al. 2004).

New problems for humans and animals create Ag in the form of nanoparticles (AgNPs). Nano-silver is used in an increasing number of products. Three major product categories are food, consumer products, and medical products. Nano-silver products are mainly manufactured in Asian countries (81 products) and in the United States (60 products); only a very small amount of these products have European countries (Wijnhoven et al. 2009).

TABLE 44.1
Silver in Foodstuffs (μg/kg FW)

Product	Content
Food of Plant Origin	
Cereals and cereal products[a]	79
Fruits and vegetables[a]	75
Meadow mushroom (*Agaricus campestris*), cultivated	10
Meadow mushroom (*A. campestris*), wild	810
Oyster mushroom (*Pleurotus ostreatus*), cultivated and wild	30
Sugar	<10
Tea	<10
Food of Animal Origin	
Meat	<50
Meat[a]	85
Offal[a]	451
Freshwater and seawater fish	<50
Shrimps	70
Fish and fish products[a]	2370
Eggs	<50
Dairy products	<50
Dairy products[a]	66

Sources: Gorbunov, A.V. et al., Assessment of human organism's intake of trace elements from staple foodstuffs in central region of Russia. Preprint of the Joint Institute for Nuclear Research, Dubna, Moscow, Russia, 2004, http://wwwinfo.jinr.ru/publish/Preprints/2004/089(D14-2004-89)_e.pdf.

[a] Millour, S. et al., *J. Food Compos. Anal.*, 25, 108–129, 2012.

FW, fresh weight.

Silver nanoparticles (particles < 100 nm in size) may have negative impact on both human and environmental health. Due to well-known bactericide properties, AgNPs are nowadays among the most commercialized nanomaterials. They may enter human and animal bodies through different routes of exposure: ingestion, inhalation, and dermal contact. They can be administered subcutaneously, intraperitonally, or intravenously. Once entering the body, some AgNPs remain in entry tissues, but generally they can be easily translocated and distributed throughout the body via bloodstream or lymphatic systems. Regardless of the routes of exposure, AgNPs are found in all major target organs: the lungs, the skin, the liver, the spleen, the brain, the adrenals, the testes, and the kidneys. AgNPs easily cross blood vessels walls and can cross the blood–brain and alveoli–capillary vessels, blood–testes and skin–blood barriers. It is highly likely that AgNPs could interact with cells and cross barriers, such as cell membranes, potentially resulting in adverse and unpredictable effects (Kruszewski et al. 2011).

Though Ag nanoparticles are often used in many medical procedures and devices, as well as in various biological fields, they have their drawbacks due to nanotoxicity. However, the downside of AgNPs is that they can induce toxicity at various degrees. Higher concentrations of AgNPs are toxic and can cause various health problems. There are also studies that prove that nanoparticles of Ag can induce various ecological problems and disturb the ecosystem if released into the environment (Prabhu and Poulose 2012). It is hypothesized that the toxic effects of nano-silver are due to a combination of the specific properties of AgNPs and the generation of ions from them (Wijnhoven et al. 2009).

45 Strontium [Sr, 38]

45.1 INTRODUCTION

Strontium (Sr) is a metal of group 2 in the periodic table of elements and has a silver appearance, before rapid oxidation, after which it turns into a yellowish color. It is relatively common in the Earth's crust, in contents within the range of 260–730 mg/kg. It is likely to concentrate in mafic rocks (140–460 mg/kg) and calcareous sediments (460–600 mg/kg). It reveals lithophilic properties, similar to Ca, with which it is often associated, and does not occur as the free element. It may be accumulated near S deposits. During weathering, Sr is easily mobile and then absorbed by clay minerals and organic matter (OM). Its contents of some coal may be elevated and vary within <300–>1000 mg/kg. Fly ash contains, on average, 720 mg/kg of Sr.

Strontium oxidation state is +2 only. It forms several minerals, of which are mainly strontianite, $SrCO_3$, and celestite, $SrSO_4$. Celestite occurs much more frequently in sedimentary deposits and is mining from these deposits. Strontianite, however, is more useful mineral. There are four naturally occurring isotopes: ^{84}Sr, ^{86}Sr, ^{87}Sr, and ^{88}Sr. During the nuclear weapon testing, ^{90}Sr, with a half-life of 28.78 years, and ^{89}Sr, with a half-live of 50 days, are released to the atmosphere. The isotope ^{90}Sr is of a real environmental concern.

Global Sr production (rounded), excluding the United States, in 2010 was 420,000 t, of which 200,000 t was produced in China, Spain—180,000, and Mexico—30,000 (USGS 2011). Strontium was previously used for television tubes. Now, its consumption is mainly in ceramics and glass manufacture. It has also been broadly applied in pyrotechnics.

Because Sr is very similar to Ca, it is incorporated in the bones, as well as its four stable isotopes. The amounts of Sr isotopes vary between geographical locations; thus, its amounts in the bone can help to determine the region it came from.

45.2 SOILS

Worldwide background mean levels of Sr in soils are estimated to be 130–240 mg/kg. However, its much higher contents are relatively common (Table 45.1). The contents of Sr are highly controlled by parent rocks and climate. The highest Sr contents are in heavy loamy soils, especially developed in hot and dry climates (Table 45.1).

In some heavy loamy soils, and in soils from industrial regions, Sr content may be above 1000 mg/kg. The range of Sr mean contents in soils of various countries are presented as follows (in mg/kg): China, 26–150; Japan, 32–130; Sweden, 112–258; the United States (California), 236–246 (Kabata-Pendias 2011). Soils of the remote region in Spitsbergen, Norway, contain Sr within the range of 0.09–19.7 mg/kg (Gulińska et al. 2003).

Strontium is concentrated mainly in residual soil fractions. It is slightly mobile and associated mainly with Fe and Mn oxides. It is moderately mobile in soils, and

TABLE 45.1
Strontium Contents of Soils, Water, and Air (mg/kg)

Environmental Compartment	Range
Soil (mg/kg)	
Light sandy	5–1000
Medium loamy	20–1000
Heavy loamy	20–3100
Calcerous (calcisols)	5–1000
Organic (histosols)	70–500
Water (μg/L)	
Rain	0.5–383
River	2–238
Sea	8000[a]
Air (ng/m³)	
Urban areas	20–50
Antarctica	0.8

Sources: Data are given for uncontaminated environments from Kabata-Pendias, A. and Mukherjee, A.B., *Trace Elements from Soil to Human*, Springer, Berlin, Germany, 2007; Reimann, C. and de Caritat, P., *Chemical Elements in the Environment*, Springer, Berlin, Germany, 1998.

[a] Mean value.

the predominated cation, Sr^{2+}, is likely to be sorbed in hydrated form by clay minerals and Fe oxides and hydroxides. In calcareous soils, Sr may be precipitated (e.g., as strontianite), which decreases its phytoavailability. When a ratio of Ca:Sr is less than 8, the Sr phytotoxicity may occur. Displacement of Sr by Ca in various compounds has practical implications in the reclamation of contaminated soils. In acidic soils, Sr is likely to be leached down, whereas in calcareous soils and soils rich in soluble OM, it may be concentrated in the upper horizons.

The main sources of Sr pollution are from coal combustions and sulfur mining and processing. Phosphorites may contain elevated amounts of Sr, up to about 2000 mg/kg; thus, P fertilizers are a source of elevated Sr amounts in soils. Also soil amendment materials may have increased Sr levels. The average Sr contents in these materials are (in mg/kg) P fertilizers, 610; limestone, 610; manure, 80; communal sludge, 75; and industrial sludge, 270.

45.3 WATERS

The median Sr concentration in seawater is calculated as 8000 μg/L. Its contents in the North Pacific are given within the range of <500–7800 μg/L (ATSDR 2004c). In rivers, Sr contents vary within the broad range of 2–238 μg/L (Table 45.1). It may be elevated to >300 μg/L in rivers of industrial regions (Szefer 2002).

Most of the aquatic organisms may accumulate Sr. This process is an inverse correlation to Ca levels in water. Mussels from the Baltic Sea may accumulate Sr up to 100 mg/kg (Szefer 2002).

The average Sr content of rainwater collected in Sweden is 1.2 μg/L and the maximum is 3.3 μg/L (Eriksson 2001a).

45.4 AIR

The common concentration of Sr in the atmosphere varies within the range of 20–50 ng/m^3. In remote regions, Sr average concentration is estimated to be 0.8 ng/m^3 (Table 45.1). In industrial regions, especially close to coal power plants, it may be significantly enriched up to 50 ng/m^3.

During thermal processes, Sr is released to air as oxide, SrO, which in contact with water may be transferred into ionic species Sr^{2+}, which is easily trapped by both biotic and abiotic environmental components.

Strontium contents of moss (*Hylocomium splendens*) from the mountains in Poland (mean 4.1 mg/kg) are lower than those of the same moss species from Alaska (mean 20.0 mg/kg). This may suggest higher Sr atmospheric deposits in Alaska than in mountains of Poland (Migaszewski et al. 2009). In mosses sampled in Norway, in the period 1990–1995, Sr content ranges from 2.8 to 51 mg/kg, mean 15 mg/kg (Berg and Steinnes 1997).

45.5 PLANTS

Although Sr is apparently not a plant's micronutrient, it is absorbed by plants similar to Ca, by mechanisms of both mass flow and exchange diffusion. It is easily taken up by plants from acidic light sandy soils. Its content of plants is highly variable, from the lowest in cereal grains (mean 1.5–2.5 mg/kg) to the highest in some vegetable leaves (mean 45–74 mg/kg).

In the European forest ecosystems, Sr contents in blueberries (*Vaccinium* sp.) range between 4.5–5.5 mg/kg in Russia and 2.9–3.09 mg/kg, in Germany (Markert and Vtorova *vide* Kabata-Pendias 2011). The mean Sr content of forest crowberries (*Empetrum nigrum*) from the Kola Peninsula is 8.9 mg/kg, whereas the mean Sr content of crowberries grown close to the Monchegorsk plant is 0.8 mg/kg, which indicates that this metallurgical industry does not emit Sr (Reimann et al. 2001).

Soil-to-plant transfer Sr ratios are relatively high and are calculated as 0.02–1.0 (ATSDR 2002b). Strontium is not readily transported from roots to shoots; thus, it is likely to be accumulated in roots. However, the distribution of Sr in plants is highly variable. However, in cereals plants, it is accumulated mainly in nonedible parts (e.g., straw), and in the cell walls of roots. Relatively high Sr contents of plant tops may be an effect of its aerial deposition.

Interactions between Sr and Ca are complex, and although they compete with each other, Sr cannot replace Ca in biochemical functions. Increased levels of Ca in growth media may both inhibit or stimulate Sr uptake, depending on several soil and plant factors. Most often, however, addition of Ca to soil decreases Sr phytoavailability. Applications to soils of other elements, such as Mg, K, and Na, also inhibit Sr uptake by plants.

Ratio of exchangeable species of Sr to Ca in soils may be a good predictor of Sr contents of some plants (Takeda et al. 2005). The interactions between Sr and P are observed, which are variable, depending on several other factors.

The mushroom Common Chanterelles (*Cantharellus cibarius*) grown in mountains of Poland contain slightly higher amounts of Sr, mean 1.1 mg/kg (range 0.51–2.6 mg/kg), than those grown in the Baltic Sea coast, mean 0.6 mg/kg (range 0.14–1.7 mg/kg), (Falandysz et al. 2012). This may indicate higher aerial Sr deposits in mountains than in low-level regions.

45.6 ISOTOPES

There are several isotopic species of Sr in the geological formations. Their natural concentrations are ^{84}Sr—0.56%, ^{86}Sr—9.87%, ^{87}Sr—7.04%, and ^{88}Sr—82.53%. ^{90}Sr is also present in the environment as a product of fission reactions, with a relatively short half-life time of about 30 years.

The ratio of ^{87}Sr to ^{86}Sr is typical for any geological formations, and therefore is used for age determination of rocks, as well as for tracing of industrial aerosol contamination (Geagea et al. *vide* Kabata-Pendias 2011). The products of several nuclear processes are ^{89}Sr and ^{90}Sr, and the second radionuclide is considered to be the most biologically hazardous in the environment. During the period 1945–1980, Sr radionuclides were released mainly from aboveground detonations of nuclear weapons. After the Chernobyl accident in 1980, the emission of ^{90}Sr becomes to be of serious environmental problems. Elevated concentrations of this isotope in soils of various regions of Russia and Europe have been reported, and the highest amounts were in soils of Belarus. Its most elevated contents, up to 340 Bq/kg, were in alluvial soils and organic-rich soils.

Biogeochemical behavior of ^{90}Sr is similar to stable Sr. The migration velocity of this isotope is calculated as 0.7–1.5 cm/yr; however, its great part is still located in the upper 10 cm soil layer (Shagalova et al. *vide* Kabata-Pendias 2011). Because ^{90}Sr is easily adsorbed by Fe hydroxides, it is also accumulated in Fe-rich soil horizons.

Plants easily uptake ^{90}Sr, similar to stable Sr, and therefore it is applied in studies of its metabolism in plants (Tsukada et al. 2005). Some plants are proposed to be used for phytoremediation of soil contaminated with Sr. Alamo switchgrass (*Panicum virginatum*) may accumulate up to 44% of the total Sr radionuclide contents of soils. The transfer factor for ^{90}Sr, expressed as its ratio in fresh weight plants to concentrations in soils, varies highly and is calculated as follows: potato tubers, 0.4; vegetable leaves, 0.8; and vegetable roots, 20 (Kabata-Pendias and Pendias 1999). In soil amendment with Ca, Mg, and K fertilizers, the phytoavailability of ^{90}Sr is inhibited.

45.7 HUMANS

The typical adult body burden of Sr is about 320 mg in the blood 30 μg/L, 35–140 mg/kg in the bone, and 120–350 μg/kg in the tissue (Emsley 2011). The distribution of absorbed Sr in the human body is similar to that of Ca, with approximately 99% found in the skeleton. Strontium can be taken into the body by eating food (Table 45.2), drinking water, or breathing air. Humans absorb some 11%–30% of ingested Sr. Its highest concentrations occur in the bones and teeth (mean 115–138 mg/kg), with

TABLE 45.2
Strontium in Foodstuffs (mg/kg FW)

Product	Content a	Content b
Food of Plant Origin		
Bread	3.7	2.27
Miscellaneous cereals	1.3	1.28
Green vegetables	1.6	2.06
Other vegetables	1.3	1.39
Potatoes	0.81	0.625
Fruits	0.79	0.859
Nuts	8.6	15.7
Sugar and preserves	1.1	1.06
Beers, various		0.14–0.74[c]
Food of Animal Origin		
Carcass meat	0.11	0.05
Offal	0.12	0.10
Fish	3.6	2.50
Egg	0.50	0.38
Milk	0.29	0.273
Dairy products	1.0	0.83

Sources: [a] Ysart, G. et al., *Food Addit. Contam.*, 16, 391–403, 1999.
[b] Rose, M. et al., *Food Addit. Contam.*, 27, 1380–1404, 2010.
[c] Pohl, P. *Food Addit. Contam.*, 25, 693–703, 2008.
FW, fresh weight.

much lower levels (0.05–0.38 mg/kg) in the muscle, brain, kidney, liver, and lung. Its concentrations in the serum/plasma/blood vary from 27 to 53 µg/L (WHO 2010).

Absorption from the lungs is rapid for soluble Sr compounds, but slow for insoluble Sr compounds. Dermal absorption of Sr compounds is also slow. It can act as an imperfect surrogate for Ca; the distribution of absorbed Sr mimics that of Ca, and Sr can exchange with Ca in the bones. Strontium uptake from the gastrointestinal tract and its skeletal retention are reduced by co-administration with Ca, phosphates or sulfates. The amount absorbed tends to decrease with age, and is higher (about 60%) in children in their first year of life, and may inhibit bone mineralization (ATSDR 2004c).

The requirement for Ca is high during the period of bone development, growth and remodelling, and thus children tend to absorb and retain Sr to a greater extent than adults. Consequently, the young are at increased risk from exposure to excess Sr. Others, who might be at an increased risk from Sr exposure, are patients with kidney failure (whose ability to excrete Sr may be limited), or with osteomalacia. Not adequate exposure to sun-light may limit Sr absorption.

Strontium can bind to proteins and, based on its similarity to Ca, probably forms complexes with various inorganic anions, such as carbonates and phosphates, carboxylic acids, citrates and lactates (WHO 2010). The Sr contents of foods showed a correlation with their Ca contents (Varo et al. 1982). Due to chemical similarity of Sr and Ca, the stable forms of Sr do not pose a significant health threat.

Strontium stimulates bone growth, increases their density. At its low contents bone fractures mainly in women may occur, escpecially at women (Cabrera et al. 1999; Genuis and Bouchard 2012). Strontium ranelate is a new orally administered agent for the treatment of women with postmenopausal osteoporosis. Strontium ranelate was chosen from among 20 different Sr compounds based on the bioavailability of Sr, gastric tolerability, and physicochemical characteristics (Blake et al. 2007). Women receiving the drug showed a 12.7% increase in bone density. Women receiving a placebo had a 1.6% decrease. It is evident that Sr-containing biomaterials have positive effects on bone tissue condition and repair (Isaac et al. 2011).

Research on the Sr toxicity has so far been focused mainly on the biological effects of radioactive ^{90}Sr, which is an important constituent of nuclear fallout. At that time dietary supplementation of stable Sr was reported to inhibit the intestinal absorption of ^{90}Sr, thereby serving as a deterrent to ingestion of radioactive Sr (Cabrera et al. 1999). Radioactive ^{90}Sr can lead to various bone disorders and diseases, including bone cancer. ^{90}Sr concentrates in bone surfaces and bone marrow, and its relatively long radioactive half-life (29 years) makes it one of the more hazardous products of radioactive fallout.

The International Agency for Research on Cancer (IARC) has determined that radioactive Sr is carcinogenic to humans, because it is deposited inside the body and emits β radiation. External and internal exposures to fission products including ^{90}Sr causes solid cancers and leukemia (IARC 2012b).

Medically, radioactive Sr probes have been used intentionally to destroy unwanted tissue on the surface of the eye or skin. Radioactive Sr may cause cancer, as a result of damage to the genetic material DNA in cells. An increase in leukemia, over time, was reported in individuals in one foreign population who swallowed relatively large amounts of ^{90}Sr (and other radioactive materials) in river water contaminated by a nuclear weapons plant.

For adult humans, the total daily intake of Sr, in many parts of the world, is estimated to be up to about 4 mg/day. Drinking water contributes about 0.7–2 mg/day, and food (mainly leafy vegetables, grains and dairy products), and another 1.2–2.3 mg/day (Table 45.2). The contribution from air is insignificant (WHO 2010). For adults in the United States, the total daily exposure to Sr has been estimated to be approximately 3.3 mg/day, made up of 2 mg/day from drinking water, 1.3 mg/day from the diet, and 0.4 μg/day from inhaled air (ATSDR 2004c). The population dietary intake in the United Kingdom was 1.3 mg/day in 1994 yr, and 1.20 in 2006 yr. The greatest contributions, from various food groups, to dietary intake of Sr are: bread (20%), vegetables (15%), and cereals (14%) (Rose et al. 2010). Cigarettes and tobacco leaves contain Sr, and individuals who smoke may be exposed to its higher levels.

46 Tantalum [Ta, 73]

46.1 INTRODUCTION

Tantalum (Ta), a blue-gray transition metal of group 5 in the periodic table of elements, is hard and highly corrosion resistant. Its mean abundance in the Earth's crust is around 2 mg/kg, within the range of 1.5–2.2 mg/kg. Its content is the highest in acidic magmatic rocks, about 4 mg/kg. In calcareous sedimentary rocks, its amounts vary within <1–2 mg/kg. Coal contains Ta up to 0.3 mg/kg.

Oxidation states of Ta are from +3 to +5, but mainly at the highest value. It reveals both chalcophilic and siderophilic affinity, and occurs in several minerals, mainly oxides. It is associated with Nb, Tl, Y, and lanthanides (LAs). Its most common mineral is tantalite/columbite $[(Fe,Mn)(Ta,Nb)_2O_6]$, also called coltan. Tantalum occurs mainly together with chemically similar Nb and may form some other minerals, such as formanite ($YTaO_4$). The chief Ta ores are tantalite, samarskite, and pyrochlore, which also contain Nb.

World mine Ta production (rounded), excluding the United States, in 2010 was 670 t, of which 180 t was mined in Brazil, 110 t in Mazambique, 100 t in Rwanda, and 80 t in Australia (USGS 2011). Tantalum is used mainly for high-temperature applications and aircraft engines, and as capacitors in electronic equipments, such as mobile phones and computers. Metallic Ta dust is a fire and explosion hazard.

Tantalum is resistant to corrosion and to several chemical attacks. For this reason, it has been employed in chemical industries, especially where strong acids are vaporized. It is rarely used as an alloying agent, because it tends to make metals brittle. Due to chemical properties, Ta may substitute Pt in various tools. Its main use is now in various electronic devices, mainly as capacitors. Due to Ta immune system to impact of body liquids, it is widely used in surgical appliances (e.g., implants).

46.2 SOILS

The mean Ta content of soils worldwide is estimated to be 1.1 mg/kg (FOREGS 2005). Its contents vary within the range of 0.2–3.9 mg/kg, but are reported to reach up to 5.3 mg/kg in soils of China (Govindaraju 1994). Agricultural soils of Sweden contain Ta from 0.5 to 1.6 mg/kg, at an average value of 1.1 mg/kg (Eriksson 2001a). The median levels of Ta in soils of Japan range from 0.7 to 1.4 mg/kg, being highest in Gleysols (Takeda et al. 2004).

46.3 WATERS

Because Ta oxide is very insoluble, it is almost not found in natural waters. Thus, there are few data on its concentration in river waters, which is within the range between 0.1 and 148 ng/L, at an average of 1.1 ng/L (Gaillardet et al. 2003). The mean Ta content of the North Pacific is <2.4 ng/L (Nozaki 2005).

Rainwater collected in Sweden, during 1999, contained Ta in the range of <0.1–0.4 ng/L, at an average value of 0.2 ng/L (Eriksson 2001b).

The median Ta concentrations in bottled water of the EU countries is <1 ng/L and is a bit lower than in tap water, estimated to be 1.8 ng/L (Birke et al. 2010).

46.4 AIR

The concentration of Ta is 10–30 pg/m^3 in the atmosphere above Greenland and up to 280 pg/m^3 in air above the United States (Reimann and de Caritat 1998).

Wet deposition of Ta in Sweden is reported to be, on average, 1.5 μg/ha/yr (Eriksson 2001a).

46.5 PLANTS

There are very limited sources of Ta in plants. Various vegetables contain Ta within the range of <1–<6 μg/kg, at a mean value of <1 μg/kg (Kabata-Pendias 2011). In cereal grains from Sweden, Ta amounts vary from 1.1 to 5 μg/kg (Eriksson 2001a).

46.6 HUMANS

Tantalum content in the human body (70 kg) is about 0.2 mg (Emsley 2011). The pure metal is, to a great degree, inert both *in vivo* and *in vitro*. The pure metal and its oxides possess low solubility and toxicity; however, its halide compounds are more biologically active. It may be concentrated in hard tissues.

Tantalum has been in clinical use since before 1940, and has found a wide range of diagnostic and implant applications. In some applications, such as for radiographic bone markers and cranial closure, Ta may well be the current material of choice; thus, metallic Ta is a promising biomaterial (Black 1994). It is inert enough to be used as an implant material for humans.

Tantalum is highly resistant to chemical attack and arouses very little adverse biological response in both reduced or oxidized forms. It is very biocompatible in a variety of applications, including bone surgery. Metals coated with Ta and Ta itself release nothing into extraction media during standardized procedures, and the surface analysis shows low impurity.

There are very little data on Ta toxicity. This probably means that Ta and its common compounds are relatively innocuous. It is an experimental tumorigenic, or tumor-causing substance. However, overall systemic industrial poisoning is apparently unknown. Tantalum and its common compounds are not listed as either presumptive or possible human carcinogens, in the United States by the International Agency for Research on Cancer (IARC 2013). However, some circumstances, following the implantation of Ta in animal tissues, it has caused cancer. Apparently, the smoothness of the implanted Ta sheet plays a role, with highly polished surfaces potentially leading to cancer in animals.

Sometimes Ta is listed as being a nonspecific, surface carcinogen. Its salts are nontoxic when taken orally, because they are poorly absorbed and quickly eliminated from mammals. Inhaled Ta oxide has caused transient bronchitis and interstitial pneumonitis, with hyperemia in mammals. No adverse effects due to Ta have been reported from industrial exposure, in places where its air content was in the range of 0.1–7.6 mg/m^3 (Chen et al. 1999; Kerwien 1996).

47 Technetium [Tc, 43]

47.1 INTRODUCTION

Technetium (Tc) is a silvery-gray radioactive transition metal of group 7 in the periodic table of elements (PTEs), without any stable isotopes. Only minute traces of ^{99}Tc, the most common radioisotope, occurs naturally in the Earth's crust. It is associated mainly with U ores and some Mo ores. Oxidation states of Tc are +4, +6, and +7, of which the last one is the most often. Its common species are insoluble cation TcO_2 and soluble anion TcO_4.

Nearly all Tc is produced synthetically, and only minute amounts are found in the nature. Its chemical properties are similar to those of Re and Mn, which are of the same PTE group.

Technetium is the only artificial metal that is used in metal processing, mainly in alloys with Mo and Nb, to obtain electrical superconductivity. It is very resistant to the oxidation and may be added to some Fe alloys to inhibit the corrosion. Its isotope is widely used in the medicine, for X-ray diagnostics, in particular thyroid disorders.

47.2 ENVIRONMENT

The source of Tc is mainly global fallout and is released from various nuclear processes. Paddy soils of Japan contain ^{99}Tc from 6.1 to 110 mBq/kg, and its uptake by rice plants is much lower from flooded soils than from nonflooded soils (Zhang et al. 2002). Apparently, it is controlled by ratio of both cationic and anionic Tc species.

Plant uptake of Tc from soils is relatively low, being lowest by tomato and the highest by spinach. Oxidized Tc species are easily available to plants than species of a lower oxidation. Due to the microbial oxidation, it becomes more phytoavailable. Because Tc is slightly mobile in plants, it is concentrated in older plant tissues.

Technetium in the form of TcO_4^- follows the behavior of NO_3^- in soils. Organic and N fertilizers inhibit Tc phytoavailability, because they stimulate changes of mobile TcO_4^- to slightly mobile TcO_2 and TcS_2. In reduced groundwaters, Tc is likely to occur as almost immobile $TcO_2 \cdot nH_2O$.

The disposal of Tc-enriched wastes is an environmental hazard, because it is easily included in the biocycle.

47.3 HUMANS

Technetium has no biological role. It does not occur naturally in the biosphere, and thus normally never presents a risk. All Tc compounds should be regarded as highly toxic, largely because of its radiological toxicity. It is studied in a few nuclear research laboratories, where its high radioactivity requires special handling techniques and precautions.

Technetium appears to have low chemical toxicity. No significant changes in the blood and body were detected in rats, which ingested, for several weeks, food with ^{99}Tc content up to 15 mg/kg (Desmet and Myttenaere 2011).

The radiological toxicity of Tc (per unit of mass) is a function of compound, type of radiation, and the isotope's half-life. All isotopes of Tc must be handled carefully. The most common isotope, ^{99}Tc, is a weak β emitter, and such radiation is stopped by the walls of laboratory glassware. The primary hazard, when working with Tc, is inhalation of dust, resulted in lung contamination, which poses a significant cancer risk.

Once in the human body, ^{99}Tc concentrates in the thyroid gland and the gastrointestinal tract. The body, however, constantly excretes ^{99}Tc, after it is ingested. The Environmental Protection Agency (EPA 2002) has established a maximum contaminant level of 4 millirem per year, for β particle and photon radioactivity, from man-made radionuclides in drinking water. As with any other radioactive materials, there is an increased chance that cancer or other adverse health effects may result from the exposure to ^{99}Tc radiation, which is carcinogenic within Group 1 (IARC 2012b). ^{99}Technetium-m labeled human serum albumin (Tc-99m HSA) is an important radiopharmaceutical for clinical applications, such as cardiac function tests or protein-losing gastroenteropathy assessment.

48 Tellurium [Te, 52]

48.1 INTRODUCTION

Tellurium (Te), a silvery-white metalloid of group 16 in the periodic table of elements, is one of the rarest stable elements in the Earth's crust, where its mean abundance is about 1 μg/kg. Its contents in sedimentary rocks, up to 5 μg/kg, are higher than in magmatic rocks. Due to the ability to be fixed by organic matter (OM), its concentration in coal is high, as well as in fly ash, within the range of 20–2000 μg/kg. Concentrations of Te in volcanic gases and volcanic deposits are also elevated.

The oxidation states of Te are –2, +2, +4, and +6, with the +4 state being most common.

There is a great similarity between Se and Te, and thus minerals and compounds are also similar. The most common Te minerals are tellurite (TeO_2), hessite, (Ag_2Te), calaverite ($AuTe_2$), sylvanite [$(Ag,Au)Te_4$], and emmonsite [$Fe_2(TeO_3)_2{\cdot}2H_2O$].

Today, most Te is obtained as a by-product of mining and refining Cu. Tellurium minerals are also associated with Fe, Ag, and Au deposits. Production of Te in 2010, in various countries, was as follows: Canada—20 t, Japan—40 t, Peru—30 t, and Russia—35 t (USGS 2011). Tellurium is primarily used as an alloying agent, mainly to make alloys easier to machine and mill, and also to increase their strength and resistance to sulfuric acid. It is also used as a semiconductor in various electronic devices and is used in rubber production to increase resistant to heat and abrasion. Tellurium is applied as a pigment to produce various colors in glass and ceramics. It is also used in optical glasses, as well as in photographic and pharmaceutical industries.

48.2 SOILS

The worldwide mean Te levels in soils are given as 0.006 mg/kg (Reimann and de Caritat 1998) and 0.03 mg/kg (FOREGS 2005). The average Te level in soils of the United States is estimated to be 0.04 mg/kg (Burt et al. 2003). The general range of Te in soils, given for various countries, is from 0.02 to 0.5 mg/kg (Kabata-Pendias 2011). The elevated Te contents of soils, up to 6.55 mg/kg, are reported for soils surrounding various smelters, as well as for soils in vicinities of Au mineralization. Soils of mining districts in the United States (Nevada State) contain Te up to 10 mg/kg (Cowgill *vide* Kabata-Pendias 2011). In soils of the remote region in Spitzbergen, its contents vary within the range of 0.00002–0.0016 mg/kg (Gulińska et al. 2003).

The elevated Te contents of upper soil layers, compared to deeper soil horizons, suggest its fixation by the OM. Because microbial metabolism of Te is similar to that of Se, bacteria that methylate Se may also methylate Te, which increases its mobility. Its anionic species, TeO_3^- and TeO_3H^-, are also mobile. Processes of reduction of tellurite, TeO_2, which is relatively common in soils, are stimulated by variety of microorganisms. Under reducing soil conditions, a stepwise change from Te^{6+} to Te^{2-} occurs. All species of Te may be adsorbed by both Fe and Mn hydroxides.

48.3 WATERS AND AIR

Tellurium concentrations in surface waters range from 0.17 to 0.9 ng/L, being highest in the Red Sea (Kobayashi 2004). In the North Pacific, it averages 0.05 ng/L (Nozaki 2005).

Rainwater sampled in 1984 in Florida contained Te within the range of 0.51–3.3 ng/L (Andrea *vide* Kobayashi 2004). The concentration of Te in rainwater of Sweden, in 1999, varied from 0.1 to 1.1 ng/L, at an average value of 0.4 ng/L. Its wet deposit was calculated as 22 mg/ha/yr (Eriksson 2001a).

The median Te concentrations in bottled waters of the EU countries is 7.98 ng/L, and it is slightly lower (9.98 ng/L) than in tap waters (Birke et al. 2010).

Tellurium in air averages 0.12 ng/m^3 in Japan, within the range of 0.35–50 ng/m^3 (Kobayashi 2004). Its concentrations of 0.78 and 0.24 ng/m^3 are reported to be in interior and exterior air samples in Au mining areas (Nozaki 2005).

48.4 PLANTS

Tellurium has no biological function, although fungi can incorporate it, in place of S and Se, into some amino acids. Plants uptake Te relatively easily and correlate positively with its contents in growth media. Seleniferous plant species (e.g., *Astragalus*) contain elevated Te amounts, up to about 6 mg/kg, especially when grown in areas of Cu mining. Several plants from mining and industrial regions, mainly of Cu, may contain Te up to 25 mg/kg (Cowgill *vide* Kabata-Pendias 2011).

Generally, Te contents of plants do not exceed 1 mg/kg. Vegetables contain Te within the range of <0.013 mg/kg fresh weight (FW) in apple fruits to >0.35 mg/kg FW in onion. The garlic odor is caused by vapors of dimethyl telluride [$(CH_3)_2Te$].

Barley and wheat grains from Sweden contain Te at the mean value of <0.001 mg/kg (Eriksson 2001a). Norwegian mosses sampled in the period 1990–1995 contained Te within the range of <0.002–0.089 mg/kg, at an average value of 0.005 mg/kg (Berg and Steinnes 1997).

48.5 HUMANS

Tellurium content in humans (70 kg) is estimated to be 0.7 mg (Emsley 2011). About 10%–20% of ingested Te appears to be absorbed by the body (Reilly 2002).

Animal studies suggest that up to 25% of orally administered Te is absorbed in the gut. It is also found in the heart, kidney, spleen, bone, and lung. Formation of dimethyl telluride [$(CH_3)_2Te$] is a characteristic feature of exposure and gives a pungent garlic-like odor to breath, excreta, and viscera. The main target sites for Te toxicity are the kidney, nervous system, skin, and fetus (hydrocephalus) (Taylor 1996).

The toxic effects associated with specific Te compounds depend on its chemical forms, such as inorganic and/or organic compounds, and often also on the oxidation state. Organotellurium compounds, with reasonably stable Te–carbon bonds, are generally considered as less toxic, compared to inorganic, such as tellurate and tellurite. This may in part be due to different pharmacological and pharmacokinetic profiles, and different metabolic conversions inside the human body (Ba et al. 2010).

Organotellurium compounds are potentially toxic and lethal to rodents at low doses. Tellurides can cause cytotoxicity, hepatotoxicity, neurotoxicity, teratogenicity, and genotoxicity. Moreover, these compounds can inhibit sulfhydryl-containing enzymes, such as the Na^+/K^+-adenosine triphosphatase (ATPase), the δ-aminolevulinic acid dehydratase, and the squalene monooxygenase. The mechanisms of toxicity by organotellurium compounds may be related to the oxidation of thiol groups of important biomolecules, the replacement of Se in selenoproteins, and the capacity of Te compounds to induce the formation of reactive oxygen species. Antioxidant properties of Te compounds, including antitumor and chemoprotective effects, have also been observed (Zemolin et al. 2013).

Despite many chemical homologies between Se and Te, a nutritional role has never been identified for Te. Applications of organic telluranes as protease inhibitors and Te applications in disease models are the most recent contribution to the scenario of its biological effects (Cunha et al. 2009).

Tellurium nanoparticles (TeNPs) can be used as antibacterial reagents against *Escherichia coli*. Compared to Ag nanoparticles that are commonly used as antibacterial

TABLE 48.1
Tellurium in Foodstuffs (μg/kg FW)

Product	Content
Food of Plant Origin	
Flour, white and wholemeal	<5
Potatoes	3.2
Apples	0.7–3.2
Tomatoes	<0.7
Lettuce	<10
Mushrooms	1.0
Onions	0.8
Garlic	0.9–<5
Curry and paprika, powder	<10
Red wine	0.5
Food of Animal Origin	
Meat (beef)	<5–10
Fish	<2
Milk	<4
Cheese (parmesan)	13

Source: Reilly, C., *Metal Contamination of Food. Its Significance for Food Quality and Human Health*, Blackwell Science, Oxford, 2002.

FW, fresh weight.

reagents, TeNPs have higher antibacterial activity and lower toxicity. Thus, TeNPs hold great practical potential as a new and efficient antibacterial agent (Lin et al. 2012).

Little information is available on Te levels in foodstuffs. Its concentration in most foods is below 10 μg/kg FW (Table 48.1). Its daily intake is between 1 and 10 μg (Reilly 2002).

49 Thallium [Tl, 81]

49.1 INTRODUCTION

Thallium (Tl), a soft gray metal of group 13 in the periodic table of elements, is discolored when exposed to air. Its average content of the Earth's crust has been estimated to be 0.7 mg/kg, within the range of 0.9–1.8 mg/kg. Its contents increase with increasing acidity of igneous rocks and with increasing clay contents of sedimentary rocks. Its amounts in coal average at about 1 mg/kg, but may be concentrated up to about 50 mg/kg. It is also associated with K-based ores.

The oxidation states of Tl are +1, +2, and +3, the most important being +1. It has lithophilic properties and is associated with several minerals, but in higher amounts it is present mainly in lorandite ($TlAsS_2$) and avicennite (Tl_2O_3). Other minerals, such as crookesite [$(Cu,Tl,Ag)_2Se$] and hutchinsonite [$(Pb,Tl)_2 As_2S_9$], do not have any mining potential. It is usually obtained as a by-product of the production of sulfuric acid, or as a by-product of refining Zn and Pb ores.

The world production (excluding the United States) of Tl in 2010 was 10,000 kg (USGS 2011). Thallium is used mainly in high-temperature superconductors. There are no uses for metallic Tl, because it quickly combines with O_2 and H_2O vapor from the atmosphere, forming a black, powdery substance. Thallium metal and its compounds are highly toxic and are strictly controlled to prevent a threat to humans and the environment.

Thallium, used in conjunction with S, Sc, and As, forms low melting glass. Its sulfate (Tl_2SO_4) is an odorless and tasteless compound, and has been used as a rat and ant poison in the United States, since 1974, but it has been banned from household use. Other thallium compounds, sulfide (Tl_2S), iodide (TlI), and bromide (TlBr), are all used in devices to detect infrared radiation. Thallium previously used in pesticides and various poisons, which have been strictly forbidden (ATSDR 2002b).

Radioisotope ^{201}Tl is used in medicine for scintigraphy and diagnosis of some diseases, including myocardial disorders and various cancer detections.

49.2 SOILS

The worldwide range of Tl in soils is 0.0014–2.8 mg/kg, at an average value of 0.5 mg/kg. Its higher levels are in cambisols and histosols, which reveals its association with both clay fraction and soluble organic matter (SOM). Its occurrences in uncontaminated soils of various countries are reported as follows (in mg/kg): Canada, 0.17–0.22; China, 0.21–2.4; France, 1.54–55; Japan, 0.39–0.50; Poland, 0.01–0.41; Sweden, <0.04–0.52; and the United States, 0.02–2.8 (Kabata-Pendias and Mukherjee 2007). Background Tl levels of soils in China are given as

<0.2 to 0.5 mg/kg (Xiao et al. 2004). Soils of the remote region in Spitzbergen contain Tl within the range of 0.0006–0.0107 mg/kg (Gulińska et al. 2003).

The elevated Tl contents, at an average value of 59 mg/kg, in the top horizon of forest soils in Sweden, are presumably affected by its strong sorption by the organic matter (Tyler 2005). Soils derived from Tl sulfide-mineralized parent materials, and soils of some industrial regions, may contain its highly elevated amounts, up to about 130 mg/kg (Xiao et al. 2004). In general, Tl contents in soils above 1 mg/kg indicate the pollution, and this content is proposed as maximum allowable concentration value for agricultural soils. Thallium and its methylated species are toxic to microorganisms, and therefore, the nitrate formation is inhibited in Tl-contaminated soils, which has an agronomic impact.

During weathering, Tl is readily mobilized, and then fixed by clay minerals, SOM, and Mn and Fe hydroxides. In most soils, its highest proportion is in residual soil fractions. It reveals a great affinity for micaceous minerals and exhibits an activity to replace K from these minerals. Under oxidation conditions, Tl may be oxidized to Tl^{3+} and become immobile. In acidic soils, especially under acid rain, Tl is highly mobile, and thus easily phytoavailable.

Anthropogenic sources of Tl are from emission of cement factories (considered to be the largest), coal-fired power plants, and sewage sludge from metal industries. Slug dumps of Tl-rich pyrite mining, in China, contain this metal within the range of 30–70 mg/kg, which is of a great risk of Tl contamination of both soils and groundwaters (Xiao et al. 2004). Landfills from Zn and Pb mines, in Poland, contain Tl at mean level of 39 mg/kg, with the maximum value of 76 mg/kg (Wierzbicka et al. *vide* Kabata-Pendias 2011).

49.3 WATERS

The average concentration of thallium in seawaters is estimated within the range of 10–15 ng/L. Its amounts in the North Pacific Ocean is reported to be 13 ng/L (Nozaki 2005). Its amounts in river waters of different countries vary from 5 to 1350 ng/L, lowest in Poland and highest in Japan (Kabata-Pendias and Mukherjee 2007). Excluding its very elevated contents, its concentrations in polluted rivers range from 5 to 56 ng/L.

The main Tl species is, most probably, $TlCl_6^{3-}$, in seawaters and Tl^{3+} in river waters. However, water pH has an impact on Tl speciation, and at acidic pH range of 4–6, it predominates $Tl(OH)^{2+}$, whereas at pH above 7 $Tl(OH)_2^+$ and $Tl\ (OH)_3$ are present (Nozaki 2005). Also the redox potential of waters governs the Tl speciation. The stability of Tl^{3+} is limited, and apparently Tl^+ is dominant species in aquatic environments.

The main source of elevated Tl concentrations in surface waters is its leaching at ore processing operations. Its content in waters of some mining areas is reported to reach 1620 μg/L. The mean Tl concentration in deep groundwaters of China is 62 μg/L, at the highest value of 1100 μg/L (Xiao et al 2004).

Compounds of Tl are readily soluble in waters, and this metal is easily fixed by bottom sediments. The content of Tl in bottom sediments of polluted rivers in Germany is about 20 mg/kg (Gunther et al. *vide* Kabata-Pendias and Mukherjee 2007).

Most of aquatic phyto- and zoo-species are capable to concentrate Tl within the range of 0.02–0.5 mg/kg. However, there are limited data on its toxicity (Lin and Nriagu *vide* Kabata-Pendias and Mukherjee 2007).

The median Tl concentration in bottled water of the EU countries is 7.98 ng/L, and it is slightly lower than in tap water, estimated to be 9.98 ng/L (Birke et al. 2010).

49.4 AIR

The mean Tl concentrations in air vary from 0.06 to 0.22 ng/m^3 (ATSDR 2002b). It enters into the atmosphere mainly from industries, such as cement plants, metallurgical plants, coal combustion, and burning solid waste. Due to physicochemical properties (volatilization), it cannot be easily precipitated from emitted gases.

49.5 PLANTS

Thallium contents of plants seem to be a function of its concentrations in soils, especially its mobile fractions. Its reference content is estimated to be 0.05 mg/kg, and it is higher in herbage and woody plants than in other plant species (Nolan et al. 2004). The common ranges of Tl amounts are given for various plants as follows (in mg/kg): herbages, 0.02–1.0; vegetables, 0.02–0.13; clover, 0.008–0.01; and meadow grass, 0.002–0.0025. Grass from Zn-Pb industrial areas contains Tl within the range of 0.98–3.46 mg/kg (Adamiec and Helios-Rybicka 2004). In industrial regions, Tl may be easily absorbed by plants, especially herbages and mushrooms, from aerial deposits.

Some plants, mainly of the Cruciferae and Gramineae families, may serve as hyperaccumulators for phytoremediation of Tl-contaminated soils. The hyperaccumulator, *Iberis intermedia*, contain Tl up to 4000 mg/kg, even when grown in soil with Tl content only 16 mg/kg (Nolan et al. 2004). Cruciferae species, grown near cement plants, accumulated Tl up to about 450 mg/kg (Kabata-Pendias 2011). Also plants grown in soils over Tl enrichment rocks contain its highly elevated amounts.

Thallium, especially of anthropogenic origin, is easily mobile in soils, and thus easily available to plants. The depletion in the rhizosphere under elevated Tl concentration, without any effect on Tl uptake by plants, indicates that its transport to roots is mainly diffusion driven (Al-Najar et al. 2002).

Increased Tl levels in plant tissues are highly toxic to all plants. Its content in roots, at about 2 mg/kg, may inhibit germination, plant growth, and chlorophyll content. Concentration of Tl in nutrient solution, at about 5 mg/L, decreases the length of roots of some plants, without any other symptoms. Especially sensitive plants to increased Tl levels are Leguminosae species, cereals, buckwheat, and tobacco.

49.6 HUMANS

Thallium content in the human body is no more than 0.5 mg, and the level in the blood is 0.5 μg/L (Emsley 2006). The average content of Tl in the human body, measured in the U.S. population, was approximately 0.1 mg, and the blood concentration of 3 μg/L (Lansdown 2013). There is no biological role for Tl. It accumulates in the body over time and mostly ends up in the skeleton. It is found in all tissues, except fat, and it can even pass the placental barrier (Emsley 2006). Thallium has a short half-life in the human body, and its intake is conveniently measured in urine, fingernails, and especially in the hair (Lansdown 2013).

Water-soluble Tl salts are readily absorbed by mucous membranes, such as the mouth, stomach, and intestines, and they can even penetrate trough the skin. Absorption of Tl by mucosa is almost complete (80%–100%) and very quick. After absorption Tl is distributed from the blood to the tissue. In humans, the kidneys display the highest concentrations, followed, in a decreasing order, by the bones, stomach, intestines, spleen, liver, muscle, lung, and brain.

Thallium can be absorbed following inhalation of contaminated air (e.g., workplace), ingestion of contaminated food and water, or dermal contact. However, the levels of Tl in air and water are very low. The greatest exposure occurs by food, mainly homegrown fruits and green vegetables, may be contaminated by Tl.

The positively charged Tl ion (Tl^+) is almost of the same size as the K ion (K^+), which is essential to living cells. Thallium mimics K so effectively that it can displace it at all sites around the body, but the most damaging is its concentration along the central nervous system (Emsley 2006). Moreover, Tl follows K distribution pathways, and in this way alters many of K-dependent processes. For example, Tl may substitute K in the Na^+/K^+-ATPase. Thallium seems to inhibit a range of enzymatic reactions and to interfere with a variety of vital metabolic processes, disrupting cell equilibrium, which in turn leads to generalized poisoning. At present, no data are available about mutagenic, carcinogenic, or teratogenic effects of Tl compounds in humans (Cvjetko et al. 2010).

Thallium is a cumulative poison, which can cause adverse health effects and degenerative changes in many organs. The effects are the most severe in the nervous system. The exact mechanism of Tl toxicity still remains unknown, although impaired glutathione metabolism, oxidative stress, and disruption of K-regulated homeostasis may play a role.

Thallotoxicosis, or Tl poisoning, is usually the result of accidental or intentional ingestion of Tl compounds. Clinical manifestations of Tl poisoning can be classified as acute, subchronic, and chronic, depending on the level of toxin, severity, and time of exposure. The most prominent feature of Tl poisoning is the loss of hair or alopecia.

Acute Tl poisoning is usually accompanied by gastrointestinal symptoms, although neurological findings predominate in chronic exposure. Other symptoms include polyneuritis; encephalopathy; tachycardia; degenerative changes of the heart, liver, and kidney; subarachnoid hemorrhage; and bone marrow depression. Gastrointestinal symptoms usually appear within the first few hours of acute poisoning. Severe abdominal pain goes along with vomiting, nausea, and bloody diarrhea. By contrast, these symptoms are mild or nonexistent in chronic exposure. Thallium poisoning can also be associated with neuropsychological symptoms and psychosis, such as depression, apathy, anxiety, confusion, delirium, and hallucinations (Cvjetko et al. 2010).

Little information is available on levels of Tl in foods. Its normal level in most food products is much less than 3 μg/kg fresh weight (FW) (Table 49.1). Low concentrations of Tl are detected in some brands of cigarette.

TABLE 49.1
Thallium in Foodstuffs (µg/kg FW)

Product	Content a	Content b
Food of Plant Origin		
Bread	1.0	0.5
Miscellaneous cereals	1.0	<0.4
Vegetables, green	3.0	1.5
Vegetables, other	1.0	0.6
Potatoes	2.0	1.4
Fruits	1.0	0.45
Nuts	2.0	1.2
Beverages	<1.0	0.15
Sugar and preserves	1.0	0.5
Food of Animal Origin		
Carcass meat	3.0	0.4
Offal	3.0	2.3
Poultry	–	2.8
Fish	1.0	1.0
Eggs	1.0	0.3
Milk	<1.0	0.08
Dairy products	1.0	<0.2

Sources: [a] Ysart, G., Miller, P., Croasdale, M. et al., *Food Addit. Contam.*, 16, 391–403, 1999.
[b] Rose, M. et al., *Food Addit. Contam.*, 27, 1380–1404, 2010.
FW, fresh weight.

The mean dietary exposure to Tl in the United Kingdom was 2 µg/day in 1994 (Ysart et al. 1999) and 0.7–0.8 µg/day in 2006 (Rose et al. 2010). The beverages and potatoes contributed the highest percentage (23% and 19%) of Tl to population dietary exposures (Rose et al. 2010). German intakes of Tl have been estimated at about 2 µg/day, whereas in Japan it was from about 40 to 64 µg/day (Reilly 2002).

50 Tin [Sn, 50]

50.1 INTRODUCTION

Tin (Sn), a silvery metal of group 14 in the periodic table of elements, occurs in the Earth's crust at an average amount of 2.5 mg/kg. Its contents are slightly elevated in acidic magmatic rocks, up to 3.6 mg/kg, and in argillaceous sediments, up to 10 mg/kg. It is easily fixed by organic matter (OM), and thus, its concentration in coal may be up to 30 mg/kg. Its content may also be increased in several other bioliths.

The oxidation states of Sn are +2 and +4, of which the last one is more common and reveals amphoteric properties. The most important of few Sn minerals is cassiterite (SnO_2). Tin is also associated with other minerals, such as stannite (Cu_2SnFeS_4), teallite (Cu_3SnS_4), and montesite ($PbSn_4S_3$).

Global Sn production (rounded), excluding the United States, in 2010 was 2,651,000 t, of which 115,000 t was produced in China, 60,000 t in Indonesia, 38,000 t in Peru, and 16,000 t in Bolivia (USGS 2011). The recovery of Sn through secondary production, or recycling of Sn scrap, is also its important source. Tin in combination with other metals forms a wide variety of useful alloys. It is most commonly alloyed with Cu. Bronze, an alloy of Sn and Cu, has been used in large quantities since 3000 BC. It is a malleable metal, not easily oxidized in air, and used to coat other metals to prevent corrosion.

Alloy of Sn–Pb is commonly used soft solder. Because of low Sn toxicity, it is also used for food packaging, including tin cans. It has a number of organometallic derivatives of an importance in various manufacturing sectors. It is also used in glass production, and as polyvinyl chloride (PVC) stabilizers and wood preservatives. It is a common component of ship paints, which may be of an environmental concern. Its major use is now estimated as follows: electrical, 28%; cans and containers, 19%; construction, 13%; transportation, 12%; and other, 28% (USGS 2011).

50.2 SOILS

The average Sn abundance in soils is 2.5 mg/kg, within the range of <0.1–>10.0 mg/kg, being a bit elevated in heavy loamy and organic soils (Table 50.1). Its contents of soils from various countries are reported as follows (in mg/kg): China, 2.5–17.7; Germany, 1.0–5.0; Japan, 2.0–3.1; Sweden, 0.4–8.6; the United States, 1.0–11.0 (Kabata-Pendias 2011). Soils derived from rocks with increased Sn levels or soils near metal smelter plants may contain increased Sn levels, up to 1000 mg/kg. Soils of the remote region in Spitzbergen contain Sn <0.41 mg/kg (Gulińska et al. 2003). The criteria for contaminated land (Dutch List 2013), following Sn concentrations in soils, are established (in mg/kg) as follows: uncontaminated, 20; medium contaminated, 50; and heavily contaminated: 300. The allowable concentration of Sn in soils is 50 mg/kg (ATSDR 2002b).

TABLE 50.1
Tin Contents of Soils, Water, and Air

Environmental Compartment	Range Mean
Soil (mg/kg)	
Light sandy	<0.1–2.0
Medium loamy	0.3–2.0
Heavy loamy	0.3–4.0
Calcerous (calcisols)	<0.1–2.0
Organic	0.2–7.9
Water (μg/L)	
River	0.005–1.2
Sea and ocean	0.003–0.05
Air (ng/m³)	
Urban areas	2–800
Industrial areas	<5000
Remote regions	2–70

Sources: Data are given for uncontaminated environments from Kabata-Pendias, A. and Mukherjee, A.B., *Trace Elements from Soil to Human*, Springer, Berlin, Germany, 2007; Reimann, C. and de Caritat, P., *Chemical Elements in the Environment*, Springer, Berlin, Germany, 1998.

The mobility of Sn during weathering is relatively high and pH dependent. The common cation, Sn^{2+}, is a strong reducing agent and occurs mainly in acidic and reducing soils. The mobile Sn species follow the behavior of Fe and Al, and remain in weathered residues with hydroxides of these metals. Inorganic Sn compounds may be methylated by microorganisms in both soils and waters. However, these compounds are also demethylated.

Tin forms several complexes with soluble OM (SOM), and thus is generally enriched in surface soil horizons and in organic soils (Table 50.1). Although Sn compounds are fixed by clay fraction and metal oxides, they are relatively easily phytoavailable. Several soil parameters, such as cation-exchange capacity, pH, SOM, salinity, and kind of minerals, have an impact on Sn fixation, as well as on its uptake by plants.

50.3 WATERS

The mean Sn concentration in the worldwide ocean waters is estimated to be 0.004 μg/L, within the range of 0.003–0.5 μg/L (Table 50.1). Its relatively low content, 0.0005 μg/L, is reported for the North Pacific Ocean (Nozaki 2005). Increased Sn levels are in waters of the Baltic Sea, close to the west coast of Sweden (Szefer 2002). Higher Sn concentrations in seawaters, up to about 3 μg/L, are also given (ATSDR 2002b).

Marine plants are important in the cycling of Sn; its contents in algae and macrophytes vary within 0.5–101 mg/kg. Algae may immobilize Sn from seawaters and stimulate the degradation of toxic methyl-Sn.

Tin is present in waters in divalent and/or tetravalent oxidation states, as well as in organic compounds (OTCs). It is present in waters as neutral or cationic species, which is highly controlled by water pH. Methylated OTCs from river estuaries and their volatile forms are significant Sn sources in the marine environment.

There are various OTCs of Sn, such as monobutylin (MBT), dibutylin (DBT), and tributylin (TBT). The main source of these compounds in waters is ship paints. Variable concentrations of OTCs are present in waters, especially in harbor regions. Their highest concentrations of harbor waters are reported (in ng Sn/L) as follows: MBT, 70, DBT, 283, and TBT, 1000. The quality standard for seawaters at 2 ng TBT/L has been recommended in the United Kingdom. In aquatic environments, all OTCs are easily adsorbed by both suspended OM and particles of bottom sediments. The highest contents are also reported for TBT; their amounts in bottom sediments of harbors may be up to 10,780 μg Sn/kg (Kabata-Pendias and Mukherjee 2007).

All OTCs are easily available to aquatic organisms and affect the growth and reproduction of both phyto- and zooplanktons. They are considered to be the most hazardous known pollutants in the aquatic environment. The amounts of TBT in various fishes differ within the range of <0.13–190 μg Sn/kg; however, their much higher amounts are reported to be in livers of otters, 4–3020 μg Sn/kg (Sudaryanto et al. 2005). At relatively low concentration, <1 ng/L, TBT is harmful to some marine biota. When fish contain Sn 10–55 μg/kg, aquatic birds accumulate TBT, within the range of 42–160 ng/kg. Although the use of OTCs has been restricted, there are still problems with these compounds in aquatic systems.

The median Sn concentration in bottled waters of the EU countries is 0.123 μg/L, and is slightly lower than in tap water, estimated to be 0.185 μg/L (Birke et al. 2010).

50.4 AIR

The concentration of Sn in the atmosphere of remote regions is <70 ng/m^3, whereas it may be up to 5000 ng/m^3 in industrial regions (Table 50.1). Very high level of Sn, up to 10,900 ng/m^3, has been reported for air from the highway tunnel in Hamburg, Germany (Dannecker et al. *vide* Kabata-Pendias and Mukherjee 2007).

The most significant sources of Sn are industrial emission and coal burning plants. The worldwide industrial Sn emission, from various countries, in the 1980 was estimated to be 1.47–10.8 kt (Nriagu and Pacyna 1988). The global emission of OTCs to the atmosphere ranges between 2000 and 6000 kg/yr. Relatively high Sn deposition, at an average of 60 g/ha/yr, is reported for agricultural soils of Sweden (Eriksson 2001a).

There are several sources of OTCs in the atmosphere, which include incineration of organic waste, use of pesticides, and antifouling paints, as well as from glass manufacturing facilities because of the tin used. Apparently, the precipitation of elemental Sn from air is about 10 times higher than that of OTCs.

Tin contents of the moss carpet in Sweden decreased from 0.49 mg/kg in 1975 to 0.09 mg/kg in 2000 (Rühling and Tyler 2004). Moss sampled in Norway, during the period 1990–1995, contained Sn within the range of <0.008–25 mg/kg, depending on the locations (Berg and Steinnes 1997).

50.5 PLANTS

There are no evidences that Sn is essential to plants, and it is considered as toxic to both higher plants and fungi. Its mobile species, mainly Sb^{2+} and Sn^{4+}, are easily available to plants, and then accumulated in roots.

The common range of Sn in food plants is between <0.04 and <0.1 mg/kg, whereas in grass it varies from 0.2 to about 2.0 mg/kg. The worldwide Sn concentrations in wheat grains are reported to be within the range of 5.6–7.9 in mg/kg (Kabata-Pendias 2011). Much lower Sn amounts, 0.01–0.12 mg/kg, are in wheat of Sweden (Eriksson 2001a). Tin contents in forest plants are variable, up to 0.3 mg/kg in *Lactarius blennius* and up to 1.8 mg/kg in *Collybia peronata* (Tyler 2004).

Plants growing in mineralized or contaminated soils accumulate Sn to high levels. Sugar beets from areas around Sn smelter contain this metal up to 1000 mg/kg, whereas various vegetables contain up to 2000 mg/kg. Some bacteria are resistant to Sn excess and may concentrate it up to 7700 mg/kg (Kabata-Pendias 2011).

50.6 HUMANS

Tin content in the human body is 30 mg, of which about 0.4 mg/L is in the blood, 1.4 mg/kg in the bones, and 0.3–2.4 mg/kg in soft tissues (Emsley 2011). Its highest concentrations occurred in the kidney, liver, lung, and bones. In the lungs, Sn appeared to increase with age.

Human exposure to Sn may occur by inhalation, ingestion, or dermal absorption. Dermal absorption is a significant route of occupational exposure for certain organotin compounds. Tin ingested in food is poorly absorbed. At most about 1% of ingested Sn enters the blood. Absorption appears to be related to the chemical form of the element, with Sn^{2+} being 4 times more readily absorbed than Sn^{4+} compounds (2.8% of Sn^{2+} compounds and 0.64% of Sn^{4+} compounds). Absorbed Sn is rapidly excreted, but its small amounts may be retained in the kidney, liver, and bones, as well as in other soft tissues.

Tin does not appear to have any biological role in the body. It affects the metabolism of other metals, such as Cu, Zn, and Fe. Humans chronically exposed to inorganic Sn (e.g., stannic oxide dust or fumes) manifest a form of pneumoconiosis, known as stannosis, which involves mainly the lower respiratory system. Gastrointestinal effects, such as nausea, vomiting, and diarrhea, have been reported, when food is contaminated with inorganic Sn. However, potential target organs for inorganic Sn toxicity have not yet been established (ATSDR 2005b).

Organotin compounds are much more toxic than inorganic forms. Triethyl- and trimethyl-tin (TET and TMT) are dealkylated rapidly to toxic trialkyl derivatives and are metabolized more slowly to less toxic di- and monoderivatives. All of these and other related compounds, especially TBT, are genotoxic. TET interferes with cellular metabolism by uncoupling oxidative phosphorylation, causing mitochondrial damage and resulting in cerebral edema, leading to neurotoxicity

(Reilly 2002). Its compounds in the liver were estimated within the ranges, as follows (in mg/kg fresh weight [FW]): MBT, 0.0003–0.018; DBT, 0.0008–0.066; and TBT, <0.0003–<0.0020 (ATSDR 2005b).

Most natural foods contain Sn in trace amounts, but its concentrations are increased due to the use of organotin pesticides and the storage of liquids in cans. In most unprocessed foods, inorganic and total Sn levels are generally <1 mg/kg (Table 50.2).

The primary sources of Sn are canned goods (Table 50.2). Elevated Sn concentrations can be, as Sn^{2+}, in canned foods due to dissolution of the Sn coating or plate. Tin levels are usually below 25 mg/kg in lacquered food cans, but may exceed 100 mg/kg in unlacquered cans. Its contents in canned foods increase with storage time and temperature, and depend largely on the type and acidity of the food. Acidic foods are more aggressive to the Sn coating in metal cans, and therefore canned acidic foods have its higher contents. Oxidizing agents (nitrates, Fe salts, Cu salts, S) accelerate detinning, whereas Sn salts, sugars, and gelatin reduce the dissolution rate.

According to the World Health Organization (WHO), the maximum limit of Sn in canned food is 250 mg/kg; the Joint FAO/WHO Expert Committee on Food Additives established a provisional tolerable weekly intake (PTWI) for Sn as 14 mg/kg bw

TABLE 50.2
Tin in Foodstuffs (mg/kg FW)

	Content		
Product	a	b	c
Food of Plant Origin			
Bread and miscellaneous cereals	0.02, 0.03	<0.02	–
Green vegetables	0.02	0.003	–
Other vegetables	0.02	0.012	–
Potatoes	0.02	0.007	–
Canned vegetables	44	36.1	<2–11
Fruit products	17	11.1	41–148
Food of Animal Origin			
Carcass meat	0.02	<0.01	–
Canned meat	–	–	<2
Fish	0.44	0.021	–
Canned fish	–	–	<2
Tuna flakes in water	–	–	0.70
Dairy products	–	0.03	<2 (canned)
Canned baby foods	–	–	<2

Sources: [a] Ysart, G. et al., *Food Addit. Contam.*, 16, 391–403, 1999;
[b] Rose, M. et al., *Food Addit. Contam.*, 27, 1380–1404, 2010;
[c] Boutakhrit, K. et al., *Food Addit. Contam.*, 28, 173–179, 2011.
FW, fresh weight.

(WHO 1989). The EU legislation suggests the maximum limit of inorganic Sn as follows: 200 mg/kg in canned foods, 100 mg/L in canned beverages, and 50 mg/kg in canned liquid foods for children (EC 2006).

The average daily Sn intake of an adult in the United States was estimated to be about 4 mg (4 mg from food and 0.003 mg from air). Its intake from drinking water is undetectable (ATSDR 2005b). In 2006, the total diet study (TDS) in the United Kingdom reported that canned vegetables group and fruit product group contributed 65% and 34% of Sn, respectively; their intake was estimated to be 1.8 mg/day. Population dietary exposure to Sn from the TDS in the United Kingdom, during the period 1976–2000, decreased from 4.4 to 1.4 mg/day (Rose et al. 2010).

In a market basket study in Japan, daily intakes of TBT and triphenyltin (TPT) in Japan were estimated to be 6.9 and 5.4 μg, respectively, in 1991, and 6.7 and 1.3 μg, respectively, in 1992. About 95% of the daily intakes of TPT was from fish, mollusk, and crustacean food groups.

Mono- and dimethyltin and mono- and dibutyltin compounds have been detected in drinking water in Canada, where PVC pipes, containing these organotin compounds, are used in the distribution of drinking water (ATSDR 2005b).

51 Titanium [Ti, 22]

51.1 INTRODUCTION

Titanium (Ti), a lustrous-silver transition metal of group 4 in the periodic table of elements, is the ninth-most abundant element in the Earth's crust, within the range of 2.7%–4.6%. Its contents of acidic magmatic rocks (about 3.5%) are higher than those in acidic rocks (about 2.5%). Sedimentary rocks contain Ti at about 4%, and in coal may be accumulated up to 7%. It is classified as a trace element, because its abundance and functions in biotic environments are similar to other trace elements.

Titanium occurs mainly at +4 oxidation state, but also at +2 and +3 states. It shows strong lithophillic and oxyphillic properties. In minerals, it occurs mainly in the tetravalent oxidation state, TiO_2, as a major component of some minerals, such as oxides, titanates, and silicates. The most important, and of an economic value are ilmenite ($FeTiO_3$), rutile (TiO_2), and, to a lesser degree, sphene (titanite, $CaTiOSiO_2$). There are also two minerals of TiO_2, anatase and brookite, but of different crystallographic coordinates.

World (without the United States) Ti production (rounded) in 2010 was 123,000 t, of which 53,000 t was produced in China, 30,000 t in Japan, 27,000 t in Russia, and 15,000 t in Kazakhstan. Titanium is also mined in various mineral forms, for example, ilmenite—5800 t and rutile—580 t (USGS 2011). Therefore, total Ti production, according to some other sources is higher, but not established exactly. Titanium is primarily used for the production of white pigments, applied in various manufactories, such as production of paints, papers, and plastics. It is added to various metallic plates, used mainly in aeronautic industries. Its special use is in alloys that are utilized in orthopedic implants and prosthesis. Titanium forms a variety of sulfides, but only TiS_2 has been used previously as cathodes in some batteries. In cosmetics it is added mainly for the therapy of skin disorders. It is also an additive to toothpaste, and is used in a variety of tablet coatings. However, $TiCl_4$ is highly irritating to the skin, eyes, and mucous membranes of humans.

51.2 SOILS

Titanium contents in soils vary from 100 to 24,000 mg/kg, being the lowest in histosols (Table 51.1), whereas soils of the remote region in Spitzbergen contain Ti within the range of 0.0006–0.011 mg/kg (Gulińska et al. 2003). Levels of Ti in soils from industrial regions near Barcelona (Spain) vary from about 12–3540 mg/kg (Bech et al. 2011). Thus, soils exposed to effluents or emissions from some industries (Ti alloys, Ti paint productions) may be contaminated by Ti; however, this element does not create any environmental problems.

TABLE 51.1
Titanium Contents of Soils, Water, and Air

Environmental Compartment	Range Mean
Soil (mg/kg)	
Light sandy	200–17,000
Medium loamy	5,000–10,000
Heavy loamy	1,000–24,000
Calcerous (calcisols)	400–8,000
Organic (histosols)	100–5,000
Water (μg/L)	
Rain	0.12–1.2
River	0.01–2.3
Ocean and sea	<0.01–9.0
Air (ng/m³)	
Urban/industrial areas	15–>100
Antarctica	0.5–2.5

Sources: Data are given for uncontaminated environments from Kabata-Pendias, A. and Mukherjee, A.B., *Trace Elements from Soil to Human*, Springer, Berlin, Germany, 2007; Reimann, C. and de Caritat, P., *Chemical Elements in the Environment*, Springer, Berlin, Germany, 1998.

Minerals of Ti are very resistant to weathering; therefore they occur, not decomposed, in soils. When Ti-bearing silicates are dissolved, Ti transformed into Ti oxides, which changes slowly into anatase or rutile. It is easily adsorbed by Fe–Mn concretions, whose content may be up to about 40% of its amounts in soils. Under reducing conditions, Fe^{2+} ions that are adsorbed on to the surfaces of Ti minerals, formed after the oxidation to Fe^{3+}, may lead to the formation of pseudorutile ($Fe_2Ti_3O_3$).

There are limited amounts of mobile Ti species, and its concentrations in soil solution are at about 0.03 mg/L. Due to low mobility, Ti contents of the top soil layers are stable, and therefore it is used in studies of soil genesis.

51.3 WATERS

Titanium, a slightly mobile metal, is present in the hydrosphere in relatively small amounts. Its worldwide mean concentration in river waters is estimated to be 0.49 μg/L, within the range between 0.02 and 2.3 μg/L (Gaillardet et al. 2003). Its amounts in some stream waters may be much higher, up to about 925 μg/L, in India (Reimann and de Caritat 1998). In the U.S. rivers, its concentrations vary from 2 to 107 μg/L. Waters of the Baltic Sea contain Ti within the range of 0.08–0.15 μg/L, whereas in waters of the drainage basin its amounts are from 1.2 to 29 μg/L (Szefer 2002). Its mean amount in the North Pacific Ocean is reported to be up to 0.007 μg/L (Nozaki 2005).

Aquatic biota, especially seaweeds, accumulate Ti within the range of 22–60 mg/kg (Szefer 2002). Increased Ti levels in some algae may introduce larger amounts of the metal into the food chain, which results in high concentrations of the metal, up to 400 mg/kg, in the hard tissues of some sea animals, which should be of environmental concern.

Rainwaters collected in Sweden, during 1999, contained Ti at an average value of 0.44 μg/L (Eriksson 2001a).

Median Ti concentrations in bottled water of the European Union (EU) countries are 0.052 μg/L, and is a bit lower than that in tap water, estimated to be 0.087 μg/L (Birke et al. 2010). Ti in drinking waters of the United States is in the range of 0.5–15 μg/L, at a mean value of 2.1 μg/L.

51.4 AIR

Titanium concentration in the atmosphere is mainly of terrestrial origin. However, anthropogenic sources of Ti are also of great importance, which is clearly demonstrated by its highly elevated amounts in air of urban and industrial regions (Table 51.1). Concentration of Ti in air of the United States range from 10 to 100 ng/m^3, but is reported to be elevated in some industrial region up to about 1000 ng/m^3 (Kabata-Pendias and Mukherjee 2007).

Lichens and mosses from Norway contain Ti within the range of 12–310 mg/kg, which is a clear evidence of Ti aerial pollution (Berg and Steinnes 1997).

51.5 PLANTS

There is no clear evidence about the biochemical role of Ti in plants. Possibly, it may have some functions in the nitrogen fixation by symbiotic microorganisms and in photooxidation of nitrogen compounds by higher plants. Thus, although it is not accepted as an essential element to plants, there are some reports on its beneficial effects on the yield of some crops. However, there are also described toxicity symptoms (necrotic and chlorotic spots on leaves) at Ti content of 200 mg/kg in bush bean (Wallace et al. *vide* Kabata-Pendias 2011). It is considered to be relatively unavailable to plants and not readily mobile in them. Concentrations of Ti decrease with age, and may be reduced by one-third in older plants, as compared to plants in early growth.

Contents of Ti in plants seem to be highly controlled by soil properties. The highest Ti contents, up to 100 mg/kg, are in plants grown on loess soils and some weathered *in situ* soils, whereas its lowest amounts (about 1 mg/kg) are in plants grown on histosols. In general, its phytoavailability is low.

Levels of Ti in plants vary considerable within the range of 0.15–80 mg/kg, and in food plants it ranges from about 0.1–7.0 mg/kg. The lowest contents are in cereal grains and fruits, and the highest in lettuce, radish, and corn. Cereal grains of Sweden contain Ti within the range of 0.17–0.25 mg/kg (Eriksson 2001a). Some weeds, especially horsetail and nettle may accumulate more Ti, up to 80 mg/kg.

51.6 HUMANS

Titanium content in human body is estimated to be 700 mg, at concentrations in blood, 50 μg/L; bones, 40 mg/kg; and other tissues, 1–2 mg/kg (Emsley 2011). Titanium does not play a significant role in any body functions and is relatively non-toxic, because the body can tolerate relatively high doses, and it is not accumulated. Titanium halogen intake causes nausea and vomiting, and acidifies the body after its resorption. Corrosion occurs at eye or skin, when it comes in contact with mucous membranes. Breathing of Ti dioxide (TiO_2), in very small particle size, may affect lung disease.

Titanium metal, in surgical implants, is well tolerated by tissues. Its compounds are generally considered to be poorly absorbed, upon ingestion and inhalation. However, detectable amounts of Ti can be found in the blood, brain, and parenchymatous organs in the general population; the highest concentrations are found in the hilar lymph nodes and the lung. Titanium crosses the blood–brain barrier, and is also transported through the placenta into the fetus. It seems to accumulate with age in the lungs, but not in other organs. The biological half-life for Ti in man has been calculated to be about 640 days. It is excreted with urine.

Titanium and its compounds presence in various environmental media, as well as in implants, are not of any health risks (WHO 1982). Small amounts of Ti, occasionally released from implants into adjacent tissues, have not caused any adverse effects. TiO_2 dust, when inhaled, has been classified by the International Agency for Research on Cancer (IARC), as a Group 2B carcinogen, which means that it is possibly carcinogenic to humans (IARC 2013).

Food is the principal source of exposure to Ti for humans. Whole grains, some vegetables and fruits, and common fish meat contained little, or not detectable Ti amount, while milled cereal grains, butter, corn oil, and lettuce may have its elevated levels. High levels of Ti in food, especially in cheese, can arise from the use of TiO_2, as a whitener in the manufacture of several cheeses (WHO 1982). The foods with the highest content of TiO_2 included candies, sweets, and chewing gums, which indicates that children are having the highest exposure to TiO_2. Typical exposure for adults in the United States may be in the order of 1 mg Ti/kg BW/day.

In several high-consumption pharmaceuticals, Ti contents range from below the detection limit (0.0001 mg Ti/kg) to a high level of 0.014 mg Ti/kg. Food-grade TiO_2 (E171) is approximately at 36%, in the particles less than 100 nm, and thus it readily disperses in water as a fairly stable colloids (Weir et al. 2012).

Conventionally, TiO_2 fine particles (FPs) have been considered as a low toxicity material. TiO_2 nanoparticles (NPs) possess different physicochemical properties, compared to their FP. TiO_2 NPs may translocate to systemic organs, from the lung and gastrointestinal tract (GIT), although the rate of translocation appears low.

In the field of nanomedicine, intravenous injection can deliver TiO_2 nanoparticulate carriers directly into the human body. Upon intravenous exposure, TiO_2 NPs can induce pathological lesions of the liver, spleen, kidneys, and brain. Long-term inhalation of TiO_2 (NPs) by rats have resulted in lung tumors (Shi et al. 2013).

Typical American diets provide approximately 300 μg of Ti daily. The 30-day mean total American dietary Ti intakes of two individuals were 370 and 410 μg/day. Daily intake of Ti at about 800 μg was reported from the United Kingdom. Its intake from drinking water is usually very low, probably <5 μg/day. Outside occupational settings, the amount of Ti absorbed via the lungs is of little significance, estimated at <1% of its total intake form food (WHO 1982).

52 Tungsten [W, 74]

52.1 INTRODUCTION

Tungsten (W), known also as wolfram, is a rare metal of group 6 in the periodic table of elements. In the Earth's crust it occurs only as chemical compounds, at an average content of about 1.5 mg/kg. Its contents of acidic magmatic rocks, <2.5 mg/kg, is a bit higher than in alkaline rocks. Coal may contain W at >1 mg/kg, some oil shales may have up to 2.5 mg/kg, and in fly ash it may be accumulated to about 4.5 mg/kg.

The oxidation states of W are from +2 to +6, and most commonly are at the highest value. Geochemical properties of W are similar to those of Mo, thus in several minerals the substitution of W for Mo takes place. Most minerals contain the anionic species of W, namely, $(WO_4)^{2-}$. Common minerals of tungsten are wolframite [$(Fe,Mn)(WO_4)$], scheelite ($CaWO_4$), and sanmartine [$(Zn,Fe)(WO_4)$], of which the first two minerals are its important ores. Under most geochemical conditions, minerals and compounds of W are slightly soluble, but some complex compounds, especially those associated with Mo are likely to be mobilized, under specific conditions.

Global W production (rounded), excluding the United States, in 2010 was 61,000 t, of which 52,000 t was produced in China; 2,500 t in Russia; 1,100 t in Bolivia; and 1,000 t in Austria (USGS 2011). The main use of W is in cemented carbide parts for cutting and wear-resistant materials, primarily in the construction, metalworking, mining, and oil- and gas-drilling industries. It is also consumed to make heavy alloys for applications requiring in high-density electrodes, filaments, wires, and several other electronic components. This metal reveals very high robustness, and in carbide (WC) form has a hardness of that of a diamond. Tungsten is also used for catalysts in automobile exhaust systems, and as pigments in dyes and inks. It is used in bullets, as a replacement for Pb, and in some tools, as a substitute for depleted U.

52.2 SOILS

Tungsten's average content in the worldwide soils is estimated to be 1.7 mg/kg, within the range of <0.17–5.0 mg/kg. The highest mean W contents, about 5.0 mg/kg, is in soils of Europe (FOREGS 2005). However, W concentration, up to 85.5 mg/kg is reported for soils in China (Govindaraju 1994). Presumably, it might be contamination effects. Range of W contents in soils of some countries are given as follows (in mg/kg): 0.6–3.4 in Brazil, 1.0–85.5 in China, 0.9–1.8 in Japan, 0.7–2.7 in Scotland, <1–51 in Slovak Republic, 0.4–2.4 in Sweden, and 0.5–5.0 in the United States. Soils surrounding the W ore-processing plant in Russia contain W within the range of 100–200 mg/kg (Kabata-Pendias 2011).

Tungsten sources in soils are from fly ash and some sludge used for soil amendments. Mean V content of sludge is given as about 8 mg/kg, whereas in slurry from dairy cows its mean amount is 0.082 mg/kg (Eriksson 2001a). Some type of ammunition may also be a source of elevated W contents in soils of certain areas. According to Thomas et al. (2009), the toxicity of W released from discharged shot was assessed against previous studies, which established a 1% toxic threshold for soil organisms.

In soils, W remains mainly in primary compounds and minerals. Some part of W is associated with Fe oxides, and with humic acid and fulvic acid. Mobile species of W, especially in soils at pH > 7, may enter groundwater. In alkaline soils, W is mainly in monomeric forms, whereas in acidic soils it is present in polymeric forms. In soil, W metal oxidizes to the W anion, which can be uptaken by some small organisms, where it may substitute for Mo in several enzymes. Presumably, it plays an essential role in the biology of microbial organisms, and is present as an active site in specific microbial enzymes.

52.3 WATERS

Although most of W minerals are slightly soluble, its global mobility is relatively high (Gaillardet et al. 2003). The median W concentration in seawaters is estimated to be 0.1 μg/L (Reimann and de Caritat 1998), whereas its content of 0.01 μg/L is reported for the North Pacific Ocean (Nozaki 2005). In seawaters it is likely to occur as oxide species, WO_4^{2-}. Stream and river waters contain W within the range of 0.01–0.05 μg/L. Exposure to water-soluble W compounds is a great threat to human and animal health.

Tungsten is readily accumulated in ocean bottom sediments, mainly in polymetallic compounds. In Fe–Mn nodules of the Baltic Sea, W contents vary from 0.4 to 284 mg/kg (Szefer 2002).

In rainwaters collected in 1999 in Sweden, the concentrations ranged from 4.2 to 23 ng/L (Eriksson 2001a).

Median W concentrations in bottled water of the European Union countries are 0.0189 μg/L, and is a bit higher than in tap water, estimated to be 0.0113 μg/L (Birke et al. 2010).

52.4 AIR

The atmosphere above Spitsbergen, Norway, contain W within the range of 0.004–0.08 ng/m^3. Its amounts in air from remote and polluted regions vary from 0.005 to 0.02 ng/m^3, respectively (Reimann and de Caritat 1998). However, the median W concentration in air was estimated to be 2.5 ng/m^3, whereas according to the recent data (ATSDR 2005c), mean W concentration in air is 10 ng/m^3.

52.5 PLANTS

Tungsten is relatively easily available to plants. A lower uptake from acidic soils suggests that W is absorbed in anionic species, WO_4. Its great proportion is accumulated in roots. Plant growing in areas around W ore bodies, may accumulate much higher W amounts, up to about 2500 μg/kg, without toxicity symptoms. Therefore, W contents in plants have been used in geochemical prospecting. Elevated W contents of these plants might be associated also with its atmospheric source.

There is no clear evidence that W may have a biological function in plants. However, it displays competitive inhibition of Mo in the enzyme nitrate reductase, reducing the enzyme's catalytic activity, what resulted in a slower N-fixation by plants. It is a component of two dehydrogenases (group of oxidoreductases) of some bacteria (e.g., *Eubacterium acidaminophilum*). Tungsten also inhibits synthesis of plant hormones, such as abscisic acid, which resulted in a lower resistance of plants to various stresses. Under drought condition this effect may be stronger.

The common range of W in terrestrial plants is established at the range of <1–150 μg/kg. However, higher contents, up to 200 μg/kg, are also reported. Vegetables growing on soil amended with fly ash, contain W within the range from 700 to 3500 μg/kg (Furr et al. *vide* Kabata-Pendias 2011). Mean contents of W in wheat and barley grains from Sweden are 6 and 5 μg/kg, respectively (Eriksson 2011a).

52.6 HUMANS

Tungsten content of human body is estimated to be 20 μg, and its concentrations in blood is 1 μg/L and in bones 0.2 μg/kg (Emsley 2011). Other sources (ATSDR 2005c) with respect to contents for adults are as follows (in μg/kg): <3–5 in kidney, <3–36 in liver, and <3–11 μg/kg in lung.

Tungsten can enter the blood through ingestion, inhalation, or injection, and is predominantly and rapidly excreted in urine, with only a very small proportion being incorporated into the kidneys, liver, spleen, and skeleton. It is deposited in bones, and may be slowly released to the blood, and eliminated mainly in the urine. Insoluble forms of orally administered W are chiefly eliminated in the feces. Under conditions of chronic uptake, it is projected that W levels in soft tissue will take up to 3 years to reach a plateau, but its level in bones will continue to increase throughout life (Voet van der et al. 2007).

Tungsten is irritating to the skin and eyes, on contact. Its inhalation will cause irritation to the lungs and mucus membrane. Irritation to the eyes will cause watering and redness. Reddening, sealing, and itching are characteristics of skin inflammation. Apparently, there is no chronic effects of exposure to W. Prolonged exposure to this metal is not known to aggravate medical conditions. All W compounds should be regarded as highly toxic (Gbaruko and Igwe 2007).

Individuals, who work in manufacturing, fabricating, and reclaiming industries, especially individuals using hard metal materials, or W carbide machining tools, may be exposed to higher levels of W and its compounds. Tungsten has been detected in hair and nail samples of hard metal workers. Pulmonary fibrosis, memory and sensory deficits, and increased mortality due to lung cancer have been attributed to occupational exposure to W dusts. However, W and its compounds are not considered very toxic for humans. Nevertheless, it may play a role in cases of pulmonary fibrosis, related to hard metal lung disease. Recently there is emerging evidence of W toxic health effects. Biomedical research indicates, at least, potential toxicity of W (Witte et al. 2012).

Recent toxicological interest in W is based on its increased use as a component of armor-piercing munitions and a replacement for Pb in other ammunition. Embedded W shrapnel represents a unique source of internal exposure to W. Military installations and areas involved in military combat operations and training may have higher concentrations of W, as a result of the use of military hardware containing this metal (ATSDR 2005c).

Acute toxic reactions were observed from W dissolved in alcoholic drinks, following a tradition of French army artillery recruits, to drink wine or beer which has rinsed a recently fired gun barrel. The symptoms of this acute intoxication were sudden nausea on ingestion, onset of seizures, rapid onset of clouded consciousness leading to coma and encephalopathy, moderate renal failure progressing to acute tubular necrosis with anuria, and hypocalcemia (Voet van der et al. 2007).

There is little information on W in foods. Its concentration is low, and range from 0.005 to 0.05 mg/kg in vegetables, and from <0.0003 to 0.004 mg/kg in rice. Wines contained W from 0.00001 to 0.001 mg/kg (Szefer and Grembecka 2007). Limited available information indicates that levels of W in food are expected to be low.

The total W intake by the general U.S. population cannot be accurately estimated, due to the lack of data regarding W content in food and drinking water (ATSDR 2005c).

52.7 ANIMALS

In animals, W compounds have the potential for causing adverse health effects involving the gastrointestinal tract and central nervous system. These effects have not been observed in humans. Chronic exposure to W compounds may result in permanent lung damage. Soluble forms of W, such as tungstate ions, are more bioavailable to fish and animals than insoluble forms.

Bioavailability of W and its compounds increases in the order of W metal < W carbide < W ions. It may preferentially occupy enzyme sites, normally reserved for the essential element, Mo, because Na tungstate may antagonize the normal metabolic action of Mo, in its role as cofactor for some enzymes, such as xanthine dehydrogenase, sulfite oxidase, aldehyde oxidase, and xanthine oxidase (secreted in animal milk).

Animals fed with W-containing diet have reduced activities of the Mo-containing enzymes, xantine oxidase and sulfite oxidase. Guinea pigs treated orally or intravenously with W suffered from anorexia, colic incoordination of movement, trembling, dyspnea, and weight loss (Gbaruko and Igwe 2007).

53 Vanadium [V, 23]

53.1 INTRODUCTION

Vanadium (V) is a silvery-gray, transition metal of group 5 in the periodic table of elements; its mean abundance in the Earth's crust is within the range of 53–60 mg/kg. It may be concentrated in alkaline mafic rocks up to 250 mg/kg. Also in argillaceous sediments, it may be accumulated up to about 130 mg/kg, whereas in limestones and sandy sediments, its content does not exceed 60 mg/kg. Its mean level in coals is about 60 mg/kg, whereas in crude oil it may reach 1300 mg/kg.

Vanadium oxidation states are from +2 to +5, and the most common is the highest value. It reveals lithophilic properties and occurs mainly in various compounds. Geochemical behavior of V strongly depends on its oxidation state and the acidity of media. Under weathering conditions, it predominates V^{5+}, which shows isomorphic relation to some cations, such as As^{5+} and Mo^{5+}, whereas V^{2+} behaves similarly to Fe^{2+}.

Vanadium occurs commonly as an admixture in about 70 minerals. Its most important minerals are vanadinite [$Pb_3Cl(VO_4)_3$], roscoelite [$KV_2(OH)_2(AlSi_3O_{10})$], and patronite of variable formulas: VS_2, VS_4, and V_2S_2. It is often associated with several minerals, such as pyroxenes, biotite, and magnetite. During weathering conditions, V mobility is controlled by host minerals and Eh–pH conditions. Mobile V species are easily fixed to the surface of clay minerals and P minerals, which resulted in its elevated contents in P fertilizers.

World production of V (rounded), excluding the United States, in 2010 was 56,000 t, of which 23,000 t was produced in China, 18,000 t in South Africa, and 14,000 t in Russia (USGS 2011). Vanadium, in various alloys, is used for a wide variety of purposes. Its great proportion (above 85%) is used in steel industry, for heat-resistant and high-strength alloys, especially for high-speed tools. It is also a significant corrosion inhibitor. It is commonly applied in various aircraft materials. Its nonmetallurgical use is in catalysts for the production of maleic anhydride and sulfuric acid. Additional use of V is in dyes for textiles, glasses and ceramics, and also photographic developers.

Vanadium is a ubiquitous trace element in the environments and is essential for all living organisms, but its increased concentration is harmful.

53.2 SOILS

Worldwide mean V level in soils is estimated to be 130 mg/kg, within the range of 70–320 mg/kg. Its contents in soils are highly related to the parent rocks. The highest V contents, up to 500 mg/kg, are heavy loamy soils derived from mafic rocks and argillaceous sediments (Table 53.1). Its lowest amounts are in organic soils, but in some peat soils it may be accumulated up to about 150 mg/kg. Vanadium

TABLE 53.1
Vanadium Contents of Soils, Water, and Air

Environmental Compartment	Range Mean
Soil (mg/kg)	
Light sandy	10–320
Medium loamy	20–330
Heavy loamy	20–500
Calcerous (Calcisols)	50–400
Organic	10–25
Water (μg/L)	
Rain	0.30–0.95
River	0.1–200
Ocean	0.1–30
Air (ng/m³)	
Urban/industrial areas	5–200
Greenland	0.8–1.5
Antarctica	0.0006–0.002

Sources: Data are given for uncontaminated environments from Kabata-Pendias, A. and Mukherjee, A.B., *Trace Elements from Soil to Human*, Springer, Berlin, Germany, 2007; Reimann, C. and de Caritat, P., *Chemical Elements in the Environment*, Springer, Berlin, Germany, 1998.

contents in soils of various countries are reported as follows (in mg/kg): Japan, 94–250; Russia, 80–100; Sweden, 28–111; and the United States, 36–150 (Kabata-Pendias 2011). The mean V content in organic surface soils of Norway is 12.3 mg/kg (Nygard et al. 2012), and in agricultural soil of Sweden it is 69 mg/kg (Eriksson 2001a). Warning V levels in soils of the EU countries are given from 100 to 340 mg/kg (Cappuyns and Slabbinck 2012), and the most common is 150 mg/kg. Soils of the remote region in Spitzbergen contain V within the range of 0.04–0.78 mg/kg (Gulińska et al. 2003).

Vanadium in soils is associated mainly with Fe hydroxides, clay fraction, and soluble organic matter. Especially clay minerals control its mobility. Vanadium is relatively mobile in soils, and thus its leaching from soil surface horizons into lower soil layers is observed. The most common in soils is vanadyl cation, VO^{2+}, which predominates in acidic soils, and is likely to be mobilized as complexes with organic acids. Its anionic species, VO_3^-, $V_4O_4^{2-}$, HV_4^{2-}, and $H_2VO_4^-$, occur mainly in solutions of neutral and alkaline soils (Kabata-Pendias and Sadurski 2004). Anionic V species are easily adsorbed by clay minerals, especially in association with Fe cations. This explains V concentration in various concretions and nodules formed in soils: 400 mg/kg in Fe-rich and 440 mg/kg in Mn-rich nodules. There is a competition between V and P for the same retention sites on surfaces of soil particles, which resulted in low adsorption of V at a high P content. Vanadium may replace Mo and Fe in nitrogenase, which play an important function in the N fixation by *Azotobacter*.

There are not many reports on V contamination of soils, but it is most likely that some industrial processing (ore smelters, cement and P-rock plants), as well as burning of coal and oil, increase its deposition on soils. Especially serious sources are combustion of fuel oils and deposition of fly ash. The highest V contaminations are reported for soils polluted by fly ash (up to 429 mg/kg) and for soils near a graphite industry (up to 840 mg/kg) (Kabata-Pendias 2011).

53.3 WATERS

The average concentration of V in seawaters is estimated to be 2.5 μg/L, within the range of 1–3 μg/L. Waters of the North Pacific contain V at an average value of 2 μg/L (Nozaki 2005). However, its higher concentration, up to 30 μg/L, is also reported (ATSDR 2012d).

Vanadium concentrations in river waters range from 0.02 to 5.8 μg/L, at the world mean of 0.71 μg/L (Gaillardet et al. 2003). Waters of the Nordic lakes from some countries contain V as follows (mean values, μg/L): Finland, 3.1; Norway, 2.2; and Sweden, 2.5 (Skjelkvale et al. 2001).

Municipal waters of the United States contain V within the range of 1–6 μg/L, whereas in waters of some U ore region its content is elevated up to 70 μg/L (ATSDR 2012d). Waters of rivers from some industrial regions of Poland contain V from 4 to 19 μg/L (Kabata-Pendias and Pendias 1999).

The highest V concentrations in surface waters, up to about 200 μg/L, were recorded in the vicinity of metallurgical plants or downstream of large cities (WHO 1988). Its anthropogenic sources account for only a small percentage of the dissolved V, reaching the oceans (IARC 2006).

Vanadium in waters occurs in several species, such as VO_2, V_2O_5, VO^{2+}, and $H_2VO_4^-$. However, V is not very mobile in aquatic environments and is rapidly fixed in bottom sediments. Thus, V contents of sediments can be an indicator for the pollution of waters. Sediments of some Baltic Sea coast regions contain V from 53 to 153 mg/kg (Szefer 2002).

Both zoo- and phytoplanktons accumulate V, and its content may reach 80 mg/kg (Szefer 2002). It stimulates photosynthesis in algae. Several species of algae produce V-bromoperoxidase (V-Br-peroxidase) that controls the behavior of Br in waters. Vanadium is involved, as a component of the enzyme (V-Br-peroxidase), in the process of reduction of hydrogen peroxide due to the alteration from V^{3+} to V^{5+}.

The median V concentration in bottled waters of the EU countries is 0.15 μg/L and is quite similar to its mean level in tap water, estimated to be 0.17 μg/L (Birke et al. 2010). According to the ATSDR (2012d), the average V concentrations in tap water are approximately 1 μg/L.

53.4 AIR

Vanadium concentrations in the atmosphere vary broadly from <0.002 to 200 ng/m^3 in industrial/urban areas (Table 53.1). Air from Spitzbergen contains V within the range of 0.4–2 ng/m^3, and from the remote regions of the United Kingdom, it varies from 5 to 10 ng/m^3 (Cawse *vide* Kabata-Pendias and Mukherjee 2007). Air in rural regions of the United States contained V, during the period 1965–1969, from <1 to 64 ng/m^3 (ATSDR 2012d).

Vanadium input to the atmosphere from natural sources (soil dust, marine aerosols) is estimated to be 65 kt/yr, and about the same amounts of V enter the atmosphere from industrial sources (Anke 2004b).

Increased V concentrations in air during cold seasons, compared with warm seasons, indicate its sources from burning fossils fuels, coal, wood, and solid wastes (ATSDR 2012d). Large amount of V-rich atmospheric particles and V dissolved in rain are its significant sources in soils. The wet deposition of V in Sweden is estimated to be 5.7 g/ha/yr (Eriksson 2001a).

Mosses collected in Norway, during 1990–1995, contained V within <10–16,000 μg/kg, at an average value of 2,800 μg/kg, with the highest contents in samples from the vicinity of the peat-fired plant (Berg and Steinnes 1997).

53.5 PLANTS

Phytoavailability of V is a function of its concentration in soil solution. Plants take up V relatively easily, especially from acidic soils. This indicates that VO^{2+} species, occurring under acidic conditions, are more easily absorbed by roots than V_3^- and HV_4^{2-} species, which predominate in neutral and alkaline soils. In general, cationic V species are more easily available to plants than anionic species. However, both species may be biotransformed and chelated by some compounds, and thus become more phytoavailable.

Vanadium is not yet considered to be an essential element for higher plants. However, there are observations that vanadate (VO_4^{3-}) and vanadyl (VO^{2+}) ions form various complexes, which have both stimulating and inhibiting impacts on several enzymes. The biotransformation of vanadate to vanadyl species may occur during uptake by plants (Selim 2012). It is a specific catalyst that may substitute Mo function in the N_2 fixation by some microorganisms, especially rhizobium bacteria.

The contents of V in food plants vary broadly and are significantly elevated in plants from V-contaminated regions. The highest V contents in the aboveground vegetables are reported to be in spinach, 533–840 μg/kg, and in lettuce, 280–710 μg/kg. The elevated V contents, 50–2000 μg/kg, are also noticed in wild mushrooms (ATSDR 2012d). The mean V contents of cereal grains are (in μg/kg) as follows: wheat, 9; winter wheat, 2.2; and barley, 3.4 (Eriksson 2001a).

There is no much evidence for the V toxicity to plants, especially under field conditions. Its increased levels, up to 3000 μg/L nutrient solution, under greenhouse conditions, reduce the growth of collard (*Brassica aleracea*). The symptoms of V toxicity (chlorosis and dwarfing) may appear at its content of 2000 μg/kg in some plants. The elevated V contents in plants, especially V^{5+} species, may inhibit the function of several enzymes, especially in some legumes.

Plants from industrial regions may contain very high V amounts. Vegetables grown in the vicinity of a thermal power station contained V within the ranges of 1900–4900 μg/kg (cabbage) and 800–1700 μg/kg (tomato) (Belsare *vide* Kabata-Pendias 2011).

53.6 HUMANS

The total amount of V in the human body is estimated to be about 20 mg, of which 0.02–0.09 μg/L is in the blood, 3 μg/kg in the bone, and 20 mg/kg in the tissues (Emsley 2011).

Absorption of V from the gastrointestinal tract is poor, not exceeding 2% (WHO 1988). Its highest concentrations occur in the liver, kidney, and lung. However, it is mainly stored in fat and serum lipids. No biochemical function of V has been yet defined in humans. Presumable, V plays a role in metabolism of carbohydrates, and also in metabolism of cholesterol and blood lipids. It controls the biosynthesis of cystine and cholesterol, depression and stimulation of phospholipid synthesis, and, at higher concentrations, inhibition of serotonin oxidation. In diabetics, V supplement may have a positive effect in regulating blood glucose levels. Possibly, V prevents heart attack, cataract development, and impaired antioxidant status (ATSDR 2012d). The potential role of V in humans is building material of bones and teeth, and stimulating muscle development (Badmaev et al. 1999). However, gastrointestinal effects (diarrhea, cramps, nausea, and vomiting) may occur in non-insulin-dependent diabetic patients, administered vanadyl sulfate or sodium metavanadate capsules, as a supplement to their diabetes treatment (ATSDR 2012d).

In general, the toxicity of V compounds is low. However, little information is available regarding the mechanism of V toxicity *in vivo*. Its pentavalent compounds and organic forms are safer, more absorbable, and able to deliver a therapeutic effect up to 50% greater than V inorganic forms.

Workers exposed to V peroxide dust suffer severe eyes, nose, and throat irritation, which have been observed especially in miners of V ores and workers in some other industries (Halliwell and Gutteridge 2007). It may have a number of effects on human health, when its uptake is too high. When V uptake takes places through air, it can cause bronchitis and pneumonia (WHO 1988).

Vanadium pentoxide has important effects on a broad variety of cellular processes. It stimulates cell differentiation, it causes cell and DNA injury via generation of reactive oxygen species, and it alters gene expression. It can pass the blood–placenta barrier. Data on genetic effects in humans exposed to vanadium pentoxide are scarce, and it is classified as possibly carcinogenic to humans (Group 2B) (IARC 2006, 2013).

Levels of V that are naturally present in food and water (Table 53.2) are not harmful. Vanadium in food is mainly ingested as VO^{2+} (vanadyl, V^{4+}), or HVO_4^{2-} (vanadate, V^{5+}). Organic V compounds are used in the treatment of diabetes and cancer. Its intakes, in supplements for individuals with diabetes, are 30–150 mg/day for vanadyl sulfate (9–47 mg V/day) and 125 mg/day for sodium metavanadate (52 mg V/day) (ATSDR 2012d).

TABLE 53.2
Vanadium in Foodstuffs (mg/kg FW)

Product	Range/Mean
Food of Plant Origin	
Cereals and cereal products	0.046 (0.010–0.316)
Bread[a]	0.01
Rice, different types[b]	0.0021–0.0061
Fruits and vegetables	0.024 (0.010–0.310)
Sweeteners, honey, and confectionery	0.057 (0.010–0.238)
Curry, powder[c]	1.1
Beer, canned and bottled[d]	0.026 and 0.039
Food of Animal Origin	
Meat and offal	0.040 (0.010–0.293)
Fish and fish products	0.010–1.310
Shellfish	0.234
Freshwater fish, varied[e]	0.02–0.36
Eggs and eggs products	0.023
Dairy products	0.009–0.53
Fat and oil	0.010–0.206
Milk	0.021
Milk[f]	0.016

Sources: Millour, S. et al., *J. Food Compos. Anal.*, 25, 108–129, 2012. Unless otherwise stated.

[a] WHO, *Environment Health Criteria*, WHO, Geneva, Switzerland, 1988.

[b] Batista, B.L. et al., *Food Addit. Contam.*, 3, 253–262, 2010.

[c] Gonzalvez, A. et al., *Food Addit. Contam.*, 1, 114–121, 2008.

[d] Rajkowska et al. (2009).

[e] Lidwin-Kaźmierkiewicz, M. et al., *Pol. J. Food Nutr. Sci.*, 59, 219–234, 2009.

[f] Gabryszuk, M. et al., *J. Elem.* 15, 259–267, 2010.

FW, fresh weight.

Daily dietary V intake, in the general population, has been estimated to be 0.01–0.03 mg per person per day, although it can reach 0.07 mg per day in some countries (IARC 2006). Dietary V intake ranges from 0.09 to 0.34 μg/kg/day in adults. A daily intake of V by adults is approximately 0.002 mg from tap water (average 0.001 mg/L). Individuals exposed to cigarette smoke may be exposed to higher than background levels of V. Approximately 0.0004 mg of V is released in the smoke of one cigarette (ATSDR 2012d).

54 Yttrium [Y, 39]

54.1 INTRODUCTION

Yttrium (Y), a silvery transition metal of group 3 in the periodic table of elements, is chemically similar to lanthanides, and often classified as a *rare earth element* (REE). It never occurs in nature as a free element. Its abundance in the Earth's crust is within the range of 20–33 mg/kg. Its levels in mafic rocks are lower (0.5–20 mg/kg) than in acidic rocks (20–33 mg/kg) and in sandstone (15–25 mg/kg). It may be accumulated in some coal, up to about 170 mg/kg. The mean Y content of fly ash is about 44 mg/kg.

Yttrium occurs at +3 oxidation state and is often associated with U ores and REE deposition, as well as with mozaite veins and pegmatites. It combines easily with various minerals, such as oxides, carbonates, silicates, and phosphates. Its most common minerals are xenotime (YPO_4) and fergusonite ($YNbO_4$).

World mine Y production (rounded), excluding the United States, in 2010 was 8900 t, of which 8800 t was produced in China, 55 t in India, and 15 t in Brazil (USGS 2011). Recycling are small quantities of Y from laser crystals and some synthetic products. Yttrium (mainly as Y_2O_3) is used in luminescence and semiconductor materials applied in various electronic devices, including lasers. It is also used in ceramic and glass industries, and in various catalysts (e.g., in production of plastics).

Yttrium has no known biological role, and exposure to its compounds can cause lung disease in humans.

54.2 SOILS

Worldwide Y content of soils is estimated to be about 15 mg/kg, within the range between 2 and 100 mg/kg. The average Y contents in soils of various countries are presented as follows (in mg/kg): China, 11–39; Japan, 5–24; California in the United States, 17–27; Alaska in the United States, 4–100; and Poland, 7–19 (Kabata-Pendias 2011). Sandy soils contain less Y than loamy ones.

Yttrium contents in materials used for soil amendments are as follows (mean, in mg/kg): NPK fertilizers, 14; P fertilizers, 114; sewage sludge, 11; pig manure, 2; and fly ash, 44.

54.3 WATERS

The mean concentrations of Y in worldwide seawaters are given as 0.009 and 0.013 μg/L (Reimann and de Caritat 1998). Its average content of water of the North Pacific is estimated to be 0.017 μg/L (Nozaki 2005). River waters of Europe contain Y within the range of 0.05–1.49 μg/L and of the north United States, 0.028–0.217 μg/L (Gaillardet et al. 2003). However, its concentration in some stream waters may be elevated up to the range of 14–98 μg/L (Reimann and de Caritat 1998).

Maximum Y contents of snow from the Kola Peninsula range from 0.01 in remote regions to 0.7 μg/L in polluted regions (Reimann and de Caritat 1998). Its concentrations in rainwater, collected in Sweden during 1998, varied from 0.003 to 0.033 μg/L, at an average value of 0.01 (Eriksson 2001a).

Yttrium is adsorbed by bottom sediments, in relatively high amounts, with the maximum values of 485, 731, and 2005 mg/kg in Germany, Scotland, and Austria, respectively (Reimann and de Caritat 1998). The top layers of bottom sediments of Swedish rivers, at the Baltic catchments, contain Y within the range of 13–23 mg/kg (Szefer 2002).

The median Y concentration in bottled waters of the EU countries is 0.014 μg/L and is similar to its mean level in tap waters, estimated to be 0.01 μg/L (Birke et al. 2010). Drinking waters contain Y usually within the range of 1–30 μg/L, with the mean value of 5 μg/L.

54.4 AIR

Atmospheric dust contains Y within the range of 0.2–2 mg/kg. Its maximum concentration, of 6.0 ng/m^3, is reported for air of industrial areas (Reimann and de Caritat 1998).

Mosses collected in Sweden, during the period 1990–1995, contained fairly similar amounts of Y from 0.04 to 2.3 mg/kg, which clearly indicates a small variation in its concentrations in air at that time (Berg and Steinnes 1997). Similar Y contents (up to 2.0 mg/kg) are reported for mosses collected earlier (1970) in Norway (Erämetsä and Viinanen *vide* Kabata-Pendias 2011). Increased Y levels in mosses and lichens, compared with higher plants, clearly indicate that its main sources are atmospheric deposition. Yttrium wet deposition in Sweden has been estimated to be 1 g/ha/yr (Eriksson 2001a).

Yttrium contents of moss (*Hylocomium splendens*) from the mountains in Poland, 0.11–0.18 mg/kg (mean 0.14 mg/kg), are slightly lower than those of the same moss species from Alaska, 0.06–0.73 mg/kg (mean 0.23 mg/kg). This may suggest relatively similar Y atmospheric deposits in both regions (Migaszewski et al. 2009).

54.5 PLANTS

Yttrium contents of higher plants vary considerably, mainly from 0.01 to 3.5 mg/kg, depending on soil and climatic factors. The reference Y contents of plants are proposed as 0.02 mg/kg (Markert and Lieth 1987). However, vegetables may contain Y within the range of 20–100 mg/kg, depending on plants and growth media. Its contents of cereal grains vary from 0.8 to 1.3 mg/kg (Eriksson 2001a).

54.6 HUMANS

Yttrium content in the human body is estimated to be 0.5 mg (Emsley 2011). It may bind weakly to various organic ligands and may compete for the binding sites of some other elements. It is toxic in excess amounts.

Mechanisms of its toxicity may include enzyme inhibition, indirect effects (i.e., binding to cofactors, vitamins, and substrates), and substitution for essential metals, which result in its imbalance (Horovitz 2000). Water-soluble compounds of Y are considered mildly toxic, whereas its insoluble compounds are nontoxic (Emsley 2011).

^{90}Yttrium is one of the most useful radioisotopes in the development of antibody-based radioimmunotherapy. The evaluation of the pharmacokinetic profile for ^{90}Y radiopharmaceuticals is usually performed by radiochemical methods (Ciavardelli et al. 2007). The radioactive isotope ^{90}Y is used in drugs for the treatment of various cancers, including leukemia, lymphoma, and bone cancer (Emsley 2011). It can also be a threat to the liver, when it accumulates in the human body. It is mostly dangerous in the working environment, due to the fact that damps and gasses can be inhaled with air. This can cause lung embolisms, especially during long-term exposure. Yttrium can also cause cancer in humans, especially lung cancer, when it is inhaled. However, it is not yet classified by the International Agency for Research on Cancer.

The mean quantity of Y ingested by preschool children in Amherst, Massachusetts, was 2.09 μg/day (Stanek et al. 1988). Its ingestion with food has been estimated to be ranged from 3 to 5 μg/day in Japanese adult males (Shiraishi et al. *vide* Stanek et al. 1988).

55 Zinc [Zn, 30]

55.1 INTRODUCTION

Zinc (Zn) is a metal of group 15 in the periodic table of elements, which also contains Cd and Hg. Its abundance in the Earth's crust is about 75 mg/kg, within the given range of 50–80 mg/kg. Its highest contents are in mafic rocks (<120 mg/kg), and it may be also concentrated, up to about 120 mg/kg, in some argillaceous sedimentary rocks. The lowest Zn amounts, 10–25 mg/kg, are in calcareous rocks. In coal its amounts vary highly, from about 50 to 19,000 mg/kg, and in fly ash it is usually around 200 mg/kg. Crude oils may contain Zn up to about 90 mg/kg.

Zinc occurs mainly at +2 oxidation state. It reveals chalcophilic affinity, and is often associated with carbonates, organic compounds, and clay minerals. Common Zn minerals are sphalerite (αZnS), wurzite (βZnS), zincite (ZnO), smithonite ($ZnCO_3$), willemite (Zn_2SiO_4), and hemimorphite [$Zn_2Si_2O_7(OH)_2 \cdot H_2O$]. All these minerals contain about 50% of Zn, present in the geosphere.

World Zn production from mines (rounded) in 2010 was 12,000 kt, of which 3,500 t was produced in China, 1,520 t in Peru, 1,450 t in Australia, 750 t in India, and 720 t in the United States (USGS 2011). Of the total Zn consumed, about 55% is used in galvanizing, 21% in Zn-based alloys, 16% in brass and bronze, and 8% in other uses, including chemical, paint, rubber, and plastic industries. It is applied in some batteries, automotive equipments, pipes, and various household devices. Some Zn compounds are also used in agriculture. Different Zn compounds have dental and medical applications.

55.2 SOILS

Average Zn contents of worldwide soils range from about 30 to 100 mg/kg (Table 55.1). Its amounts are, in general, closely associated with soil texture, and usually the lowest are in light sandy soils. It may be elevated in some calcareous and organic soils. However, mean Zn content in organic surface soils of Norway is 60.8 mg/kg, and is a bit lower than the mean value (62.4 mg/kg) for all investigated soils (Nygard et al. 2012).

The ranges of mean Zn contents in soils of various countries are as follows (in mg/kg) Brazil, 4–20; Japan, 60100; Poland, 10–5800; South Africa, 12–115; Slovak Republic, 45–14,925; Sweden, 6–152; the United States, 3–265; and Vietnam, 22–237. Its mean amounts in soils of uncontaminated areas of the EU countries vary within the range of about 7–90 mg/kg (Kabata-Pendias 2011). The elevated Zn contents are in soils from industrial regions. Soils of the remote region in Spitzbergen contain Zn from 0.5 to 12.7 mg/kg (Gulińska et al. 2003).

TABLE 55.1
Zinc Contents of Soils, Water, and Air

Environmental Compartment	Range
Soil (mg/kg)	
Light sandy	30–50
Medium loamy	35–65
Heavy loamy	35–75
Calcareous (calcisols)	50–100
Organic	50–100
Water (µg/L)	
Rain	1.2–6.6
River (EU countries)	<5–40
Ocean	0.5–5.0
Air (ng/m³)	
Urban/industrial areas	550–16,000
Village regions	10–200
Antarctica	0.002–0.05

Sources: Data are given for uncontaminated environments from Kabata-Pendias, A. and Mukherjee, A.B., *Trace Elements from Soil to Human*, Springer, Berlin, Germany, 2007; Reimann, C. and de Caritat, P., *Chemical Elements in the Environment*, Springer, Berlin, Germany, 1998.

The criteria for contaminated land (Dutch List 2013), following Zn concentrations in soils and groundwaters, are established (in mg/kg and µg/L) as follows: uncontaminated, 200 and 50; medium contaminated, 500 and 200; and heavily contaminated, 3000 and 800.

Agricultural practices are known to increase Zn contents of surface soils. However, there are observed rather small variations in Zn contents, from 57 to 82 mg/kg, between cereals and vegetables fields. Zinc balance in surface soils of various ecosystems shows that its atmospheric input exceeds its output due to both leaching and production of biomass. Only in unpolluted forest regions of Sweden Zn discharge with water flux higher than its atmospheric input (Tyler 2005).

Zinc status in soils depends highly on parent materials, soil formation processes, soluble organic matter (SOM) contents and clay fractions, especially when are composed of vermiculite and gibbsite. Both clay fraction and SOM reveal a great sorption capacity for Zn, and thus have an impact on its accumulation in surface soil horizons. Zinc fixation is relatively slow and highly controlled by soil pH and forms of metal added. The amounts of fixed Zn increase with time. However, under increased temperature Zn became more mobile (Li et al. 2008).

The most common and mobile Zn species in soils form simple and complex ions that are present mainly in soil solutions, such as $ZnHCO_3^+$ and $Zn(OH)_3^-$ (Kabata-Pendias and Sadurski 2004). Complex Zn compounds are also likely to be formed in soil solution: $ZnFe_2O_4$, $Zn(PO_4)_2{\cdot}4H_2O$, $Zn(PO_4)_2$, and $ZnCO_3$ (Barsova and Mozutova 2012). Depending on the techniques used for obtaining soil solution, Zn concentrations vary between about 20 and 570 μg/L. In very acidic soils (pH < 4), Zn concentrations average 7,137 μg/L, whereas in contaminated soils its contents may reach 17,000 mg/L (Kabata-Pendias 2011).

Although Zn is very mobile in most soils, clay fractions, SOM, and P compounds are capable of holding Zn quite strongly, especially in neutral and alkaline soils (Kumpiene et al. 2008). SOM is known to be capable of bonding Zn in relatively stable forms, therefore its accumulation in organic soil horizons and in some peat is observed. However, the stability of Zn–organic matter (OM) compounds are relatively low. The addition of sewages sludge to soil increases the contents of two Zn species, mobile and exchangeable. In general, organic complexes of Zn increase its uptake by roots. Carbonates added to contaminated chermozems increase Zn forms bound to Ca compounds and significantly decrease its amounts adsorbed on Fe and Mn hydroxides (Minkina et al. 2010).

Oxides and hydroxides of Al, Fe, and Mn also play important functions in Zn binding, in some soils. Especially goethite, αFeOOH, reveals a great affinity for Zn sorption. Microorganisms may partly reduce this process, due to the accumulation of Zn in living cells (Perelomov and Kandeler 2006). Also dissolved OM is an important factor affecting Zn mobility in soils, in particular, soils with alkaline pH range of 7–7.5 (Wong et al. *vide* Kabata-Pendias 2011). At higher pH values, Zn organic complexes may also account for its solubility, whereas in acidic (mainly sandy) soils SOM seems to be the most important soil component in binding this metal. The most important factors in controlling Zn mobility in most soils are clay minerals, hydrous oxides, and pH. However, in some soils, especially with high OM contents (e.g., peaty soils), Zn organic species are highly responsible for its behavior. Apparently, there are two different mechanisms of Zn sorption: in acidic media it is associated with cation exchange sites, whereas in alkaline media it is mainly chemisorption, highly influenced by organic ligands.

In acidic sandy soils, Zn is always very mobile and is often easily leached down the soil profiles. The immobilization of Zn occurs in soils rich in Ca, P, and S compounds, as well as in clay minerals, especially allophane, imogolite, and montmorillonite. Zinc deficiencies, especially its phytoavailable species, result, most often, from management practices used during crop production (e.g., overliming, P fertilization, and OM amendment).

Zinc contamination of soils is relatively common. Anthropogenic Zn sources are related to several industrial processes, and agricultural practices. Soils around nonferric smelters contain Zn up to about 1115 mg/kg (ATSDR 2005d). Soils surrounding the Zn smelter, in Poland, contain this metal within the range of 202–4832 mg/kg (Diatta et al. *vide* Kabata-Pendias 2011). Along high traffic roads, Zn content may be up to about 900 mg/kg (Niesiobędzka 2012).

Agricultural Zn sources are mainly from various fertilizers, which contain its variable amounts, as follows (in m/kg): triple superphosphate, 142–624; NPK fertilizers, 60–200; cow slurry, 580; pig slurry, 920; poultry slurry, 495; and sewage sludge (of EU countries), 131–1670 (Kabata-Pendias 2011).

In soils heavily contaminated with Zn, the formation of Zn pyromorphite [$Zn_5(PO_4)_3OH$] is observed mainly at the surface of grass roots, which suggests the impact of the rhizosphere on this process. Some soil microorganisms (*Rhizopus arrhizus*) accumulate Zn up to 28,000 mg/kg, at the pH range of 4–7.5 (Perelomov et al. 2013). The elevated Zn amounts may be toxic to some microorganisms, especially in acidic soils. Zinc is relatively easily taken up by earthworms, and the content of this metal, within the range of 114–369 mg/kg, depends on various soil amendments (Siebielec et al. 2013). In Zn-contaminated soils, earthworms may contain this metal up to 2000 mg/kg (Kabata-Pendias 2011).

The fate of Zn from various sources (e.g., atmospheric deposition, fertilizers, pesticides, sewage sludge, ashes) differs, depending on its chemical species and their affinity for various soil components. The contents of Zn in crop-farm soils increase (in g/ha/yr) from 115 in Germany to 838 in France (Eckel et al. 2005). Sewage sludges applied to agricultural land of EU countries contain Zn, at average values, from 606 to 1628 mg/kg (ICON 2001). Long-term irrigation (above 100 years) with municipal wastewaters influences Zn speciation in soils, increasing its fraction bound to amorphous Fe oxides (Kabala et al. 2011). Decontamination of Zn from wastewaters is proposed by electrochemical technology (Chen 2013).

Half life-time of Zn as a pollutant in soils may be quite long and has been estimated for about 70–80 years (Kitagishi and Yamane *vide* Kabata-Pendias 2011). Amelioration of Zn-contaminated soils is commonly based on controlling its bioavailability by application of lime or OM, or both. Addition of a brown coal preparation (composed of coal, fly ash, and peat) decreases Zn phytoavailability (Kwiatkowska-Malina and Maciejewska 2013). However, various organic Zn complexes may also increase its uptake by roots (Degryse et al. 2006).

55.3 WATERS

The median Zn concentrations in ocean and seawaters are estimated within the range of 0.5–5.0 µg/L. The mean Zn amount in the North Pacific is 0.35 µg/L (Nozaki 2005), whereas in the Baltic Sea it ranges from 1.4 to 5.4 (Szefer 2002).

The worldwide mean Zn content of river waters is calculated for 0.6 µg/L, and its riverine flux for 23 kt/yr (Gaillardet et al. 2003). In river waters its amounts may be quite elevated, and range (in EU rivers) from <5 to 40 µg/L (Table 55.1). Terrestrial waters are often contaminated by domestic wastewaters, in which Zn concentrations range between 0.1 and 1.0 mg/L, but may be elevated up to 35 mg/L in the United Kingdom, and up to 50 in France (ICON 2001).

Zinc in aquatic media dominates at divalent state, Zn^{2+}. However, its speciation varies with pH. At pH between 4 and 6, it exists in freshwater, mainly as aqua ion (hydrated ion), $Zn(H_2O)_6^{2+}$; at pH about 7 it dominates free cation, Zn^{2+}; and at pH 9 it occurs mainly as monohydroxide ion, $HOZn^+$.

In lakes, rivers, and estuaries, Zn is bound by hydroxides, clay minerals, and other sediment materials. Fine granulometric fraction of sediments in polluted rivers may contain elevated amounts of Zn, for example, up to around 355 mg/kg (Bojakowska *vide* Kabata-Pendias and Mukherjee 2007). Much higher Zn accumulation, up to 2,000 and 14,000 mg/kg, in sediments of selected regions of Vistula and Odra rivers (Poland), respectively, are also reported (Helios-Rybicka et al. *vide* Kabata-Pendias and Mukherjee 2007).

Zinc contents in bottom sediments of streams in National Park, Montgomery (Pennsylvania) were, in 1995, within the range of 80–5400 mg/kg (Reif and Sloto 1997).

Sediments of Sergipe river estuary (Brazil) contain Zn from 7.4 to 89.7 mg/kg (Garcia et al. 2011). The Zn concentration in the wetland sediments from the effluent at the Savannah River Site (Aiken, SC) is 602 mg/kg, and its highest amounts, 425.9 mg/kg, are fixed in SOM fraction. High amounts of Zn (114.4 mg/kg) in these sediments occur as amorphous oxides (Knox et al. 2006). Zinc content of surface-bottom sediments of harbor in Klaipeda (Lithuania) depends also on granulometric composition and is (in mg/kg, average and maximum, respectively): in sand, 28.6 and 79.8, and in mud, 40.5 and 185.2 (Galkus et al. 2012).

The assessment limits for Zn in sediments are established as follows (in mg/kg): effect range low—150, effect range median—410, probable effect level (PEL)—315, severe effect level—820, and toxic effect threshold—540 (EPA 2000, 2013). The Environment Canada sediment-quality guidelines (USGS 2001) gave the following values for Zn in the lake-bottom sediments (in mg/kg): threshold effect level—123, PEL—315, threshold effect concentration—121, and probable effect concentration—459.

Zinc may enter into river systems from numerous sources, such as mine drainage, industrial and municipal wastes, urban runoff, and soil erosion waters. The largest, up to about 30% Zn discharge to aquatic environments in the EU countries, is from the manufacturing of basic industrial chemicals. The estimated total Zn load (in 1990) into the Baltic Sea was 168 t/yr, of which about 80% was from anthropogenic sources (Matschullat 1997).

The median Zn concentration in bottled waters of the EU countries is 1.2 μg/L and is much lower than in tap waters, estimated to be 23.5 μg/L (Birke et al. 2010). The guide value for Zn in the United States is 3 mg/L, and a maximum level is 5 mg/L (EPA 1999). The World Health Organization (WHO) has not yet established a guideline value for Zn in drinking water, because it is not of health concern. However, drinking water containing Zn at levels >3 mg/L may not be acceptable to consumers (WHO 2011a).

Zinc contents of aquatic organisms vary greatly and are influenced mainly by sites of sampling. Fish of the Baltic Sea contain Zn within the range of 1.4–48 mg/kg fresh weight (FW), whereas soft tissues of mussels contain from 50 to 600 mg/kg (Szefer 2002).

Zinc in waters is not very toxic to the biota; however, concentrations above 240 μg/L may have adverse effects on some sensitive organisms, for example, salmons. The limit for safe Zn concentration to aquatic biota is established to be 180–570 μg/L, for soft and very hard water, respectively (Peganova and Eder *vide* Kabata-Pendias and Mukherjee 2007).

Rainwaters collected in Sweden, during 1999, contain Zn at an average value of 11 μg/L, within the range of 1.6–24 μg/L. Its wet deposition has been estimated to be 110 g/ha/yr (Eriksson 2001a).

55.4 AIR

General levels of Zn in the atmosphere are relatively fairly constant. Its worldwide median concentrations are estimated to be 0.04 and 900 ng/m^3, for remote and polluted regions, respectively. However, air of some industrial areas may contain up to 16,000 ng/m^3 (Table 55.1). Aerosol samples from over the North Sea contain Zn within the range of 3–320 ng/m^3, at the mean value of 67 ng/m^3 (Injuk et al. *vide* Kabata-Pendias and Mukherjee 2007).

Zinc enters the atmosphere in vapor forms, as well as particles, from industrial processes, incinerations of waste, cement plants, and fuel-fired power plants. The coal combustion for heat and electricity production is considered as the highest Zn source in the atmosphere.

Apparently, the wet deposition dominates and can be up to about 90% of the total Zn fallout. The global Zn deposition in the 1980s varied between 683 and 1954/yr (Nriagu and Pacyna 1988). In the Netherlands Zn fallout, at that period, was estimated to be 575 t/yr.

Speciation of Zn in street dust is quite variable and its amounts are in forms (in mg/kg): exchangeable, 1.2; absorbed, 21.7; carbonate-bound, 138.7; OM-bound, 52.4; and sulfide-bound, 145.7 (Mirosławski et al. 2006).

Zinc contents of moss (*Hylocomium splendens*) from the mountains in Poland, 32.5–56.1 mg/kg (mean 41.8 mg/kg), are slightly lower than those of the same moss species from Alaska, 12.0–39.6 mg/kg (mean 20.2 mg/kg). This suggests higher atmospheric deposits of Zn in Poland than in Alaska (Migaszewski et al. 2009).

55.5 PLANTS

Mobile species of Zn are readily available to plants, and its uptake is reported to be linear with metal concentration in nutrient solutions and in soils. Organic Zn complexes increase its bioavailablity (Degrys et al. 2006). The composition of soil solution, especially contents of Ca, has a significant impact on its uptake. However, various plants reveal different capability to Zn uptake, which is illustrated by its variable contents among plants from the contaminated region; the highest Zn content, mean 12.38 mg/kg, was in aboveground Spanish foxglove (*Digitalis thapsi*), and the lowest in oat grass (*Arrhenatherum album*), mean 0.87 mg/kg (Garcia-Salgado et al. 2012).

There is still a disagreement whether Zn uptake is an active or a passive process. It is most probably that both processes may occur, depending on both soils and plants. However, there are many suggestions that Zn uptake is mostly metabolically controlled. In Zn-deficient soils, exudates of roots of cereal plants may mobilize this metal from various stable compounds (Fussuo Zhang *vide* Kabata-Pendias 2011). Plants uptake mainly two species: hydrated Zn and Zn^{2+}. However, several complex Zn ions and their organic complexes may also be readily absorbed. In general, Zn is the most easily adsorbed and transported metal within plants, compared with other trace metals.

Zinc is mainly bound to mobile low-molecular proteins; however, Zn phytates and other insoluble Zn complexes are also present in plants. A high part of Zn bound to light organic compounds influenced its high mobility in plants. It is likely to be concentrated in old leaves, and accumulated in chloroplasts. Under Zn-deficiency conditions, some plants may mobilize Zn from old leaves and transport to generative organs.

Zinc plays essential metabolic functions in plants, of which the most important is its activity in several enzymes, such as dehydrogenases, proteinases, peptidases, and phosphohydrolases. Therefore, Zn functions in plants are related to metabolisms of several important compounds: proteins, carbohydrates, phosphates, as well as ribosomes and RNA. Presumably, Zn stimulates the resistance of plants to dry and hot weather, and also to bacterial and fungal diseases, due to its positive impact on levels of proteins, chlorophyll, and abscisic acid. (Zengin *vide* Kabata-Pendias 2011).

The bioavailability of Zn is a function of several plants and soils properties. The most important soil factors limiting its uptake are (1) low soil Zn content, (2) calcareous soils and pH > 7, (3) low OM level, (4) microbial inactivation of Zn, and (5) antagonistic effects of other chemical elements. Phytotoxic threshold levels of Zn in soils extracted with 1 M HCl was established in the experiment with spring wheat at 65 mg/kg for acidic media (pH < 4.6) and at 144 mg/kg for alkaline media (pH > 6.6) (Stanisławska-Głupiak and Korzeniowska 2014).

There are several observed and reported interactions of Zn and other elements, of which the most common and important are as follows:

- Zn–P interaction is relatively common, especially after phosphate and lime applications. Addition of P decreases Zn concentration in soil solution, and thus its availability to plants, and may induce its deficiency. Sometimes, there is also an observed impact of Zn on P uptake by plants. The balanced P and Zn amounts in plants nutrition seem to be essential for the proper activity of *Rhizobium* and N fixation.
- Zn–N interaction is mainly a secondary effect, resulting from increased biomass under heavy N treatment. The elevated content of proteins and amino acids in roots may disturb Zn transport to plant tops.
- Zn–Ca and Zn–Mg interactions vary for a given plant and media and depend on several other factors. Therefore, there are both observed antagonistic and synergistic effects. In general, the elevated amounts of Ca and Mg decrease Zn phytoavailability.
- Zn–Fe antagonism is widely observed. There are two possible mechanisms of this interaction: (1) the competition between Zn^{2+} and Fe^{2+} in uptake processes and (2) the interference in chelation processes. Iron decreases Zn absorption and toxicity.
- Zn–Cd interaction may be both antagonistic and synergistic in uptake and transport processes. However, most often there are observed synergistic effects in field conditions, especially at low pH values. Some plants with high Zn content reveal a tolerance to increased Cd levels.
- Zn–As interaction is observed in plants with high content of As, whose toxic effects decrease at elevated Zn amounts.

Zinc contents of plants vary considerably, reflecting impact on both ecosystems and plant genotypes. Nevertheless, its concentrations in some plants, mainly cereal grains and pasture herbage, do not differ widely. The mean Zn contents (rounded) of cereal grains from various EU countries are (in mg/kg) as follows: wheat, 25; barley, 26; and oats, 33. The background Zn contents in fodder plants, throughout the world, are also relatively similar and vary in grasses from 12 to 47 mg/kg and in clovers from 24 to 45 mg/kg.

Pollution of both air and soils greatly influences the Zn content of plants. Plants cultivated in nonferric metal-mining region contain this metal up to 530 and 710 mg/kg, in lettuce leaves and onion bulbs, respectively. Zinc concentration in plants from areas of various metal industries was the highest in lettuce leaves, 393 mg/kg, and carrot tops, 458 mg/kg. Also plants grown in sludged soils may contain increased Zn levels, up to 85 mg/kg in oat grains, and up to 114 mg/kg in soybean seeds. However, plants grown in Zn-contaminated soils accumulated a great proportion of the metal in roots (Kabata-Pendias 2011). Plants grown on Zn–Pb waste deposits containing Zn up to 171,790 mg/kg, have this metal, in top parts, up to 902 mg/kg (Wójcik et al. 2014). Some bacteria species (e.g., *Rhizobium leguminosarum*) may promote the growth of plant, *Brassica juncea*, on Zn-contaminated sites, due to an impact on Zn speciation and its accumulation in roots (Adediran et al. 2013). Some Zn-hyperaccumulating plants (e.g., *Thlaspi caerulescens*, Brassicaceae) contain this metal, in top parts, up to 10,000 mg/kg.

Mushrooms easily absorb Zn from aerial deposits. Common chanterelles (*Cantharellus cibarius*) grown in mountains of Poland contain higher amounts of Zn, from 74 to 130 mg/kg, than those grown in the Baltic Sea coast, from 52 to 67 mg/kg (Falandysz et al. 2012).

55.6 HUMANS

Zinc is one of the most abundant trace metals in humans and its total content of the body (70 kg) is in the range 2–3 g, of which about 7 mg/L is in the blood, 70–100 in the bones, and 50 mg/kg in the muscle (Emsley 2011). Zinc is an essential nutrient for human health and is vital for many biological functions, supporting normal growth and development in pregnancy, childhood, and adolescence.

Sources of exposure to Zn include ingestion of food, drinking water, polluted air, tobacco products, and occupational exposure. Food is its primary source. Humans are currently exposed to various levels of Zn, as food supplements and additives, medicines, and disinfectants, and in antiseptic and deodorant preparations, and dental cement.

After ingestion, Zn in humans is initially transported to the liver and then distributed throughout the body. Zinc is absorbed at about 20%–30% of ingested quantities. The highest concentrations of Zn in humans are in the liver, kidney, pancreas, prostate, and eyes. Zinc is also present in plasma, erythrocytes, and leukocytes. It is mostly bound to albumin (60%–80%), and to a lesser extent to α-2-macroglobulin and transferrin (WHO 2001d).

Zinc plays specific and important catalytic, cocatalytic, and structural roles in enzymes, and in many other proteins and biomembranes. Over 300 Zn metalloenzymes have been identified, including alkaline phosphatase, lactate dehyrogenase, carbonic anhydrase, carboxypeptides, and DNA and RNA polymerases. It is vital for enzymes, activating growth (height, weight, and bone development), growth and cell division, immune system, fertility, taste, smell and appetite, skin, hair and nails, and vision (WHO 2001d).

Zinc deficiency in the humans is associated with numerous health effects. Its deficiency, especially in infants and young children under 5 years of age, has received global attention. Zinc deficiency is the fifth leading cause of death and disease, in the developing world. According to the WHO, about 800,000 people die annually, due to Zn deficiency, of which 450,000 are children under the age of 5. Globally, around 2 billion people are affected by Zn deficiency. It is also estimated that 60%–70% of the population in Asia and Sub-Saharan Africa could be at risk of low Zn intake (Prasad 2006).

Some of the reported symptoms of Zn deficiency in humans, especially in infants and young children, include dwarfism (growth retardation), dermatitis (alopecia), impaired neurology, decreased immune function, infections, and death. In industrialized countries, cases of mild Zn deficiency can be observed. The most common symptoms include dry and rough skin, dull hair, brittle finger nails, white spots on nails, reduced taste and smell, loss of appetite, mood swings, reduced adaptation to darkness, frequent infections, delayed wound healing, dermatitis, and acne (Black 2008). Zn toxicity has been seen in both acute and chronic forms. In most cases, dermal exposure to Zn, or its compounds, does not result in any noticeable toxic effects.

Intakes of 150–450 mg of Zn per day have been associated with low Cu status, altered Fe function, reduced immune function, and reduced levels of high-density lipoprotein (Fraga 2005). A disproportionate intake of Zn in relation to Cu may induce Cu deficiency, resulting in increased *Cu requirements*, increased Cu excretion, and impaired Cu status. Pharmacological intakes of Zn have been associated with effects ranging from leukopenia and/or hypochromic microcytic anemia. These conditions were reversible upon discontinuation of Zn therapy, together with Cu supplementation. Both Cu and Zn appear to bind to the same metallothionein protein; however, Cu may displace Zn from these proteins.

Following longer term exposure to low Zn doses (about 0.5–2 mg Zn/kg/day), the absorption of Cu from the diet decreased, leading to symptoms of its deficiency. High-dose Zn administration has also resulted in reductions of leukocyte number and function. Long-term consumption of excess Zn resulted in decreased Fe stores (ATSDR 2005d).

Poisoning incidents, with symptoms of gastrointestinal distress, nausea, and diarrhea, may occur after a single or short-term exposure to concentrations of Zn in water or beverages, at concentrations of 1000–2500 mg/L. Similar symptoms, occasionally leading to death, follow the inadvertent intravenous administration of large Zn doses. Kidney dialysis patients, exposed to Zn through the use of water stored in galvanized units, have developed symptoms of Zn toxicity, which are reversible when the water is subjected to activated carbon filtration (WHO 2001d).

In humans, the absorption of Zn in the diet ranges widely. Bioavailability can be affected by abnormalities in the gastrointestinal tract, in transport ligands or in substances that interfere with Zn absorption. A decreased absorption was noted for elderly persons. A significantly reduced absorption of Zn in humans and laboratory animals was observed, after oral uptake of phytate (from grain and vegetable components), owing to the formation of insoluble Zn–phytate complexes in the upper gastrointestinal tract. The absorption of Zn decreases with increasing gastric pH. Other components that may reduce its availability are casein and its phosphopeptides in dairy products; and Ca in the diet. Zinc is less bioavailable from whole grains and legumes, due to inhibitory effects of phytic acid, fiber, and ligninon. It is relatively high bioavailable in meat, eggs, and seafood, and is present in all food groups (Table 55.2).

In 1982, the Joint FAO/WHO Expert Committee on Food Additives proposed a provisional maximum tolerable daily intake for Zn of 1 mg/kg bw. The daily requirement for adult man is 15–20 mg/day (WHO 2001d). The U.S. recommended dietary allowance for Zn is as follows: for adult men, 11 mg/day, and for women, 8 mg/day.

TABLE 55.2
Zinc in Foodstuffs (mg/kg FW)

Product	Content
Food of Plant Origin	
Bread	9.9
Miscellaneous cereals	9.4
Rice, brown[a]	44.2
Rice, different species[b]	15.0
Green vegetables	3.26
Other vegetables	2.62
Fresh fruits	0.61
Nuts	31.0
Food of Animal Origin	
Carcass meat	64.8
Meat products	46.5
Offal	23.0
Poultry	16.3
Fish	7.67
Eggs	11.4
Milk	3.71
Dairy products	9.66

Sources: Rose, M. et al., *Food Addit. Contam.*, 27, 1380–1404, 2010. Unless otherwise stated.

[a] Batista, B.L. et al., *Food Addit. Contam.*, 3, 253–262, 2010.

[b] Jorhem, L. et al., *Food Addit. Contam.* 25, 841–850, 2008b.

FW, fresh weight.

Infants, children, adolescents, and pregnant and lactating women are at an increased risk of Zn deficiency (FNB/IOM 2001).

In the United States, the average daily intake of Zn by humans is on the order of 5.2–16.2 mg/day (ATSDR 2005d). The estimated average daily intakes of Zn were reported to be (in mg/day) 14, 11, 14, and 13.2 in France, Spain, Sweden, and Belgium, respectively (Biego et al. 1998). The average daily intakes of Zn for residents of Japan were estimated to be 8.7 and 8.5 mg/day, for 1991 and 1992, respectively (Tsuda et al. 1995). In the United Kingdom, population dietary exposure was 8.4 mg/day in 1995 and 8.8 mg/day in 2006 (Rose et al. 2010).

55.7 ANIMALS

Zinc is required in biological processes of animals. Its deficit, or excess, could be harmful to animal health. The zinc content in animal body varies, depending on tissues species, and is in relation to their surrounding habitat (Table 55.3).

TABLE 55.3
Zinc in Animal Tissues (mg/kg FW)

	Tissues		
Animal	**Muscle**	**Liver**	**Kidney**
Cattle, agricultural area[a]	43.3	40.3	18.3
Cattle, industrial area[a]	41.2	53.9	22.0
Cattle[b]	36.6	34.3	19.3
Pigs[b]	23.0	43.7	22.8
Hunting animals[b]	30.8	38.8	37.1
Carps[b]	5.6	58.6	–

Sources: [a] Waegeneers, N. et al., *Food Addit. Contam.*, 26, 326–332, 2009a.
[b] Szkoda, J. and Żmudzki, J., *Zinc in the Environment*, Polish Academy of Sciences, Warsaw, Poland, 2002.
FW, fresh weight.

Zinc deficiency in animals is characterized by reduction in growth, cell replication, adverse reproductive effects, adverse developmental effects, which persist after weaning, and reduced immune responsiveness.

Increases in Zn concentrations in experimental animals, exposed to Zn, are accompanied by reduced levels of Cu. Moreover, exposure to Zn alters the levels of other essential metals, including Fe. Some signs of Zn toxicity observed in animals can be alleviated by the addition of Cu or Fe to the diet. Toxic effects of Zn in rodents, following short-term oral exposure, include weakness, anorexia, anemia, diminished growth, loss of hair, and lowered food utilization, as well as changes in the levels of liver and serum enzymes, morphological and enzymatic changes in the brain, and histological and functional changes in the kidney (WHO 2001d).

56 Zirconium [Zr, 40]

56.1 INTRODUCTION

Zirconium (Zr), a lustrous, grayish-white metal of group 4 in the periodic table of elements, is soft, ductile, and malleable, and exhibits both oxyphilic and lithophilic tendencies. Its abundance in the Earth's crust is within the range of 100–250 mg/kg and is distributed fairly similar to that in various rocks. Its contents in coal are from about 30 to 50 mg/kg.

The oxidation state of zirconium is mainly +4, but it may also be +2 and +3. It occurs in several silicates and its principal source is the mineral zircon, $ZrSiO_4$. Also, baddeleyite (called zirconia), ZrO_2, has technological importance and is used as a diamond simulant in jewelry. It is associated with Ti and Hf minerals, and may occur in several minerals, such as pyroxenes, amphiboles, micas, and ilmenites.

World mine Zr production (rounded), excluding the United States, in 2010 was 1190 kt, of which 481 kt was produced in Australia, 390 kt in South Africa, 140 kt in China, 60 kt in Indonesia (USGS 2011). Zirconium deposits are often enriched in U and Th. It is also obtained as a by product of Ti mining.

The principal use of Zr is for atomic energy purposes. It is highly resistant to corrosion, especially due to impacts of alkalis and acids; thus, it is an important component of various alloys. However, at high temperature (~900°C), Zr properties are changed, which may result in meltdown of some equipments, especially of nuclear reactors, which happened in 1979 at Three Miles Island, central Pennsylvania, and in 2011 at Fukushima, Japan, after the Tōhoku earthquake and tsunami (Emsley 2014). Zirconium is applied in glass and ceramic productions, as well as in various thermal installations, dyes, and pyrotechnics. It is added to paints to replace small amounts of Pb compounds, which are still used (Emsley 2014).

Zirconium application in medicine is relatively broad, mainly as an addition to implant materials. It is also a component of some cosmetics and toothpastes.

56.2 SOILS

Zirconium contents of soils are mainly inherited from parent rocks, and therefore, its distribution in soil profiles is often used in studies of soil genetic. Lower Zr amounts are in glacial drift soils (mean 140 mg/kg), and higher are in residual soils, especially derived from Zr-rich rocks (mean 305 mg/kg). Increased Zr content in coarse fraction of desert dusts reflects the presence of detrital zircon (Castillo et al. 2008; Jaworska and Dąbkowska-Naskręt 2006).

Soils contain Zr within the range of 90–850 mg/kg, and the highest concentration is in heavy loamy soils (Table 56.1). Soils of various countries contain Zr as follows (mean and range, in mg/kg): Argentina, 100–550; Australia, 350; China, 219–500; Denmark, 115; Russia, 200–550; the United States, 184–760 (Kabata-Pendias 2011).

TABLE 56.1
Zirconium Contents of Soils, Water, and Air

Environmental Compartment	Range
Soil (mg/kg)	
Light sandy	90–100
Medium loamy	200–550
Heavy loamy[a]	330–850
Organic	200–550
Water (μg/L)	
Rain	0.01–0.04
River	0.03–90
Ocean	0.015–0.05
Air (ng/m³)	
Urban areas	3.0
Contaminated regions	<27

Sources: Data are given, in general, for uncontaminated environments from Kabata-Pendias, A. and Mukherjee, A.B., *Trace Elements from Soil to Human*, Springer, Berlin, Germany, 2007; Reimann, C. and de Caritat, P., *Chemical Elements in the Environment*, Springer, Berlin, Germany, 1998.

[a] Soils derived from basalts and andesites.

Zirconium is slightly mobile in soils, and its movement and bioavailability are controlled by organic acids, and thus its mobile species are mainly in acidic soils. It is strongly fixed by soluble organic matter and Mn hydroxides. Migration of Zr within soil profiles serves as an indicator of a biogeochemical cycle balance for large ecosystems (Smith and Carson *vide* Kabata-Pendias 2011).

Materials used for soil amendments contain relatively low amounts of Zr; its mean content is 1.6 mg/kg in slurry from dairy cows and 53 mg/kg in sludge (Eriksson 2001a).

56.3 WATERS

The median zirconium concentration in seawaters is estimated to be 0.03 μg/L (Reimann and de Caritat 1998), and its content in the North Pacific is calculated as 0.015 μg/L (Nozaki 2005). The concentration of Zn in stream and river waters is more variable, within the range of 0.03–90 μg/L (Table 56.1), but much higher values, up to 200 μg/L, are also reported (Kabata-Pendias and Mukherjee 2007).

Due to low solubility of Zr compounds and minerals, its concentration in surface waters is relatively low. Water-soluble Zr species are converted into zirconia (ZrO_2) at a broad range of water pH 4–9.5. Zirconium may reduce the availability of P compounds to phytoplanktons.

Rainwaters collected in Sweden, during 1999, contained Zr at an average value of 0.024 μg/L within the range of 0.01–0.04 μg/L. Its wet deposition has been estimated to be 0.25 g/ha/yr (Eriksson 2001a).

The median Zr concentration in bottled waters of the EU countries is 0.015 μg/L and is higher than in tap waters, estimated to be 0.009 μg/L (Birke et al. 2010).

Zirconium in stream sediments is likely to accumulate, up to about 2433 mg/kg, with median values within the range of 256–818 mg/kg (Reimann and de Caritat 1998).

56.4 AIR

The average Zr concentration in air collected in urban areas is about 3 ng/m^3, and its contents from air of contaminated regions vary between 0.7 and 27 ng/m^3 (Table 56.1).

56.5 PLANTS

Although most soils contain relatively high amounts of Zr, its availability to plants, presumably in anionic form, $Zr(OH)_5^-$, is greatly limited. Its availability is controlled by its chemical species, and the easiest taken up are Zr–organic ligands.

Zirconium contents are usually higher in roots than in tops of plants (especially of legumes). Higher accumulation of Zr in roots than in tops is confirmed in the experiment with tomato plants, which contain Zr at levels 0.56 and 7.96 mg/kg in aerial parts and roots, respectively (Vera Tome et al. 2003). However, when aerial dust is enriched in Zr, it may be easily absorbed by plant tops. Increased Zr level in mosses, at an average value of 0.35 mg/kg, indicates its sources from dust and rainwater (Berg and Steinnes 1997).

Zirconium contents of food plants vary from 0.005 to 2.5 mg/kg, being lowest in cereal grains and highest in peanuts and beans. Winter wheat and barley grains grown in Sweden contain Zr at the mean values of 0.008 and 0.013, respectively (Eriksson 2001a). Some herbages, especially legumes, trees, and shrubs, may concentrate more Zr than other plants. Plants grown along highways in Germany contain Zr within the range of 0.07–0.92 mg/kg, being highest in dandelion (Djingova et al. 2003).

Although the toxic effects of Zr on plants, and especially on root growth, are observed, its stimulating effects on growth and protein synthesis of some microorganisms are also reported (Ferrand et al. 2006).

56.6 HUMANS

The total amount of Zr in the human body is 1 mg, of which 0.01 mg/L is in the blood, <0.1 mg/kg in the bone, and average 0.1 mg/kg in other tissues (Emsley 2011). Most of ingested Zr passes through the gut, without being adsorbed, and that which is adsorbed tends to accumulate slightly more in the skeleton than in tissue. Zirconium has no known biological role or toxicity and is biocompatible and is used for surgical implants and prosthetic devices.

Zirconium is able to cross the blood–brain barrier and the placental barrier. Zr oxychloride is mutagenic *in vivo.* Skin granulomas have occurred from the use of deodorants containing Zr salts, and lung granulomas have been reported in Zr workers (EU 1998).

Little information is available on Zr level in foodstuffs. However, it may be present in almost all foods. Very low levels of Zr are in vegetables (0.3–30 μg/kg fresh weight [FW]), and wine (0.01–1.0 μg/kg FW) (Szefer and Grembecka 2007). There are no indications that Zr in food is a cause for concern.

The mean quantity of Zr ingested by preschool children in Amherst, Massachusetts, was 7 μg/day (Stanek et al. 1988). Daily adult Zr ingestion with food has been estimated in the United Kingdom to be 53 μg/day (Hamilton and Minski 1972).

57 Lanthanides

57.1 INTRODUCTION

Lanthanides (LAs), often called rare earth elements (REEs), comprise 15 elements, of which only one, promethium [Pm, 61], does not occur naturally in the Earth's crust. The term REE is related to LA series, distinguished in the periodic table of elements. However, the International Union of Pure and Applied Chemistry proposed to add to the REEs other elements, such as Sc and Y. To avoid any confusion, the term LAs will be used for the elements from La to Lu. Two subgroups of LAs are distinguished: the first one (light, LLAs) obtains more basic and more mobile light elements, from La to Gd, than the elements of the second group (heavy, HLAs), from Tb to Lu.

LAs are relatively abundant in rocks and soils (Tables 57.1 and 57.2). Their terrestrial distribution is peculiar; their contents decrease with an increase of atomic weights and atomic number. According to the Oddo–Harkins rule, the element with the even atomic number is more frequent than the next element with an odd atomic number. This rule governs the distribution of all elements in the universe, but is much less pronounced than in the case of LAs.

During weathering processes, LAs are fractionated and are likely to accumulate in weathered materials. Almost all LAs occur preferable as +3 cations, exhibit an affinity for oxygen, and are likely to be concentrated in acidic igneous rocks and argillaceous sediments (Table 57.1). Cerium is the most frequent element in both the Earth's crust (60 mg/kg) and rocks (>250 mg/kg), whereas lutetium is the less abundant element (0.4 and >1.2 mg/kg, respectively).

The global mobility of LAs, calculated as the ratio of water-dissolved concentrations to those of the upper continental crust is relatively low, at about 0.01 ratio value (Gaillardet et al. 2003). However, during longer period, the mobility of LAs affects their distribution.

LAs are constituents of several minerals, such as monazite [$(La,Ce,Th)PO_4$], bastnasite [$(Ce,F)CO_3$], and cheralite [$(Ce,La,Y,Th)PO_4$]. Lanthanum is also associated with xenotime (YPO_4). All these minerals are likely to be concentrated in phosphorites, which resulted in elevated amounts of some LAs in P fertilizers. In addition, some calcite rocks may contain higher amounts of LAs.

Global production of LAs in 2003 was estimated to be about 75 Mt (WMSY 2004). Two of their major sources, and especially of Ce, are monazite and bastnasite. Monazite is associated with granites and gneisses, and due to its great resistance to weathering, it is concentrated in weathered materials and often accumulated in beach sands.

Elevated amounts of LAs are reported for fly ash and fire clay. LLAs in fly ash are within the range of 424–493 mg/kg, and HLAs are within the range of 55–61 mg/kg (Hower et al. 2013). Therefore, there are suggestions to extract them from coal ash.

TABLE 57.1
Lanthanides in Geosphere (mg/kg)

Element	Oxidation State[a]	Earth's Crust	Igneous Rock	Sedimentary Rock
Lanthanum [La, 57]	+3	18.3	2–150	4–90
Cerium [Ce, 58]	**+3**, +4	60.0	4–250	3–90
Praseodymium [Pr, 59]	**+3**, +4	8.2	1–30	1–10
Neodymium [Nd, 60]	+2, **+3**, +5	28.0	2–80	5–48
Samarium [Sm, 62]	+2, **+3**	4.7	0.1–11	1–10
Europium [Eu, 63]	+2, **+3**	1.2	0.01–4	0.2–2
Gadolinium [Gd, 64]	+1, +2, **+3**	5.4	0.1–10	1.3–10
Terbium [Tb, 65]	**+3**, +4	0.6	0.1–2.5	0.3–2
Dysprosium [Dy, 66]	+2, **+3**, +4	4.0	0.05–8	0.8–7
Holmium [Ho, 67]	+3	1.0	0.1–2	0.05–2
Erbium [Er, 68]	+3	2.8	0.1–4.7	0.4–6
Thulium [Tm, 69]	+2, **+3**	0.5	0.1–0.7	0.2–0.7
Ytterbium [Yb, 70]	+2, **+3**	2.5	0.1–4.5	0.3–4.4
Lutetium [Lu, 71]	+3	0.4	0.1–1.2	0.003–1.2

[a] Valences in bold indicate the main oxidation states.

TABLE 57.2
Lanthanides in Soils (mg/kg)

Element	A	B	C	D
La	26.1	8.4–46.9	35.2	3.5–33.2
Ce	48.7	15.8–64.4	97.4	11–68
Pr	7.6	1.5–5.0	8.4	1.3–7.5
Nd	1.9	7.6–28.6	29.3	0.4–53
Sm	3.3	1.8–4.0	5.5	0.9–4.6
Eu	4.8	0.44–1.43	0.8	0.22–0.83
Gd	1.23	1.77–4.54	3.4	1.0
Tb	6.03	0.27–0.92	0.6	0.15–0.65
Dy	0.71	0.68–4.68	2.9	0.9–3.74
Ho	1.08	0.36–0.95	0.5	0.2–0.74
Er	1.58	1.1–2.72	1.4	0.63–2.2
Tm	0.46	0.16–0.4	0.2	0.09–0.33
Yb	2.06	1.11–2.64	1.1	0.6–2.3
Lu	0.34	0.16–0.4	0.2	0.09–0.34

A: Mean contents, data compiled by Kabata-Pendias and Pendias (1999) for various soils, mainly of Europe.
B: Range of means for 77 soils of Japan. Yoshida, S. et al., *Environ. Int.*, 24, 275–286, 1998.
C: Mean for 27 soils of China. Zhu, J. et al., *Fourth International Conference Biogeochemistry Trace Elements*. Abstract. Berkeley, CA, 1997.
D: Range for 30 forest soils of Sweden. Tyler, G., *Plant Soil*, 267, 191–206, 2004.

The use of LAs is relatively broad in various industries, from glass production (mainly as colorant) to electronic devices. They are also applied in catalytic converters, batteries, radars, and so on, and are added to some metallic alloys. Cerium is added to diesel fuel for lowering of the soot ignition temperature and is relatively slightly trapped by filter (Ulrich and Wicher *vide* Kabata-Pendias 2011). Especially Nd is broadly used as a component of the alloys of high strength and applied in various magnets, such as microphones, computer hard disks, and high-energy lasers. Recently, it has been utilized in alloys used in the production of green energy.

57.2 SOILS

The mean contents of LAs in various soils range from the highest amounts of La (26 mg/kg) and Ce (49 mg/kg) to the lowest levels of Yb (2 mg/kg) and Lu (0.3 mg/kg). Relatively high LA concentrations are reported for soils of China (Table 57.2). Their occurrence in soils shows a relation to the geologic origin and mineral composition of parent rocks. Usually their lowest amounts are in sandy soils and the highest in loamy soils. Fine granulometric fraction reveals a great capacity for LAs, especially for their heavy fraction. Usually elements of the group LLAs are more abundant in soils than elements of the group HLAs. However, HLAs are likely to be concentrated more in clay fraction than LLAs. In general, LAs are accumulated in residual fraction, associated with oxides, phosphates, carbonates, and silicates.

Organic compounds play a significant role in the distribution of LAs and affect their increased levels in humus horizons of soils and in forest soil litter. Relatively high binding activity of soluble organic matter (SOM) toward the LAs is controlled by pH values. Due to a high capacity of organic matter to bound LAs, their contents of peats are about 10 times higher than in the surrounding mineral soils (Stern et al. 2007). In most soils, LAs are likely to be concentrated in amorphous Fe oxide-bound fraction (Hu et al. 2006). According to these authors, LAs with an odd atomic number are likely to predominate in water-soluble and exchangeable fractions.

All LAs are slightly mobile in soils, and their behavior is under the impact of microorganisms. Some bacteria increase their adsorption to about 1.5% and 4% for quartz and goethite, respectively (Perelomov and Yoshida 2008). The authors concluded that increased amounts of LAs, in nonexchangeable forms, on mineral surface are due to the formation of low-soluble complexes with organic substances produced by bacteria.

The mobility of LAs in soils is variable and controlled by both soil pH and climatic factors (Tyler 2004). Their distribution in soil profiles is affected by weathering and leaching processes. Soil solutions contain relatively high amounts of ALs, exceeding 100 times those present in rainwater. Their concentrations in soil solutions vary for La to Lu from 188 to 1.3 ng/L (Table 57.3).

Fertilizers enriched in LAs have been commonly used in China, since 1980, to stimulate seed germination, seedling growth, and chlorophyll content. However, their effects on plant metabolism are variable, and environmental effects have not been investigated (d'Aquino et al. 2009). There is an estimation that 50–100 million tons of LAs were added to the agricultural systems of China in 2001 (He et al. *vide*

TABLE 57.3
Lanthanides in Waters (Mean, ng/L)

Element	Seawater[a]	River Water[a]	Rainwater[b]	Soil Solution[c]	Bottled Water[d]	Tap Water[d]
La	5.6	120	1.68	188.4	3.4	0.2
Ce	0.7	263	1.38	114.5	2.3	1.6
Nd	3.3	152	0.76	58.0	2.9	2.2
Sm	0.6	36	0.63	17.6	1.4	1.5
Eu	0.2	10	0.19	3.9	0.96	2.0
Gd	0.9	40	0.86	19.8	1.6	2.0
Tb	0.2	6	0.22	3.6	0.19	0.2
Dy	1.3	30	1.48	19.0	1.2	1.1
Ho	0.4	7	0.3	3.2	0.54	0.4
Er	1.2	20	0.83	8.7	0.99	8.7
Tm	0.2	3	–	–	0.19	0.2
Yb	1.2	17	0.96	10.9	1.1	1.1
Lu	0.2	2	0.15	1.3	0.23	2.3

Sources: [a] Gaillardet, J. et al., *Treatise on Geochemistry*, Elsevier, Oxford, 2003.
[b] Rainwater from Sweden, collected during 1999. Eriksson, J.E., Swedish EPA Report 5159, Stockholm, Sweden, 2001a.
[c] Soils of the Vosgens Mountains region. Aubert, D. et al., *Geochim. Cosmochim. Acta*, 19, 3339–3350, 2002.
[d] Birke, M. et al., *J. Geochem. Expl.*, 107, 217–226, 2010.

d'Aquino et al. 2009). Phosphate fertilizers and some amending materials are their serious sources (Tables 57.4 and 57.5).

57.3 WATERS

The concentration of LAs in waters vary highly; however, the general rule of their distribution is observed. The highest average content is for Ce (26 ng/L) and the lowest for Lu (2 ng/L) (Table 57.3). According to Gaillardet et al. (2003), pH and dissolved organic carbon of waters control LA concentrations and behavior. River waters contain higher levels of LAs than seawaters. Their highest amounts are in soil solutions (Table 57.3). LA concentrations in snow are higher than in rainwaters (Aubert et al. 2002).

57.4 AIR

Data on LAs in air are very limited. The concentrations of some LAs in air of the remote regions (Antarctica or Greenland) are as follows (in pg/m^3): Ce, 0.8–4.0; Eu, 0.004–0.02; La, 0.2–110; Sm, 0.03–12; and Tb, 1–5 (compiled by Kabata-Pendias and Pendias [1999]). Amounts of these elements in air from urban and industrial regions are increased by a factor of 10–10,000, which clearly indicate their anthropogenic sources.

TABLE 57.4
Lanthanides in Sewage Sludge and Farmyard Manure (mg/kg)

	Sewage Sludge		Stable Manure	
Element	**A**	**B**	**A**	**B**
La	2.7	16	24	110
Ce	41.9	24	55	180
Pr	4.3	2.8	11	22
Nd	2.5	11.3	2.5	78
Sm	3.5	1.8	5.2	14
Eu	3.7	0.3	0.7	2.9
Gd	6.8	2	1.5	13
Tb	1.4	0.3	0.3	2
Dy	0.7	1.7	1	10
Ho	0.3	0.4	0.4	2.1
Er	1.2	1	0.7	5.9
Tm	0.4	0.2	0.1	0.8
Yb	0.6	1.1	1.8	5.4
Lu	0.1	0.2	0.6	0.8

A: Data for the United States. Furr, A.K. et al., *Environ. Sci. Technol.*, 10, 683–695, 1976.
B: Data for Sweden. Eriksson, J.E., Swedish EPA Report 5159, Stockholm, Sweden, 2001a.

TABLE 57.5
Contents of Some Lanthanides in Surface Soils and Amending Materials (Mean, mg/kg)

Element	Soil	P Fertilizer	Sewage Sludge	Fly Ash
Dysprosium	3.6	35	1.7	9.5
Europium	1.2	25	0.3	–
Lanthanum	29	422	16	57
Ytterbium	2.5	5	1.1	4.8

Source: Various sources, Kabata-Pendias, A. *Trace Elements in Soils and Plants*, 4th ed., CRC Press, Boca Raton, FL, 2011.

Mosses collected during the last century from the Central Barents region contain La up to 35.2 mg/kg, whereas its content in mosses from Norway was up to 4.5 mg/kg (Steinnes et al. *vide* de Caritat et al. 2001). LA contents in mosses also follow the general rule. La is at the highest amount (1100 μg/kg) and Lu is at the lowest amount (4 μg/kg) (Berg and Steinnes 1997).

57.5 PLANTS

A close relationship between LA contents of plants and their occurrence in soils has been always observed, and their concentration orders, similar to terrestrial materials, decrease with the increase of atomic numbers.

The concentration of LAs in food plants is of a special concern. LA amounts in vegetable are very variable, apparently due to their various contents in soils (Table 57.6). Wheat grains contain relatively low amounts of LAs and are smaller than wheat shoots. Elevated LA contents in wheat grains from China, compared to their contents in grains from Sweden, clearly indicate that soils and farming habits have an impact.

The first attempts to use LAs in agriculture were made in China in 1970. Later, such attempts were also made in Western countries (Hanczakowska and Hanczakowski 2010). In China, both yield increases and quality improvements were achieved in multiple plant species, including cereals, fruits, and vegetables after their application. Recommended application rates vary with the crop species, the application technique (soil, foliar, or seed dressing) as well as the timing. Soil dressing with LAs used in China had visible effect on their contents in wheat grains. Also other food plants cultivated in China contain elevated amounts of some LAs (Liang et al. 2005). Relatively high

TABLE 57.6
Lanthanides in Terrestrial Plants (μg/kg)

Element	Vegetable[a]	Wheat Grain[b]	Wheat Grain[c]	Mushroom[d]
La	4–2000	1.7	16.8	27
Ce	1–50	3.4	26.9	56.0
Pr	1–2	0.4	1.9	6.5
Nd	10	0.9	6.8	23.0
Sm	0.2–100	<0.1	1.1	7.3
Eu	0.04–70	0.2	0.3	1.9
Gd	<2	<0.1	0.2	8.1
Tb	0.1–1	0.1	0.1	1.3
Dy	–	<0.1	0.7	3.7
Ho	0.06–0.1	<0.1	0.1	1.5
Er	1.1–2	<0.1	1.8	2.9
Tm	0.2–4	<0.1	1.8	0.9
Yb	0.08–20	<0.1	1.5	2.2
Lu	0.01–60	<0.1	1.5	–

Sources: [a] Kabata-Pendias and Pendias (1999).
[b] Grains from Sweden. Eriksson, J.E., Swedish EPA Report 5159, Stockholm, Sweden, 2001a.
[c] Grains from China. Liang, T. et al., *Environ. Geochem. Health*, 27, 301–311, 2005.
[d] *Boletus edulis*. Falandysz et al. (2012).

LA amounts contain mushrooms, which may suggest their aerial sources (Falandysz et al. 2012).

Plants sampled in 1999 along the highways in Germany contained the following amounts of some LAs (in μg/kg): La, 38; Ce, 820; and Nd, 330. The highest concentrations were always in dandelion (Djingova et al. 2003). Variable absorption of LAs from a forest ecosystem is affected by their organic complexes and changes in the oxidation state. The plant-to-soil ratio for almost all LAs in the needles of Norway spruce was about 5×10^{-3}, but revealed great variation among various trees (Wyttenbach *vide* Kabata-Pendias 2011).

The main soil parameters controlling LA bioavailability are pH, SOM, and amorphous Fe oxides. There are also evidences that both microorganisms and earthworms increase their bioavailability (Aouad et al. 2006). Also micorrhizal fungi may influence the translocation of some LAs from roots to shoots (Boulois de et al. 2008). Soybean roots reveal an ability to enhance mobility of LAs and to increase their uptake by plants (Nakamaru et al. 2006).

There is no clear evidence on the toxic effects of LAs on plants; however, they have impaired impact on cell membranes of vascular plants and on Ca metabolism in microorganisms. Although LAs may regulate some biological functions, their overdose has inhibitory effects on all organisms. Although LAs are toxic to cell metabolism, there are some elements (e.g., La, Pr, Nd, Eu, Tb) that regulate Ca accumulation by mitochondria of microorganism cells. There are observations that La increases the accumulation of several elements (e.g., Se, Co, Rb, V) in chloroplast of cucumber (Shi et al. 2006).

Although there are reports on the stimulating effect of LAs on several processes in some plants, such as seed germination, root growth, nodulation, and chlorophyll production, these elements have not yet been proved to be essential to plants.

57.6 HUMANS

The contents of some LAs in the human body (70 kg) are as follows (in mg): La, <1.0; Ce, 40; and Sm, 0.05 (Emsley 2011). Their concentrations in the rib bones are as follows (in mg/kg): La, 0.02; Ce, 0.029; Pr, 0.0032; Sm, 0.0014; Gd, 0.0015; Tb, 0.00041; Dy, 0.002; Er, 0.0011; Tm, ≤0.00006; Yb, 0.00072; and Lu, ≤0.00024. The concentrations of Nd, Eu, and Ho are under the detection limit. In humans, the amounts of Ce, Dy, Er, Gd, La, Nd, Pr, Sm, Tb, and Yb increase with age (Zaichik et al. 2011).

LAs have no known biological role in living organisms, but they are present in human bones, and to a lesser extent, in the liver and kidneys. Compared with other metals, their toxicity is relatively low, especially when they are complexed by ligands.

LA ions are very close to Ca^{2+} in properties and their pharmacological effects. In general, their toxicity decreases with increasing atomic number, probably due to greater solubility, ionic stability, and smaller radius.

The chemical forms of LA compounds determine deposition and retention of LAs, following different ways of exposure. The clearance of chelated LAs from the body depends on the stability of the complexes. The chelated LAs are excreted rapidly via urine, whereas unchelated ionic LAs easily form colloids in blood, which are taken up by phagocytic cells of the liver and spleen. Nevertheless, the bones are one of the target organs of LAs. Because chemical properties of LAs are very similar, it is plausible that their binding affinities to biomolecules, metabolism, and toxicity in the living system are also very similar (Hirano and Suzuki 1996).

LAs entering the human body, due to exposure to various industrial works, can affect metabolic processes. Trivalent LA ions, especially La^{3+} and Gd^{3+}, can interfere with Ca channels in human and animal cells. They may also alter or even inhibit the action of various enzymes. LA ions present in neurons can regulate synaptic transmission, as well as block some receptors (e.g., glutamate receptors). Numerous enzymes are affected by LAs: Dy^{3+} and La^{3+} block Ca^{2+}-ATPase and Mg^{2+}-ATPase, whereas Eu^{3+} and Tb^{3+} inhibit calcineurin. In neurons, LA ions regulate the transport and release of synaptic transmitters, and block some membrane receptors, for example, gamma-aminobutyric acid and glutamate receptors. It is likely that LA ions significantly and uniquely affect biochemical pathways, thus altering physiological processes in the cells of humans, animals, and plants (Pałasz and Czekaj 2000).

The use of LAs for medical purposes first began with their prescription as antiemetics. Several of the LAs are well known for their antimicrobial and anticoagulant properties, and have been employed as antithrombic drugs (Jakupec et al. 2005).

Lanthanum carbonate is an efficacious noncalcium, nonresin phosphate binder that is being increasingly used in patients with chronic kidney disease stage dialysis (Di Iorio and Cucciniell 2010). Cerium compounds are known for their uses in topical burn treatments due to their bacteriostatic and bactericidal effects. Antiseptic effects of Ce chloride, nitrate, and sulfate demonstrated particular susceptibility of both gram-negative and gram-positive bacteria (which tend to coat burn wounds) (Jakupec et al. 2005). Soluble LA salts are mildly toxic, whereas insoluble ones are not (Emsley 2011).

LA nanoparticles (NPs) and nanorods have been widely used for diagnostic and therapeutic applications in biomedical nanotechnology, due to their fluorescence properties and pro-angiogenic to endothelial cells, respectively. Some LA elements (Eu, Tb, Gd) as NPs play important roles in biology and medicine due to their unique physical, chemical, and electronic properties (Patra et al. 2009).

LAs have the ability to accumulate in tumor tissues. As they provide various radioisotopes, emitting α, β, or γ radiation, some of them even possess paramagnetic properties and have become useful for anticancer diagnosis and therapy.

Human diets have not been monitored for LA contents, so it is not known how much the average human takes in, but estimations show that the amount is only about several micrograms per year, all coming from tiny amounts taken by plants.

57.7 ANIMALS

The distribution and excretion of LAs in laboratory animals is well known. When LAs are administrated intravenously as salts, the main part of the dose (around 60%–80%) accumulates in the skeleton and the liver. It has been shown that the higher the atomic number of the LAs, the higher the distribution ratio of the liver versus the bone. The toxicity of LAs in rodents was dependent on the route of administration, their chemical forms, and the used animal model (Zielhuis et al. 2005).

As feed additives, LAs improve body weight gain and feed conversion, in nearly all categories of farming animals (chickens, pigs, ducks, cattle). Additionally, improvements in milk production of dairy cows, egg production in laying hens, and output and survival rate of fish and egg hatching of shrimps were noticed.

At present, no definitive statement on optimum LA levels can be made. However, a dose dependency was observed in several trials, and better effects have been achieved when the mixture of LAs was applied, instead of a single element. Feed additives used contain predominantly LAs (La, Ce, Pr, Nd). Both organic and inorganic LA feed additives are commercially available, but organic ones are claimed to provide better results. This is probably ascribable to different chemical characteristics, which lead to variations in both absorption and bioavailability (Redling 2006).

The highest LA concentrations in organs of fattening bulls was measured in the liver followed by the kidneys and rib bones. In the liver, the concentration amounted to (in µg/kg) 22–482 for La, 37–719 for Ce, and 4–73 for Pr. The muscle tissue, playing an important role to evaluate food safety, showed the lowest La, Ce, and Pr concentrations with [in µg/kg dry matter (DM)] 3–5, 5–7, and 0.5–0.7, respectively. The health risk to humans consuming edible tissues of LA-supplemented animals is very low (Schwabe et al. 2012).

It has been suggested that LAs may promote growth, by influencing the development of undesirable bacterial species within the gastrointestinal tract. For example, La may bind to the surface of bacteria. Another explanation for the growth-promoting effects of LAs is due to improvements in nutrient digestibility and availability. LAs have several properties that make them attractive alternatives to antibiotics. Generally, absorption of orally applied LAs is low, with more than 95% being recovered in the feces of animals. As a result, the chances of LA residues present in meat are low and similar to their levels in muscle tissues of animals fed with commercial diets. There have been no reports on the development of bacterial resistance in treated animals (Thacker 2013).

58 Actinides

58.1 INTRODUCTION

Actinides (ACs) is the term used for the elements from actinium [Ac, 89] to lawrencium [Lr, 103]. The International Union of Pure and Applied Chemistry recommended term actinoids. Sometimes they are also included in the group of rare earth elements.

The group of ACs contains 14 various elements, but only 5 of them are present in nature. These are actinum [Ac, 89], thorium [Th, 90], proactinum [Pa, 91], uranium [U, 92], and plutonium [Pu, 94]. Only two elements exist in relatively large quantities as long-lived nuclides: two isotopes of U (^{235}U and ^{238}U) and one of Th (^{232}Th). Thorium is several times more abundant than U. Other ACs, especially Pu, may also occur in the environment, as a result of natural nuclear reaction of U, but in very negligible amounts. Soils in the vicinity of the Tomsk–Seversk facility (Siberia, Russia) contain elevated amounts of $^{239+240}Pu$ (up to 5900 Bq/m^2) and ^{241}Am (up to 1220 Bq/m^2) (Gauthier-Lafaye et al. 2008).

All ACs are radioactive and release energy upon decay. Naturally occurring U and Th, and synthetically produced Pu are the most abundant ACs on the Earth. They are used in nuclear reactors and nuclear weapons. Both U and Th have relatively diverse uses. Thorium may replay U in nuclear reactors.

ACs share the following common properties:

- All are radioactive.
- All are highly electropositive.
- The metals tarnish readily in air.
- They are very dense metals and may form several allotropes.
- They react with boiling water or dilute acids to release hydrogen gas.
- All combine directly with most nonmetals.

ACs, as well as lanthanides, are relatively easily taken up by plants, and thus its concentrations in plants may identify the place of growth. Their content in pumpkin seed oil has been applied for assessing the geographic origin of plants (Bandoniene et al. 2013).

58.2 THORIUM [Th, 90]

58.2.1 INTRODUCTION

Thorium is a silvery-white metal and its luster tarnishes in air due to the oxidation. The degree of oxidation greatly influences its physical properties. Its occurrence in the Earth's crust is estimated within the range of 3.6–9.6 mg/kg and is several times more abundant than all isotopes of U. Its isotope, ^{232}Th, decays very, very slowly.

It is likely to accumulate in acidic igneous rocks, 10–23 mg/kg, and may also be concentrated, up to 10–22 mg/kg, in some argillaceous sedimentary rocks. It occurs in several minerals, of which the most common are thorite ($ThSiO_4$), thioranite, ($ThO_2 + UO_2$), and monazite [(Sm,Gd,Ce,Y,Th)PO_4].

Data on Th production are limited, and according to Reimann and Caritat (1998), it was 0.7 Mt in 1984. Its economically available reserves in the world are estimated to be 1913 Mt of ThO_2, of which 1000 Mt is present in Canada, 963 Mt in India, 440 Mt in the United States, and 300 Mt in Australia (Cotton 1991).

Thorium is used in several electronic equipments, mainly as a component of various electrodes. It is also applied in special optic and scientific instrumentations. It is used in aircraft equipment, nuclear reactors, nuclear weapons, and rockets to impart high strength and resistance to elevated temperature. It is commonly used as a fuel in nuclear power stations. It produces a radioactive ^{220}Rn. Thorium (as well as U) is the only radioactive element with major commercial applications.

58.2.2 Soils

Thorium contents of soils in the EU countries are estimated within the range of 0.2–53.2 mg/kg, and in the United States, 6.1–7.6 mg/kg (Haneklaus and Schung *vide* Kabata-Pendias 2011). Elevated Th amounts are in soils around AC ore bodies (e.g. 76–96 mg/kg in Brazil). The background Th content of soils from the wetland in South Carolina (USA) is given as 1.3 mg/kg, whereas in contaminated sites its content is 166 mg/kg (Hinton et al. 2005). Various Th mean contents of soils are reported for several countries: the lowest is for Poland, 3.4 mg/kg, and the highest for Russia, 13.4 mg/kg (Kabata-Pendias 2011).

The hydroxides of Th^{4+} are dominant species in soils and are responsible for its mobility over a broad range of pH. Especially some organic acids may increase its solubility. Its mobility may be limited due to the formation of slightly soluble compounds (carbonates, phosphates, oxides) as well as adsorption by clays, some minerals, and soluble organic matter (OM). The increased use of apatite, as a soil amendment, results in the enhanced desorption of Th in soils and sediments. Nevertheless, Th predominates in a residual fraction of both apatite and soil amended with apatite (Kaplan and Knox 2004).

The range of Th concentration in groundwaters is 0.08–2.0 μg/L, and it is higher in soils of areas of hot and dry climates (Kabata-Pendias and Mukherjee 2007).

58.2.3 Waters

The worldwide average Th concentration in river water is estimated to be 0.041 μg/L, within the range of 0.0005–1.054 μg/L (Gaillardet et al. 2003). Waters of the Baltic Sea contain Th from 0.0005 to 0.0077 μg/L (Szefer 2002). Some bacteria strains (*Streptomyces levoris*) may adsorb relatively high amounts of Th from water, at pH 3.5 (Tsuruta 2004). The levels of Th in seaweeds range from 0.02 to 0.62 mg/kg (Szefer and Ostrowski 1980).

The median Th concentration in bottled waters of the EU countries is 0.00062 μg/L and is slightly lower than in tap water, estimated to be 0.00077 μg/L (Birke et al. 2010).

In rainwaters collected during 1999 in Sweden, the mean Th content was 0.0023 µg/L and its wet deposition was 0.08 g/ha/yr (Eriksson 2001a). Atmospheric Th deposition in city regions of the United Kingdom is estimated to be <0.5–1 g/ha/yr (Cawse 1987).

58.2.4 Air

The concentration of Th in air over Antarctica is within the range of 0.02–0.08 pg/m^3, whereas in air over cities of various continents its amounts vary from 1,000 to 13,000 pg/m^3 (Reimann and de Caritat 1998).

Mosses, collected during the last century, from Spitsbergen, Norway, contain Th up to 1.14 mg/kg, and its fairly similar maximum content was in mosses from Norway, up to 1.8 mg/kg (Steinnes et al. *vide* de Caritat et al. 2001).

58.2.5 Plants

Phytoavailability of Th from soils is variable and controlled by various factors; however, it is often accumulated in roots (Shtangeeva 2010). Thorium uptake by netted chain fern (*Woodwardia areolata*) from contaminated soils was very high (329 mg/kg), compared with plant grown in greenhouse soil (6.4 mg/kg). This plant is proposed for the phytoremediation of Th-contaminated soils (Knox et al. 2008). Red maple (*Acer rubrum*) grown in background soil contains Th at 0.03 mg/kg, whereas in plants from contaminated soil its content is 2.6 mg/kg (Hinton et al. 2005). The addition of Th to the germination medium increased its contents from <0.02 to 0.17 mg/kg, and from <0.02 to 2.4 mg/kg, in leaves and roots of wheat, respectively. Elevated Th contents in wheat seedling resulted in the decrease of Ca in leaves, as well as in the uptake of other nutrients (Shtangeeva et al. *vide* Kabata-Pendias 2011).

Mobile Th species in soils are easily available to plants. Its contents in pumpkin seed oil vary, depending on its geographic origin (median value, in mg/kg): 0.04 in Austria, 0.04 in Russia, and 0.34 in China (Bandoniene et al. 2013). Thorium amount in sweet gum (*Liquidambar styraciflua* L.) from uncontaminated soil is 0.04 mg/kg, whereas it is 5.3 mg/kg from polluted soil.

Among various native plants grown in the polluted area of the Northern Europe (Kola Peninsula), crowberry (*Empetrum nigrum*) contains Th <0.02 mg/kg and moss (*Hylocomium splendens*) about 0.05 mg/kg (Reimann et al. 2001).

58.2.6 Humans

The total Th content in the human body is 40 µg (Emsley 2011). Thorium can be taken into the body by food, drinking water, and inhalation of Th-contaminated dust. However, its amounts in air are so small that the uptake through air can usually be ignored (ATSDR 1990b).

Most Th that is inhaled or ingested is excreted within a few days, with only a small fraction being absorbed into the bloodstream, about 0.02%–0.05% of the ingested amount. Gastrointestinal absorption from food or water is the principal source of internally deposited Th in humans. Of the amount entering the blood, about 70% deposits in the bones and 4% deposits in the liver, where it is retained, with a biological half-life of about 22 years and of 700 days, respectively. Its small proportion, about 16%, is uniformly distributed to all other organs and tissues. Most of the remaining 10% is directly excreted (Argonne 2007). Thorium binds to aspartic and glutamic acids, and bone glucoprotein. It inhibits certain digestive enzymes.

Thorium is generally a health hazard. It is odorless and tasteless. Excessive exposure to Th has been linked to diseases of the lung, liver, bone, kidney, pancreas, and blood. External γ exposure is not a major concern, because Th emits only a small amount of γ radiation. The main health concern for environmental exposures is generally bone cancer (Argonne 2007). ^{232}Th and its decay products are classified as carcinogenic to humans in Group 1 (International Agency for Research on Cancer [IARC] 2001, 2013). Because Th is radioactive and may be stored in the bone for a long time, bone cancer is also a potential concern for people exposed to Th. Liver diseases and effects on the blood occur in people injected with Th in order to take special X-rays. Several types of cancer developed in people many years after Th was injected into their bodies. Although some forms of Th can stay in the lungs for long periods of time, in most cases, small amounts of Th leave the body in the feces and urine within days (ATSDR 1990b).

58.3 URANIUM [U, 92]

58.3.1 Introduction

Uranium is a silvery-white metallic element, weekly radioactive, but very stable. Its most common isotopes are ^{238}U (99.28% natural abundance) and ^{235}U (0.72%). The isotope ^{234}U is very rare (<0.006%). Its contents in the Earth's crust are estimated at the mean value of 2.7 mg/kg, and its highest concentrations are in igneous rocks, within the range of 3–5 mg/kg. It may also be accumulated in some sedimentary rocks: in black shales, up to 1244 mg/kg, and in phosphate rocks, up to 300 mg/kg. It occurs naturally, in low concentrations of a few mg/kg, in soils, rocks, and waters. The mobility of U depends highly on the host minerals, which after the dissolution are likely to be quickly precipitated as oxides, carbonates, phosphates, and other compounds. Thus, the distribution of U in the lithosphere is highly controlled by the Eh–pH system. The most common forms of U oxides are U_3O_8, UO_2, and U_3O_8. They are relatively stable over a wide range of environmental conditions. The oxidation state of U may vary from +2 to +6, which is the most common.

Uranium ore, carnotite, contains U, at an average of 11%. Uranium in minerals is associated mainly with Si; these minerals are soddyite [$(UO_2)_2SiO_4{\cdot}2H_2O$], kasolite [$Pb(UO_2)(SiO_4){\cdot}H_2O$], uranosilite ($UO_3{\cdot}7SiO_2$), and sklodovskite [$Mg(UO_2)_2(HSiO_4)_2{\cdot}H_2O$]. A major commercial mineral, and the primary source of U, is pitchblende (uraninite) of the composition: $UO_2 + UO_3$.

The worldwide production of U in 2009 amounted to about 50 Mt, of which 27% was mined in Kazakhstan, 20% in Canada, and 16% in Australia (USGS 2011). The major application of U is in the military sector, as a high-density penetrator. This ammunition consists of depleted U (DU), alloyed with 1%–2% of other elements. However, DU contains low amounts of ^{235}U, and thus the use of DU in munitions is controversial, because of questions about potential long-term health effects. Tank armor and other removable vehicle armor are hardened with DU plates. Most DU arises as a by-product of the production of U for the use in nuclear reactors and nuclear weapons.

The first use of munitions with DU was in Sarajevo war in 1992, and the second time in Iraq war in 2003. DU distributed with munitions and various war vehicles remains a long time in the environment and is taken by people with water, plants, and aerial dust. some amounts of DU in food are still noticed in Bosna and Hercegovina (Cubadda et al. 2009). Since 2008, the addition of DU to munitions has been forbidden.

DU is also applied as a shielding material in some containers used to store and transport radioactive materials. Uranium density is about 70% higher than that of Pb, but less dense than that of Au and W. Although the metal itself is radioactive, its high density makes it more effective than Pb in halting radiation from strong sources, such as Ra. Other uses of DU include counterweights for aircraft control surfaces, as ballast for missiles, and as a shielding material. It is also used as a colorant in U-glass, producing orange-red to lemon yellow hues.

58.3.2 Soils

The worldwide uncontaminated soils contain U within the range of <0.4–96.0 mg/kg. The lowest U contents (mean 0.6 mg/kg) are reported for sand dunes and peat soils of Japan, whereas agricultural soils contain median U, within the range of 1.4–2.4 mg/kg (Takeda et al. 2004). Uranium contents of soils in Spitsbergen, Norway, are within the range of 0.0032–0.0765 mg/kg (Gulińska et al. 2003), whereas in soils around Sankt Petersburg (Russia) its amounts vary from 0.5 to 5.5 mg/kg (Shtangeeva 2010). Various U mean contents of soils are reported for several countries: the lowest is for Poland, 0.79 mg/kg, and the highest for the United States, 3.70 mg/kg (Kabata-Pendias 2011).

Elevated U contents of soils are from various sources, mainly from fossil fuel power plants, nuclear wastes, and also P fertilizer works (Paulo and Pratas 2008). Extremely high levels of U are in P fertilizers, within the range of 80–1300 mg/kg, depending upon the kind of phosphate rocks. Thus, these fertilizers are the important source of U, which is likely to accumulate in the upper layer of soils, up to about 200 mg/m^2, after 61 years of cultivation period (Takeda et al. 2006). Some amounts of U in soils may also be from other fertilizers; its content is 0.22 mg/kg in slurry from dairy cows and 10 mg/kg in sludge (Eriksson 2001a). Uranium may be fixed by OM, and thus it is slightly mobile in peat soils. Changing the oxidation state from U^{6+} to U^{4+} by some bacteria reduces its mobility, as well as adding calcite materials, for example, apatite. Also melanin produced by various bacteria enhanced U immobilization in soils (Turick et al. 2008).

Soils of the United States show a relatively small variation in U concentration among soil types, with the exception of radionuclide ^{238}U, which is likely to be concentrated in soils close to aquatic systems (Meriwether *vide* Kabata-Pendias 2011). Uranium in seepage pore water of contaminated soil occurs mainly as sulfate species $UO_2SO_{4(aq)}$ (Baumann et al. 2012). Variable parameters (pH and Eh) of groundwater may control the reactivity of U species. The range of U concentration in groundwaters is 2–12 μg/L, and it is higher in soils in areas of hot and dry climates.

58.3.3 Waters

The worldwide average U concentration in river waters is estimated to be 0.372 μg/L, within the range of 0.02–4.94 μg/L (Gaillardet et al. 2003). The Baltic Sea waters contains U from 0.67 to 1.65 μg/L (Szefer 2002), and its mean concentration in the North Pacific is 3.2 μg/L (Nozaki 2005).

Radioactive ^{238}U of water basins in Poland was at the level 8.37 mBq/L, during the period 1998–1991, and has not changed after the Chernobyl accident (Skwarzec 1997). Annual discharge of ^{238}U with water of Vistula river system to the Baltic Sea is estimated to be 240 GBq (Kabat et al. 2013).

The median U concentration in bottled waters of the EU countries is 0.209 μg/L and is slightly lower than in tap waters, estimated to be 0.307 μg/L (Birke et al. 2010). The World Health Organization (WHO) established 30 μg/L as a provisional guideline value for U because of scientific uncertainties about U toxicity (WHO 2011a). Levels of U in drinking waters are generally less than 1 μg/L; however, higher U amounts are reported for drinking waters (Table 58.1). Also concentrations as high as 700 μg/L have been measured in some private supplies (WHO 2011a). In drilled well water from Helsinki region (with bed rocks of various granites), Finland, U concentration is high, from >1,000 up to 14,870 μg/L (Asikainen and Kahlos *vide* Kabata-Pendias 2011).

Algal biomass may accumulate up to 95% of the total U concentration in seawater. Its contents in marine plankton are within the range of 0.01–1.8 mg/kg (Szefer and

TABLE 58.1
Uranium in Tap and Bottled Waters of Various EU Countries (μg/L)

Country	Mean	Median	Maximum
	Tap Water		
France	3.12–3.33	0.43–0.71	112.51
Germany	0.82–0.83	0.50	10.50
Hungary	2.06	1.90	4.20
Sweden	1.79–2.75	0.00–1.20	40.80
Switzerland	2.12	0.73	92.02
The United Kingdom	0.12–0.83	0.00–0.80	1.69
	Bottled Water		
France	8.74	1.89	152.70
Germany	1.08–1.29	0.15–0.50	35.0
Italy	0.72	0.33	7.15
Portugal	1.03	0.26	13.90
Switzerland	2.54	1.02	30.35
The United Kingdom	3.31–3.32	1.69	14.86

Source: EFSA, *Eur. Food Saf. Auth. J.* 1018, 1–59, 2009b.

Ostrowski 1980). Some bacteria strains (*S. levorice*) may adsorb relatively high amounts of U from water, at pH 3.5 (Tsuruta 2004).

Rainwaters collected in Sweden during 1999 contain U at an average value of 0.0022 μg/L, within the range of 0.0008–0.0054 μg/L. Its wet deposition in Sweden was 0.021 g/ha/yr (Eriksson 2001a). Annual deposition of U on soils of the agricultural region of Croatia is estimated to be 3.1 Bq/m^2 for ^{235}U and 67 Bq/m^2 for ^{238}U (USGS 2011).

58.3.4 Air

The mean uranium concentration in air of Europe is estimated to be 0.002 ng/m^3 and increases up to about 0.5 ng/m^3 above the urban regions (Reimann and de Caritat 1998). Elevated amounts of U in air are associated mainly with coal combustion and nuclear power plants, and are considered the most important risk to the environment and health. Volcanic eruptions are significant sources of U in the atmosphere. Its contents in aerial dust in some industrial regions vary from <2 up to about 10 mg/kg (Manecki and Skowroński *vide* Kabata-Pendias 2011). Limit values given by the Agency for Toxic Substances and Disease Registry (ATSDR 2011) are 50 μg/m^3 for soluble U compounds and 250 μg/m^3 for its insoluble compounds.

Mosses collected during the last century from the Central Barents region contain U up to 0.45 mg/kg, whereas its content in mosses from Norway was up to 0.89 mg/kg (Steinnes et al. *vide* Caritat et al. 2001).

58.3.5 Plants

Uranium contents in vegetable plants vary within the range (rounded) of 0.0014–0.0025 mg/kg fresh weight (FW) and are higher than in fruits, 0.0004–0.0011 mg/kg FW (Shtangeeva 2008). The mean U content in winter wheat grain is 0.0001 mg/kg (Eriksson 2001a).

Among various native plants grown in the polluted area of the Northern Europe (Kola Pennisula), U content was <0.005 mg/kg in crowberry (*E. nigrum*) and 0.034 mg/kg in moss (*H. splendens*) (Reimann et al. 2001). Forest mushrooms (grown in Spain) may contain elevated amounts of U, up to 2.0 mg/kg (Campos et al. 2012). Uranium contents of pumpkin seed oil vary, depending on its geographic origin (median value, in mg/kg): 0.08 in Austria, 0.13 in Russia, and 0.42 in China (Bandoniene et al. 2013).

In regions with U ores, the contents of this element in some herbs increase up to the range of 0.1–1 mg/kg (Kovalevskyj and Kovalevskaja 2010). Uranium content of netted chain fern (*W. areolata*) from contaminated soil was 44 mg/kg, whereas its content of this plant grown in greenhouse was 6.4 mg/kg. Use of this plant for phytoremediation has been considered (Knox et al. 2008).

58.3.6 Humans

The total body burden of U in humans is 40 μg, of which 40% is in the muscles, 20% in the skeleton, 10% in the blood, 4% in the lungs, 1% in the liver, and 0.3% in the kidney (WHO 2005b). However, there are also other estimations: the adult body burden is 90 μg, of which 66% is in the bones (ATSDR 2011), and the body burden (70 kg) is 100 μg, of which up to 70 μg/kg is in bones (Emsley 2011).

Uranium can be taken into the body by food, drinking water, or breathing air. People who work with materials and products that contain U may be exposed to its elevated amounts. This includes workers who mine, mill, or process U, or make items that contain U (e.g., phosphate fertilizers).

After ingestion, most U is excreted within a few days and never enters the bloodstream. The small fraction (0.2%–5%), which is absorbed into the bloodstream, is deposited preferentially in the bones (about 22%) and kidneys (about 12%), with the rest being distributed throughout the body (12%) and excreted. Most of what goes to the kidneys leaves within a few days (in urine), whereas that deposited in the bones can remain for a long time; the half-life of U in bones is 70–200 days. Uranium deposited in other tissues leaves the body in 1–2 weeks. After inhalation, generally only its small fraction penetrates to the lung's alveolar region, where it can remain for years, and from which it can also enter the bloodstream (Argonne 2007). The main sites of long-term retention for inhaled insoluble compounds that are deposited in the deep respiratory tract are the lungs and pulmonary lymph nodes. In the bloodstream, U is associated with red cells, and its clearance is relatively rapid. Although renal toxicity is a major adverse effect of U, the metal has toxic effects also on the cardiovascular system, liver, muscle, and nervous system (Taylor and Taylor 1997).

The bioavailability of U after oral intake is relatively low. Depending on the doses and solubility, U gastrointestinal absorption ranges from 0.1% to 6% of the ingested dose. Oral bioavailability is limited, and only up to 1%–2% of soluble U and 0.2% of insoluble U are absorbed. Toxicity of ingested U is related to the solubility of the U compounds (EFSA 2009b).

Little information on the chronic health effects in humans due to the exposure to environmental U is available (WHO 2011a). Several international agencies (IARC, National Toxicology Program, Environmental Protection Agency) have not classified natural U or DU, with respect to carcinogenicity. Natural U and DU have identical chemical impact on the human body, and their health effects are due to chemical properties and not to radiation (ATSDR 2011). The tolerable daily intake (TDI) for U is estimated to be 60 μg/day (WHO 2011a). For most people, food (Table 58.2) and drinking water (Table 58.1) are the main sources of U exposure. Water may contribute about 50% of the total U exposure (EFSA 2009b).

Uranium intake with food and water varies from about 1.0 to about 5 μg/day in uncontaminated regions, and from 13 to above 18 μg/day in U mining areas (Taylor and Taylor 1997). The average overall dietary exposure to U in European countries was estimated to range from 0.05 to 0.09 μg/kg body weight (bw)/day. High dietary exposure to U varies between 0.09 and 0.14 μg/kg bw/day. These dietary exposure estimates are all well below the TDI of 0.6 μg/kg bw/day. Human daily intake of U has been estimated to range from 0.9 to 1.5 μg/day (0.6–1.0 pCi/day) (ATSDR 2011). However, there are also other data that its intake through food is between 1 and 4 μg/day. Intake of U through drinking water is normally extremely low, but when it is present in drinking water sources, the majority of intake can be through drinking water (WHO 2011a).

TABLE 58.2
Uranium in Foodstuffs (μg/kg FW)

Group of Products	*n*	Contents
Food of Plant Origin		
Cereal and cereal products	171	0.08–2.93
Sugar and sugar products	38	0.63–3.96
Vegetables	678	1.37–2.54
Starchy roots or potatoes	79	0.55–1.94
Fruits	167	0.35–1.07
Fruit and vegetable juices	11	0.30–1.33
Food of Animal Origin		
Meat and meat products	67	0.46–1.71
Fish and seafood	22	0.14–0.87
Eggs	17	0.00–0.60
Milk and dairy-based products	15	0.30–2.10

Source: EFSA, *Eur. Food Saf. Auth. J.* 1018, 1–59, 2009b.
All data are from Germany.
FW, fresh weight.

The daily U intake from water in Finland has been estimated to be 2.1 μg. The daily intake from drinking water in Salt Lake City, Utah, is estimated to be 1.5 μg. The daily intake of U from drinking water in Ontario, Canada, is 0.8 μg. The average, per capita, intake of U in food is 2–3 μg/day in the United States and 1.5 μg/day in Japan (WHO 2005).

58.4 PLUTONIUM [Pu, 94]

58.4.1 Introduction

Plutonium is a transuranic radioactive metal of a silvery-gray appearance, which in air tarnishes and becomes coating when oxidized. It exhibits six allotropes (different forms) and four oxidation states, of which +4 is the most common. Plutonium has several isotopes, of which ^{239}Pu, ^{240}Pu, and ^{241}Pu have a very long half-life time and are the most important, because they are applied in nuclear weapons and nuclear reactors. Pu easily reacts with some elements, especially with C, N, Si, O, and H. When exposed to moist air, it forms oxides and hydrides, which may spontaneously ignite. During fission of Pu, a fraction of the binding energy, which holds the nucleus together, is released as a large amount of electromagnetic and kinetic energy, which results in explosion.

Plutonium is mostly a by-product of nuclear reactions, when due to released neutrons ^{238}U is converted into Pu. Two radionuclides, ^{239}Pu and ^{241}Pu, are easily fissible, can sustain nuclear reactions, and thus are applied in radioisotope thermoelectric generators, are used in some spacecrafts as reactor fuel, and may be a component of some nuclear weapons and reactors.

Currently, most of the Pu found in the Earth's environment results from human activities, in particular, the testing of nuclear weapons in the atmosphere. There is an estimation that about 3 t of ^{239}Pu and ^{241}Pu was released into the atmosphere, of which about 80% are in the Northern Hemisphere (United Nations Scientific Committee on the Effects of Atomic Radiation). Disposal of Pu waste from nuclear power plants and nuclear weapons is of environmental concern. Earlier aboveground nuclear tests are still its source in some places.

58.4.2 Environment

Plutonium is discharged from nuclear facilities mainly as PuO_2, and when in soils, its behavior is controlled by various soil factors. Its migration down rural soils is relatively low, and it is likely to accumulate in 10 cm top layer (Roussel-Debet 2005). In soils of a buffered zone near Denver, Colorado, which was contaminated with Pu in 1960, about 99% of its radionuclides remain in the top layer (Little et al. *vide* Kabata-Pendias 2011).

Elevated concentrations of Pu radionuclides in soils, bottom sediments, and aquatic plants along the Yenisei river, Siberia, were reported, as an effect of a long-term impact of weapon plants and mining activities (Bolsunovsky and Bondareva *vide* Kabata-Pendias 2011). The average fallout of $^{239+240}Pu$ on soils in the temperate zone of the United States, was calculated as 7.5×10^4 Bq/km^2 and of ^{238}Pu as 1.9×10^3 Bq/km^2 (ATSDR 2002b). The levels of ^{238}Pu in soils of the United Kingdom were detected within the range of 0.2–18 nCi/kg and of $^{239+240}Pu$ in the range of 0.8–83 nCi/kg (ATSDR 2002b). Soils of Kazakhstan, sampled in 1992, contained $^{239+240}Pu$ from 10 to 93 mCi/km^2 (Panin 2004). Soils of Iran contain $^{239+240}Pu$ from 80 to 360 mBq/kg (Aliabadi *vide* Kabata-Pendias 2011). Meadow soils around Heidelberg, Germany, contain both Pu isotopes within the range of 290–450 mCi/km^2 (Jakubick *vide* Kabata-Pendias 2011).

Plutonium is known to bind to soil particles very strongly, especially it is adsorbed by hematite in larger amounts than by silica and montmorillonite. It is easily complexed by fulvic and humic acids, and is fixed by mineral particles in organic forms. Low pH value, low clay minerals and OM contents, as well as microbial activity may enhance Pu mobility. However, it may also be accumulated in some microorganisms. Its nitrate compound, $Pu(NO_3)_4$, is mobile, and thus may migrate in soils and is available to plants. Mobile Pu species (organic and nitrate compounds) are easily available to plants. However, Pu is more likely to be fixed to the root surface than to be transported into plant tops. Observed absorption of Pu on the plant surface indicates predomination of its aerial sources.

PuO_2^{2+} is tightly bound in soils, whereas $Pu(NO_3)_4$ is easily mobile and thus available to plants. Also Pu complexes with organic compounds are phytoavailable. However, even in areas where soils contained elevated amounts of Pu (53 mCi/m^2), its concentrations in plants were relatively low (Little et al. *vide* Kabata-Pendias 2011).

Radionuclides, $^{239+240}Pu$, were present in waters of the southern Baltic Sea, during the period 1980–1991, at the level about 0.0035 MBq/L, and have not changed after the

Chernobyl accident (Skwarzec 1997). Annual discharge of $^{239+240}$Pu with water of the Vistula river system to the Baltic Sea is estimated to be 130 MBq (Kabat et al. 2013).

The contents of $^{239+240}$Pu in fish and shellfish sampled in the Irish Sea were reported to be up to 2pCi/g (ATSDR 2002b). Mesoplankton collected from about 200 m depth contain this radionuclide up to 452 mBq/kg, whereas the content of ^{238}Pu was much smaller, 13 mBq/kg. Mesoplankton from deeper layer contain much less of these elements (Sanchez-Cabeza et al. 2003).

Concentration of both Pu radionuclides, $^{239+240}$Pu, in air near Heildelberg, Germany, during the period 1954–1975 varied from 0.086 to 0.75 fCi/m^3, and their total cumulative deposition was calculated to range between 0.054 and 2.10 mCi/km^2 (Jabick *vide* Kabata-Pendias 2011). Volcanic eruptions are significant sources of Pu.

58.4.3 Humans

Humans may be exposed to Pu by breathing air, drinking water, and eating food containing Pu. However, the levels of Pu in the environment are generally very low and of little health consequence. Exposure to its higher levels could occur from an accidental release during its use. Workers at nuclear facilities using Pu may be exposed to its higher levels.

Little Pu (about 0.05%) is absorbed from the gastrointestinal tract after ingestion, and little is absorbed through the skin following dermal contact. When it is inhaled, a significant fraction can move from the lungs through the blood to other organs, depending on the solubility of Pu compounds. After leaving the intestine or lung, about 10% of Pu clears the body. The rest of Pu compounds enter the bloodstream deposits, about equally in the liver and skeleton, where it remains for long time, with biological retention half-lives of about 20 and 50 years. Plutonium metabolism in physiological systems consists, primarily, of hydrolytic reactions and formation of complexes with proteins and nonprotein ligands.

The Pu amounts deposited in the liver and skeleton increase with age. However, it is possible that the bones of children could be more severely affected by Pu than the bones of adults. Most of the body burden of Pu resides in the skeleton and liver, and following inhalation exposures, in the lung and lung-associated lymph nodes. Risks for adverse outcomes of Pu exposures are strongly dependent on radiation doses received by specific tissues and organ systems.

The main health effect of exposure to Pu is cancer (mainly the lung, bones, and liver), which may occur years after the exposure. Especially workers exposed to Pu in air at high levels are at risk. The IARC classified ^{239}Pu and its decay products (^{240}Pu and other isotopes) in Group 1 as carcinogenic to humans (IARC 2001, 2013). As for other radionuclides, the risk coefficients for tap water are about 80% of those for dietary ingestion (Argonne 2007).

Plutonium generally poses a health hazard only if it is taken into the body because all isotopes, except ^{241}Pu, decay by emitting the α particles, whereas the β particles emitted by ^{241}Pu are of low energy. Inhaling airborne Pu is the primary concern for all isotopes, and cancer resulting from the ionizing radiation is the health effect of concern. The ingestion hazard associated with common forms of Pu is much lower than the inhalation hazard, because absorption into the body after ingestion is quite low.

The Pu levels, to which most people are exposed, are very low and of little health consequence. Levels of all anthropogenic radionuclides (including ^{238}Pu and $^{239,240}Pu$) in beverages available to the public in the United Kingdom were below the limits of detection (MAFF 1997a *vide* ATSDR 2010b). Also 20 Emmental cheese samples collected from six EU regions contained these radionuclides below the detection limit (ATSDR 2010b).

The content of $^{239,240}Pu$ in daily diet samples, collected from the hospital in Białystok, Poland (during March 1987 to May 1992), was investigated. The estimated annual intake of Pu was 0.774 Bq/yr (20.9 pCi/yr), in the first year after the accident at Chernobyl, and after six years, the daily intake was 0.088 Bq/yr (2.4 pCi/yr) (Pietrzak-Flis and Orzechowska 1993).

References

Abu-Surrah, A.S., Al-Sa'doni, H.H., Abdalla, M.Y. 2008. Palladium-based chemotherapeutic agents: Routes toward complexes with good antitumor activity. *Cancer Ther.* 6:1–10.

Abyar, H., Safahieh, A., Zolgharnein, H., Zamani, I. 2012. Isolation and identification of *Achromobacter denitrificans* and evaluation of its capacity in cadmium removal. *Pol. J. Environ. Stud.* 21:1523–1527.

Adamiec, E., Helios-Rybicka, E. 2004. Study of thallium content in the soil and grass at the Zn-Pb industrial area in Bukowno. *Geologia/AGH, Krakow* 30:141–152 (in Polish).

Adediran, G., Ngwenya, B., Heal, K. et al. 2013. Bacterial induced increase in Zn bioavailability and toxicity attenuation in the hyperaccumalating plant *Brassica juncea. 7th International Workshop on Chemical Boioavailability in the Terrestrial Environment.* Nottingham, November 3–6. Abstr. 79–80.

Adriano, D.C., Wenzel, W.W., Vangronsveld, J., Bolan, N.S. 2004. Role of assisted remediation in environmental cleanup. *Geoderma* 122:121–142.

Ahmed, Z.U., Panaullah, G.M., Gauch, Jr, H. et al. 2011. Genotype and environment effect on rice (*Oryza sativa* L.) grain arsenic concentration in Bangladesh. *Plant Soil* 338:367–382.

Alam, L., Mohamed, C.A.R. 2011. A mini review on bioaccumulation of 210Po by marine organisms. *Int. Food Res. J.* 18:1–10.

Allen, H.E. 1993. The significance of trace metal speciation for water, sediments and soil quality criteria and standards. *Sci. Total Environ. Suppl.* 1:23–45.

Al-Najar, H., Schultz, R., Rőmheld, V. 2002. Plant availability of thallium in the rhizosphere by hyperaccumator plats: a key factor for assessment of phytoextraction. *Plant Soil* 249:97–105.

Al Rmalli, S.W., Haris, P.I., Harrington, C.F., Ayub, M. 2005. A survey in foodstuffs on sale in the United Kingdom and imported from Bangladesh. *Sci. Total Environ.* 337:23–30.

Alshatwi, A.A., Subbarayan, P.V., Ramesh, E. et al. 2013. Aluminum oxide nanoparticles induce mitochondrial-mediated oxidative stress and alter the expression of antioxidant enzymes in human mesenchymal stem cells. *Food Addit. Contam.* 30:1–10.

Álvarez, E., Monterroso, C., Fernandez-Marcos, M.L. 2002. Aluminum fractionation in Galician (NW Spain) forest as related to vegetation and parent material. *Forest Ecol.* 166:103–206.

Amoroso, M.J., Abate, C.M. 2012. Bioremediation of copper, chromium and cadmium by actinomycetes from contaminated soils. In: *Bio-Geo Interactions in Metal-Contaminated Soils*, eds. E. Kothe, A. Varma, 349–364. Berlin, Germany: Springer.

An, B., Zhao, D. 2012. Immobilization of As(III) in soil and groundwater using a new class of polysaccharide stabilized Fe-Mn oxide nanoparticles. *J. Hazard. Mater.* 211/212:332–341.

Anbar, A. 2004. Molybdenum stable isotopes: Observations, interpretations and directions. In: *Reviews in Mineralogy & Geochemistry*, eds. C.M. Johnson, B.L. Beard, and F. Albarede, Vol. 55, 429–454. Washington, DC: Mineralogical Society of America.

Anderson, D.L., Kitto, M.E., McCarthy, L., Zollar, W.H. 1994. Sources of atmospheric distribution of particulate and gas phase boron. *Atmos. Environ.* 28:1401–1410.

Anderson, R.A. 1981. Nutritional role of chromium. *Sci. Total Environ.* 17:13–29.

Andrade, R.P., de Mello, J.W.V., Windmöller, C.C. et al. 2012. Evaluation of arsenic availability in sulfide materials from gold mining areas in Brazil. *Water Air Soil Pollut.* 223:4679–4686.

Andren, A.W., Bober, T.W. (eds.). 2002. *Silver, Environment, Transport, Fate and Effects*. SETAC Press: Pensacola, FL.

Angelova, M., Asenova, S., Nedkova, V., Koleva-Kolarova, R. 2011. Copper in the human organism. *Trakia J. Sci.* 9:88–98.

Anke, M. 2004a. Essential and toxic macro, trace and ultra elements in the nutrition of man. In: *Elements and Their Compounds in the Environment*. Eds. E. Merian, M. Anke, M. Ihnat, M. Stoeppler, 2nd edn., 343–367, Wiley-VCH: Weinheim, Germany.

Anke, M. 2004b. Vanadium. In: *Elements and Their Compounds in the Environment*. Eds. E. Merian, M. Anke, M. Inhnat, M. Stoeppler, 1171–1191. Wiley-VCH: Weinheim, Germany.

Anke, M., Arnhold, W., Gleich, M. et al. 1995. Essentiality and toxicity of lithium. Lithium in trophic chain, soil-plant-animal-man, ed. T. Kośla, *Proceedings of International Symposium*, Warsaw, Poland, 19–42.

Antonarakis, E.S., Emadi, A. 2010. Ruthenium-based chemotherapeutics: Are they ready for prime time? *Cancer Chemother. Pharm.* 66:1–9.

Aouad, G., Stille, P., Crovisier, J.L. et al. 2006. Influence of bacteria on lanthanide and actinide transfer from specific soil components (humus, soil minerals, and vitrified municipal solid waste incinerator bottom ash) to corn plants: Sr-Nd isotope evidence. *Sci. Total Environ.* 370:545–551.

Argonne. 2007. *Radiological and Chemical Fact Sheets to Support Health Risk Analyses for Contaminated Areas*. Argonne National Laboratory Environmental Science Division. University of Chicago: Chicago, IL.

Argyrou, M., Valassi, A., Andreou, M., Lyra, M. 2013. Rhenium-188 production in hospitals, by W-188/Re-188 generator, for easy use in radionuclide therapy. *Int. J. Mol. Imaging*. Article ID 290750, http://dx.doi.org/10.1155/2013/290750.

Armienta, M.A., Ongley, L.K., Rodriquez, R. et al. 2008. Arsenic distribution in mesquite (*Prosopis laevigata*) and huzaiche (*Acacia farnesiana*) in the Zimapan mining area, Mexico. *Geochem. Explor. Environ. Anal.* 8:191–197.

Asami, T. 1988. Environmental pollution by beryllium in Japan. In: *Contaminated Soil*, eds. K. Wolf, W. van den Brink, and F.J. Colon, 261–272. Dordrecht, The Netherlands: Kluwer Academic Publication.

Ashowrth, D.J., Alloway, B.J. 2004. Soil mobility of sewage sludge-derived dissolved organic matter, copper, nickel and zinc. *Environ. Pollut.* 127:137–144.

ATSDR (Agency for Toxic Substances and Disease Registry). 1990a. *Toxicological Profile for Radium*. Atlanta, GA: ATSDR.

ATSDR (Agency for Toxic Substances and Disease Registry). 1990b. *Toxicological Profile for Thorium*. Atlanta, GA: ATSDR.

ATSDR (Agency for Toxic Substances and Disease Registry). 2002a. *Draft Toxicological Profile for Several Trace Elements*. Atlanta, GA: ATSDR.

ATSDR (Agency for Toxic Substances and Disease Registry). 2002b. *Toxicological Profile for Beryllium*. Atlanta, GA: ATSDR.

ATSDR (Agency for Toxic Substances and Disease Registry). 2003. *Toxicological Profile for Fluorides, Hydrogen Fluoride, and Fluorine*. Atlanta, GA: ATSDR.

ATSDR (Agency for Toxic Substances and Disease Registry). 2004a. *Toxicological Profile for Cesium*. Atlanta, GA: ATSDR.

ATSDR (Agency for Toxic Substances and Disease Registry). 2004b. *Toxicological Profile for Cobalt*. Atlanta, GA: ATSDR.

ATSDR (Agency for Toxic Substances and Disease Registry). 2004c. *Toxicological Profile for Strontium*. Atlanta, GA: ATSDR.

ATSDR (Agency for Toxic Substances and Disease Registry). 2005a. *Toxicological Profile for Nickel*. Atlanta, GA: ATSDR.

ATSDR (Agency for Toxic Substances and Disease Registry). 2005b. *Toxicological Profile for Tin and Tin Compounds*. Atlanta, GA: ATSDR.

ATSDR (Agency for Toxic Substances and Disease Registry). 2005c. *Toxicological Profile for Tungsten*. Atlanta, GA: ATSDR.

ATSDR (Agency for Toxic Substances and Disease Registry). 2005d. *Toxicological Profile for Zinc*. Atlanta, GA: ATSDR.

ATSDR (Agency for Toxic Substances and Disease Registry). 2007a. *Toxicological Profile for Barium and Barium Compounds*. Atlanta, GA: ATSDR.

ATSDR (Agency for Toxic Substances and Disease Registry). 2007b. *Toxicological Profile for Lead*. Atlanta, GA: ATSDR.

ATSDR (Agency for Toxic Substances and Disease Registry). 2008. *Toxicological Profile for Aluminum*. Atlanta, GA: ATSDR.

ATSDR (Agency for Toxic Substances and Disease Registry). 2010a. *Toxicological Profile for Chlorine*. Atlanta, GA: ATSDR.

ATSDR (Agency for Toxic Substances and Disease Registry). 2010b. *Toxicological Profile for Plutonium*. Atlanta, GA: ATSDR.

ATSDR (Agency for Toxic Substances and Disease Registry). 2011. *Toxicological Profile for Uranium*. Atlanta, GA: ATSDR.

ATSDR (Agency for Toxic Substances and Disease Registry). 2012a. *Toxicological Profile for Chromium*. Atlanta, GA: ATSDR.

ATSDR (Agency for Toxic Substances and Disease Registry). 2012b. *Toxicological Profile for Manganese*. Atlanta, GA: ATSDR.

ATSDR (Agency for Toxic Substances and Disease Registry). 2012c. *Toxicological Profile for Radon*. Atlanta, GA: ATSDR.

ATSDR (Agency for Toxic Substances and Disease Registry). 2012d. *Toxicological Profile for Vanadium*. Atlanta, GA: ATSDR.

Aubert, D., Stille, P., Gather-Lafaye, A. et al. 2002. Characterization and migration of atmospheric REE in soils and surface waters. *Geochem. Cosmochim. Acta* 19:3339–3350.

Azarenko, Y. 2007. The boron content in soil of solonetzic complexes in the Irtysh Region of Omsk oblast and boron resistance of plants. *Eurasian Soil Sci.* 40:512–521.

Ba, L.A., Döring, M., Jamier, V., Jacob, C. 2010. Tellurium: An element with great biological potency and potential. *Org. Biomol. Chem.* 8:4203–4216.

Badmaev, V., Prakash, S., Majeed, M. 1999. Vanadium: A review of its potential role in the fight against diabetes. *J. Altern. Complement Med.* 5:273–291.

Bahaminyakamwe, L., Simunek, J., Dane, J.H. et al. 2006. Copper mobility in soils as affected by sewage sludge and low molecular weight organic acids. *Soil Sci.* 171:29–39.

Bakirdere, S., Örenay, S., Korkmaz, M. 2010. Effect of boron on human health. *Open Miner. Process. J.* 3:54–59.

Bandomiere, D., Zettle, D., Meisel, F., Maneko, M. 2013. Suitability of elemental fingerprinting for assessing the geographic origin of pumpkin (*Curcubita pepo* var. *styriaca*) seed oil. *Food Chem.* 136: 1533–1543.

Bang, S., Pena, M.E., Patel, M. et al. 2011. Removal of arsenate from water by adsorbents: A comparative case study. *Environ. Geochem. Health* 33:133–141.

Banuelos, G.S. 2001. The green technology of selenium phytoremediationn. *Biofactors* 14:255-260.

Barány, E., Bergdahl, I.A., Bratteby, L.-E. et al. 2005. Iron status influences trace element levels in human blood and serum. *Environ. Res.* 98:215–223.

Barker, A.V., Pilbaem, D.J. 2007. *Handbook of Plant Nutrition*. Boca Raton, FL: Taylor & Francis.

Barregard, L., Sallsten, G., Conradi, N. 1999. Tissue levels of mercury determined in a deceased worker after occupational exposure. *Int. Arch. Occup. Environ. Health* 72:169–173.

Barsova, N., Mozutova, G. 2012. Transport parameters and sorption-desorption of Zn and Cu Sin soddy soils of the Upper Volga. In: *Competitive Sorption and Transport of Heavy Metals in Soils and Geological Media*, ed. H.M. Selim, 233–266. Boca Raton, FL: CRC Press.

Basta, N.T., Ryan, J.A., Chaney, R.L. 2005. Trace element chemistry in residual-treated soils: Key concepts and metal bioavailability. *J. Environ. Qual.* 34:49–63.

Batista, B.L., De Oliveira Souza, V.C., Da Silva, F.G., Barbosa, Jr, F. 2010. Survey of 13 trace elements of toxic and nutritional significance in rice from Brazil and exposure assessment. *Food Addit. Contam.* 3:253–262.

Baumann, N., Arnold, T., Lonschinski, M. 2012. Speciation of uranium in seepage and pore waters of heavy metal-contaminated soil. In: *Bio-Geo Interactions in Metal-Contaminated Soils*, eds. E. Kothe, A. Varma, 131–141. Berlin, Germany: Springer.

Bech, J., Suarez, M., Revrter, T. et al. 2010. Selenium and other trade elements in phosphate rock of Bayovar-Sechura (Peru). *J. Geochem. Expl.* 107:136–145.

Bech, J., Reverter, F., Tune, P. et al. 2011. Pedogeochemical mapping of Al, Ba, Pb, Ti and V in surface soils of Barcelona province (Catalonia, NE Spain): relationships with soil physicochemical properties. *J. Geochem. Explor.* 109:26–37.

Becquer, T., Quantin, C., Sicit, M., Boudot, J.P. 2003. Chromium availability in ultramafic soils from New Caledonia. *Sci. Total Environ.* 301:367–373.

Bennett, J.P., Wetmore, C.M. 2003. Elemental chemistry of four lichen species from Apostle Island, Wisconsin, 19987, 1995 and 2001. *Sci. Total Environ.* 305:46–48.

Berg, D., Steinnes, E. 1997. Recent trends in atmospheric deposition of trace elements in Norway as evident from the 1995 moss survey. *Sci. Total Environ.* 208:197–206.

Bergamo, A., Sava, G. 2011. Ruthenium anticancer compounds: Myths and realities of the emerging metal-based drug. *Dalton Trans.* 40:7817–7823.

Bernstein, L.R. 1998. Mechanisms of therapeutic activity for gallium. *Pharmacol. Rev.* 50:665–682.

Bertine, K.K. Koide, M. Goldberg, E.D. 1993. Aspects of rhodium, marine chemistry. *Marine Chem.* 42:199–210.

Battacharya, P., Chakrabari, K., Chakraborty, A. 2008. Cobalt and nickel uptake by rice and accumulation in soil amended with municipal solid waste compost. *Ecotoxicol. Environ. Saf.* 69:506–512.

Battacharya, P., Frisbie, S.H., Smith, E. et al. 2002. Arsenic in the environment: A global perspective. In: *Handbook of Heavy Metals in the Environment*, ed. B. Sarkar, 147–215. New York: Marcel Dekker.

Bialek, B., Roland, A., Diaz-Bone, R.A. et al. 2011. Toxicity of methylated bismuth compounds produced by intestinal microorganisms to bacteroides thetaiotaomicron, a member of the physiological intestinal microbiota. *J. Toxicol.* Article ID 60834,9

Biego, G.H., Joyeux, M., Hartemann, P. et al. 1998. Daily intake of essential minerals and metallic micropollutants from foods in France. *Sci. Total Environ.* 217:27–36.

Bini, C., Maleci, L., Romanin, A. 2008. The chromium issue in soils of the leather tannery district in Italy. *J. Geochem. Explor.* 96:194–202.

Birke, M., Reimann, C., Demetriades, A. et al. 2010. Determination of major and trace elements in European bottled water—Analytical methods. *J. Geochem. Explor.* 107:217–226.

Black, J. 1994. Biological performance of tantalum. *Clin. Mater.* 16:167–173.

Black, R.E. 2008. Zinc deficiency, infectious disease and mortality in the developing world. *J. Nutr.* 133:1485–1489.

Blake, G.M., Lewiecki, E.M., Kendler, D.L., Fogelman, I. 2007. A review of strontium ranelate and its effect on DXA scans. *J. Clin. Densitom.* 10:113–119.

Bojakowska, I., Gliwicz, T., Wolkowicz, S. 2007. Trace elements in bottom sediments of Zegrzynski Lake. *Bug and Narew rivers. WSEiZ.* 8: 123–135, Warsaw, Poland: University of Ecology and Management (in Polish).

Borowska, K., Koper, J., Grabowska, N. 2012. The effects of farmyard manure in relation to its on storage of selenium and its phytoavailable fractions content in soil in the background of catalase activity. *Protect. Environ. Natural Resour.* 54:141–149 (in Polish).

Boulois de, H.D., Jones, E.J., Leyval, C. et al. 2008. Role and influence of mycorrhizal fungi on radiocesium accumulation by plants. *J. Environ. Radioact.* 99:785–800.

Boutakhrit, K., Crisci, M., Bolle, F., Van Loco, J. 2011. Comparison of four analytical techniques based on atomic spectrometry for determination of total tin in canned foodstuffs. *Food Addit. Contam.* 28:173–179.

Bratakos, S.M., Lazou, A.E., Bratakos, M.S., Lazos, E.S. 2012. Aluminum in food and daily dietary intake estimate in Greece. *Food Addit. Contam.* 5:33–44.

Broberg, K., Concha, G., Engström, K. et al. 2011. Lithium in drinking water and thyroid function. *Environ. Health Persp.* 119:827–830.

Brokbartold, M., Wischermann, M., Marschner, B. 2012. Plant availability and uptake of lead, zinc, and cadmium in soils contaminated with anti-corrosion paint from pylons in comparison to heavy metal contaminated urban soils. *Water Air Soil Pollut.* 223:199–213.

Burd, G.I., Dixon, D.G., Glick, B.R. 2000. Plants growth-promoting bacteria that decrease heavy metal toxicity in plants. *Can. J. Microbiol.* 46:237–245.

Burges, W., Ahmed, K.M. 2006. Arsenic in aquifers of the Bengal Basin. In: *Managing Arsenic in the Environment: From Soil to Human Health*, eds. R. Naidu, G. Smith, G. Owens, P. Bhattacharya, and P. Nadebrum, 31–56, Collingwood, VI: CSIRO Publication.

Burt, R., Wilson, M.A., Mays, M.D., Lee, C.W. 2003. Major and trace elements of selected pedons in the USA. *J. Environ. Qual.* 32:2109–2121.

Butterwick, I., de Oude, N., Rayminf, K. 1989. Safety assessment of boron in aquatic and terrestrial environment. *Ecotoxicol. Environ. Saf.* 17:339–371.

Cabrera, W.E., Schrooten, I., De Broe, M.E., D'Haese, P.C. 1999. Strontium and bone. *J. Bone Miner. Res.* 14:661–668.

Cain, D.J., Luoma, S.N., Carter, J.N., Fend, S.V. 1992. Aquatic insects as bioindicators of trace element contamination in cobble-bottom rivers and sediments. *Can. J. Fish Aquat. Sci.* 49:2141–2154.

Campos, V., de Toro, J.A., Perez de los Reyes, C. et al. 2012. Lifestyle influence on the content of copper, zinc and rubidium in wild mushrooms. *Appl. Environ. Soil Sci.* 12, Article ID 687160.

Canavese, C., DeConstanzi, E., Branciforte, L. 2001. Rubidium deficiency in dialysis patients. *J. Nephrol.* 14:169–175.

Cappuyns, V., Slabbinck, E. 2012. Occurrence of vanadium in Belgian and European alluvial soils. *Appl. Environ. Soil Sci.* Article ID 97501.

Carraher, Jr, C.E., Pittman, Jr, C.U. 2004. Organometallic compounds in biomedical applications, In: *Macromolecules Containing Metal and Metal-Like Elements*, Biomedical Applications, eds. A.S. Abd-El-Aziz, C.E. Carraher, Jr, C.U. Pittman, Jr, J.E. Sheats, and M. Zeldin, Vol. 3, 2–18. Hoboken, NJ: Wiley.

Casado, M., Anavar, H.M., Garcia-Sanchez, A., Santa-Regina, I. 2007. Antimony and arsenic uptake by plants in abandoned mining area. *Common. Soil Sci. Plant Anal.* 38:1255–1275.

Castillo, S., Moreno, T., Querol, X. et al. 2008. Trace element variation in size-fractionated African desert dusts. *J. Arid Environ.* 71:1031–1042.

Castronovo, F.P., Wagner, H.N. 1971. Factors affecting the toxicity of the element indium. *Br. J. Exp. Pathol.* 52:543–559.

Catalani, S., Leone, R., Rizzetti, M.C. et al. 2011. The role of albumin in human toxicology of cobalt: Contribution from a clinical case. *ISRN Hematol.* 1:1–6, Article ID 690620.

Cawse, P.A. 1987. Trace and major elements in the atmosphere at rural locations in Great Britain, In: *Pollutant Transport and Fate in Ecosystems*, eds. J.P. Coughtrey, M.H. Marti, and M.H. Unsworth, 1972–1981, Oxford: Black Science Publication.

Cefalu, W.T., Hu, F.B. 2004. Role of chromium in human health and in diabetes. *Diabetes Care* 27:2741–2751.

Cejnar, R., Mestek, O., Dostálek, P. 2013. Determination of silicon in Czech beer and its balance during the brewing process. *Czech J. Food Sci.* 13:166–171.

Chen, G.-T., Yun, S.-T., Mayer, B. et al. 2007. Fluorine geochemistry in bedrock groundwaters of South Korea. *Sci. Total Environ.* 385:272–283.

Chaney, R.L., Maliok, M., Li, Y.M. et al. 2005. Using hyperaccumulator plants to phytoextract soil Ni and Cd. *Z. Natuforsch.* 60C:190–198.

Chaudri, A., McGrath, S., Gibbs, P. et al. 2007. Cadmium availability to wheat grain in soils treated with sewage sludge or metal salt. *Chemosphere* 66:1423–1451.

Chen, J.P. 2013. *Decontamination of Heavy Metals: Processes, Mechanisms, and Applications.* Boca Raton, FL: CRC Press.

Chen, Y., Yin, X., Ning, G. 1999. Effects of tantalum and its oxides on exposed workers. Zhonghua Yu Fang, Yi Xue Za Zhi. *Chin. J. Prev. Med.* 33:234–235 (in Chinese).

Chen, W., Krage, N., Weu, L. et al. 2008. Ferilizer applications and trace elements in vegetable production soils of California. *Water Air Soil Pollut.* 190:209–214.

Chitambar, C.R. 2010. Medical applications and toxicities of gallium compounds. *Int. J. Environ. Res. Public Health* 7:2337–2361.

Choi, A.L., Sun, G., Zhang, Y., Grandjean, P. 2012. Developmental fluoride neurotoxicity: A systematic review and meta-analysis. *Environ. Health Persp.* 120:1362–1368.

Chung, S., Chan, A., Xiao, Y. et al. 2013. Iodine content in commonly consumed food in Hong Kong and its changes due to cooking. *Food Addit. Contam.* 6:24–29.

Chung, S.W., Lam, C., Chan, B.T. 2014. Total and inorganic arsenic in foods of the first Hong Kong total diet study. *Food Addit. Contam.* 31:650–657.

Chung, S.W.C., Kwong, K.P., Yan, J.C.W., Wong, W.W.K. 2008. Dietary exposures to antimony, lead and mercury of secondary school students in Hong Kong. *Food Addit. Contam.* 25:831–840.

Chung, J.B., Sa, T.M. 2001. Chromium oxidation potential and related soils characteristics in arable upland soils. *Commun. Soil Sci. Plant Anal.* 32:1719–1733.

Ciavardelli, D., D'Anniballe, G., Nano, G. et al. 2007. An inductively coupled plasma mass spectrometry method for the quantification of yttrium-antibody based drugs using stable isotope tracing. *Rapid Commun. Mass Spectrom.* 21:2343–2350.

Ciba, J., Zołotajkin, M. 2001. Chemical forms of iron in the composts obtained from municipal solid waste contaminated with pulverized metallic Fe, Fe_2O_3 or FeS. *Arch. Environ. Prot.* 27:101–114.

Cieśla, W., Borowska, K. 1994. The selenium and other trace elements content in muscle and organs of hares from different habitats. In: *Arsenic and Selenium in Environment. Ecological Analytical Problems*, eds. A. Kabata-Pendias, B. Szteke, 126–129, Warsaw, Poland: The Polish Academy of Sciences (in Polish).

Clarkson, T.W., Magos, L., Myers, G.J. 2003. The toxicology of mercury-current exposures and clinical manifestations. *N. Engl. J. Med.* 349:1731–1737.

Coenen, M. 2004. Chlorine. In: *Elements and Their Compounds in the Environment*, 2nd ed., eds. E. Merian, M. Anke, M. Ihnat, and M. Stoeppler, 1423–1443, Weinheim, Germany: Wiley.

Concha, G., Eneroth, H., Hallström, H., Sand, S. 2013. Contaminants and minerals in foods for infants and young children. *Part 2: Risk and Benefit Assessment.* Livsmedelsverkets rapportserie, 1/2013. Uppsala, Sweden: National Food Agency, Sweden Livsmedels Verket.

Conklin, S.D., Ackerman, A.H., Fricke, M.W. et al. 2006. In vitro biotransformation of an arsenosugar by mouse anaerobic cecal microflora and cecal tissue as examined using IC-ICP-MS and LC-ESI-MS/MS. *Analyst* 131:648–655.

Cornell, R.M., Schwertmann, U. 2003. *The Iron Oxides: Structure, Properties Reactions, Occurrences and Uses*, 2nd ed., Munich, Germany: Wiley.

Cotton, S.A. 1991. *Lanthanides and Actinides.* Macmillan Publ.: London.

Coughlin, J.R., Nielsen, F.H. 1999. Advances in boron essentiality research: Symposium summary. *J. Trace Elem. Exp. Med.* 12:171–284.

Cubadda, F., Aureli, F., Ciardullo, S. et al. 2009. Determination of depleted uranium in food samples from Bosnia and Herzegovina. *3rd International IUPAC Symposium on Trace Elements in Food*. October 10–13, Rome, Italy, Abstr. 56.

Cunha, R.L.O.R., Gouvea, I.E., Juliano, L. 2009. A glimpse on biological activities of tellurium compounds. *An. Acad. Bras. Cienc.* 81:393–407.

Čurlik, F., Šefčik, P. 1999. *Geochemical Atlas of the Slovak Republic*. V. Soils. Rome, Italy: Ministry of the Environment of the Slovak Republic.

Cvjetko, P., Cvjetko, I., Pavlica, M. 2010. Thallium toxicity in humans. *Arh. Hig. Rada. Toksikol.* 61:111–119.

Dąbkowska-Naskręt, H. 2009. Natural and synthetic iron oxides as adsorbents of trace elements in soils. *Ochr. Srod. Zasob. Nat.* 41:631–639 (in Polish).

d'Aquino, L., Concetta de Pinto, M., Nardi, L. et al. 2009. Effect of some light rare earth elemenys on seed germination, seedlings growth and antioxidation metabolism in *Triticum durum. Chemisphere* 75:900–905.

Darnerud, P.O. 2003. Toxic effects of brominated flame retardants in man and in wildlife. *Environ. Int.* 29:841–853.

Davies, B.E., Vaugham, J., Lalor, G.C., Vutchkov, M. 2003. Cadmium and zinc adsorption maxima of geochemically anomalous soils (Oxisols) in Jamaica. *Chem. Spec. Bioavailab.* 15:59–66.

de Caritat, P., Reimann, C., Bogatyrev, I. et al. 2001. Regional distribution of Al, B, Ba, Ca, K, La, Mg, Mn, Na, P, Rb, Si, Sr, Th, U, and Y in terrestrial moss within a 188 000 km^2 area of the central Barents region: Influence of geology, sea spray and human activity. *Appl. Geochem.* 16:137–159.

Degryse, F., Smolders, E., Parker, D.R. 2006. Metal complexes increase uptake of Zn and Cu by plants: Implications for uptake and deficiency studies in chelator-buffered solutions. *Plant Soil* 289:171–185.

Dell'Amino, E., Cavalca. L., Andreoni, V. 2008. Improvement of Brassica napus growth under cadmium stress by cadmium-resistance rhizobacteria. *Soil Biol. Biochem.* 40:74–84.

Deng, G.-F., Li, K., Ma, J. et al. 2011. Aluminum content of some processed foods, raw materials and food additives in China by inductively coupled plasma-mass spectrometry. *Food Addit. Contam.* 4:348–353.

Deschamps, E., Matschullat, J. 2011. *Arsenic: Natural and Anthropogenic Arsenic in the Environment*. Boca Raton, FL: CRC Press.

Desmet, G., Myttenaere, C. 2011. *Technetium in the Environment*. Springer: the Netherlands.

Dhillon, S.K., Dhillon, K.S. 2009. Phytoremediation of selenium-contaminated soils: the efficiency of different cropping systems. *Soils Use Managem.* 25:441–453.

Diduch, M., Polkowska, Z., Namieśnik, J. 2011. Chemical quality of bottled waters: A review. *J. Food Sci.* 76:178–196.

Di Iorio, B., Cucciniell, E. 2010. Lanthanum carbonate is not associated with QT interval modification in hemodialysis patients. *Clin. Pharmacol.* 2:89–93.

Dimitrova, M., Petrova, E., Gluhcheva, Y. 2013. Neurodegenerative changes in rat produced by lithium treatment. *J. Toxicol. Environ. Health A* 76:304–310.

Dinesh, R., Ramanathan, G., Singh, H. 1995. Influence of chlorine and sulphate ions on soil enzymes. *J. Agric. Crop Sci.* 175:129–133.

Djingova, R., Kovacheva, P., Wagner, G., Markert, B. 2003. Distribution of platinum group elements and other traffic related elements among different plants along some highways in Germany. *Sci. Total Environ.* 30:235–246.

Dong, J., Wu, F.B., Huang, R.G., Zang, G.P. 2007. Chromium-tolerant plants growing in Cr-contaminated. *Int. J. Phytoremediation* 9:167–179.

Dradrach, A., Karczewska, A. 2013. Mercury in soils of municipal lawns in Wrocław, Poland. *Fresenius Environ. Bull.* 22:968–972.

Drake, P.L., Kyle, J., Hazelwood, K.J. 2005. Exposure-related health effects of silver and silver compounds: A review. *Ann. Occup. Hyg*. 49:575–585.

Duke, J.A. 1970. Ethnobotanical observations on the Choco Indians. *Econ. Bot*. 23:344–350.

Dutch List. 2013. Soil and ground water criteria used in the Netherlands for contaminated land. http://www.epd.gov.hk/eia/register/report (Accessed: April 2, 2013).

EC (European Commission). 2001. Commission recommendation of 20 December 2001 on the protection of the public against exposure to radon in drinking water supplies. (2001/928/Euratom). *Off. J. Eur. Commun.* L:344/85.

EC (European Commission). 2006. Commission regulation (EC) No 1881/2006 of 19 December 2006 setting maximum levels for certain contaminants in foodstuffs. *Off. J. Eur. Union* L:364/5.

EC (European Commission). 2008. Commission regulation (EC) No 629/2008 of 2 July 2008 amending regulation (EC) No 1881/2006 setting maximum levels for certain contaminants in foodstuffs. *Off. J. Eur. Union* L:173/6.

ECETOC (European Centre for Ecotoxicology and Toxicology of Chemicals). 1997. *Ecotoxicology of Some Inorganic Borates—Interim Report*. Special Reports 11, Brussels, Belgium: ECETOC.

Eckel, H., Roth, U., Döhler, H. et al. 2005. *Assessment and Reduction of Heavy Metal Input into Agro-Ecosystems*. Final report EU-Concerted Action. Aromis. Darmstadt, Germany: KTBL-Schrift.

Edwardson, J.A., Moore, P.B., Ferrier, I.N. et al. 1993. Effect of silicon on gastrointestinal absorption of aluminum. *Lancet* 342:211–212.

EFSA (European Food Safety Authority). 2008. Safety of aluminum from dietary intake. *Eur. Food Saf. Authority J*. 754:1–34.

EFSA (European Food Safety Authority). 2009a. Scientific opinion. Potassium molybdate as a source of molybdenum added for nutritional purposes to food supplements. *Eur. Food Saf. Authority J*. 1136:1–21.

EFSA (European Food Safety Authority). 2009b. Scientific opinion. Uranium in foodstuffs, in particular mineral water. *Eur. Food Saf. Authority J*. 1018:1–59.

EFSA (European Food Safety Authority). 2010. Scientific opinion on the substantiation of health claims related to molybdenum. *Eur. Food Saf. Authority J*. 8:1745.

EFSA (European Food Safety Authority). 2012a. Cadmium dietary exposure in the European population. *Eur. Food Saf. Authority J*. 10:2551–2558.

EFSA (European Food Safety Authority). 2012b. Scientific opinion on safety and efficacy of cobalt compounds (E3) as feed additives for all animal species. *Eur. Food Saf. Authority J*. 10:2791.

EFSA (European Food Safety Authority). 2013. EFSA NDA, scientific opinion on dietary reference values for manganese. *Eur. Food Saf. Authority J*. 11:3419.

Eisler, R. 1998. *Copper Hazard to Fish, Wildlife, and Invertebrates: A Synoptic Review*. National Biological Service, Washington, DC: USDI.

Eisler, R. 2004. Gold concentration in abiotic materials, plants, and animals: A synoptic review. *Environ. Monit. Assess*. 90:73–88.

El Samad, O., Baydoun, R., El Jeaid, H. 2010. Activity concentrations of polonium-210 and lead-210 in Lebanese fish. *Lebanese Sci. J*. 11:39–45.

Emsley, J. 2006. *The Element of Murder: A History of Poison*. University Press: Oxford.

Emsley, J. 2011. *Nature's Building Blocks: An A–Z Guide to the Elements*. Oxford: Oxford University Press.

Emsley, J. 2014. The A–Z of zirconium. *Nat. Chem*. 6:254.

Ensminger, A.H., Ensminger, M.E., Konlade, J.E., Roson, J.R.K. 1995. *The Concise Encyclopedia of Foods and Nutrition*, 2nd ed. Boca Raton, FL: CRC Press.

EPA (Environmental Protection Agency). 1986. *Quality Criteria for Water*. Washington, DC: EPA.

EPA (Environmental Protection Agency). 1999. *National Recommended Water Quality Criteria Correction*. EPA-822-Z-99-001. Washington, DC: EPA.

EPA (Environmental Protection Agency). 2000. *Prediction of Sediment Toxicity Using Consensus-Based Freshwater Sediment Quality Guidelines*. USEPA GLNPO. Washington, DC: EPA.

EPA (Environmental Protection Agency). 2002. *Facts About Technetium-99*. Washington, DC: EPA.

EPA (Environmental Protection Agency). 2013. http://www.epa.gov/emap/maia/html/docs/Est5.pdf (Accessed: April 7, 2013).

Erickson, B.E. 2007. The therapeutic use of radon: A biomedical treatment in Europe; an “alternative” remedy in the United States. *Dose Response* 5:48–62.

Eriksson, J.E. 2001a. *Concentration of 61 Trace Elements in Sewage Sludge, Farmyard Manure, Mineral Fertilizers, Precipitation and in Oil and Crops*. Swedish EPA Report 5159. Stockholm, Sweden: Swedish Environmental Protection Agency.

Eriksson, J.E. 2001b. Critical load set to “no further increase in Cd content of agricultural soils” consequences. *Proceedings of International Expert Group on Effect-Based Critical Limits for Heavy Metals*, September 15–17, Bratislava, Slovakia.

Ermakov, V.V. 2004. The halogens. Fluorine. In: *Elements and Their Compounds in the Environment*, 2nd ed., eds. E. Merian, M. Anke, M. Ihnat, and M. Stoeppler, 1415–1421, Weinheim, Germany: Wiley.

EU (European Union). 1998a. Council directive of 3 November 1998 on the quality of water intended for human consumption. EU Directive 98/83/EC. *Off. J. Eur. Union*. 330:32–54.

EU (European Union). 1998b. Opinion of the scientific committee on cosmetic and non-food products intended for consumers concerning zirconium and compounds, adopted by the plenary session of the SCCNFP of 21 January 1998. http://ec.europa.eu/health/scientific_committees.

EU (European Union). 2014. Commission regulation (EU) No488/2014 of 12 May 2014 amending regulation (EC) No 1881/2006 as regards maximum levels of cadmium in foodstuffs. *Off. J. Eur. Union*. L:138/75.

Fahey, N.S.C., Karagatzides, J.D., Jayasinghe, R., Tsuji, L.J.S. 2008. Wetland soil and vegetation bismuth content following experimental deposition of bismuth pellets. *J. Environ. Monit*. 10:951–954.

Falandysz, J., Drewnowska, M., Jarzyńska, G. et al. 2012. Mineral constituents in common chanterelles and soils collected from a high mountain and lowland sites in Poland. *J. Mt. Sci*. 9:697–705.

Fan, J.-L., Ziadi, N., Belenger, G. et al. 2009. Cadmium accumulation in potato tubers produced in Quebec. *Can. J. Soil Sci*. 89:435–443.

Farago, M.E., Kavanagh, P.O., Blanks, R.P. et al. 1998. Platinum concentration in urban dust and soil, and in blood and urine in the United Kingdom. *Analyst* 23:451–454.

Farmer, J.G., Graham, M.C., Thomas, R.P. et al. 1999. Assessment and modeling of the environmental chemistry and potential for remediation treatment of chromium-contaminated land. *Environ. Geochem. Health* 21:331–337.

Faurschou, A., Manné, T., Johansen, J.D., Thyssen, J.P. 2011. Metal allergen of the 21st century—A review on exposure, epidemiology and clinical manifestations of palladium allergy. *Contact Dermatitis* 64:185–195.

Felix, H.R., Kayse, A., Schulin, R. 1999. Phytoremediation, field trails in the years 1993–1998. *5th ICOBTE International Conference*. October 20–22, Vienna, Austria.

Fendorf, S., La Force, M.J., Li, G. 2004. Temporal changes in soil partitioning and bioaccessability of arsenic, chromium, and lead. *J. Environ. Qual*. 33:2049–2055.

Feng, X.H., Zhai, L.M., Tan, W.F. et al. 2007. Adsorption and redox reactions of heavy metals on synthesized Mn oxide minerals. *Environ. Pollut*. 147:366–373.

Fernandez-Calvino, D., Rodriques-Suares, J.A., Lopez-Periago, E. et al. 2008. Copper contents of soils and river sediments in vinegrowing area, and its distribution among soil or sediments compartments. *Geoderma* 145:91–97.

Ferrand, E., Benedetti, M.F., Leclerc-Cessac, E., Dumat, C. 2006. Study of the mechanisms involved in the rhizosphere for the absorption of zirconium by vegetables. *Difpolmine Conference*. Le Corum: Montpellier, France, December, 12–14, 1–4.

Ferrari, C.P., Hong, S., van de Velde, K. et al. 2000. Natural and anthropogenic bismuth in Central Greenland. *Atmos. Environ.* 34:941–948.

Filella, M., Benzile, N., Chen, Y.-W. 2000. Antimony in the environment: A review focused on natural waters. *Earth Sci. Rev.* 57:125–176.

Finkelman, R.B. 1999. Trace elements in coal. Environmental and health significance. *Biol. Trace Elem. Res.* 67:197–204.

Fjällborg, B., Dave, G. 2003. Toxicity of copper in sewage sludge. *Environ. Int.* 28:761–769.

Flachowsky, G. 2007. Iodine in animal nutrition and iodine transfer from feed into food of animal origin. *Lohmann Inf.* 42:47–59.

Flanagan, S.V., Johnston, R.B., Zheng, Y. 2012. Arsenic in tube well water in Bangladesh: Health and economic impacts and implications for arsenic mitigation. *Bull. World Health Organ.* 90:839–846.

Flora, S.J.S., Das Gupta, S. 1994. Toxicology of gallium arsenide: An appraisal. *Defence Sci. J.* 44:5–10.

Flynn, H.C., Meharg, A.A., Bowyer, P.K., Paton, G.I. 2003. Antimony bioavailability in mine soils. *Environ. Pollut.* 124:93–100.

FNB/IOM (Food and Nutrition Board/Institute of Medicine). 2001. *Dietary Reference Intakes for Vitamin A, Vitamin K, Arsenic, Boron, Chromium, Copper, Iodine, Iron, Manganese, Molybdenum, Nickel, Silicon, Vanadium, and Zinc: Food and Nutrition Board. Institute of Medicine*. Washington, DC: National Academy Press.

Fołta, M., Bartoń, H. 2011. Lithium levels in mineral water and household drinking water from Kraków and Southern Poland. *Bromat. Chem. Toksykol.* 44:754–759 (in Polish).

Font, Q., Querol, X., Lopez-Soler, A. et al. 2005. Ge extractions from gasification fly ash. *Fuel* 84:1384–1392.

Fordyce, F. 2005. Selenium deficiency and toxicity in the environment. In: *Essentials of Medical Geology*. Eds. O. Selinus, B.J. Alloway, B.J. Centeno et al. 373–415, Elsevier: Amsterdam, the Netherlands.

FOREGS (Forum of the European Geological Survey). 2005. *Forum of the European Geological Survey Directors*. Geological Atlas of Europe, Survey of Finland. Espoo, Finland.

Forlenza, O.V., de Paula, V.J., Machado-Vieira, R. et al. 2012. Does lithium prevent Alzheimer's disease? *Drugs Aging* 29:335–342.

Fraga, C.G. 2005. Review. Relevance, essentiality and toxicity of trace elements in human health. *Mol. Aspects Med.* 26:235–244.

Fuge, R. 1988. Source of halogens in the environment, influence on human and animal health. *Environ. Geochem. Health* 10:51–61.

Furr, A.K., Lawrence, A.W., Tong, S.S.C. et al. 1976. Multielement and chlorinated hydrocarbon analysis of municipal sewage sludge of American cities. *Environ. Sci. Technol.* 10:683–695.

Gabryszuk, M., Słoniewski, K., Metera, E., Sakowski, T. 2010. Content of mineral elements in milk and hair of cows from organic farms. *J. Elementol.* 15:259–267 (in Polish).

Gać, P., Pawlas, K. 2011. Blood selenium concentration in various populations of healthy and sick people-review of literature from the years 2005–2010. *Med. Środowisk.* 14:93–104 (in Polish).

Gaillardet, J., Viers, J., Dupré, B. 2003. Trace elements in river waters. In: *Treatise on Geochemistry*, eds. H.D. Holland, K.K. Turekian, Vol. 5, 225–227. Oxford: Elsevier.

Gajewska. E., Skłodowska, M. 2008. Different biochemical response of wheat shoots and roots to nickel stress: Antioxidative reactions and proline accumulation. *Plant Growth Regul.* 54:179–188.

Gal, J., Hursthouse, A., Cuthbert, S. 2007. Bioavailability of arsenic and antimony in soils from an abandoned mining area, Glendinning (SW Scotland). *J. Environ. Sci. Health Part A* 42:1263–1274.

Galkus, A., Joksas, K., Stakeniene, R., Lagunaviciene, L. 2012. Heavy metal contamination of harbor bottom sediments. *Pol. J. Environ. Stud.* 21:1583–1594.

Garcia, C.A.B., Passos, E.A., Alves, J.P.H. 2011. Assessment of trace metals pollution in estuarine sediments using SEM-AVS and ERM-ERL predictions. *Environ. Monit. Assess.* 181:385–395.

Garcia-Salgado, S., Garcia-Casillas, D., Quijano-Nieto, M.A., Bonilla-Simon, M.M. 2012. Arsenic and heavy metal uptake and accumulation in native plant species from soil polluted by mining activities. *Water Air Soil Pollut.* 223:559–572.

Gauthier-Lafaye, F., Pourcelot, L., Eikenberg, J. et al. 2008. Radioisotope contamination from releases of the Tomsk-Seversk nuclear facility (Siberia, Russia). *J. Environ. Radioact.* 99:680–693.

Gbaruko, B.C., Igwe, J.C. 2007. Tungsten: Occurrence, chemistry, environmental and health exposure issues. *Global J. Environ. Res.* 1:27–32.

Ge, Y., MacDonald, D., Sauve, S., Hendershot, W. 2005. Modeling of Cd and Pb speciation in soil solutions by VinHumicV and NICA-Donnan model. *Environ. Model. Software* 20:353–359.

Geebelen, W., Bijnens, O.A., Claeys, N. et al. 2005. Transplanted lichens as biomonitors for atmospheric fluoride pollution near two fluoride points sources in Flanders (Belgium). *Belgian J. Bot.* 138:141–151.

Geelhoed, J.S., Meeussen, C.L., Roe, M.J. et al. 2003. Chromium remediation or release? Effect of iron (III) sulfate addition on chromium (VI) leaching from columns of chromite ore processing residue. *Environ. Sci. Technol.* 32:3206–3213.

Gensemer, R.W., Playle, R.C. 1999. The bioavailability and toxicity of aluminum in aquatic environments. *Crit. Rev. Environ. Sci. Technol.* 29:315–450.

Genuis, S.J., Bouchard, T.P. 2012. Combination of micronutrients for bone (COMB) study: Bone density after micronutrient intervention. *J. Environ. Public Health.* Article ID 354151, doi:10.1155/2012/354151.

Goldbach, H., Wimmer, A. 2007. Boron in plants and animals: Is there a role beyond cell-wall structure? *J. Plant Nutr. Soil Sci.* 170:39–48.

Gomez, B., Palacios, M.A., Gomez, M. et al. 2002. The rare platinum metals. *Sci. Total Environ.* 299:1–19.

Gonzalvez, A., Armenta, S., De la Guardia, M. 2008. Trace elements composition of curry by inductively coupled plasma optical emission spectrometry (ICP-OES). *Food Addit. Contam.* 1:114–121.

Gopal, J., Hasan, N., Manikandan, M., Hui-Fen Wu, H.-F. 2013. Bacterial toxicity/compatibility of platinum nanospheres, nanocuboids and nanoflowers. *Sci. Rep.* Article ID 1260, doi:10.1038/srep 01260.

Gorbunov, A.V., Lyapunov, S.M., Okina, O.I. et al. 2004. Assessment of human organism's intake of trace elements from staple foodstuffs in central region of Russia. Preprint of the Joint Institute for Nuclear Research, Dubna, Moscow, Russia. http://wwwinfo.jinr.ru/publish/Preprints/2004/089(D14-2004-89)_e.pdf.

Gough, L.P., Severson, R.C., Shacklette, H.T. 1988. Element concentration in soils and other surficial materials of Alaska. *U.S. Geol. Prof. Pub.* 53.

Govindaraju, K. 1994. Compilation of working values and sample description for 383 geostandards. *Geostand. Newslett.* 18:1–158.

Gramowska, H., Siepak, J. 2002. The effect of fluoride level on the state of leaves and needles of trees in Poznan city and its vicinities. *Roczn. Ochrona Srod.* 4:445–477 (in Polish).

Gray, C.W., McLarre, R.G., Roberts, H.G. 2003. Atmospheric accession of heavy metals to some New Zealand pastoral soils. *Sci. Total Environ.* 305:105–115.

Gregurek, D., Melcher, F., Niskavaara, H. et al. 1999. Platinum-group elements (Rh, Pt, Pd) and Au distribution in snow samples from the Kola Peninsula (NW Russia). *Atmosph. Environ.* 33:3281–3290.

Griethuysen von, C., de Lange, H.J., van der Heuij, M. et al. 2006. Temporal dynamics of AVS and SEM in sediment of shallow fresh floodplain lakes. *Appl. Geochem.* 21:632–642.

Grunfeld, J.P., Rossier, B.C. 2009. Lithium nephrotoxicity revisited. *Nat. Rev. Nephrol.* 5:270–276.

Guilarte, T.R. 2010. Manganese and Parkinson's disease: A critical review and new findings. *Environ. Health Persp.* 118:1071–1080.

Gulińska, J., Rachlewicz, I.E., Szczuciński, W. et al. 2003. Soils contamination in high Arctic area of human impact, Central Spitsbergen, Svalbard. *Pol. J. Environ. Stud.* 12:701–707.

Gundersen, V., Bechmann, I.E., Behrens, A., Stürup, S. 2000. Comparative investigation of major and trace elements in organic and conventional Danish agricultural crops. 1. Onions and peas. *J. Agric. Food Chem.* 48:6094–6102.

Haldimann, M., Alt, A., Blanc, A. et al. 2013. Migration of antimony from PET trays into food simulant and food: Determination of Arrhenius parameters and comparison of predicted and measured migration data. *Food Addit. Contam.* 30:587–598.

Hallberg, L. 1981. Bioavailability of dietary iron in man. *Ann. Rev. Nutr.* 1:123–147.

Hallberg, L., Hulthén, L., Gramatkovski, E. 1997. Iron absorption from the whole diet in men: How effective is the regulation of iron absorption? *Am. J. Clin. Nutr.* 66:347–356.

Halliwell, B., Gutteridge, J. 2007. *Free Radicals in Biology and Medicine*. 4th ed. Oxford: Oxford University Press.

Hamilton, E.I., Minski, M.J. 1972/1973. Abundance of the chemical elements in man's diet and possible relations with environmental factors. *Sci. Total Environ.* 1:375–394.

Hammel, W., Debus, R., Steubing, I. 2000. Mobility of antimony in soil and its availability to plants. *Chemosphere* 41:1791–1798.

Hamon, R., McLaughlin, M., Lombi, E. (eds.) 2006. *Natural Attenuation of Trace Elements Availability in Soils*. Boca Raton, FL: Taylor & Francis.

Hanczakowska, E., Hanczakowski, P. 2010. Rare earth elements in farm animal feeding. *Wiad. Zootech.* 48:15–20 (in Polish).

Hashimoto, Y. 2007. Citrate sorption and biodegradation in acid soils with implication for aluminum rhizotoxicity. *Appl. Geochem.* 22:2861–2871.

Hassanein, M., Anke, M., Hussein, L. 2000. Determination of iodine content in traditional egyptian foods before and after a salt iodination programme. *Pol. J. Food Nutr. Sci.* 9/50:23–25.

He, M., Yang, J. 1999. Effects of different forms of antimony on rice during the period of germination and growth and antimony concentration in rice tissue. *Sci. Total Environ.* 243/244:149–155.

Health Canada. 1991. *Boron Guideline, 09/90*. Ottawa, ON: Publications Health Canada.

Health Canada. 1999. *Antimony—Guidelines for Canadian Drinking Water Quality: Supporting Documentation*. Ottawa, ON: Publications Health Canada.

Health Canada. 2010a. Draft screening assessment for the challenge. Antimony oxide. Environment Canada, Health Canada. http://www.hc-sc.gc.ca/ewh-semt/pubs/contaminants/chms-ecms/section8-eng.php#tphp.

Health Canada. 2010b. *Report on Human Biomonitoring of Environmental Chemicals in Canada*. Ottawa, ON: Publications Health Canada.

Hécho, L.L., Tolu, J., Thiry, Y. et al. 2012. Influence of selenium speciation and fractionation on its mobility in soils. In: *Competitive Sorption and Transport of Heavy Metals in Soils and Geological Media*, ed. H.M. Selim, 215–232. Boca Raton, FL: CRC Press.

Heitkemper, D.T., Kubachka, K.M., Halpin, P.R. et al. 2009. Survey of total arsenic speciation in US-produced rice as a reference point for evaluating change and future trends. *Food Addit. Contam. Part B*. 2:112–120.

Hendrix, P., Van Cauvenbergh, R., Robberecht, H., Deelstra, H. 1997. Daily dietary rubidium intake in Belgium using duplicate portion sampling. *Z. Lebesm. Forsch. A* 204:165–167.

Hinton, T.G., Knox, A.S., Kaplan, D.I., Bell, N.C. 2005. Phytoremedation potential of native trees in a uranium and thorium contaminated wetlands. *J. Radioanal. Nucl. Chem.* 264:417–422.

Hirano, S., Suzuki, K.T. 1996. Exposure, metabolism, and toxicity of rare earths and related compounds. *Environ. Health Persp.* 104S:85–95.

Hlušek, J. 2000. Beryllium and its effect on the quality of lettuce. *Probl. Post. Nauk Roln.* 472:305–310.

Höll, W.H. 2011. Mechanisms of arsenic removal from water. *Environ. Geochem. Health* 32:287–290.

Holmgren, G.G.S., Meyer, M.W., Chaney, R.L., Davies, R.B. 1993. Cadmium, lead, zinc, copper and nickel in agricultural soils of the United States of America. *J. Environ. Qual.* 22:335–348.

Hope, S.-J., Daniel, K., Gleason, K.L. et al.2006. Influence of tea drinking on manganese intake, manganese status and leucocyte expression of MnSOD and cytosolic aminopeptidase. *P. Eur. J. Clin. Nutr.* 60:1–8.

Hoppstock, K., Sures, B. 2004. Platinum group metals. In: *Elements and Their Compounds in the Environment*, 2nd ed., eds. E. Merian, M. Anke, M. Ihnat, and M. Stoeppler, 1047–1086, Weinheim, Germany: Wiley.

Horovitz, C.T. 2000. *Biochemistry of Scandium and Yttrium*. Part 2. New York: Kluwer Academic Publication.

Hou, H., Swennen, R., Deckers, J., Maquill, R. 2005. Concentration of Ag, In, Sn, Sb and Bi, and their chemical fractionations in typical soils in Japan. *Eurasian J. Soil Sci.* 57:214–277.

Houba, V.J.G., Uittenbogaard, J. 1994. *Chemical Composition of Various Plant Species*. Wageningen, the Netherlands: Agriculture University Press.

Howe, A., Fung, L.H., Lalor, R. et al. 2005. Elemental composition of Jamaican foods 1: A survey of five food crop categories. *Environ. Geochem. Health* 27:19–30.

Hower, J., Dai, Sh., Serdin, V. et al. 2013. Lanthanides in coal combustion fly ash. *World of Coal Ash Conference*. Lexington, KY, April 22–24.

Hsu, C.-W., Chang, Y.-J., Chang, C.-H. et al. 2012. Comparative therapeutic efficacy of rhenium-188 radiolabeled-liposome and 5-fluorouracil in ls-174 human colon carcinoma solid tumor xenografts. *Cancer Biother. Radio* 27:481–489.

Hsu, P.-C., Guo, Y.L. 2002. Antioxidant nutrients and lead toxicity. *Toxicology* 30:33–44.

Hu, Z., Haneklaus, S., Sparovek, G., Schnung, E. 2006. Rare earth elements in soils. *Commun. Soil Sci. Plant Anal.* 37:1381–1420.

Huang, J.-H., Hu, K.-N., Ilgen, J., Ilgen, G. 2012. Occurrence and stability of inorganic and organic arsenic species in wines, rice wines and beers from Central Europe market. *Food Addit.Contam.* 20:85–93.

Hunt, C.D., Shuler, T.R., Mullen, L.M. 1991. Concentration of boron and other elements in human foods and personal-care products. *J. Am. Diet. Assoc.* 91:558–568.

Hunt, G.J., Rumney, H.S. 2007. The human alimentary tract transfer and body retention of environmental polonium-210. *J. Radiol. Prot.* 27:405–426.

Huq, S.M.I., Naidu, R. 2005. Arsenic in groundwater and contamination of food chain: Bangladesh scenario. In: *Natural Arsenic in Groundwater: Occurrence, Remediation and Management*, eds. J. Bundschuh, P. Bhattacharya, and D. Chandrasekharam, 451–502, Leiden, The Netherlands: A.A. Balkema Publication.

IARC (International Agency for Research on Cancer). 1988. Man-made mineral fibres and radon. *Monographs on the Evaluation of Carcinogenic Risks to Humans*. Vol. 43. Lyon, France: IARC.

IARC (International Agency for Research on Cancer). 1989. Summaries and evaluations. Antimony trioxide and antimony trisulfide. *Monographs on the Evaluation of Carcinogenic Risks to Humans*. Vol. 47. Lyon, France: IARC.

IARC (International Agency for Research on Cancer). 1990. Chromium, Nickel and welding. *IARC Monographs on the Evaluation of Carcinogenic Risks to Humans*, Vol. 49. Lyon, France: IARC.

IARC (International Agency for Research on Cancer). 1993. Beryllium, Cadmium, Mercury, and exposures in the Glass Manufacturing Industry. *IARC Monographs on the Evaluation of Carcinogenic Risks to Humans*. Vol. 58. Lyon, France: IARC.

IARC (International Agency for Research on Cancer). 2001. Ionizing radiation, part 2: Some internally deposited radionuclides. *IARC Monographs on the Evaluation of Carcinogenic Risks to Humans*. Vol. 78. Lyon, France: IARC.

IARC (International Agency for Research on Cancer). 2006. Cobalt in hard metals and Cobalt Sulfate, Gallium Arsenide, Indium Phosphide and Vanadium Pentoxide. *IARC Monographs on the Evaluation of Carcinogenic Risks to Humans*. Vol. 85. Lyon, France: IARC.

IARC (International Agency for Research on Cancer). 2012a. A review of Human Carcinogens: Arsenic, Metals, Fibres, and Dusts. *IARC Monographs on the Evaluation of Carcinogenic Risks to Humans*. Vol. 100C. Lyon, France: IARC.

IARC (International Agency for Research on Cancer). 2012b. A review of Human Carcinogens: Radiation. *IARC Monographs on the Evaluation of Carcinogenic Risks to Humans*. Vol. 100D. Lyon, France: IARC.

IARC (International Agency for Research on Cancer). 2013. Agents classified by the IARC monographs. *List of Classification*. Vols. 1–108. Lyon, France: IARC.

ICON. 2001. *Pollutants in Urban Waste Water and Sewage Sludge*. Final Report. London: IC Consultants Ltd.

Ikem, A., Nwankwoala, A., Odueyungbo, S. et al. 2002. Levels of 26 elements in infant formula from USA, U.K. and Nigeria by microwave digestion and ICP-OES. *Food Chem.* 77:439–447.

Inter Clinical Laboratories. 2011. Bismuth, germanium and zirconium. *Newsletter* 15, http://www.interclinical.com.au/newsletters/nl2011v15n1.pdf.

Isaac, J., Nohra, J., Lao, J. et al. 2011. Effects of strontium-doped bioactive glass on the differentiation of cultured osteogenic cells. *Eur. Cells Mater.* 21:130–143.

Isaure, M.P., Fraysse, A., Deves, G. et al. 2005. Micro-chemical imaging of cesium distribution in *Arabidopsis thaliana* plants and its interaction with potassium and essential trace elements. *Biochemie* 88:1538–1590.

Iwegbue, C.M.A. 2011. Concentration of selected metals in candies and chocolates consumed in southern Nigeria. *Food Addit. Contam.* 4:22–27.

Jacks, G., Bhattacharya, P., Chandhary, V., Singh, K.P. 2005. Controls on the genesis of some high-fluoride groundwaters in India. *Appl. Geochem.* 20:221–228.

Jagoe, C.H., Chesser, R.K., Smith, M.H. et al. 1997. Levels of cesium, mercury and lead in fish, and cesium in pond sediments in an inhabited region of the Ukraine near Chernobyl. *Environ. Pollut*. 98:223–232.

Jakupec, M.A., P. Unfried, P., Keppler, B.K. 2005. Pharmacological properties of cerium compounds. *Rev. Physiol. Biochem.* 153:101–111.

Jankiewicz, B., Adamczyk, D. 2007. Assessing heavy metals contents in soils surrounding Lodz EC4 power plant, Poland. *Pol. J. Environ. Stud.* 16:933–938.

Jaritz, M. 2004. Barium. In: *Elements and Their Compounds in the Environment*, 2nd ed., eds. E. Merian, M. Anke, M. Ihnat, and M. Stoeppler, 627–634, Weinheim, Germany: Wiley.

Jaworska, H., Dąbkowska-Naskręt, H. 2006. Zirconium in alfisols of different granulometric composition within soil profiles. *Pol. J. Environ. Stud.* 15:312–315.

Jędrzejczak, R. 2002. Determination of total mercury in foods of plant origin in Poland by cold vapour atomic absorption spectrometry. *Food Addit. Contam.* 19:996–1002.

Jedynak, L., Kowalska, J., Leporowska, A. 2012. Arsenic uptake and phytochelatin synthesis by plants from two arsenic-contaminated sites in Poland. *Pol. J. Environ. Stud.* 21:1629–1633.

Jefferson, R.D., Goan, R.E., Blain, P.G., Thomas, S.H. 2009. Diagnosis and treatment of polonium poisoning. *Clin. Toxicol. (Phila).* 47:379–392.

Jeng, M.S., Jeng, W.L., Hung, T.C. et al. 2000. Mussel watch: A review of Cu and other metals in various marine organisms in Taiwan, 1991–98. *Environ. Pollut.* 110:207–215.

Jenny-Burri, J., Haldimann, M., Dudler, V. 2010. Estimation of selenium intake in Switzerland in relation to selected food groups. *Food Addit. Contam.* 27:1516–1531.

Jorhem, L., Estrand, C., Sunström, B. et al. 2008a. Elements in rice from the Swedish market: Part 1. Cadmium, lead and arsenic (total and inorganic). *Food Addit. Contam.* 25:284–292.

Jorhem, L., Strand, C., Sunström, B. et al. 2008b. Elements in rice from the Swedish market: Part 2. Chromium, copper, iron, manganese, platinum, rubidium, selenium and zinc. *Food Addit. Contam.* 25:841–850.

Jugdaohsingh, R. 2007. Silicon and bone health. *J. Nutr. Health Aging* 11:99–110.

Jugdaohsingh, R., Hui, M., Simon, H.C. et al. 2013. The silicon supplement "Monomethylsilanetriol" is safe and increases the body pool of silicon in healthy pre-menopausal women. *Nutr. Metab.* 10:37–42.

Julshamn, K., Nilsen, B.M., Frantzen, S. 2012. Total and inorganic arsenic in fish samples from Norwegian waters. *Food Addit. Contam. Part.* 5:229–235.

Kabala, C., Singh, B.R. 2001. Fractionation and mobility of copper, lead and zinc in soil profiles in the vicinity of a copper smelter. *J. Environ. Qual.* 30:485–492.

Kabat, K., Skwarzec, B., Astel, A. 2013. Discharge of polonium, uranium and plutonium from Vistula and Odra river systems. *Analityka* 14:28–31 (in Polish).

Kabata-Pendias, A. 2011. *Trace Elements in Soils and Plants*. 4th ed. Boca Raton, FL: CRC Press.

Kabata-Pendias, A., Mukherjee, A.B. 2007. *Trace Elements from Soil to Human*. Berlin, Germany: Springer.

Kabata-Pendias, A., Pendias, H. 1999. *Biogeochemistry of Trace Elements*. 2nd ed. Warsaw, Poland: Wydawnictwo Naukowe PWN (in Polish).

Kabata-Pendias, A., Sadurski, W. 2004. Trace elements and compounds in soils. In: *Elements and Their Compounds in the Environment*, 2nd ed., eds. E. Merian, M. Ihnat, and M. Stoeppler, 79–99, Weinheim, Germany: Wiley.

Kagan, L.M., Kadatsky, V.B. 1996. Depth migration of Chernobyl originated 137Cs and 90Sr in soils of Belarus. *J. Environ. Radioact.* 33:27–39.

Kaneko, Y., Thoendel, M., Olakanmi, O. et al. 2007. The transition metal gallium disrupts *Pseudomonas aeruginosa* iron metabolism and has antimicrobial and antibiofilm activity. *J. Clin. Invest.* 117:877–888.

Kapala, J., Karpinska, M., Mnich, Z. et al. 2008. The changes in the contents of 137Cs in bottom sediments of some Masuria lakes during 10–15 y observation. *Radiat. Prot. Dosimetry* 130:178–185.

Kaplan, D.I., Gergely, V., Dernovics. M., Fodor, P. 2005. Cesium-137 partitioning to wetland sediments and uptake by plants. *J. Radioanal. Nucl. Chem.* 264:393–399.

Kaplan, D.I., Knox, A.S. 2004. Enhanced contaminant desorption induced by phosphate mineral additions to sediments. *Environ. Sci. Technol.* 38:3153–3160.

Kapusta, N.D., Mossaheb, N., Etzersdorfer, E. et al. 2011. Lithium in drinking water and suicide mortality. *Br. J. Psychiat.* 198:346–350.

Karczewska, A. 2004. Function of iron and manganese oxides in the sorption of heavy metals in polluted soils as related to the sequential extraction. *Sci. Heft. Agr. Univ.* 432:1–59. (in Polish)

Karczewska, A., Gałka, B., Gersztyn, L., Popielas, K. 2013. Effects of forest litter on copper and zinc solubility in polluted soils—Examined in a pot experiment. *Fresenius Environ. Bull.* 22:949–954.

Karolewski, P., Siepak, J., Gramowska, H. 2000. Response of Scots pine, Norway spruce and Douglas fir needles to environment pollution with fluorine compounds. *Dendrology* 45:41–46.

Kashparov, V., Colle, C., Zvarich, S. et al. 2005. Soil to plant halogen transfer studies-2. Root uptake of radiochlorine by plant. *J. Environ. Radioact.* 79:233–253.

Kekli, A., Aldaham, A., Meili, M. et al. 2003. 129I in Swedish rivers: Distribution and sources. *Sci. Total Environ.* 309:161–172.

Kerwien, S.C. 1996. Toxicity of tungsten, molybdenum, and tantalum and the environmental and occupational laws associated with their manufacture, use, and disposal. Special Publication ARAED-SP-96002. http://www.dtic.mil/cgi-bin/GetTRDoc?AD=ADA310298.

Khanal, D.R., Knight, A.P. 2010. Selenium: Its role in livestock health and productivity. *J. Agric. Environ.* 11:101–106.

Khellat, N., Zerdaouri, M. 2010. Growth repsonse of duckweed *Lemna gibba* L. to copper and nickel phytoaccumulation. *Ecotoxicology* 19:1363–1368.

Khlebtsov, N., Dykman, L. 2011. Biodistribution and toxicity of engineered gold nanoparticles: A review of in vitro and in vivo studies. *Chem. Soc. Rev.* 40:1647–1671.

Kielhorn, J., Melber, C., Keller, D., Mangelsdorf, I. 2002. Palladium—A review of exposure and effects to human health. *Int. J. Hyg. Environ. Health* 205:417–432.

Kitano, Y. 1992. Water chemistry. In: *Encyclopedia of Earth System Science*, ed. W.A. Nierenberg, 449–470, San Diego, CA: Academic Press.

Knox, A.S., Kaplan, D.I., Hinton, T.G. 2008. Elevated uptake of Th and U by netted chain fern (*Woodwardia areolata*). *J. Radioanal. Nucl. Chem.* 277:169–173.

Knox, A.S., Paller, M.H., Nelson, E.A. et al. 2006. Metal distribution and stability in constructed wetland sediments. *J. Environ. Qual.* 35:1948–1959.

Kobayashi, R. 2004. Tellurium. In: *Trace Elements and Their Compounds in the Environment*. Eds. E. Merian, M. Anke, M. Ihnat, M. Stoeppler, 1407–1414. Wiley-VCH: Weinheim, Germany.

Koch, I. 1998. Arsenic and Antimony Species in the Terrestrial Environment. Ph.D. Dissertation, Vancouver, BC: University of British Columbia.

Korobova, E., Chizhikova, N. 2006. Concentration levels, distribution and mobility of the Chernobyl 137Cs in the alluvial soil profiles in relation to physical and chemical properties of the soil horizons (a case study). *Geophys. Res. Abstr.* 8:91–96.

Kostiainen, E., Turtainen, T. 2013. Artificial radioactivity of Finnish vegetables in the 2000s. *Food Addit. Contam.* 30:1316–1321.

Kot, B., Bobrowska-Grzesik, E. 2006. Contents of some microelements in sunflower and pumpkin seeds. *Bromat. Chem. Toksykol.* 44:176–181 (in Polish).

Kothe, E., Ajit, V. (eds.) 2012. *Bio-Geo Interactions in Metal-Contaminated Soils*. Berlin, Germany: Springer.

Kovalevskyj, A.L., Kovalevskaja, O.M. 2010. *Biogeochemistry of Uranium Ores and Methods of Prospecting*. Novosybirsk, Russia: Akad. Izd. Geo. (in Russian).

Kruszewski, M., Brzoska, K., Brunborg, G. et al. 2011. Toxicity of silver nanomaterials in higher eukaryotes. *Adv. Mol. Toxicol.* 5:179–218.

Krylova, L., Nosov, A.V., Kiseleva, V.P. 2011. Analysis of accumulation factor of cesium-137 in bottom sediments of surface water bodies. *Russian Meteor. Hydro.* 36:33–39.

Kucharczak, E., Moryl, A., Szyposzyński, K., Jopek, Z. 2005. Influence of environment on content of arsenic in tissues of roes and wild pigs. *Acta Sci. Pol. Medicina Veterinaria* 4:141–151 (in Polish).

Kumpiene, J., Bränvall, E., Taraskievićius, R. et al. 2011. Spatial variability of soil contamination with trace elements in preschools in Vilnius, Lithuania. *J. Geochem. Eplor.* 108:15–20.

Kumpiene, J., Ores, S., Mench, M., Maurice, C. 2005. Remediation of CCA contaminated soils with zero valent iron. *ICOBTE 8th International Conference*. Adelaide, Australia, pp. 264–265.

Kwiatkowska-Malina, J., Maciejewska, A. 2013. Uptake of heavy metals by darnel multiflora (*Lolium multiflorum*, Lam.) at diverse soil reactions and organic matter. *Soil Sci. Ann.* 64:19–23.

Kyzioł, J. 2002. Sorption and Binding Force of Organic Substances (Peat) for Selected Cations of Heavy Metals. Ph.D. Thesis. Zabrze, Poland: The Polish Academy of Sciences (in Polish).

Laborga, F., Goriz, M.P., Bolea, E., Castillo, J.R. 2007. Mobilization and speciation of chromium in compost: A methodological approach. *Sci. Total Environ.* 373:383–390.

Langford, N.J., Ferner, R.E. 1999. Toxicity of mercury. *J. Human Hypertens.* 13:651–656.

Lansdown, A.B.G. 2010. A pharmacological and toxicological profile of silver as an antimicrobial agent in medical devices. *Adv. Pharm. Sci.* Article ID 910686. http://dx.doi.org/10.1155/2010/910686.

Lansdown, A.B.G. 2013. *The Carcinogenicity of Metals: Human Risk through Occupational and Environmental Exposure*. Cambridge: Royal Society of Chemistry.

Laxen, D.P.H. 1985. Trace metal adsorption/coprecipitation on hydrous ferric oxide under realistic conditions. *Water Res.* 19:1229–1233.

Leblanc, J.C., Guérin, T., Noël, L. et al. 2005. Dietary exposure estimates of 18 elements from the 1st French Total Diet Study. *Food Addit. Contam.* 22:624–641.

Lee, S.-H., Jung, W., Jeon, B.-H. et al. 2011. Abiotic subsurface behaviors of As(V) with Fe(II). *Environ. Geochem. Health* 33:13–22.

Leikin, J.B., Paloucek, F.P. 2008. *Poisoning and Toxicology Handbook*. New York: Informa Healthcare.

Li, Y.-H., Zhao, Q.-L., Huang, M.-H. 2005. Cathodic adsorptive voltametry of the gallium-alizarin red complex at a carbon paste electrode. *Electroanalysis* 17:343–347.

Liang, T., Zhang, S., Wang, L. et al. 2005. Environmental biochemical behavior of rare earth elements in soil-plant systems. *Environ. Geochem. Health* 27:301–311.

Liang, F., Li, Y., Zhang, G,. Tan, M. et al. 2010. Total and speciated arsenic levels in rice from China. *Food Addit.Contam.* 27:810–816.

Lidwin-Kaźmierkiewicz, M., Pokorska, K., Protasowicki, M. et al. 2009. Content of selected and toxic metals in meat of freshwater fish from West Pomerania, Poland. *Pol. J. Food Nutr. Sci.* 59:219–234.

Limbong, D., Kumanpung, J., Rimper, J. et al. 2003. Emission and environmental implications of mercury from artisal gold mining in North Sulawesi, Indonesia. *Sci. Total Environ.* 302:227–236.

Limeback, H. 2001. Recent studies confirm old problems with water fluoridation: A fresh perspective. *Fluoride* 34:1–6.

Lin, Z.-H., Lee, C.-H., Chang, H.-Y., Chang, H.-T. 2012. Antibacterial activities of tellurium nanomaterials. *Chemistry* 7:930–934.

Liu, H., Probst, A., Liao, B. 2005. Metal contamination of crops affected by the Chenzhou lead/zinc mine spill (Hunan, China). *Sci. Total Environ.* 339:153–166.

Llorens, A.W., Fernandez, J.L., Querol, X. 2000. The fate of trace elements in a large coal-fired power plant. *Environ. Geol.* 40:409–446.

Lombeck, I., Kasperek, K., Feinendegen, L.E., Bremer, H.J. 1980. Rubidium—A possible essential trace element: 1. The rubidium content of whole blood of healthy and dietetically treated children. *Biol. Trace Elem. Res.* 2:193–198.

Lourekari, K., Mäkelä-Kurto, R., Pasanen, J. et al. 2000. Cadmium in fertilizers. Risk to human health and the environment. *Min. Agric. Forestry Publ.* 4/2000. Helsinki, Finland.

Lourencetti, C., Grimalt, J.O., Marco, E. et al. 2012. Trihalomethanes in chlorine and bromine disinfected swimming pools: Air-water distributions and human exposure. *Environ. Int.* 45:59–67.

Lucho-Constantino, C.A., Priesto-Garci, F., Del Razo, L.M. et al. 2005. Chemical fractionation of boron and heavy metals in soil irrigated with wastewater in central Mexico. *Agric. Ecosyst. Environ.* 108:57–71.

Luttrell, W.E., Giles, C.B. 2007. Toxic tips: Osmium tetroxide. *J. Chem. Health Saf.* 14:40–41.

Lyons, M.P., Papazyan, T.T., Surai, P.F. 2007. Selenium in food chain and animal nutrition: Lessons from nature—Review. *Asian-Aust. J. Anim. Sci.* 20:1135–1155.

Ma, L.Q., Komar, K.M. Zhang, T. et al. 2001. A fern that hyperaccumulate arsenic: A hardly, versatile, fast-growing plant helps to remove arsenic from contaminated soils. *Nature* 409:579.

Ma, Y.B., Lombi, E., Nolan, A.L., McLaughlin, M. 2006. Short-term natural attenuation of copper in soils: Effect of time, temperature, and soil characteristics. *Environ. Toxicol. Chem.* 25:652–658.

Maćkowiak, C.L., Grossal, P.R., Bugbee, B.G. 2003. Biogeochemistry of fluoride in a plant-soil system. *J. Environ. Qual.* 32:2230–2237.

Maggiorella, L., Barouch, G., Devaux, C. et al. 2012. Nanoscale radiotherapy with hafnium oxide nanoparticles. *Future Oncol.* 8:1167–1181.

Maján, G.Y., Kozák, M., Püspöki, Z. et al. 2001. Environmental geological examination of chromium contamination in eastern Hungary. *Environ. Geochem. Health* 23:229–233.

Makris, K.C., Andra, S.S., Herrick, L. et al. 2013. Association of drinking-water source and use characteristics with urinary antimony concentrations. *J. Exp. Sci. Environ. Epidemiol.* 23:120–127.

Makuch, I. 2012. Forms of lead in profiles of different use soil. *Soil Sci. Ann.* 63:41–45.

Malik, M., Chaney, R., Brewer, E. et al. 2000. Phytoextraction of soil cobalt using hyperaccumulator plants. *Int. J. Phytoremediation* 2:319–329.

Malinowska, E., Szefer, B., Falandysz, J. 2004. Metal bioaccumulation by bay bolete, *Xerocomus badius*, from selected sites in Poland. *Food Chem.* 84:196–204.

Mallela, V.S., Ilankumaran, V., Rao, N.S. 2004. Trends in cardiac pacemaker batteries. *Indian Pacing Electrophysiol. J.* 4:201–212.

Manaka, M. 2006. Amount of amorphous materials in relation to arsenic, antimony, and bismuth concentrations in a brown forest soil. *Geoderma* 136:75–86.

Manceau, A., Nagy, K.L., Marcus, M.A. et al. 2008. Formation of metallic copper nanoparticles at the soil-root interface. *Environ. Sci. Technol.* 42:1766–1772.

Marengo, E., Aceto, M. 2003. Statistical investigation of the difference in the distribution of metals in Nebbolo-based wines. *Food Chem.* 81:621–630.

Marin, C.M.D.C., Oron, G. 2007. Boron removal by duckweed *Lemna gibba*: A potential method for the remediation of boron-polluted waters. *Water Res.* 41:4579–4584.

Markert, B., Lieth, H. 1987. Element concentration cadaster in Swedish biotope. *Fresenius Z. Anal. Chem.* 327:716–718.

Markert, B., Vtorova, V.N. 1995. Concentration cadasters of chemical elements in plants of eastern European forest ecosystem. *Biol. Bull.* 22:453–460.

Markert, B. 1992. Multi-element analysis in plant materials–Analytical tools and biological questions, In: *Biochemistry of Trace Elements*. Ed. D.C. Adriano, 401–428. Boca Raton, FL: Lewis Publishing.

Markiewicz, K., Nowak-Polakowska, H., Markiewicz, E. et al. 2006. Contents of selected macro- and microminerals and toxic elements in walnuts. *Bromat. Chem. Toksykol.* 39:237–241 (in Polish).

Marschner, H. 2005. *Mineral Nutrition of Higher Plants*. 2nd ed. Amsterdam, The Netherlands: Academic Press.

Martens, R.J., Miller, N.A., Cohen, N.D. et al. 2007. Chemoprophylactic antimicrobial activity of allium maltolate against intracellular *Rhodococcus equi*. *J. Equine Vet. Sci.* 27:341–345.

Marzec, Z., Marzec, A., Zaręba, S. 2004. Daily food rations as a source of iron and manganese for adults. *Roczn. PZH.* 55S:29–34 (in Polish).

Masironi, R., Shaper, A.G. 1981. Epidemiological studies of health effects of water from different sources. *Ann. Rev. Nutr.* 1:375–400.

Matschullat, J. 1997. Trace elements fluxes to the Baltic Sea: Problems of input budgets. *Ambio* 26:363–368.

Matschullat, J. 2000. Arsenic in the geosphere—A review. *Sci. Total Environ.* 249:297–312.

Matsuda, R., Sasaki, K., Saito, Y. 1994. Determination of total bromine in foods by ECD gas chromatograph. *Esei Shikenjo Hokoku* 112:108–111 (in Japanese).

Matsumoto, H., Hinkley, T.K. 2001. Trace metals in Antarctic pre-industrial ice are consistent with emissions from quiescent degassing of volcanoes worldwide. *Earth Planet. Sci. Lett.* 186:3–43.

Matsumoto, H., Yamamoto, Y., Devi, S.R. 2001. Aluminum toxicity in acid soils: Plants response to aluminum stress. In: *Metals in the Environment: Analysis by Biodiversity*, ed. M.N.V. Prasad, 289–320. New York: Marcel Dekker.

Matusik, J., Bajda, T., Manecki, M. 2008. Immobilization of aqueous cadmium by addition of phosphates. *J. Hazard. Mater.* 125:1332–1339.

Mboringong, M.N., Brown, M.E.A., Ashano, E.C., Olasehinde, A. 2013. Assessment of lead, zirconium and iron concentration in soils in parts of Jos Plateau, North Central Nigeria. *Earth Resour.* 1:48–53.

McHard, J.A., Foulk, S.J., Winefordner, J.D. 1979. A comparison of trace element contents of Florida and Brazil orange juice. *J. Agric. Food Chem.* 27:1326–1328.

McMahon, M., Regan, F., Hughes, H. 2006. The determination of total germanium in real food samples including Chinese herbal remedies using graphite furnace atomic absorption spectroscopy. *Food Chem.* 97:411–417.

Meacham, S., Karakas, S., Wallace, A., Altun, F. 2010. Boron in human health: Evidence for dietary recommendations and public policies. *Open Miner. Process. J.* 3:36–53.

Meler, J., Meler, G. 2006. Fluoridation of drinking water-advantages and disadvantages. *J. Elementol.* 11:379–387 (in Polish).

Melo de, W.J. 2012. Mercury sorption and desorption by tropical soils. In: *Competitive Sorption and Transport of Heavy Metals in Soils and Geological Media*, ed. H.M. Selim, 147–214. Boca Raton, FL: CRC Press.

Méplan, C. 2011. Trace elements and ageing, a genomic perspective using selenium as an example. *J. Trace Elem. Med. Biol.* 25:S11–S16.

Mertz, W. 1993. Chromium in human nutrition: A review. *J. Nutr.* 123:626–633.

Mesjarz-Przybyłowicz, J., Balkwill, K., Przybyłowicz, W.J., Annegarn, H.J. 1994. Proton microprobe and X-ray fluorescence investigations of nickel distribution in serpentine flora from South Africa. *Nucl. Instrum. Methods Phys. Res. B* 89:208–212.

Messer, A. 2010. Mini-review: Polybrominated diphenyl ether (PBDE) flame retardants as potential autism risk factors. *Physiol. Behav.* 100:245–249.

Meyers, D.E.R., Achterlonide, G.J., Webb, R.I., Wood, B. 2008. Uptake and localization of lead in the root system of *Brassica juncea*. *Environ. Pollut.* 153:323–332.

Mezhibor, A., Arbuzov, S., Rikhvanov, L., Gauthier-Lafaye, F. 2011. History of the pollution in Tomsk region (Siberia, Russia) according to the study of high-moor peat formation. *Int. J. Geosci.* 2:493–501.

Michalke, K., Schmidt, A., Huber, B. et al. 2008. Role of intestinal microbiota in transformation of bismuth and other metals and metalloids into volatile methyl and hydride derivatives in humans and mice. *Appl. Environ. Microbiol.* 74:3069–3075.

Migaszewski, Z.M., Gałuszka, A., Crock, J.G. et al. 2009. Interspecies and interregional comparison of the chemistry of PAHs and trace elements in mosses *Hylocomium splendens* (Hedw.) B.S.G. and *Pleurozium schreberi* (Brid.) Mitt. from Poland and Alaska. *Atmos. Environ.* 43:1464–1473.

Millour, S., Noël, L., Chekri, R. et al. 2012. Strontium, silver, tin, iron, tellurium, gallium, germanium, barium and vanadium levels in foodstuffs from the Second French Total Diet Study. *J. Food Compos. Anal.* 25:108–129.

Minervino, A.H.H., Barreto, Jr, R.A., Ferreira, R.N.F. et al. 2009. Clinical observations of cattle and buffalos with experimentally induced chronic copper poisoning. *Res. Vet. Sci.* 87:473–478.
Minkina, T.M., Motusova, G.V., Nazarenko, O.G., Mandzhieva, S.S. 2010. *Heavy Metal Compounds in Soil: Transformation upon Soil Pollution and Ecological Significance.* New York: Nova Science Publication Inc.
Mirosławski, J., Kwapuliński, J., Paukszto, R. et al. 2006. Speciation of metals in street dusts. *Pol. J. Environ. Stud.* 15:426–430.
Molina, J.-A., Oyarzun, R., Esbri, J.-M., Higuera, P. 2006. Mercury accumulation in soils and plants in the Almaden mining district, Spain: One of the most contaminated sites on Earth. *Environ. Geochem. Health*, 51:487–498.
Morgado, O., Pereira, V., Pinto, M.S. 2001. Chromium in Portuguese soils surrounding electroplating facilities. *Environ. Geochem. Health* 23:225–228.
Moskalyk, R.R. 2004. Review of germanium processing worldwide. *Miner. Eng.* 17:393–402.
Mukherjee, A.B., Bhattacharya, P. 2001. Arsenic in ground water in the Bengal delta plain: Slow poisoning in Bangladesh. *Environ. Rev.* 9:189–220.
Mukherjee, A.B., Zevenhoven, R., Brodersen, J. et al. 2004. Mercury in waste of the European Union: Sources, disposal methods and risks. *Res. Conserv. Recycling* 42:155–182.
Nachtigall, G.R., Nogueiro, R.C., Alleori, L.R.F., Cambri, M.A. 2007. Copper concentration in vineyard soils as a function of pH variation and addition of poultry litter. *Braz. Arch. Biol. Technol.* 50:941–948.
Nakamaru, Y., Tagami, K., Uchida, S. 2006. Effect of nutrient uptake by plant roots on the fate of REEs in soil. *J. Alloy. Comp.* 408:413–416.
Nassem, M.G., Abdalla, Y.H. 2003. Cobalt status in the North Western Coast soils of Egypt in relation to cobalt content of barley for ruminants. *Proceedings of the 1st International Symposium, Environment Biogeochemistry.* October 10–15, Edinburgh, Scotland.
Neathery, M.W., Miller, W.J. 1975. Metabolism and toxicity of cadmium, mercury, and lead in animals: A review. *J. Dairy Sci.* 58:1767–1781.
Nersyan, G.S. 2007. Studies of accumulation of chlorine in plants of Yerevan City. *Proceedings of International Conference Mountain Areas—Ecological Problem of Cities.* September 5–7, Yerevan, Armenia (in Russian).
Nicholson, F.A., Smith, S.R., Alloway, B. et al. 2003. An inventory of heavy metals inputs to agricultural soils in England and Wales. *Sci. Total Environ.* 311:205–219.
Niedzielski, P., Siepak, J., Kowalczuk, Z. 2000. Speciation analysis for determinations of arsenic, antimony and selenium in water samples from Lednickie Lake. *Arch. Environ. Prot.* 26:73–82.
Niedzielski, P., Siepak, J., Pelechaty, M., Burchardt, L. 2000. Concentrations of arsenic, antimony and selenium in water of lake Wielkopolski National Park. *Morena* 7:69–77 (in Polish).
Nielsen, F.H. 1986. Other elements: Sb, Ba, B, Br, Cs, Ge, Rb, Ag, Sr, Sn, Ti, Zr, Be, Bi, Ga, Au, In, Nb, Se, Tl, W. In: *Trace Elements in Human and Animal Nutrition*, 5th ed., ed. W.M. Orlando, 415–463, New York: Academic Press.
Nielsen, F.H. 2008. Is boron nutritionally relevant? *Nutr. Rev.* 66:183–191.
Niesiobędzka, K. 2012. Transfer of copper, lead and zinc in soil-grass ecosystem in aspect of soil properties in Poland. *Bull. Environ. Contam. Toxicol.* 88:627–633.
Nocito, F.F., Epsen, L., Crema, B. et al. 2008. Cadmium induces acidosis in maize root cells. *New Phytol.* 179:700–711.
Noël, L., Leblanc, J.-C., Guérin, T. 2003. Determination of several elements in duplicate meals from catering establishments using closed vessel microwave digestion with inductively coupled plasma mass spectrometry detection: Estimation of daily dietary intake. *Food Addit. Contam.* 20:44–56.
Nolan, A., Schaumlőffel, D., Lombi, E. et al. 2004. Determination of $Tl_{(I)}$ and $Tl_{(III)}$ by IC-ICP-MS, and application to Tl species in the Tl hyperaccumulator plant *Iberis intermedia. J Anal. At. Spectro.* 19:757–761.

Nowak-Winiarska, K., Wróbel, S., Sienkiewicz-Cholewa, U. 2012. Application of sequential analysis with the BRC method in the estimation of effects of chemical remediation of soils polluted with copper. *Chem. Spec. Bioavailab.* 24:53–59.

Nozaki, Y. 2005. A fresh look at element distribution in the North Pacific, AGU. http://www.agu.org/eos.elec/97025e.html.

NRC. 2001. *Nutrient Requirements of Dairy Cattle*. 7th rev. ed. Washington, DC: National Research Council, National Academy of Sciences.

Nriagu, J.O. 1980. Human influence on the global cadmium cycles. In: *Cadmium in the Environment*, ed. J.O. Nriagu, 81–89, New York: Wiley.

Nriagu, J.O., Pacyna, J.M. 1988. Quantitative assessment of worldwide contamination of air, water and soils by trace metals. *Nature* 1669:134–139.

Nygard, T., Steinnes, E., Royset, O. 2012. Distribution of 32 elements in organic surface soils: Contribution from atmosphere transport of pollutants and natural sources. *Water Air Soil Pollut.* 223:699–713.

OEHHA (Office of Environmental Health Hazard Assessment). 2000. http://www.oehha.ca.gov.

Offem, B.O., Ayotunde, E.O. 2008. Toxicity of lead to freshwater invertebrates (water fleas: *Daphnia magma* and *Cyclop* sp) in fish ponds in a tropical floodplain. *Water Air Soil Pollut.* 192:39–46.

Onishi, Y., Yokuda, A.R. 2013. *Annual Report: Simulate and Evaluate the Cesium Transport and Accumulation in Fukushima-Area Rivers by the TODAM Code*. 2012 PNNL-22364. U.S. Department of Energy.

Oosterhuis, F.H., Brouwer, F.M., Wijnants, H.J. 2000. *A Possible EU Wide Charge on Cadmium in Phosphate Fertilizers: Economic and Environmental Implications*. Final Report to the European Commission. European Commission Report No E-00/02. Amsterdam, The Netherlands: Vrije Universiteit.

Outola, L., Pehrman, R., Jaakola, T. 2003. Effects of industrial pollution on the distribution of 137Cs in soils and soil-to-plant transfer in a pine forest in SW Finland. *Sci. Total Environ.* 303:221–230.

Ozel, H.U. 2012. Biosorption of Cd(II) ions by nordmann fir cones. *Fresenius Environ. Bull.* 21:2527–2535.

Pacha, J., Galimska-Stypa, R. 1988. Mutagenic properties of selected tri- and hexavalent chromium compounds. *Acta Biol. Sil.* 9:30–36.

Pachocki, K.A., Wierzbowski, K., Bekas, M., Różycki, Z. 2009. Occurrence of radon, ^{222}Rn, in mineral water. *Ann. PZH*, 60:129–136 (in Polish).

Pacyna, J.M., Pacyna, E.G. 2001. An assessment of global and regional emissions of trace elements to the atmosphere from anthropogenic sources. *Environ. Rev.* 9:269–298.

Pakkanen, T.A., Loukkola, K., Korhonen, C.H. et al. 2002. Sources and chemical composition of atmospheric fine and coarse particles in the Helsinki area. *Atmos. Environ.* 35:5381–5391.

Pałasz, A., Czekaj, P. 2000. Toxicological and cytophysiological aspects of lanthanides action. *Acta Biochim. Pol.* 47:1107–1114.

Pampura, T., Groenenberg, J.E., Lofts, S., Priturine, I. 2007. Validation of transfer functions predicting Cd and Pb free metal ion activity in soil solution as a function of soil characteristic and reactive metal contents. *Water Air Soil Pollut.* 184:217–234.

Pan, J., Plant, J.A., Voulvouli, N. et al. 2010. Cadmium levels in Europe: Implications for human health. *Environ. Geochem. Health* 32:1–12.

Panin, M.S. 2004. Main sources of pollution in Kazakhstan territory with radionuclides. *Proceedings of the 3rd International Conference Heavy Metals, Radionuclides and Elements-Biofills in Environ*. September 8–10, Semipalatynsk, Kazakhstan (in Russian).

Papić, P., Cuk, M., Todorović, M. et al. 2012. Arsenic in tap water of Serbia's South Pannonian Basin and arsenic risk assessment. *Polish J. Environ. Sci.* 21:1783–1790.

Pappas, A.C., Zoidis, E., Georgiou, C.A. et al. 2011. Influence of organic selenium supplementation on the accumulation of toxic and essential trace elements involved in the antioxidant system of chicken. *Food Addit. Contam.* 28:446–454.

Park, J.-S., Jung, S.-Y., Son, Y.-J. et al. 2011. Total mercury, methylmercury and ethylmercury in marine fish and marine fishery products sold in Seoul, Korea. *Food Addit. Contam.* 4:268–274.

Paschalis, C., Jenner, F.A., Lee, C.R. 1978. Effects of rubidium chloride on the course of manic-depressive illness. *J. Royal Soc. Med.* 71:343–352.

Pasieczna, A. 2012. Content of antimony and bismuth in agricultural soils of Poland. *Pol. J. Agron.* 10:21–29 (in Polish).

Patra, C.R., Moneim, S.S.A., Wang, E.E. et al. 2009. In vivo toxicity studies of europium hydroxide nanorods in mice. *Toxicol. Appl. Pharmacol.* 240:88–98.

Paulíková, I., Kováč, G., Bíreš, J. et al. 2002. Iodine toxicity in ruminants. *Vet. Med.-Czech.* 47:343–350.

Paulo, C., Pratas, J. 2008. Environmental contamination control of water drainage from uranium mines by aquatic plants. In: *Trace Elements as Contaminants and Nutrients*, ed. M.N.V. Prasad, 623–651, Hoboken, NJ: Wiley.

Pavelka, S. 2004a. Bromine. In: *Elements and Their Compounds in the Environment*, 2nd ed., eds. E. Merian, A.M. Ihnat, and M. Stoeppler, 1445–1455. Weinheim, Germany: Wiley.

Pavelka, S. 2004b. Metabolism of bromide and its interference with the metabolism of iodine. *Physiol. Res.* 53. Suppl. 1:81–90.

Pechova, A., Pavlata, L. 2007. Chromium as an essential nutrient: A review. *Veter. Med.* 52:1–18.

Pekka, I., Ingri, J., Widerlund, A. et al. 2004. Geochemistry of the Kola River, northwest Russia. *Appl. Geochem.* 19:1975–1995.

Peng, K.J., Luo, C.L., You, W.X. et al. 2008. Manganese uptake and interactions with cadmium in the hyperaccumulator—*Phytolacca americana* L. *J. Hazard. Mater.* 154:674–681.

Pennington, J.A.T. 1991. Silicon in foods and diets. *Food Addit. Contam.* 8:97–118.

Pennington, J.A.T., Jones, S.W. 1998. Aluminum in foods and diets. *Food Addit. Contam.* 5:161–232.

Perelomov, L.V., Chulin, A.N. 2013. Molecular mechanisms of interaction between microelements and microorganisms in the environment. Direct biological transformation of microelements compounds. *Uspechy Sovr. Biol.* 133:425–470 (in Russian).

Perelomov, L.V., Cozzolino, V., Pigna, M., Violante, A. 2011. Adsorption of Cu and Pb on goethite in the presence of low-molecular mass aliphatic acids. *Geomicrobiol. J.* 28:582–589.

Perelomov, L.V., Kandeler, E. 2006. Effects of soil microorganisms on the sorption of zinc and lead compounds by goethite. *J. Plant Nutr. Soil Sci.* 169:95–100.

Perelomov, L.V., Perelomova, L.V., Pinskij, D.D. 2013. Molecular mechanisms of interaction between microelements and microorganisms in the complex biotic/abiotic systems (biosorption and bioaccumulation). *Agrochimia* 3:80–94 (in Russian).

Perelomov, L.V., Yoshida, S. 2008. Effect of microorganisms on the sorption of lanthanides by quarts and goethite at the different pH values. *Water Air Soil Pollut.* 194:217–225.

Petrunic, B.M., MacQuarrie, K.T.B., Al, T.A. 2005. Reductive dissolution of Mn oxides in river-recharged aquifers: A laboratory column study. *J. Hydrol.* 301:163–181.

Pietrzak-Flis, Z., Orzechowska, G. 1993. Plutonium in daily diet in Poland after the Chernobyl accident. *Health Phys.* 65:489–492.

Pietrzak-Flis, Z., Rosiak, L., Suplińska, M.M. et al. 2001. Daily intake of 338U, 234U, 232Th, 230Th and 236Ra in the adult population of central Poland. *Sci. Total Environ.* 273:163–169.

Pirrone, N., Cinnirella, S., Feng, X. et al. 2009. Global mercury emission to the atmosphere from natural and anthropogenic sources. In: *Mercury Fate and Transport in the Global Atmosphere*. Eds. N. Pirrone, R. Masson, 3–49. Springer: New York.

Pluta, I. 2001. Barium and radium discharged from coal mines in the Upper Silesia, Poland. *Environ. Geol.* 40:345–348.

Pocock, S.J., Shaper, A.G., Walker, M. et al. 1983. Effects of tap water lead, water hardness, alcohol, and cigarettes on blood lead concentrations. *J. Epidemiol. Commun. H.* 37:1–7.

Pohl, P. 2008. Determination and fractionation of metals in beer. A review. *Food Addit. Contam.* 25:693–703.

Polak-Juszczak, L. 2008. Mineral elements content in smoked fish. *Roczn. PZH.* 59:187–196 (in Polish).

Polec-Pawlak, K., Ruzik, R., Lipiec, E. 2007. Investigation of Pb(II) binding to pectin in *Arabidopsis thaliana. J. Anal. At. Spectrom.* 22:968–972.

Poletti, J., Pozebon, D., de Fraga, M.V.B. et al. 2014. Toxic and micronutrient element in organic, brown and polished rice in Brazil. *Food Addit. Contam.* 7:63–69.

Ponizovsky, A.A. et al. 2006. Effect of soil properties on copper release in soil solution at low moisture content. *Environ. Toxic. Chem.* 25:671–682.

Powell, J.J., McNaughton, S.A., Jugdaohsingh, R. et al. 2005. A provisional database for the silicon content of foods in the United Kingdom. *Br. J. Nutr.* 94:804–812.

Prabhu, S., Poulose, E.K. 2012. Silver nanoparticles: Mechanism of antimicrobial action, synthesis, medical applications, and toxicity effects. *Int. Nano Lett.* 2:32.

Prasad, R. 2006. Zinc in soils and in plant, human and animal nutrition. *Indian J. Fertli.* 2:103–119.

Price, C.T., Koval, K.J., Langford, J.R. 2013. Review article. Silicon: A review of its potential role in the prevention and treatment of postmenopausal osteoporosis. *Int. J. Endocrinol.* Article ID 316783.

Puig, S., Mira, H., Dorcey, E. 2007. Higher plants posses two different types of ATX1-like copper chaperones. Biochem. *Biophys. Res. Commun.* 354:385–390.

Rajkowska, M., Holak, M., Protasowicki, M. 2009. Macro- and microelements in some selected assortments of beer. *Żywność. Nauka. Technol. Jakość* (Poland) 2:112–118 (in Polish).

Ramírez, G., Rodil, S.E., Arzate, H. et al. 2011. Niobium based coatings for dental implants. *Appl. Surf. Sci.* 257:2555–2559.

Ramirez-Diaz, M.L., Diaz-Perez, C., Vargas, E. et al. 2008. Mechanisms of bacterial resistance to chromium compounds. *Biometals* 21:321–332.

Randall, P., Chattopadhyay, S. 2004. Influence of pH and oxidation-reduction potential (Eh) on the dissolution of mercury-containing mine wastes from the Sulphur Bank Mercury Mine. *Miner. Metallur. Process.* 21:93–98.

Rasmussen, P.E., Subramanian, K.S., Jessiman, B.J. 2001. A multi-element profile of home dust in relation to exterior dust and soils in the city of Ottawa, Canada. *Sci. Total Environ.* 267:125–140.

Ravindra, T., Bencs, I., van Grieken, R. 2004. Platinum group elements in the environment and their health risk. *Sci. Total Environ.* 318:1–43.

Ravindra, K., Bencs, L., Van Grieken, R. 2004. Platinum group elements in the environment and their health risk. *Sci. Total Environ.* 318:1–43.

Ręczajska, W., Jędrzejczak, R., Szteke, B. 2005. Determination of chromium content in food and beverages of plant origin. *Pol. J. Food Nutr. Sci.* 55:183–188.

Redling, K. 2006. Rare earth elements in agriculture with emphasis on animal husbandry. Ph.D. dissertation. Munich, Germany: Ludwig-Maximilians-Universität München, Tierärztliche Fakultät.

Reif, A.G., Sloto, R.A. 1997. *Metal, Pesticides, and Semivolatile Organic Compounds in Sediment in Valley Forge Historical Park, Montgomery County, Pennsylvania.* USGS Water-Resources Investigations Report. P. 1–18.

Reilly, C. 2002. *Metal Contamination of Food. Its Significance for Food Quality and Human Health.* 3rd ed. Oxford: Blackwell Science Ltd.

Reimann, C., de Caritat, P. 1998. *Chemical Elements in the Environment*. Berlin; Heidelberg, Germany: Springer-Verlag.

Reimann, C., Halleraker, J.H., Kashulina, G., Bogatyrev, I. 1999. Comparison of plant and precipitation chemistry in catchments with different level of pollution on the Kola Peninsula, Russia. *Sci. Total Environ.* 243/244:169–191.

Reimann, C., Koller, G., Kashulina, G. et al. 2001. Influence of extreme pollution on the inorganic chemical composition of some plants. *Environ. Pollut.* 115:239–252.

Reimann, C., Niskavaara, H. 2006. Regional distribution of Pd, Pt and Au emissions from the nickel industry on the Kola Peninsula, NW-Russia, as seen in moss and humus samples. In: *Palladium Emissions in the Environment–Analytical Methods, Environmental Assessment and Health Effects*. Eds. F. Zereini and F. Alt, 53–70, Springer:Heidelberg, Germany.

Reith, F., McPhail, D.C. 2007. Mobility and microbially mediated mobilization of gold and arsenic in soils from two gold mines in semi-arid and tropical Australia. *Geochim. Cosmochim. Acta* 71:1183–1196.

Renaud, P., Pourcelot, L., Metivier, J.-M., Morello, M. 2003. Mapping of 137Cs deposition over eastern France 16 years after the Chernobyl accident. *Sci. Total Environ.* 309:257–264.

Ristić, M., Popović, I., Pocajt, V. et al. 2011. Concentration of trace elements in mineral and spring bottled waters on the Serbian markets. *Food Addit. Contam.* 4:6–14.

Robinson, B.H., Bischofberger, S., Stoll, A. et al. 2008. Plant uptake of trace elements on a Swiss military shooting range. *Environ. Pollut.* 153:668–676.

Robinson, B.H., Green, S.R., Chancerel, B. et al. 2008. Poplar for phytoremanagement of boron contaminated sites. *Environ. Pollut.* 150:225–233.

Rodrigues, J.A., Nans, N., Grau, J.M. et al. 2008. Multiscale analysis of heavy metal contents in Spanish agricultural topsoils. *Chemosphere* 70:1085–1096.

Rojas, F.S., Ojeda, C.B., Pavón, J.M.C. 2007. Determination of rhodium and platinum by electrothermal atomic absorption spectrometry after preconcentration with a chelating resin. *J. Braz. Chem. Soc.* 18:1270–1275.

Rose, M., Baxter, M., Brereton, N., Baskaran, C. 2010. Dietary exposure to metals and other elements in the 2006 UK Total Diet Study and some trends over the last 30 years. *Food Addit. Contam.* 27:1380–1404.

Rosenberg, E. 2009. Germanium: Environmental occurrence, importance and speciation. *Environ. Sci. Biotechnol.* 8:29–57.

Rossmann, M.D. 2004. Beryllium. In: *Elements and Their Compounds in the Environment*, 2nd ed., eds. E. Merian, A. Anke, A., M. Ihnat, and M. Stoeppler, 575–586. Weinheim, Germany: Wiley.

Roussel-Debet, S. 2005. Experimental values for 241Am and 249+240Pu Kd's in French agricultural soils. *J. Environ. Radioact.* 79:171–185.

Ruan, J., Ma, L., Shi, Y. 2006. Aluminum in tea plantations—Mobility in soils and plants, and the influence of nitrogen fertilization. *Environ. Geochem. Health* 28:519–528.

Rudy, M. 2009. Correlation of lead, cadmium and mercury levels in tissue and liver samples with age in cattle. *Food Addit. Contam.* 26:847–853.

Rühling, A., Tyler, G. 2004. Changes in the atmospheric deposition of minor and trace elements between 1975 and 2000 in South Sweden, as measured by moss analysis. *Environ. Pollut.* 131:417–423.

Sabadell, J.E., Axtmann, R.C. 1975. Heavy metal contamination from geothermal sources. *Environ. Health Persp.* 12:1–7.

Sadler, P.J., Guo, Z. 1998. Metal complexes in medicine: Design and mechanism of action. *Pure Appl. Chem.* 70:863–887.

Sager, M. 2010. Analysis of less bioactive elements in green plants, food and feed samples (Sc-Y-La-Ce-Rb-Cs-Ti). *Ecol. Chem. Engin.* 17:289–295.

Sager, M. 2012. Chocolate and cocoa products as a source of essential elements in nutrition. *J. Nutr. Food. Sci.* 2:123.

Sanchez-Cabeza, J.-A., Merino, J., Masqué, O. et al. 2003. Concentration of plutonium and americium in plankton from western Mediterranean Sea. *Sci. Total Environ.* 311:233–245.

Santini, J.M. (ed.) 2012. *The Metabolism of Arsenite*. Arsenic in the Environment. Boca Raton, FL: CRC Press.

Satarug, S., Garett, S.H., Sens, M.A., Satarug, D. 2010. Cadmium, environmental exposure and health outcomes. *Environ. Health Persp.* 118:182–190.

SCF. 2003. *Opinion of the Scientific Committee on Food on the Tolerable Upper Intake Level of Copper*. SCF/CS/NUT/UPPLEV/57 Final.

Schäfer, J., Puchelt, H. 1998. Platinum group elements (PGM) emitted from automobile catalytic converters and their distribution in roadside soil. *J. Geochem. Exp.* 64:307–314.

SCHER. 2012. Assessment of the tolerable daily intake of barium. Scientific Committee on Health and Environmental Risk. http://ec.europa.eu/health/scientific_committees/environmental_risks/index_en.html.

Schrauzer, G. 2002. Lithium: Occurrence, dietary intakes, nutritional essentiality. *Am. Coll. Nutr.* 21:14–21.

Schümann, K. 2001. Safety aspects of iron in food. *Ann. Nutr. Metab.* 45:91–101.

Schwabe, A., Meyer, U., Grün, M. et al. 2012. Effect of rare earth elements (REE) supplementation to diets on the carry-over into different organs and tissues of fattening bulls. *Livest. Sci.* 143:5–14.

Sekomo, C.B., Rousseau, D.P.L., Lens, P.N.L. 2012. Use of Gisenyi volcanic rock for adsorptive removal of Cd(II), Cu(II), Pb(II), and Zn(II) from wastewater. *Water Air Soil Pollut.* 223:533–547.

Selim, H.M. 2012 Competitive sorption of heavy metals in soils: Experimental evidence. In: *Competitive Sorption and Transport of Heavy Metals in Soils and Geological Media*, ed. H.M. Selim, 1–48. Boca Raton, FL: CRC Press.

Sembratowicz, I., Rusinek-Prystupa, E. 2012. Content of cadmium, lead, and oxalic acid in wild edible mushrooms harvested in places with different pollution levels. *Pol. J. Environ. Stud.* 21:1825–1830.

Sereno, M.L., Almeida, R.S., Nishimura, D.S., Figuera, A. 2007. Response of sugarcane to increasing concentration of copper and cadmium and expression of metallothioneien genes. *J. Plant Physiol.* 164:1499–1515.

Sharma, A.D., Brar, M.S., Malhi, S. 2005. Critical toxic range of chromium in spinach plants and in soils. *J. Plant Nutr.* 28:1555–1568.

Shi, H., Castranova, R.M., V., Zhao, J.J. 2013. Titanium dioxide nanoparticles: A review of current toxicological data. *Part. Fibre Toxicol.* 10:15.

Shi, P., Huang, Z.W., Chen, G.C. 2006. Influence of lanthanum on the accumulation of trace elements in chloroplast of cucumber leaves. *Biol. Trace Elem. Res.* 109:181–188.

Shiller, A.M. 2003. Dissolved gallium in the Atlantic Ocean. *Mar. Chem.* 61:87–99.

Shiraishi, K. 2005. Dietary intakes of eighteen elements and 40K in eighteen food categories by Japanese subjects. *J. Radioanal. Nucl. Chem.* 266:61–69.

Shorrocks, V.M. 1997. The occurrence and correction of boron deficiency. In: *Boron in Soils and Plants. Reviews,* eds. B. Dell, P.H. Brown, R.W. Bell, and W. Shotyk, 121–148. London: Kluwer Academic Publication.

Shotyk, W., Chen, B., Krachler, M. 2005. Lithogenic, oceanic and anthropogenic sources of atmospheric Sb to a maritime blanket bog, Myramar, Faroe Islands. *J. Environ. Monit.* 7:1148–1154.

Shtangeeva, I. 2008. Uranium and thorium accumulation in cultivated plants. In: *Trace Elements as Contaminants and Nutrients*, ed. M.N.V. Prasad, 295–342. Hoboken, NJ: Wiley.

Shtangeeva, I. 2010. Uptake of uranium and thorium by native and cultivated plants. *J. Environ. Radioact.* 101:458–463.

Shtangeeva, I., Ayrault, S., Jain, J. 2004. Scandium bioaccumulation and its effect on uptake macro- and trace elements during initial phase of plant growth. *Soil Sci. Plant Nutr.* 50:877–883.

Shtiza, A., Swennen, R., Tashko, A. 205. Chromium and nickel distribution in soils, active river, overbank sediments and dust around the Burrel chromium smelter (Albania). *J. Geomech. Explor.* 87:92–108.

Siebielec, G., Smreczak, B., Klimkowicz-Pawlas, A. et al. 2012. *Monitoring of Chemical Properties of Agricultural Soils in Poland in 2010–2012 Years.* Warsaw, Poland: Libr. Monit. Environ (in Polish).

Siebielec, G., Ukalska, A., Gałązka, R. 2013. Trace elements accumulation by earthworms in amended soils. *7th International Workshop on Chemical Boiavailability in the Terrestrial Environment.* Nottingham, November 3–6. Abstr: 58–59.

Siepak, M., Niedzielski, P., Przybyłek, J. 2003. Investigations on trace elements in ground water, using analysis of speciation. *Contemp. Problems Hydrogeol. Part 2.* 11:305–313 (in Polish).

Silanpää, M., Jansson, H. 1992. Status of cadmium, lead, cobalt and selenium in soils and plants of thirty countries. In: *Soil Bull*, Vol. 65. Rome, Italy: FAO.

Simon, L., Balazsy, S., Balogh, A., Pais, I. 1990. Phytoextraction of heavy metals from a galvanic mud contaminated soils. In: *Soil Pollution*, ed. G. Filep. 892–899. Debrecen, Hungary: Agriculture University Press.

Simonescu, C.M., Ferdes, M. 2012. Fungal biomass for Cu(II) uptake from aqueous systems. *Pol. J. Environ. Stud.* 21:1831–1839.

Singh, G., Brar, M.S., Malhhi, S.S. 2007. Decontamination of chromium by farm yard manure application in spinach grown in two texturally different Cr-contaminates soils. *J. Plant Nutr.* 30:289–308.

Sinha, R., Saxena, R., Sing, S. 2002. Comparative studies on accumulation of Cr from metal solution and tannery effluent under repeated metal exposure by aquatic plants: Toxicity effects. *Environ. Monit. Assess.* 80:7–31.

Skibniewska, K., Smoczyński, S.S., Wiśniewska, I. 1993. Content of radioactive cesium in selected food products. II. Radioactive cesium in daily food rations of selected population groups. *Roczn. PZH.* 44:367–371 (In Polish).

Skjelkvale, B.L., Anderssen, T., Fjeld, E. et al. 2001. Heavy metal survey in Nordic lakes; concentration, geographic patterns and relation to critical limits. *Ambio* 30:2–10.

Škribić, B., Onjia, A. 2007. Multivariate analyses of microelement contents in wheat cultivated in Serbia. *Food Control* 18:338–345.

Skwarzec, B. 1997. Polonium, uranium and plutonium in selected components of the southern Baltic Sea. *Ambio* 26:113–117.

Skwarzec, B., Strumińska, D.I., Boryło, A. 2001. The radionuclides ^{234}U, ^{238}U and ^{210}Po in drinking water in Gdańsk agglomeration (Poland). *L. Radioanal. Nuclear Chem.* 250: 315–318.

Slikkerver, A., de Wolff, F.A. 1989. Pharmacokinetics and toxicity of bismuth compounds. *Med. Toxicol. Adverse Drug Exp.* 4:303–323.

Smith, A.H., Lingas, E.O., Rahman, M. 2000. Contamination of drinking-water by arsenic in Bangladesh: A public health emergency. *Bull. World Health Organ.* 78:1093–1103.

Smith, E., Naidu, R., Alston, A.M. 1998. Arsenic in the soil environment: A review. *Adv. Agron.* 64:149–195.

Speziali, M., Orvini, E., Rizzio, E. et al. 1989. Gallium distribution in several human brain areas. *Biol. Trace Elem. Res.* 22:9–15.

Sposito, G., Page, A.L. 1984. Cycling of metal ions in the soil environments. In: *Metal Ions in Biological Systems*, ed. H. Sigel, 287–298. New York: Dekker.

Srikanth, R., Reddy, S.R.-J. 1991. Lead, cadmium and chromium levels in vegetables grown in urban sewage sludge—Hyderabad, India. *Food Chem.* 40:229–234.

Sripanyakorn, S., Jugdaohsingh, R., Thompson, R.P.H., Powell, J.J. 2005. Dietary silicon and bone health. *Nutr. Bull.* 30:222–230.

Stahl, T., Taschan, H., Brunn, H. 2011. Aluminum content of selected foods and food products. *Environ. Sci. Eur.* 23:37.

Stanek, E.J., Calabrese, E., Barnes, R.M. et al. 1988. Ingestion of trace elements from food among preschool children: Al, Ba, Mn, Si, Ti, V, Y and Zr. *J. Trace Elem. Exp. Med.* 1:179–190.

Stanisławska-Głupiak, E., Korzeniowska, J. 2014. Phytotoxic thresholds for Zn in soil extracted with 1 M HCl. *J. Food Agric. Environ.* 12:146–149.

Stern, B.R., Solioz, M., Krewski, D. et al. 2007. Copper and human health: Biochemistry, genetics, and strategies for modeling dose-response relationships. *J. Toxicol. Environ. Health B* 10:157–222.

Stern, J.C., Sonke, J.E., Salters, V.J.M. 2007. A capilary electrophoresis-ICP-MS study of rare elements complexation by humic acids. *Chem. Geol.* 246:170–180.

Stoecker, B. 2004. Chromium. In: *Elements and Their Compounds in the Environment*, 2nd ed., eds. E. Merian, M. Anke, M. Ihnat, and M. Stoeppler, 709–729, Weinheim, Germany: Wiley.

Su, Y., Chen, L.-J., He, J.-R. et al. 2011. Urinary rubidium in breast cancers. *Clin. Chim. Acta* 412:2305–2309.

Sudaryanto, A., Takahashi, S., Iwata, H. et al. 2005. Organotin residues and the role of anthropogenic tin sources in the coastal environment of Indonesia. *Marine Poll. Bull.* 50:226–235.

Sun, J.H., Ji, J.H., Park, J.D. et al. 2011a. Subchronic inhalation toxicity of gold nanoparticles. *Part. Fibre Toxicol.* 8:16.

Sun, J.-F., Wang, C.-N., Wu, B.-J. et al. 2011b. Long term dietary exposure to lead of the population of Jangsu Province, China. *Food Addit. Contam.* 28:107–114.

Sundar, S., Chakravarty, J. 2010. Antimony toxicity. *Int. J. Environ. Res. Public Health* 7:4267–4277.

Sundby, H., Phipippe, M., Gobeil, C. 2004. Comparative geochemistry of cadmium, rhenium, uranium, and molybdenum in continental margin sediments. *Geochim. Cosmochim. Acta* 68:2485–2493.

Sures, B., Zimmermann, S., Messerschmidt, J. et al. 2001. First report on the uptake of automobile catalyst emitted palladium by European eels (*Anguilla anguilla*) following experimental exposure to road dust. *Environ. Pollut.* 113:341–345.

Suttle, N.F. 2012. Copper imbalances in ruminants and humans: Unexpected common ground. *Adv. Nutr.* 3:666–674.

Świetlik, R. 1998. Speciation analysis of chromium in waters. *Pol. J. Environ. Stud.* 7:257–266.

Szabó, S.A. 2009. Minerals in foodstuffs. Part XLV. Rhenium in foodstuffs. *Élelmezési Ipar.* 63:159–160 (in Hungarian).

Szefer, P. 2002. *Metals, Metalloids and Radionuclides in the Baltic Sea Ecosystem*. Amsterdam, The Netherlands: Elsevier.

Szefer, P., Grembecka, M. 2007. Mineral components in food crops, beverages, luxury food, spices, and dietary food. In: *Mineral Components in Food*, eds. P. Szefer and J.O. Nriagu, 163–230, Boca Raton, FL: CRC Press.

Szefer, P., Ostrowski, S. 1980. On the occurrence of uranium and thorium in the biosphere of natural waters. Part I. Uranium and thorium in plankton and inherent plants. *Oceanology* 13:35–44.

Szilagyi, M. 2004. Hafnium. In: *Elements and Their Compounds in the Environment*, 2nd ed., eds. E. Merian, M. Anke, S. Ihnat, and M. Stoeppler, M., 795–800. Weinheim, Germany: Wiley.

Szkoda, J., Nawrocka, A., Kmiecik, M., Żmudzki, J. 2011. Monitoring study of toxic elements in food of animal origin. *Ochr. Srod. Zasob. Nat.* 48:475–484 (in Polish).

Szkoda, J., Żmudzki, J. 1996. Levels of copper in animal tissues, milk, eggs and feeds. In: *Copper and Molybdenium in the Environment. Ecological and Analytical Problem*, eds. A. Kabata-Pendias and B. Szteke, 216–220, Warsaw, Poland: The Polish Academy of Sciences (in Polish).

Szkoda, J., Żmudzki, J. 2000. Alive test for evaluation of cadmium contamination in horses. In: *Cadmium in the Environment. Ecological and Analytical Problems*, eds. A. Kabata-Pendias and B. Szteke, 403–409, Warsaw, Poland: The Polish Academy of Sciences (in Polish).

Szkoda, J., Żmudzki, J. 2002. Zinc in animal tissues and food of animal origin. In: *Zinc in the Environment. Ecological and Analytical Problem*, eds. A. Kabata-Pendias and B. Szteke, 457–463, Warsaw, Poland: The Polish Academy of Sciences (in Polish).

Szkoda, J., Żmudzki, J., Grzebalska, A. 2004. Iron in animal tissues and milk. *Roczn. PZH*. 55S:61–66 (in Polish).

Szopka, K., Karczewska, A., Jezierski, P., Kabała, C. 2013. Spacial distributioon of lead in the surface layers of mountain forest soils, an example from the Karkonosze National Park, Poland. *Geoderma* 192:259–268.

Szteke, B. 2006. Monitoring of trace elements in edible plants. *Pol. J. Environ. Stud.* 15(2a):189–194.

Szteke, B., Jędrzejczak, R., Ręczajska, W. 2004. Iron and manganese in selected edible plants. *Roczn. PZH*. 55S:21–27 (in Polish).

Takeda, A., Kimura, K., Yamasaki, S.-I. 2004. Analysis of 57 elements in Japanese soils, with special references to soil groups and agricultural use. *Geoderma* 119:291–307.

Takeda, A., Tsukada, H., Nanzyo, M. et al. 2005. Effects of long term fertilizer applications on the concentration and solubility of major and trace elements in a cultivate andosol. *Soil Sci. Plant Nutr*. 51:251–260.

Takeda, A., Tsukada, H., Takaku, Y. et al. 2006. Accumulation of uranium derived from long-term fertilizers applications in a cultivated Andisol. *Sci. Total Environ*. 367:924–931.

Taylor, A. 1996. Biochemistry of tellurium. *Biol. Trace Elem. Res*. 5:231–239.

Taylor, M.D., Percival, H.J. 2001. Cadmium in soils solution from a transect of soils away from a fertilizer bin. *Environ. Pollut*. 113:35–40.

Taylor, D.M., Taylor, S.K. 1997. Environmental uranium and human health. *Rev. Environ. Health* 12:147–157.

Telford, K., Maher, J.L., Krikowa, F., Foster, S. 2008. Measurement of total antimony and antimony species in mine contaminated soils by ICPMS and HPLC-ICPMS. *J. Environ. Monit*. 10:136–140.

Thacker, P.A. 2013. Alternatives to antibiotics as growth promoters for use in swine production: A review. *J. Anim. Sci. Biotechnol*. 4:35.

Thomas, F., Bialek, B., Hensel, R. 2012. Medical use of bismuth: The two sides of the coin. *J. Clin. Toxicol*. S3:004.

Thomas, V.G., Roberts, M.J., Harrison, P.T.C. 2009. Assessment of the environmental toxicity and carcinogenicity of tungsten-based shot. *Ecotoxicol. Environ. Saf.* 72:1031–1038.

Thomson, C.D. 1998. Selenium speciation in human body fluids. *Analyst* 123: 827–831.

Thomson, A., Chadwick, O.A., Rancourt, D.G., Chorover, J. 2006. Iron-oxide crystallinity increases during soil redox oscillations. *Geochimica* 70:1710–1727.

Tome, F.V., Rodriquez, M.P.B., Lozano, J.C. 2003. Soil-to-plant transfer factors for natural radionuclides and stable elements in Mediterranean area. *J. Environ. Radioact*. 65:161–175.

Torres-Escribano, S., Vélez, D., Montoro, R. 2010. Mercury and methylmercury bioaccessibility in swordfish. *Food Addit. Contam*. 27:327–337.

Tsuda, T., Inoue, T., Kojima, M. et al. 1995. Market basket and duplicate portion estimation of dietary intakes of cadmium, mercury, arsenic, copper, manganese, and zinc by Japanese adults. *J. AOAC Int*. 78:1363–1367.

Tsukada, H., Hasekawa, H., Hisamau, S. 2002. Distribution of alkali and alkaline earth metals in several agricultural plants. *Radioprot. Colloq.* 37:C1535–C1540.

Tsukada, H., Nakamura, Y. 1999. Transfer of 137Cs and stable Cs from soil to potato in agricultural fields. *Sci. Total Environ.* 228:111–120.

Tsukada, H., Shibata, H., Sugiyama, H. 1998. Transfer of radiocesium and stable cesium from substrata to mushrooms in a pine forest in Rokkaho-mura, Saomoriu, Japan. *J. Environ. Radioact.* 39:149–160.

Tsukada, H., Takeda, A., Takabashi, T. et al. 2005. Uptake and distribution of Sr-90 and stable Sr in rice plants. *J. Environ. Radioact.* 81:221–231.

Tsuruta, T. 2004. Cell-associated adsorption of thorium and uranium from aqueous system using various microorganisms. *Water Air Soil Pollut.* 159:35–47.

Tumi, A.F., Mihailović, N., Gajić, B.A. 2012. Comparative study of hyperaccumulation of nickel by Alyssum murales. l. Population from the ultramafics of Serbia. *P. J. Environ. Stud.* 21:1855–1866.

Turick, Ch.E., Knox, A.S., Leverette, Ch.L., Kritzas, Y.G. 2008. In situ uranium stabilization by microbial metabolites. *J. Environ. Radioact.* 99:890–899.

Tyler, G. 2004. Rare earth elements in soils and plant systems—A review. *Plant Soil* 267:191–206.

Tyler, G. 2005. Changes in the concentrations of major, minor and rear-earth elements during leaf senescence and decomposition in a Fagus sylvatica forest. *Forest Ecol. Manag.* 206:167–177.

Udeh, K.O., Targowski, Z., Kais-Samborska, M. 2001. Determination of iodine in food products. *Pol. J. Food Nutr. Sci.* 10/51:35–38.

Ursinyova, M., Hladikova, V. 2000. Cadmium in the environment of Central Europe. In: *Trace Elements in the Environment*, eds. B. Markert and K. Friese, 87–101. Oxford: Elsevier.

USDA (U.S. Department of Agriculture). 2005. *National Fluoride Database of Selected Beverages and Foods*. Nutrient Data Laboratory Beltsville, Human Nutrition Research Center Agricultural Research Service, U.S. Department of Agriculture.

USDA. 2009. Mineral commodity summaries. USGS. http://minerals.usgs.gov/minerals/pubs/mcs (Accessed: November 20, 2009).

USGS. 1997. Water-Resources Investigations. Report 97–4120.

USGS. 2001. http://pubs.usgs.gov/fs/2001/fs-068-01/FS_068-01.htm (Accessed: April 7, 2013).

USGS. 2010. *Minerals Yearbook.* Salt. U.S. Department of Interion.

USGS. 2011. *Mineral Commodity Summaries*. U.S. Department of Interion. http://minerals.usgs.gov/minerals/pubs/commodity/ (Accessed: March 15, 2012).

Usydus, Z., Szlinder-Richert, J. 2009. Iodine and fluorine in fish products. *Bromat. Chem. Toksykol.* 42:822–826 (in Polish).

Vaessen, H.A.M.G., Szteke, B. 2000. Beryllium in food and drinking water—A summary of available knowledges. *Food Addit. Contam.* 17:149–159.

Vaessen, H.A.M.G., van Ooik, A. 1989. Speciation of arsenic in Dutch total diets: Methodology and results. *Z. Lebensm. Unters. F.* 189:232–235.

Van Cauwenbergh, R., Hendrix, P., Robberecht, H., Deelstra, H., Daily dietary chromium intake in Belgium, using duplicate portion sampling. 1996, *Z. Lebensm. Unters. F.* 203:203–206.

Varo, P., Saari, E., Paaso, A., Koivistoinen, P. 1982. Strontium in Finnish foods. *Int. J. Vitam. Nutr. Res.* 52:342–350.

Vaughan, G.T., Florence, T.M. 1992. Platinum in the human diet, blood, hair and excreta. *Sci. Total Environ.* 111:47–58.

Veiga, M.M., Baker, R.F. 2004. *Protocols for Environmental and Health Assessment of Mercury Realized Artisanal and Small Scale Gold Miners*. Global Mercury Project. Vienna, Austria: UNIDO.

Verma, P., George, K.V., Singh, H.V., Singh, R.N. 2007. Modeling cadmium accumulation in radish, carrot, spinach and cabbage. *Appl. Math. Model.* 32:1652–1661.

Vernay, P., Gauthier-Moussard, C., Hitmi, A. 2007. Interaction of bioaccumulation of heavy metal chromium with water relation, mineral nutrition and photosynthesis in developed leaves of *Lolium perenne* L. *Chemosphere* 68:1563–1575.

Veselý, J., Norton, S.A., Skřivan, P. et al. 2002. Environmental chemistry of beryllium. *Rev. Miner. Geochem.* 50:291–317.

Vietrov, V.A. 2002. Eko-geochemical description of Baikal at the end of XX century as base for assessment of technogenic pollution of the lake. *2nd International Conference on Heavy Metals, Radionuclides and Elements—Biofills in the Environment.* October 20–22, Semipalatynsk, Kazakhstan, 1:85–94 (in Russian).

Violante, A. 2013. Elucidating mechanisms of competitive sorption at the mineral/water interface. In: *Advances in Agronomy*, ed. D.L. Sparks, Vol. 118, 111–176. Amsterdam, The Netherlands: Elsevier.

Voet van der, G.B., Todorov, T.I., Centeno, J.A. et al. 2007. Metals and health: A clinical toxicological perspective on tungsten and review of the literature. *Mil. Med.* 172:1002–1005.

Voigt, H.R. 2004. Concentration of Mercury (Hg) and Cadmium (Cd) and the Condition of Some Coastal Baltic Fishes. Ph.D. thesis. University of Helsinki, Finland.

Von Glasow, R., von Kuhlmann, R., Lawrence, M.G. et al. 2004. Impact of reactive bromine chemistry in the troposphere. *Atmos. Chem. Phys.* 4:2481–2497.

Vrooman, V., Waegeneers, N., Cornelis, C. et al. 2010. Dietary cadmium intake by the Belgian adult population. *Food Addit. Contam.* 27:1665–1673.

Waegeneers, N., Pizzolon, J.-C., Hoenig, M., De Temmerman, L. 2009a. Accumulation of trace elements in cattle from rural and industrial areas in Belgium. *Food Addit. Contam.* 26:326–332.

Waegeneers, N., Pizzolon, J.-C., Hoenig, M., De Temmerman, L. 2009b. The European maximum level for cadmium in bovine kidneys is in Belgium only realistic for cattle up to 2 years of age. *Food Addit. Contam.* 26:1239–1248.

Wan, A.T., Conyers, R.A.J., Coombs, C.J., Masterton, J.P. 1991. Determination of silver in blood, urea and tissues of volunteers and burn patients. *Clin. Chem.* 37:1683–1687.

Wang, X., Nan, Z., Liao, Q. et al. 2012. Fractions and bioavailability of cadmium and nickel to carrot crops in oasis soil. *Pol. J. Environ. Stud.* 21:1867–1874.

Wang, Y.P., Li, Q.B., Shi, J.Y. et al. 2008. Assessment of microbial activity and bacteria community composition in the rhizosphere of a copper accumulator and non-accumulator. *Soil Biol. Biochem.* 40:1167–1177.

Wangstrand, H., Erikskon, J., Öborn, I. 2007. Cadmium concentration in winter wheat as affected by nitrogen fertilization. *Eur. J. Agron.* 25:209–214.

Watmough, S.A., Eimers, M.C., Dillon, P.J. 2007. Manganese cycling in central Ontario forest: Response to soil acidification. *Appl. Geochem.* 22:1241–1247.

Wegenke, M., Junge, M., Diemer, J., Nittka, J. 2005. Monitoring of antimony in the environment of Bavaria. *1st Workshop; Antimony in the Environment*, September 25–27, Heidelberg, Germany.

Weggler, K., McLaughlin, J., Graham, R.D. 2004. Effect of chlorine in soils solution on the plant availability of biosolid-borne cadmium. *J. Environ. Qual.* 33:495–505.

Weir, A., Westerhoff, P., Fabricius, L. et al. 2012. Titanium dioxide nanoparticles in food and personal care products. *Environ. Sci. Technol.* 46:2242–2250.

Welch, S.A., Green, E.G., Banfield, J.F. 2004. *Geochemistry and Biochemistry of Ga, Ge, and Ti during Weathering*. Copenhagen, Denmark: Goldschmidt.

Welle, F., Franz, R. 2011. Migration of antimony from PET bottles into beverages. *Food Addit. Contam.* 28:115–126.

West, K. 2008. *The Elements-Bromine*. New York: Marshall Cavendish Corporation.

White, P., Bowen, M., Hommond, C. et al. 2003. The mechanism of cesium uptake by plants. *International Symposium Radioecology and Environmental Dosimetry*. October 3–5, Rokkasho, Aomori, Japan. 255–262.

Whiteley, J.D., Murray, F., 2003. Anthropogenic platinum group element (Pt, Pd and Rh) concentrations in road dusts and roadside soils from Perth, Western Australia. *Sci. Total Environ.* 317:121–135.

WHO (World Health Organization). 1982a. *Titanium.* Environmental Health Criteria, 24. Geneva, Switzerland: WHO.

WHO (World Health Organization). 1982b. *Toxicological Evaluation of Certain Food Additives: Copper.* Food Additives Series, 17. Geneva, Switzerland: WHO.

WHO (World Health Organization). 1988. *Vanadium.* Environmental Health Criteria, 81. Geneva, Switzerland: WHO.

WHO (World Health Organization). 1989. *Toxicological Evaluation of Certain Food Additives and Contaminants.* Food Additives, 24. Geneva, Switzerland: WHO.

WHO (World Health Organization). 1991a. *Inorganic Mercury.* Environmental Health Criteria, 118. Geneva, Switzerland: WHO.

WHO (World Health Organization). 1991b. *Platinum.* Environmental Health Criteria, 125. Geneva, Switzerland: WHO.

WHO (World Health Organization). 1995. *Methyl bromide.* Environmental Chemistry Criteria, 166. Geneva, Switzerland: WHO.

WHO (World Health Organization). 1998. *Boron.* Environmental Health Criteria, 204. Geneva, Switzerland: WHO.

WHO (World Health Organization). 2001a. Barium and barium compounds. *Concise International Chemistry.* Assess. Doc. 33. Geneva, Switzerland: WHO.

WHO (World Health Organization). 2001b. Beryllium and beryllium compounds. *Concise International Chemistry.* Assess. Doc. 32. Geneva, Switzerland: WHO.

WHO (World Health Organization). 2001c. *Evaluation of Certain Food Additives and Contaminants. Tin.* Food Additives Series, 46, Geneva, Switzerland: WHO.

WHO (World Health Organization). 2001d. *Zinc.* Environmental Health Criteria, 221. Geneva, Switzerland: WHO.

WHO (World Health Organization). 2002a. *Fluorides.* Environmental Health Criteria, 227. Geneva, Switzerland: WHO.

WHO (World Health Organization). 2002b. *Palladium.* Environmental Health Criteria, 226. Geneva, Switzerland: WHO.

WHO (World Health Organization). 2003. *Antimony in Drinking-Water. Background Document for Preparation of WHO Guidelines for Drinking-Water Quality.* Geneva, Switzerland: WHO.

WHO (World Health Organization). 2004. *Vitamin and Mineral Requirements in Human Nutrition*, 2nd ed. Geneva, Switzerland: WHO.

WHO (World Health Organization). 2005a. *Nickel in Drinking-Water.* WHO/SDE/WSH/05.08/55. Geneva, Switzerland: WHO.

WHO (World Health Organization). 2005b. *Uranium in Drinking-Water.* WHO/SDE/WSH/03.04/118. Geneva, Switzerland: WHO.

WHO (World Health Organization). 2006a. Cobalt and inorganic cobalt compounds. *Concise International Chemistry.* Assess. Doc. 69. Geneva, Switzerland: WHO.

WHO (World Health Organization). 2006b. *Elemental Speciation in Human Health Risk Assessment.* Environmental Health Criteria, 234. Geneva, Switzerland: WHO.

WHO (World Health Organization). 2007. *Evaluation of Certain Food Additives and Contaminants.* Sixty-seventh Report of the JECFA. WHO Technical Report Series, 940. Geneva, Switzerland: WHO.

WHO (World Health Organization). 2009. *WHO Handbook on Indoor Radon. A Public Health Perspective.* Geneva, Switzerland: WHO.

WHO (World Health Organization). 2010. Strontium and strontium compounds. *Concise International Chemistry.* Assess. Doc. 77. Geneva, Switzerland: WHO.

WHO (World Health Organization). 2011a. *Evaluation of Certain Food Additives and Contaminants. Arsenic and Mercury*. Technical Report 959. Geneva, Switzerland: WHO.

WHO (World Health Organization). 2011b. *Evaluation of Certain Food Additives and Contaminants. Cadmium and Lead*. WHO Technical Report Series 960. Geneva, Switzerland: WHO.

WHO (World Health Organization). 2011c. *Guidelines for Drinking-Water Quality*. 4th ed., Geneva, Switzerland: WHO.

WHO (World Health Organization). 2011d. *Molybdenum in Drinking Water*. WHO/SDE/WSH/03.04/11/Rev/1. Geneva, Switzerland: WHO.

WHO (World Health Organization). 2011e. *Selenium in Drinking-Water*. WHO/HSE/WSH/10.01/14. Geneva, Switzerland: WHO.

Wijnhoven, S.W.P., Peijnenburg, W.J.G.M., Herbert, C.A. et al. 2009. Nano-silver—A review of available data and knowledge gaps in human and environmental risk assessment. *Nanotoxicology* 3:109–138.

Willey, J.D., Inscore, M.T., Kieber, R.J., Skrobal, S.A. 2009. Manganese in coastal rainwater: Speciation, photochemistry and deposition to seawater. *J. Atmos. Chem.* 62:31–43.

Witte, M.L., Sheppard, P.R., Witten, B.L. 2012. Tungsten toxicity. *Chem. Biol. Interact.* 196:87–88.

WMSY (World Metal Statistic Yearbook). 2004. *World Metal Statistic Yearbook*. London: World Bureau of Metal Statistic.

Wojciechowska-Mazurek, M., Mania, M., Starska, K. et al. 2011. Noxious elements in edible mushrooms in Poland. *Bromat. Chem. Toksykol.* 44:143–149 (in Polish).

Wójcik, M., Sugier, P., Siebielec, G. 2014. Metal accumulation strategies in plants spontaneously inhabiting Zn—Pb waste deposits. *Sci. Total Environ.* 497:313–322.

Wong, W.W.K., Chung, S.W.C., Kwong, K.P. et al. 2010. Dietary exposure to aluminum of the Hong Kong population. *Food Addit. Contam.* 27:457–463.

Wróbel, S. 2012. Lettuce yields as an indicator of remediation efficiency of soils contaminated with trace metals emitted by copper smelter. *J. Food Agric. Environ.* 10:828–832.

Wu, W.H., Xie, Z.M., Xu, J.M. 2007. Distribution of aluminum and fluoride in subtropical hilly red soils of tea plantation and its influential factors. In: *Biogeochemistry of Trace Elements: Enviromental Protection, Remediation and Human Health*, eds. Y. Zhu, N. Lepp, and R. Naidu, 136–137. Beijing, China: Tsinghua University Press.

Wyszkowska, J., Boros, E., Kucharski, J. 2007. Effects of interaction between nickel and other heavy metals on soil microbiological properties. *Plant Soil Environ.* 53:544–552.

Wyszkowska, J., Boros, E., Kucharski, J. 2008. Enzymes activity of nickel-contaminated soils. *J. Elementol.* 13:139–151.

Wyszkowska, J., Kucharski, J. 2004. Effects of chromium on multiplications of bacteria in artificial media. *J. Elementol.* 9:165–174 (in Polish).

Xiao, T., Guha, J., Boyle, D. et al. 2004. Environmental concersn related to high thallium levels in soils and thallium uptake by plants in southewest Guizhou, China. *Sci. Total Environ.* 318:223–244.

Yokel, R.A. 2004. Aluminum. In: *Elements and Their Compounds in the Environment*, 2nd ed., eds. E. Merian, M. Anke, S. Ihnat, and M. Stoeppler, 635–658. Weinheim, Germany: Wiley.

Yokoi, K., Kimura, M., Itokawa, Y. 1996. Effect of low dietary rubidium on plasma biochemical parameters and mineral levels in rats. *Biol. Trace Elem. Res.* 51:199–208.

Yordanova, I., Stanova, D., Zlatev, A. et al. 2007. Study of the radiocesium content in Bulgarian mushrooms for the year of 2005. *J. Environ. Prot. Ecol.* 8:934–939.

Yoshida, S., Muramatsu, Y., Tagami, K., Uchida, S. 1998. Concentration of lanthanide elements, Th and U in 77 Japanese surface soils. *Environ. Int.* 24:275–286.

Ysart, G., Miller, P., Croasdale, M. et al. 1999. Dietary exposure estimates of 30 elements from U.K. total diet study. *Food Addit. Contam.* 16:391–403.

Yuita, K. 1983. Iodine, bromine and chlorine in soils and plants of Japan. *Soil Sci. Plant Nutr.* 9:403–407.

Zaichick, S., Zaichik, V., Karandasshev, V.K., Moskvina, I.R. 2011. The effect of age and gender on 59 trace-element contents in human rib bone investigated by inductively coupled plasma mass spectrometry. *Biol. Trace Elem. Res.* 43:41–57.

Zayed, A.M., Terry, N. 2003. Chromium in the environment: Factors affecting biological remediation. *Plant Soil* 249:139–156.

Zayed, A.W., Lytle, C.M., Qian, J.H., Terry, N. 1998. Chromium accumulation, translocation and chemical speciation in vegetable crops. *Planta* 206:239–299.

Zdrojewicz, Z., Strzelczyk, J. 2006. Radon treatment controversy. *Dose Response* 4:106–118.

Zemolin, A.P.P., Farina, M., Dafre, A.L. et al. 2013. Sub-acute administration of (S)-dimethyl2-(3-(phenyltellanyl) propanamido) succinate induces toxicity and oxidative stress in mice: Unexpected effects of N-acetylcysteine. *SpringerPlus* 2:182. http://www.springerplus.com/content/2/1/182.

Zereini, F., Skerstupp, B., Rankenburg, K. et al. 2001. Anthropogenic emission of platinum-group elements into the environment. Concentration, distribution and geochemical behavior in soils. *J. Soil Sediment* 1:444–449.

Zevenhoven, R., Mukherjee, A.B., Bhattacharya, P. 2006. Arsenic flows in the environment of the Europe Union: A synoptic review. In: *Arsenic in Soils and Environment: Biological Interactions, Health Effects and Remediation.* Eds. P. Bhattacharya, A.B. Mukherjee, J. Bundshuh, R. Zevenhoven, and R.H. Loeppert, 527–548. Amsterdam, The Netherlands: Elsevier.

Zhang, W. 2003. Nanoscale iron particles for environmental remediation: An overview. *J. Nanopart. Res.* 5:323–332.

Zhang, P.-Ch., Krumhand, J.L., Brady, P.V. 2002. Introduction to properties, sources and characteristics of soil radionuclides. In: *Geochemistry of Soils Radionuclides*. Eds. P.-Ch. Zhang, P.V. Brady, 1–20. SSSA Special Publ.: Madison,WI.

Zhiyanski, M., Bech, J., Sokolovaska, M. et al. 2006. Analysis of Cs-137 spatial distribution in forest soils of mountainous region in Bulgaria. *Geophys. Res. Abstr.* 8:87–95.

Zhu, Y.-G., Liu, W.-J., Cheng, Z., Geng, N.-C., 2005. The role of iron in controlling the dynamics of arsenic in the rhizosphere of rice plants. *ICOBTE 8th International Conference*, Adelaide, Australia, pp. 208–209.

Zhu, J., Zhong, G., Kennedy, M. et al. 1997. The distribution of rare earth elements (REEs) in Chinese soils. *4th International Conference on Biogeochemistry Trace Elements.* November 10–13, Berkeley, CA, Abstr.

Zielhuis, S.W., Nijsen, J.F.W., Seppenwoolde, J.H. et al. 2005. Lanthanide bearing microparticulate systems for multi-modality imaging and targeted therapy of cancer. *Curr. Med. Chem.-Anti-Cancer Agents.* 5:303–313.

Zimmermann, M.B. 2011. The role of iodine in human growth and development. *Semin. Cell Dev. Biol.* 22:645–652.

Żmudzki, J., Szkoda, J. 1994. Arsenic and selenium in animal tissues in Poland. In: *Arsenic and Selenium in Environment—Ecological and Analytical Problems*. Eds. A. Kabata-Pendias and B. Szteke, 120–125. Warsaw, Poland: The Polish Academy of Sciences (in Polish).

Żmudzki, J., Szkoda, J. 2000. Cadmium in animal tissues and food of animal origin. In: *Cadmium in the Environment—Ecological and Analytical Problems*, eds. A. Kabata-Pendias and B. Szteke, 381–389. Warsaw, Poland: The Polish Academy of Sciences (in Polish).

Index

Note: Locators followed by "*f*" and "*t*" denote figures and tables in the text

C

D

E

F

G

J

K

L

M

N

O

P

T

Printed in the United States
by Baker & Taylor Publisher Services